Biology of Extracellular Matrix

Volume 8

Extracellular matrix (ECM) biology, which includes the functional complexities of ECM molecules, is an important area of cell biology. Individual ECM protein components are unique in terms of their structure, composition and function, and each class of ECM macromolecule is designed to interact with other macromolecules to produce the unique physical and signaling properties that support tissue structure and function. ECM ties everything together into a dynamic biomaterial that provides strength and elasticity, interacts with cell-surface receptors, and controls the availability of growth factors. Topics in this series include cellular differentiation, tissue development and tissue remodeling. Each volume provides an in-depth overview of a particular topic, and offers a reliable source of information for post-graduates and researchers alike.

All chapters are systematically reviewed by the series editor and respective volume editor(s).

"Biology of Extracellular Matrix" is published in collaboration with the American Society for Matrix Biology and the International Society for Matrix Biology.

More information about this series at http://www.springer.com/series/8422

Florence Ruggiero
Editor

The Collagen Superfamily and Collagenopathies

Springer

Editor
Florence Ruggiero
Institut de Génomique Fonctionnelle de
Lyon, ENS de Lyon
UMR CNRS, Université Lyon 1
LYON CEDEX 07, France

ISSN 0887-3224 ISSN 2191-1959 (electronic)
Biology of Extracellular Matrix
ISBN 978-3-030-67591-2 ISBN 978-3-030-67592-9 (eBook)
https://doi.org/10.1007/978-3-030-67592-9

This Springer imprint is published by the registered company Springer Nature Switzerland AG.
The registered company address is: Gewerbestrasse 11, 6330 Cham, Switzerland

Preface

Collagen is the most abundant extracellular matrix protein in organisms. It is known for more than one hundred years for its fascinating capacity to yield glue when denatured by boiling collagen-rich tissues such as bones, skin, and tendons. This simple "*do-it-yourself*" procedure to obtain an extra-strong glue was extensively used in the past by the Greeks and Romans for assembling furniture and mosaic floors which have stood the test of time. This is why the name of collagen was given to this protein as, in Greek, it literally means "*a substance that generates glue*". Sensu stricto, collagen is the glue that holds tissues and organs together in organisms.

Collagen I was described at the molecular level in the 1930s, and it is when a second collagen was discovered in cartilage almost 40 years; after that collagens started to be numbered with Roman numerals. At that time, it was certainly not suspected that collagens will form a superfamily of proteins that today counts up to 28 genetically distinct members in vertebrates; otherwise using Roman numerals for collagen types (from type I to XXVIII) would not have been considered as an option. Yet, this is still the current nomenclature for collagen proteins.

Obviously, collagens have long been established not just as sticky proteins. Apart from their primary function to provide structural support and strength to cells to maintain biomechanical integrity of tissues, they act as extracellular modulators of cell signaling events. Cells in tissues are structurally and functionally integrated via numerous dynamic connections with their surrounding extracellular matrix in which collagens play critical roles. In addition to direct interactions with cells, collagens are engaged in specific interactions with other extracellular matrix macromolecules adding in this way to a broad and complex repertoire of their functions in tissues. In that respect, collagens have been regarded for decades now as a superfamily of multifaceted scaffolding and signaling proteins. Their functions are nevertheless far from being completely elucidated due to their high number, complex nature, and involvement in multiple biological processes.

By the end of the last century, the importance of type I collagen has been well exemplified with the discovery that collagen I genes are the molecular cause of a severe disorder of bone fragility called osteogenesis imperfecta, a disease known for thousand years. Since then, mutations in a growing number of collagen genes have been implicated in a broad spectrum of heterogenous heritability-related diseases that affect various organs and tissues. They are now known as "collagenopathies." Yet, the most common mutation remains the replacement of a glycine (Gly) residue in a (*Gly*–Pro–Hyp) *repeating* sequence of the triple collagen helix by one of any other residues. This results in the destabilization of the trimeric structure of the molecule and eventually leads to the loss of extracellular matrix integrity and altered cell functional properties. Notably, various mutations of genes encoding collagens yield abnormal molecules that fold improperly. The intracellular retention of these abnormal collagen molecules has recently been identified as an important cause of collagenopathies by provoking sustained endoplasmic reticulum (ER) stress and ultimately cell death. Efforts are underway to develop approaches to target ER stress mechanisms for therapy.

Built on decades of research, the understanding of collagen biosynthesis and function and their deep implication in development and tissue homeostasis is clearly the result of incessant back and forth discussions between pure biologists and medical scientists that obviously exist in other fields but, in my view, are particularly intense in collagen research. This is something you feel when reading the chapters of this book. The last few decades have seen impressive progress on the understanding of collagen functions and how mutations in collagen genes affect them.

This book provides timely insights into the mechanisms underlying collagen-related diseases and thereby documents the remarkable diversity of collagen functions. The first chapter addresses readers who are not familiar with collagens and provides a context for harnessing collagenopathies. In the following chapters introduced in more detail in Chap. 1, issues related to collagenopathies and the associated defective collagens are presented and critically discussed. They include osteogenesis imperfecta (collagen I), specific subtypes of the Ehlers–Danlos syndrome (collagens I, III, and V), different types of chondrodysplasias and skeletal dysplasia (collagens II, IX, and XI), Alport syndrome and other collagen IV-related vascular diseases (collagen IV), Bethlem myopathy and Ullrich congenital muscular dystrophy also known as collagen VI-related myopathies (collagen VI), epidermolysis bullosa (collagen VII), Knobloch syndrome (collagen XVIII), and other recently identified nervous system-related diseases involving various collagen genes.

Chapters have been written by renowned experts in collagen biology who have made key discoveries in the molecular mechanisms of collagenopathies and participated in the development of therapeutic strategies to cure patients. I warmly thank all the authors for their contribution to this book endeavor despite the difficult working conditions caused by the COVID-19 pandemic.

My gratitude also goes to Prof. Nikos Karamanos, Series Editor, and Dr. Miriam Latuske, Associate Editor, and her colleagues who contributed to the whole process of preparing this book.

I hope you will enjoy reading this book as much as we all enjoyed writing it!

Lyon, France Florence Ruggiero

Contents

About the Editor

Florence Ruggiero is Director of Research at the CNRS and Director of the Institute of Functional Genomics of Lyon at the "Ecole Normale Supérieure de Lyon." Her primary research interest, as a young investigator, was the analysis of cell–collagen interactions and collagen receptors. Since then, the collagen superfamily became "her family," and she has authored over a hundred publications in the matrix biology field. Her lab developed new methods to produce recombinant collagens and derived domains in various expression systems including plants to identify binding partners and to resolve extracellular protein networks. Her lab is currently investigating the functional role of unconventional collagens in development, regeneration, and disease using zebrafish as a model organism. She is a past councilor of the International Society for Matrix Biology and again serves as a councilor of the French Society for Matrix Biology. She was honored to serve as Chair for the Collagen Gordon Research Conference in 2017.

Chapter 1
The Collagen Superfamily: Everything You Always Wanted to Know

Mélanie Salamito, Pauline Nauroy, and Florence Ruggiero

Abstract Collagens are the most abundant extracellular matrix proteins in multicellular organisms. In humans, the collagen superfamily counts 28 proteins encoded by 44 genes whose functions are far from being completely elucidated. Their primary function is to provide structural support and strength to cells in order to maintain the biomechanical integrity of tissues. However, collagens are no longer considered just as structural proteins and there is an extensive literature that documents the high diversity of collagens' functions in cell behavior, integrity, and tissue homeostasis. A number of common diseases are directly linked to an imbalance between collagen synthesis and degradation that leads to tissue dysfunction. In addition, collagens are associated with a broad spectrum of heritability-related diseases known as "collagenopathies" that affect a multitude of organs and tissues including sensory organs. However, the particular complexity and diversity of the collagen nomenclature, structure, biosynthesis, and molecular assembly often repel scientists and clinicians who are not in the field. This chapter aims at providing a glossary of the collagen superfamily that allows readers to locate the piece of information they need without alienating them.

Abbreviations

BM	Basement membrane
COL	Collagenous domain
ECM	Extracellular matrix
EDS	Ehlers–Danlos syndrome
ER	Endoplasmic reticulum

M. Salamito · F. Ruggiero (✉)
Institut de Génomique Fonctionnelle de Lyon, ENS de Lyon, UMR CNRS 5242, Université Lyon 1, Lyon Cedex 07, France
e-mail: florence.ruggiero@ens-lyon.fr

P. Nauroy
Department of Dermatology, Medical Center – University of Freiburg, Freiburg, Germany

F. Ruggiero (ed.), *The Collagen Superfamily and Collagenopathies*, Biology of Extracellular Matrix 8, https://doi.org/10.1007/978-3-030-67592-9_1

FACIT	Fibril-associated collagens with interrupted triple helices
MACIT	Membrane-associated collagens with interrupted triple helices
NC	Non-collagenous domain
OI	Osteogenesis imperfecta
P4H	Prolyl-4-hydroxylase
UCMD	Ullrich congenital muscular dystrophy

1.1 Introduction

Cells use a thousand genes to produce, assemble, and remodel tissue-specific extracellular matrices (ECM) that show a remarkable versatility and permeate all aspects of cellular function in health and disease. Among them, 44 genes code for collagens in humans. Collagens account for one-third of the total protein mass in mammals and are the prevalent component of ECM. The collagen superfamily comprises 28 trimeric proteins that display large diversity in their structure, supramolecular organization and function. Initially, collagens have been given Roman numerals (I–XXVIII) in the order of their discovery. Reaching the number XXVIII, which was certainly not expected when the first collagen types were discovered, has become somewhat annoying for easy identification of samples and easy reading of papers. The classic (and still recommended) nomenclature has the tendency to include now Arabic numbers not only for collagen genes but also the proteins (for example, Col28 instead of ColXXVIII). In some animal models, the mutant names are still used instead of the gene name (for example, *stumpy* in zebrafish is referring to *col19a1* gene). For didactic learning of collagen structure and nomenclature, the reader is referred to a collagen lexicon in Fig. 1.1.

Collagens were first described as banded fibrils and famous for providing structural and tensile properties in connective tissues such as skin or cartilage. They are also key components of the basement membrane (BM), a specialized ECM structure that physically separates and anchors cell sheets to the underlying connective tissue and ensures molecular filtration and tissue homeostasis. Collagens act as extracellular modulators of signaling events and serve critical regulatory roles in various cell functions during embryonic development and adult homeostasis. The later identification of novel members in the early twenty-first century, after the completion of the human genome sequencing together with the discovery that bioactive fragments are released from most of the collagen molecules, led to new exciting biological functions including cell differentiation, behavior and integrity as well as biological processes such as tissue regeneration, wound healing, inflammation, angiogenesis, and many others.

A number of common diseases (microenvironmentally-driven tumors, fibrotic disorders, neurodegenerative diseases, etc.) are directly linked to an imbalance between collagen synthesis and degradation that leads to tissue dysfunction. Thus, not surprisingly, these multifunctional and broadly expressed proteins are also

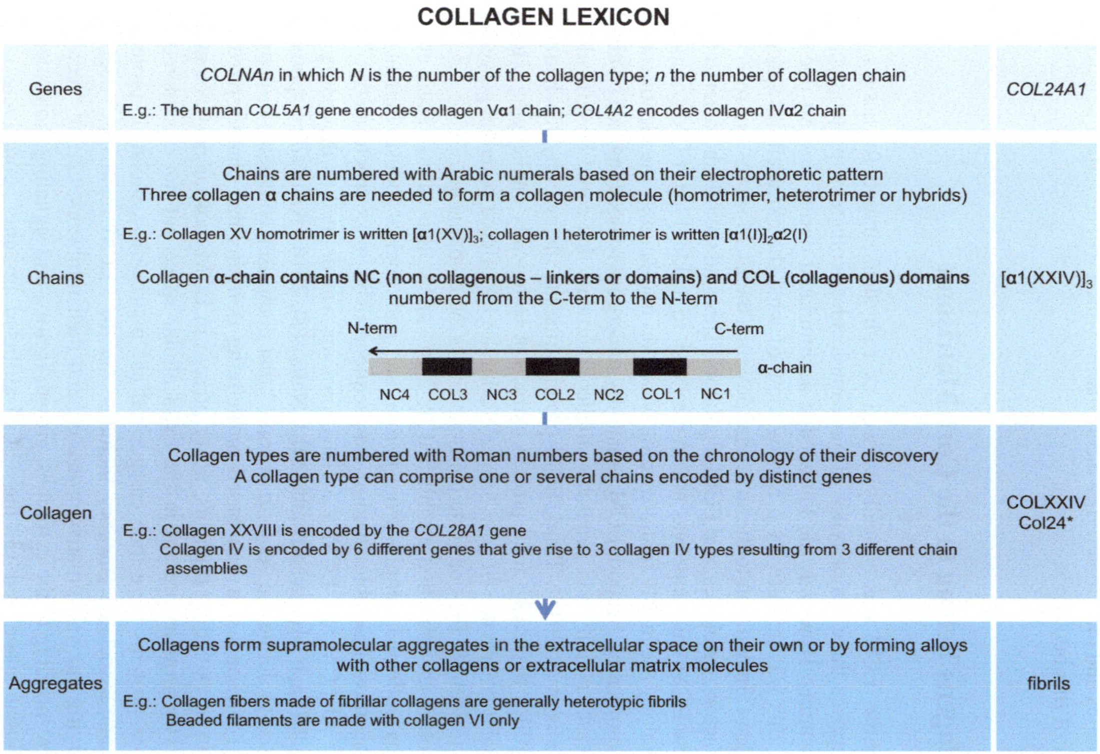

Fig. 1.1 The collagen lexicon. The collagen lexicon shows the different levels of collagen organization from the synthesis of individual α-chains to the supramolecular aggregates found in the extracellular space, and indicates the commonly used nomenclature for each organizational level (left panel uses collagen XXIV as an example). *There is now a tendency to use Arabic numbers not only for collagen genes but also for the proteins. *N-term* N-terminus, *C-term* C-terminus

associated with a wide spectrum of heritability-related diseases known as "collagenopathies." Thousands of mutations in more than half of the collagen genes affect a multitude of organs and tissues including sensory organs. This chapter aims at providing an overview of the fascinating collagen superfamily and giving specific insights into the collagen molecules that are critical for the understanding of collagenopathies and their underlying mechanisms, which will be further detailed in the chapters of this book.

1.2 Collagens Belong to the Core Matrisome

For a long time, ECM was typically viewed as a physical support for tissues that chiefly comprises three major types of macromolecules: collagens, proteoglycans, and glycoproteins. However, the knowledge accumulated during the last decades told us that this was definitively a narrow view of what the extracellular space is exactly made of. In addition to the structural proteins, the ECM serves as a reservoir for growth factors, cytokines, and ECM regulators. This said the full characterization of the diverse ECM landscape was not an easy task to tackle. First, ECM composition and topology are highly tissue-specific which do not facilitate the identification of all ECM molecules. Second, ECM constantly undergoes remodeling as a result of matrix proteases activity and generates smaller fragments that could be difficult to identify (Wells et al. 2015). Third, ECM proteins display distinct physicochemical property profiles that make total protein extraction technically challenging. However, the recent emergence of new "omics" technologies offered potential powerful approaches to identify the complete landscape of ECM components.

ECM proteomes were first established by Naba and collaborators (Naba et al. 2012a). They collected samples from normal and pathological mouse, human tissues, and the most insoluble fraction was sequenced by mass spectrometry. The ensemble of all ECM and ECM-associated proteins was defined as the matrisome (Naba et al. 2012a, b; Hynes and Naba 2012; Naba et al. 2016). An in silico approach was then used to define the "core matrisome" that comprises collagens, proteoglycans, and glycoproteins. It was based on defining domains that characterized an ECM protein or excluding domains, never present in ECM molecules (Naba et al. 2012a, b). The proteins present in the extracellular space but not considered as "pure" matrix proteins were then categorized as "matrisome-associated" proteins. The matrisome-associated category includes: (1) the "ECM regulators" such as ECM-modifying enzymes, matrix proteases, and their inhibitors; (2) the "secreted factors" corresponding to the cytokines, growth factors, and signaling molecules of the extracellular space and (3) the "ECM-affiliated proteins" containing other secreted factors and proteins that were regularly found in ECM-enriched preparations (Naba et al. 2016).

The characterization of the matrisome revealed that organisms used up to a thousand genes to build tissue-specific ECMs. The human matrisome is composed of 1027 genes while 1110 genes encode the mouse matrisome (Naba et al. 2016).

Both the mouse and human core matrisome comprises 274 genes but the number of matrisome-associated genes varies considerably between the two species, even between the different subcategories (Naba et al. 2016). Although human and mouse both have 44 collagen genes (Table 1.1), the *COL21A1* gene does not exist in mice and the *COL6A4* gene is not functional in humans due to a chromosome inversion (Fitzgerald and Bateman 2004; Gara et al. 2008). It is worth noting that ECM receptors such as integrins or Discoidin Domain Receptors (DDR) were not included in the matrisome. ECM transmembrane receptors are responsible for signaling cross-talks between ECM and cells and, thereby, trigger specific cell responses as diverse as cell adhesion, migration, proliferation, survival, and death, to name but a few. The complete characterization of the integrin adhesion complex components (adaptors, enzymes, and cytoskeleton) was termed adhesome (Zaidel-Bar et al. 2007; Zaidel-Bar 2013; Winograd-Katz et al. 2014) (www.adhesome.org).

1.3 Collagen, a Singular Protein that Should Be Read as Plural

In essence, collagens are large trimeric molecules containing at least one triple helical domain and secreted in the extracellular space (Fig. 1.2a). However, the latter definition does not entirely include transmembrane collagens that exist as full-length transmembrane proteins with soluble ectodomains that are proteolytically released from cell surface by sheddases. Collagen trimers consist in the association of three polypeptide chains, called α-chains (Fig. 1.2a). Each of the so-called α-chains is composed of Gly-Xaa-Yaa triplet repeats, in which Xaa is often a proline and Yaa a hydroxyproline. This motif allows the formation and the stabilization of the triple helix. Interchain hydrogen bonds and electrostatic interactions are holding the three α-chains together. Glycines (Gly) are strictly present as every third residue to fit into the triple helix as it is the least bulky amino acid. As such, a single substitution of a Gly residue into any other residue can lead to a distortion of the triple helix. The generation of pathogenic triple helix micro-unfoldings can further provoke cascading pathological scenario by causing disturbance of collagen stability, biosynthesis, and supramolecular aggregates thus impacting tissue biomechanics and/or interactions with cells or binding partners.

The triple helical region of collagens, named collagenous domain (COL), can be either large and uninterrupted in some molecules (e.g., collagens I, III, and VII) or short and interrupted (e.g., collagens IV and XII, and the transmembrane collagen XIII). Consequently, the COL domains represent 96% of the collagen I structure whereas they represent less than 10% in collagen XII (Fig. 1.2b). Besides this structural collagen hallmark, collagens also possess non-triple helical domains, called non-collagenous domains (NC) (Fig. 1.2b, c). The NC domains are globular domains when located at the collagen molecule extremities and/or short linker regions of about 20 residues when located between domains. The NC domains

Table 1.1 Atlas of the 28 members of the collagen superfamily: genes, chain assembly, bioactive domains and supramolecular assemblies

Collagen type	Gene	α-chains	Released domain/ Matricryptins	α-chains assembly	Supramolecular assembly
Fibrillar collagens					
Collagen I	***COL1A1*** ***COL1A2***	α_1(I) α_2(I)	Trimer C-propeptide (C3) N-propeptide	$[\alpha_1(I)]_2\ \alpha_2(I)$ $[\alpha_1(I)]_3$	Collagen fibril
Collagen II	***COL2A1***	α_1(II)	Chondrocalcin N-propeptide	$[\alpha_1(II)]_3$	
Collagen III	***COL3A1***	α_1(III)	N-propeptide C-propeptide	$[\alpha_1(III)]_3$	
Collagen V	***COL5A1*** ***COL5A2*** *COL5A3*	α_1(V) α_2(V) α_3(V)	TSPN[a]	$[\alpha_1(V)]_2\alpha_2(V)$ $[\alpha_1(V)]_3$ $\alpha_1(V)\alpha_2(V)\alpha_3(V)$	
Collagen V/XI [b]	***COL5A1*** ***COL11A1*** ***COL2A1***	α_1(V) α_2(XI) α_3(XI)[b]		$\alpha_1(XI)\alpha_1(V)\alpha_3(XI)$	
Collagen XI [b]	***COL11A1*** ***COL11A2*** ***COL2A1***[b]	α_1(XI) α_2(XI) α_3(XI)[b]	TSPN[a]	$\alpha_1(XI)\alpha_2(XI)\alpha_3(XI)$	
Collagen XXIV	*COL24A1*	α_1(XXIV)		$[\alpha_1(XXIV)]_3$	
Collagen XXVII	***COL27A1***	α_1(XXVII)		$[\alpha_1(XXVII)]_3$	
Fibril-associated collagens with interrupted triple helices (FACIT)					
Collagen IX	***COL9A1*** ***COL9A2*** ***COL9A3***	α_1(IX) α_2(IX) α_3(IX)	TSPN[a]	$\alpha_1(IX)\alpha_2(IX)\alpha_3(IX)$	Fibril-Associated Collagens
Collagen XII	***COL12A1***	α_1(XII)		$[\alpha_1(XII)]_3$	
Collagen XIV	*COL14A1*	α_1(XIV)		$[\alpha_1(XIV)]_3$	
Collagen XVI	*COL16A1*	α_1(XVI)		$[\alpha_1(XVI)]_3$	
Collagen XIX	*COL19A1*	α_1(XIX)	NC1 α_1(XIX)	$[\alpha_1(XIX)]_3$	
Collagen XX	*COL20A1*	α_1(XX)		$[\alpha_1(XX)]_3$	
Collagen XXI	*COL21A1*	α_1(XXI)		$[\alpha_1(XXI)]_3$	
Collagen XXII	*COL22A1*	α_1(XXII)		$[\alpha_1(XXII)]_3$	

Network forming collagens					
Collagen IV	***COL4A1*** ***COL4A2*** ***COL4A3*** ***COL4A4*** ***COL4A5*** ***COL4A6***	α_1(IV) α_2(IV) α_3(IV) α_4(IV) α_5(IV), α_6(IV)	Arresten α1(IV) 1263-1277 α1(IV) CB3 α1(IV) 531-543 Canstatin Tumstatin Oncothanin Tetrastatin 1-3 Pentastatin 1-3 Hexastatin 1-2 NC1 α6(IV)	[α_1(IV)]$_2\alpha_2$(IV) α_3(IV)α_4(IV)α_5(IV) [α_5(IV)]$_2\alpha_6$(IV)	Hexagonal network
Collagen VI [c]	***COL6A1*** ***COL6A2*** ***COL6A3*** *COL6A5* *COL6A6*	α_1(VI), α_2(VI) α_3(VI) α_5(VI) α_6(VI)		α_1(VI)α_2(VI)α_3(VI) α_1(VI)α_2(VI)α_5(VI) α_1(VI)α_2(VI)α_6(VI)	Beaded filaments
Collagen VII	***COL7A1***	α_1(VII)		[α_1(VII)]$_3$	Anchoring fibrils
Collagen VIII	*COL8A1* ***COL8A2***	α_1(VIII), α_2(VIII)	Vastatin	[α_1(VIII)]$_3$ [α_2(VIII)]$_3$ [α_1(VIII)]$_2\alpha_2$(VIII) α_1(VIII)[α_2(VIII)]$_2$	Hexagonal network
Collagen X	***COL10A1***	α_1(X)		[α_1(X)]$_3$	
Transmembrane collagens or MACIT for Membrane-Associated Collagens with Interrupted Triple helices)					
Collagen XIII	***COL13A1***	α_1(XIII)		[α_1(XIII)]$_3$	
Collagen XVII	***COL17A1***	α_1(XVII)		[α_1(XVII)]$_3$	
Collagen XXIII	*COL23A1*	α_1(XXIII)		[α_1(XXIII)]$_3$	
Collagen XXV	***COL25A1***	α_1(XXV)		[α_1(XXV)]$_3$	
Multiplexins					
Collagen XV	*COL15A1*	α_1(XV)	Restin	[α_1(XV)]$_3$	Basement membrane associated-collagens
Collagen XVIII	***COL18A1***	α_1(XVIII)		[α_1(XVIII)]$_3$	

(continued)

Table 1.1 (continued)

Collagen type	Gene	α-chains	Released domain/ Matricryptins	α-chains assembly	Supramolecular assembly
			Endostatin Neostatin 7 and 14 Frizzled module		
Others					
Collagen XXVI	*COL26A1*	α_1(XXVI)		$[\alpha_1(XXVI)]_3$	
Collagen XXVIII	***COL28A1***	α_1(XXVIII)		$[\alpha_1(XXVIII)]_3$	

Collagen types and genes for which mutations have been associated with human diseases are in bold

[a] TSPN is released from the N-terminal extremity of the collagen molecules by BMP1 activity and has no known function so far

[b] α_3(XI) is encoded by *COL2A1* gene. The α_3(XI) incorporated in the collagen XI molecule is overmodified compared to α_1(II) that forms the collagen II molecule

[c] In contrast to its murine ortholog, the human *COL6A4* gene has been converted into pseudogene by a chromosomal break

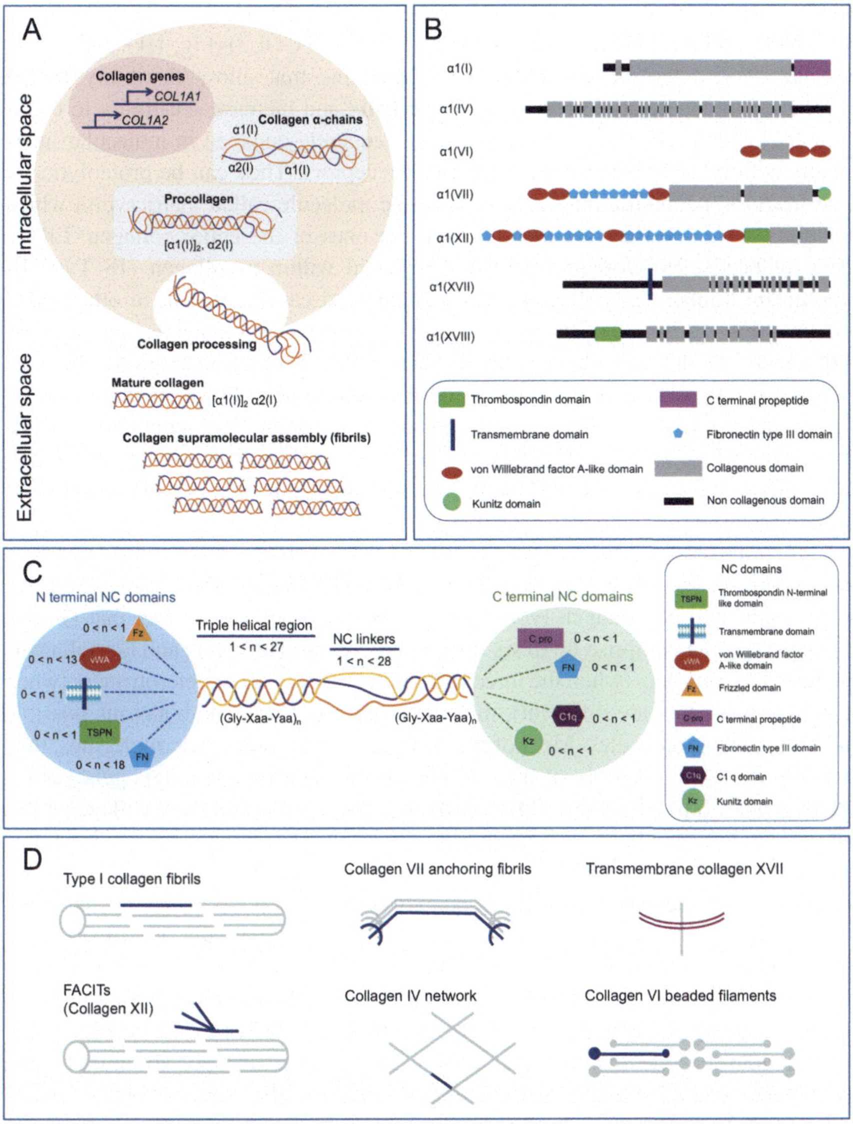

Fig. 1.2 The collagen superfamily: biosynthesis, molecular structure and extracellular aggregates. (**a**) Collagen I biosynthesis as a textbook case. *COL1A1* and *COL1A2* genes are translated respectively in proα1(I) and proα2(I) collagen chains that start to assemble as a triple helix through their C-terminal domain in the endoplasmic reticulum. Procollagen [α1(I)] α2(I) transits through the Golgi apparatus and is secreted into the extracellular space for C- and N-terminal processing. Mature [α1(I)] α2(I) collagen assembles in supramolecular assembly to form collagen fibrils. (**b**) Examples of the high structural diversity of collagens that contain collagenous (COL) and non-collagenous (NC) domains of varying length and number. (**c**) Prototype of a collagen molecule. All collagens contain at least one and up to 27 triple helical regions (also named collagenous domain - COL) that are composed of (Gly-Xaa-Yaa) repeats where Gly is a glycine, Xaa often a proline and

alternate with COL domains and are numbered starting from the C-terminus (e.g., NC1, NC2, and NC3 alternate with COL1, COL2, and COL3) (Fig. 1.1). NC linkers provoke short interruptions within COL domains that allow flexibility to the rod-shaped molecule conferred by the triple helix and increase sensibility to extracellular enzymes. NC domains are the most frequently engaged in molecular interactions with other ECM proteins and/or cell receptors. They can be proteolytically shed from the parental collagen as a bioactive molecule called matricryptin which functions, once released, are distinct from the ones of the entire collagen. Like a string of beads, NC domains are often repeated within a collagen (18 Type III Fibronectin domains in collagen XII α1 chain) and can be present in other ECM proteins (Fig. 1.2b, c). The most frequent domains in collagens are the von Willebrand like domain which is involved in protein-protein interactions; Type III Fibronectin domains and the thrombospondin N-terminal like domain while the Kunitz and Frizzled domains are only present in collagen VII and collagen XVIII respectively (Fig. 1.2b, c). The functions of these domains are not yet fully elucidated and what is even more obscure is how important is the number and order of NC domain repetition in collagen molecules.

Further diversity of the collagens occurs with the different existing subunits for a given collagen type. In humans there are 44 collagen genes, encoding for 44 α-chains, and 28 collagen types, that can be explained by the fact that (1) one collagen type can comprise up to six different α-chains (Table 1.1) and (2) collagens can form homotrimers (when the three α-chains are identical) or heterotrimers (when at least two α-chain are different), making several α-chains combinations possible and thus several molecules. Most of the collagens are homotrimeric molecules, thus resulting from the assembly of three α1-chains encoded by one single gene, as for collagen types II $[\alpha 1(II)]_3$ and III $[\alpha 1(III)]_3$. It is also the case for most collagen types with the highest numbers (collagen types XII to XVIII—Table 1.1). On the top of that, rare hybrid collagen molecules resulting from the association of α-chains from two different collagen types can be found in tissues. So far it has been described only for collagen V and the cartilaginous collagen XI, two structurally and functionally closely related collagens. An explanation for this rare event is, as previously suggested (Fichard et al. 1995), that collagen V/XI α-chains were initially wrongly attributed to two distinct collagen types due to the exclusive tissue-specific chain combinations despite the fact that they were found as closely structurally and functionally related. Finally, some collagens possess glycosaminoglycans chains and are thus by definition collagen/proteoglycan hybrids. Collagen IX carries chondroitin sulfate chains, collagen XVIII possesses heparan sulfate chains whereas

Fig. 1.2 (continued) Yaa often a hydroxyproline. Non-collagenous (NC) regions are intercalated between COL domains. Several others NC domains can be present at the C- or N-terminal part of the molecule and repeated up to 18 times as for the Fibronectin type III domain. (**d**) Collagen supramolecular aggregates. Mature collagens assemble into specific homotypic or heterotypic supramolecular aggregates to fulfil their functions. Some collagens do not exhibit any particular supramolecular organization

collagen XV is decorated by chondroitin sulfate chains and, to a less extent, heparin sulfate chains. This high molecular diversity results in the formation of a variety of supramolecular aggregates (Fig. 1.2d) that are responsible for tissue-specific mechanical properties such as stiffness, deformability, tensile force, or shear resistance but also to a wide range of biological functions through chemical and/or mechanical signals.

With the expansion of the collagen superfamily, collagen types have been classified into subfamilies according to similarities in their molecular or supramolecular organizations. This classification gave rise to six subsets of the collagen superfamily (Table 1.1 and Fig. 1.2d). The "supramolecular organization" subsets comprise the "fibril-forming collagens" (7 members) and the "network-forming collagens" (5 members). This subset includes collagen VI that assembles into "beaded filaments" and collagen VII that forms "anchoring fibrils." The "molecular organization" subsets include the FACIT (Fibril-Associated Collagens with Interrupted Triple helices) comprising 8 members; the MACIT (Membrane-Associated Collagens with Interrupted Triple helices) that are transmembrane collagens (4 members) and the multiplexins (for collagens with "*multiple* triple helix domains with interruptions"; 2 members). Two collagens remain unclassified, collagens XXVI and the last discovered collagen type XXVIII. Whereas the function of collagen XXVI is still unclear, type XXVIII collagen was found to be chiefly expressed in the peripheral nervous system and mutations in *COL28A1* are responsible for the Charcot-Marie-Tooth disease, a disease that primary impairs nerve myelinization (Grimal et al. 2010). For further and more in-depth analysis on the collagen subfamilies, we refer to previous comprehensive reviews (Ricard-Blum and Ruggiero 2005; Gordon and Hahn 2010; Ricard-Blum 2011).

It is here worthy to note that besides the "genuine" collagen molecules, a sizeable number of proteins that do contain one or several collagenous domains were not included in the collagen superfamily (reviewed in Ricard-Blum 2011). This holds true for the members of the collectin family, the acetylcholinesterase, emilins, adiponectin, C1q, ficolin, and more recently the gliomedin and ectodysplasin A, many of them just because they were first reported by "non-matrix biologists".

1.4 Collagen Biosynthesis: A Complex and Multistep Process

Fibrillar collagen biosynthesis has been extensively studied but regarding the extreme diversity of collagens, it would be naïve to assume that their biosynthesis, which includes supramolecular assembly, is identical for the other collagen family subtypes. They do share similarities in some early post-translational modifications, but even the initiation of the triple helix formation differs from one type to another. Yet, collagens IV, VI and VII display important biosynthetic specificities. For further information about these collagens biosynthesis, we invite the reader to refer to the corresponding chapters in this book. Here, we will discuss the biosynthesis of fibril-forming collagens only as it represents a textbook case of collagen biosynthesis. The full elucidation of this multistep process is crucial for the understanding of the pathogenesis of collagenopathies and further development of effective treatment for affected patients.

The biosynthesis of fibrillar collagens comprises a succession of intracellular and extracellular events from the formation of the triple helix and its secretion to the supramolecular organization into fibrils in the extracellular space. As for all secreted proteins, fibrillar collagens are synthetized as proα-chains by membrane-bound ribosomes and are then translocated into the lumen of the endoplasmic reticulum (ER) where they assemble into triple helix molecules (Fig. 1.2a). Collagen pro-α-chains then undergo numerous and complex post-translational modifications in the ER among which the most important are: (1) cleavage of the signal peptides; (2) hydroxylation of most of the prolines by the Prolyl-4-hydroxylases (P4H) at the Yaa position of the Gly-Xaa-Yaa triplets. Prolyl-3-hydroxylases also hydroxylate selected prolines residues in several collagens but not all (Rappu et al. 2019); (3) hydroxylation of some lysines and (4) glycosylation of hydroxylysines generated in step (3). These post-translational modifications are followed by an important step of collagen biosynthesis, that is the specific assembly of the newly synthesized collagen proα-chains into a triple helix. This occurs in a zipper-like manner, starting from the C-terminus of the molecule and propagating towards the N-terminus (Fig. 1.2a). Collagen C-propeptides play a major role in the selection of the pro-α-chains and procollagen folding. C-propeptides association depends on specific recognition sequences and protein disulfide isomerase (PDI) activity to create disulfide bonds between three individual C-propeptides that facilitate the initiation of the triple helix formation (Doege and Fessler 1986; Bulleid et al. 1997; Lees et al. 1997). In vivo, the presence of hydroxyprolines (about 10% in fibrillar collagens) allows the binding of HSP47 chaperon protein and the subsequent procollagen trimer stabilization together with the formation of interchain disulfide bonds within the C-propeptides discussed above (Makareeva and Leikin 2007).

The molecule is further translocated to the Golgi apparatus where accurate procollagen glycosylation and trafficking are ensured. During their transport to the extracellular space, procollagens start to aggregate laterally within Golgi to plasma membrane carriers. When procollagens reached the plasma membrane, carriers connect to the ECM through deep projections of the plasma membrane, called "fibripositors" (Canty and Kadler 2005; Kadler 2017). The precise nature of these vesicles is still a matter of considerable debate. Specifically, the question of how large rod-shape molecules of up to 700 nm in length can fit into secretory vesicles remains puzzling. Collagen trafficking is far from being completely understood and we refer the reader to comprehensive reviews on this topic (Canty and Kadler 2005; Banos et al. 2008; McCaughey and Stephens 2019).

Once secreted in the extracellular space, N- and C-propeptides of fibrillar procollagens are entirely (e.g., collagens I and III) or partially (e.g., collagens V and XI) removed by specific matrix enzymes including BMP-1, mTLD, and TLL-1 for the C-propeptide and ADAMTS family members for the N-propeptide. Mature collagens spontaneously self-assembled into fibrils that result from end-overlap alignments of collagen molecules, in a quarter-staggered packing pattern (Fig. 1.2a, d). Fibrils structure is strengthened by intra- and intermolecular crosslinks catalyzed by lysyl oxidases between lysine and hydroxylysine residues (Eyre et al. 1984; Mäki and Kivirikko 2001; Mäki et al. 2001; Vallet and Ricard-Blum 2019). Heterotypic fibrils made of two or three different fibrillar collagen types are the most common structure found in tissues (e.g., collagen I, III, and V in skin, collagen II and XI in cartilage) while homotypic fibrils made of collagen III only can be found in blood vessel walls (Kuivaniemi and Tromp 2019) and homotypic collagen V fibrils were described in embryonic tissues and in specific location in skin and cornea (Fichard et al. 1995; Bonod-Bidaud et al. 2012). The fact that banded collagen fibers could be formed in vitro directly from purified fibril-forming collagens without the contribution of cells has been overlooked and is still disputed.

1.5 Collagens Across Species

Although mice were historically used as a preferred model organism for human connective tissue diseases studies and pre-clinical trials, matrix scientists could not reasonably conduct in vivo research using only one animal model. Therefore, additional organisms such as zebrafish, drosophila, and *Caenorhabditis elegans* have rapidly emerged as reliable model organisms to study collagen function and are now also extensively used to investigate the pathogenesis and molecular mechanisms of collagenopathies. Accordingly, their matrisomes were established thanks to fruitful collaborative work between Naba's lab, the matrisome's maker, and ECM experts in zebrafish (our lab, (Nauroy et al. 2018)); *C. elegans* (Ewald's lab, (Teuscher et al. 2019)) and drosophila (Horne-Badovinac's lab, (Davis et al. 2019)). Complete matrisome gene lists are available on The Matrisome Project website (http://matrisomeproject.mit.edu).

As the zebrafish underwent an additional whole genome duplication compared to mammals, a large number of collagen genes are duplicated and referred as "a" and "b" in the nomenclature. The in silico approach used to define the zebrafish matrisome retrieved 58 collagen genes (Nauroy et al. 2018). Among them, the orthologue of the human *COL21A1* gene, whose function is still unknown, is present in the zebrafish genome but absent in the mouse genome as mentioned before. A second gene encoding for an additional chain of collagen XVIII that was lost during evolution in mammals exists in zebrafish (Gebauer et al. 2016). The duplicated collagen genes in zebrafish are submitted to different evolutionary processes such as non-functionalization of one copy, sub-functionalization, neofunctionalization, or dosage selection. Previous work that aimed at deciphering the function of collagen genes using zebrafish revealed gene specification of collagen paralogues (reviewed in Bretaud et al. 2019).

To establish the *C. elegans* matrisome, the authors combined gene orthology and protein structure analysis such as, for example, the presence of Gly-Xaa-Yaa repeats (the structural hallmark of collagens). The *C. elegans* matrisome contains 719 proteins including 181 collagens (Teuscher et al. 2019). However, except eight collagen genes, most of them are restricted to the cuticle and do not share orthology with human collagen genes. To clarify the *C. elegans* collagen classification, the authors created four different groups based on the structural organization of the molecules. The first group contains the vertebrate-like collagens that are conserved in the nematode including the BM collagens *emb-9* and *let-2* (orthologs of human collagen IV genes); a single multiplexin gene *cle-1* and a transmembrane collagen (MACIT), *col-99*. The second group is composed of proteins that contain a collagenous domain and resemble mammalian gliomedins and collectins. Non-cuticular *C. elegans* collagens with no orthology with mammalian genes are in the third group and 173 cuticular collagens with unknown function form the last group that is itself divided into five different clusters.

Using similar biocomputational approaches, the drosophila matrisome was established and contains 641 genes including only 4 collagens that are all BM collagens (Davis et al. 2019). Three of them possess mammalian orthologs while only one (pericardin) is specific to the drosophila. Contrary to the mammalian matrisomes, the drosophila matrisome has an additional category that is called "Apical Matrix" representing about 50% of the total drosophila matrisome and containing chitins-based ECM proteins secreted by the apical side of epithelial tissues.

1.6 An Introduction to Collagenopathies

Collagens are the most abundant proteins in humans and are broadly expressed in tissues. They display diverse structural features, as discussed above, and are known to fulfill diverse mechanical and biological functions. It is thus not surprising that defects in collagen synthesis and assembly lead to a large variety of common and genetic disorders that affect various tissues and organs (Fig. 1.3). A number of common diseases (microenvironmentally-driven tumors, fibrotic disorders, neurodegenerative diseases, etc.) are directly linked to an imbalance between collagens synthesis and degradation that lead to tissue dysfunction. In addition, collagens are associated with a broad spectrum of heterogeneous genetic diseases now known as "collagenopathies" that affect a multitude of organs and tissues including sensory organs. Mutations in genes of a single collagen type, such as collagen IV genes, can be responsible for different diseases; and conversely tissues as skin or cartilage can be affected by mutations in different collagen types. Both of these cases will be illustrated in the coming chapters.

Among the 44 human collagen genes, up to 29 genes coding for 17 distinct collagen types are associated with collagenopathies. Intuitively, it would have been assumed that genes responsible for collagenopathies would be the ones highly expressed in tissues. But genes responsible for collagenopathies are often expressed at low levels in the affected tissues or do not affect at all the tissues in which the

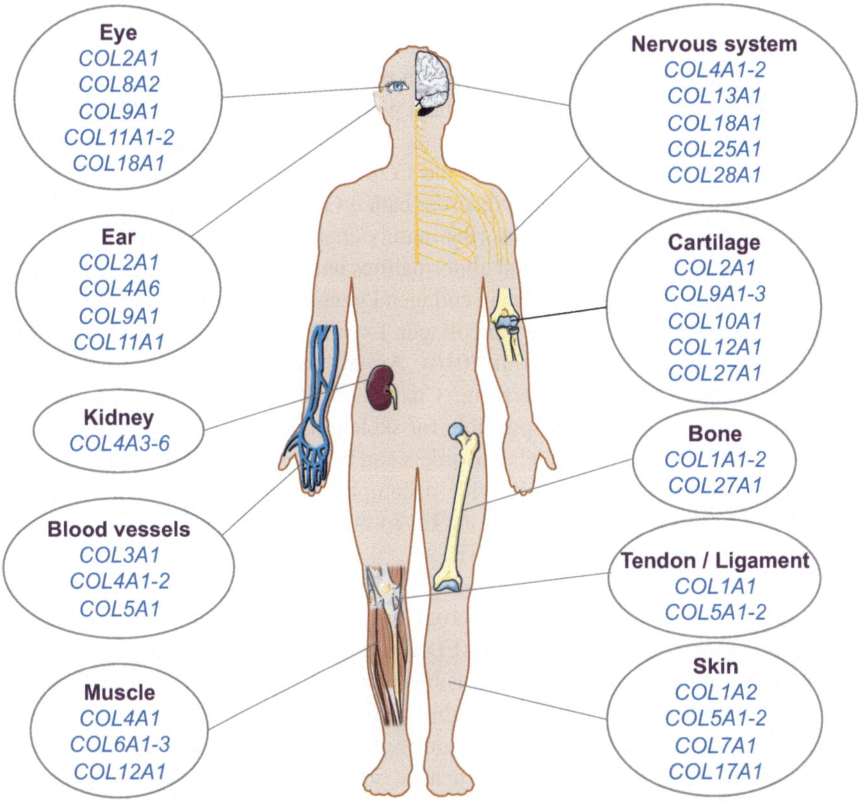

Fig. 1.3 Major tissues and organs of the human body affected by collagenopathies. For clarity, only primary affected functions of tissues and organs by a given collagen mutation are highlighted. This figure was created using Servier Medical Art under a Creative Commons Attribution 3.0 Unported License

corresponding protein is highly present. For instance, collagen V is highly expressed in cornea and minorly expressed in skin, but mutations in *COL5A1* or *COL5A2* genes primary result in defects in skin and joints (Ricard-Blum and Ruggiero 2005). Collagen XVIII isoforms are localized in various BM and the long isoform is highly expressed in liver, however, mutations in *COL18A1* are responsible for a rare condition called Knobloch syndrome characterized by severe vision problems (high myopia, retinal detachment) as well as skull defect (occipital encephalocele) (Sertié et al. 2000). In addition, there is growing evidence for diseases with overlapping symptoms and clinical presentation between different conditions. Collagen XII was first discovered in the early 1990s and knock-out mice for this gene showed a bone phenotype (Izu et al. 2011). Mutations in *COL12A1* was reported in 2013 with patients displaying a novel Ehlers–Danlos syndrome (EDS)/myopathy syndrome, named myopathic EDS in the recent classification of EDS (Zou et al. 2014; Hicks et al. 2014). Clinical and morphological overlaps between the Ullrich congenital muscular dystrophy (UCMD) caused by collagen VI gene mutations and classic EDS have also been reported (Kirschner et al. 2005; Delbaere et al. 2020).

Among collagens, collagen I is by far the most abundant collagen in vertebrates particularly in bone and skin providing them with structural support. This is thus not surprising that mutations in collagen I genes are associated with several connective tissue disorders. Mutations in *COL1A1* and *COL1A2* genes have been early reported to be responsible for a severe bone disorder called Osteogenesis imperfecta (OI). OI is an inherited connective tissue disorder mainly characterized by bone fragility and deformity leading to severe skeletal abnormalities and growth deficiency. In the last 15 years, it appeared that mutations in collagen I genes explain only 85%–90% of OI and that other molecules involved in collagen I synthesis, processing, or maturation can cause OI (Forlino and Marini 2016). As collagen I-related OI was already extensively reviewed in the literature, Chap. 2 by Omari et al. focuses on the newly discovered molecules responsible for skeletal disorders. Based on the recent discovery that OI and other skeletal disorders can be caused by defects in the protein secretory pathway, the authors provide a complete review on the importance of procollagen folding, trafficking from the ER to the Golgi, and degradation for bone and other connective tissue homeostasis.

Together with the fibril-forming collagens III and V, collagen I deficiency was also identified as responsible for a heterogeneous group of connective tissue disorders named Elhers–Danlos syndrome (EDS) as briefly introduced above. The EDS common features are fragile hyperextensible skin with abnormal wound healing and joint hypermobility but the important clinical heterogeneity of EDS prompted the clinicians to classify them into subcategories. An update of the Villefranche nosology was provided in 2017 and resulted in a list of 13 different EDS types that involve mutations in 20 different ECM genes (Malfait et al. 2017). In Chap. 3 Malfait and collaborators discuss the impact of mutations in other ECM proteins or collagen-modifying enzymes that were recently found to cause new EDS types or to augment the list of genes causing one of the already characterized EDS types. The authors have gathered the more recent information obtained from EDS patients, cells, and animal models of EDS to provide the reader with a clear review of the various and complex molecular defects causing the different subtypes of EDS and their supramolecular consequences.

Collagen II is the major component of the hyaline cartilage and the territorial matrix contains not less than six different collagens. Contrary to most of the members of the collagen superfamily, collagens II, IX, and XI are expressed only in cartilage with the exception of the vitreous in the eye. They co-polymerize into collagen II/IX/XI heterotypic fibrils in which collagen II is the major component while collagens IX and XI are minor components. Nevertheless, these minor components play important role in fibril formation and structure by regulating fibrillogenesis. Mutations in collagens II and XI, two fibrillar collagens, and in collagen IX, a member of the FACIT collagen subfamily, have been associated with a large spectrum of cartilage disorders among which osteoarthrosis. Common clinical features of cartilage disorders are disproportionate stature, dwarfism, short limbs, and premature osteoarthrosis. As some of the cartilaginous collagens are also expressed in the vitreous in the eye, patients with cartilage-related disorders also often have eye defects. In Chap. 4, Uwe Hansen outlines all aspects of the physiological and pathological roles of cartilaginous collagens and cartilage-related disorders.

Collagen IV is together with laminins, perlecan, and nidogen, one of the major components of all BM, a.k.a. BM tool-kit (Randles et al. 2016). BM are thin and highly organized ECM structures that underline all epithelia and endothelium, and surround a few individual cell types as for instance muscle cells, adipocytes, and Schwann cells (Jayadev and Sherwood 2017). Contrary to interstitial connective tissues, they are conserved across species and can be found in most multicellular organisms. Thereby, mutations in the six human genes encoding collagen IV lead to a wide spectrum of clinical signs and diseases ranging from vascular defects to eye and kidney disorders. Chapter 5 by Dean and Van Agtmael starts with an in-depth description of the distinctive organization of collagen IV genes, the complex biosynthesis of collagen IV molecules, and their expression patterns. The authors then carefully described the different genetic collagen IV related diseases among which Alport and Goodpasture syndromes are surely the most well-known. Importantly they also gave insights into the underlying mechanisms of those diseases and reported on the development of novel therapeutic strategies.

ECM components, and more specifically collagens, are critical for the development and homeostasis of the neuromuscular system. Twelve different collagens are expressed in skeletal muscle, including the neuromuscular junction and the myotendinous junction. Among them, collagen VI is definitely the most studied collagen and its deficiency underlies the so-called collagen VI-related myopathies that encompass a spectrum of disorders ranging from the milder Bethlem myopathy to the severe UCMD (Lamandé and Bateman 2018). These disorders are characterized by muscle weakness and atrophy, distal joint hyperlaxity, and proximal joint contractures. Chapter 6 by Tonelotto et al. carefully describes collagen VI and collagen VI-related disorders, but also highlights the important contribution that mouse and zebrafish models brought to the understanding of collagen VI-related collagenopathies and describes the role of other collagens in muscle-associated disorders.

The skin contains a large number of different collagen types, and consequently this tissue is highly subjected to collagenopathies. In Chap. 7, Nyström et al. emphasize the role of collagens VII and XVII on skin homeostasis. Both collagens are critical components of the dermo-epidermal junction as they ensure the cohesion of the epidermis with the underlying dermis either by forming anchoring fibrils for collagen VII or by being a component of hemidesmosomes for collagen XVII. Mutations in *COL7A1* or *COL17A1* genes lead to severe genetic and autoimmune skin blistering disorders such as epidermolysis bullosa (Has et al. 2018). The authors give a comprehensive overview of collagen VII and XVII physiology and of diseases associated with *COL7A1* or *COL17A1* mutations. They also described the different therapeutic approaches that are currently under evaluation to treat patients affected with these collagenopathies.

The past decades have seen the emergence of the role of ECM in the nervous system. Chapter 8 by Heikkinen et al. emphasizes the critical role of collagens in the development, homeostasis, and repair of the different morphological compartments of the central and peripheral nervous systems. Unexpectedly, mutations in collagens, that originally were not even thought to be expressed in the nervous system, were found to cause various neurodevelopmental, degenerative, and psychiatric disorders. Through the description of in vitro and in vivo data from several mouse models, this chapter gives a thorough and comprehensive overview of this expanding type of collagenopathies.

1.7 Outlook

There is little question that the number of collagen superfamily members will further increase, the identification of *COL6A5*, a novel chain of collagen VI, being wrongly attributed to the discovery of collagen XXIX (Fitzgerald et al. 2008). But the full landscape of collagen functions is far from being completely elucidated. This is already evidenced by the new findings and concepts in which collagens are involved in cell stem niche, cell signaling, growth factor bioavailability, mechanotransduction, regulation of metabolism, etc.

Mice have been instrumental in furthering our understanding of collagenopathies and for pre-clinical studies but zebrafish has recently joined the mouse as a powerful new model (Bretaud et al. 2019). The development of diverse animal models can be very helpful in the understanding of the disease mechanisms and in the identification of candidate genes for unresolved collagenopathies. Zebrafish model has pinpointed a critical role of collagen XXII and XV in the development of the neuromuscular system, and, as such, has revealed *COL15A1* and *COL22A1* as candidate genes for orphan neuromuscular disorders (Charvet et al. 2013; Pagnon-Minot et al. 2008; Guillon et al. 2016).

Nonetheless, there is already evidence that the list of collagenopathies will rapidly increase. Mutations in *COL8A2* are responsible for Fuchs' corneal dystrophy, a progressive degenerative disease of cornea (Mok et al. 2009), recessive mutations in *COL25A1* were shown to cause Congenital Cranial Dysinnervation Disorder (Shinwari et al. 2015), and mutations in *COL27A1* were described in patients with Steel syndrome (Gonzaga-Jauregui et al. 2015; Pölsler et al. 2020). *COL21A1*, a collagen gene remaining poorly documented, could be associated with nonsyndromic orofacial cleft (Mohamad Shah et al. 2019). With the widespread use of whole exome sequencing and whole genome sequencing in clinical diagnostics, the identification of new mutations involved in collagen-related diseases will definitely be shortly unraveled, an issue of outmost importance for the development of therapies for various collagenopathies.

1.8 Concluding Remarks

Collagens are the most abundant protein in mammals and form a superfamily of 28 large trimeric ECM proteins that display diverse composition and complex biosynthesis and spatial supramolecular organization. These multifaceted proteins ensure essential biological functions during development, homeostasis, repair, and disease. It thus should come as no surprise that out of the 44 human collagen genes, more than half are associated with collagenopathies so far. These genetic disorders that affect various tissues and organs generally show poor correlation between genotype and phenotype due in part to clinical heterogeneity and phenotypic overlap but it also highlights incomplete knowledge of collagen functions. A proper understanding on how these proteins influence matrisome function as a whole and the topology of the ECM in tissues will necessitate the use of various model organisms and integrated "omics" methodologies. This will undoubtedly help to shape new strategies for the development of innovative therapies for collagenopathies.

References

Banos CC, Thomas AH, Kuo CK (2008) Collagen fibrillogenesis in tendon development: current models and regulation of fibril assembly. Birth Defects Res Part C – Embryo Today Rev 84:228–244

Bonod-Bidaud C, Roulet M, Hansen U, Elsheikh A, Malbouyres M, Ricard-Blum S, Faye C, Vaganay E, Rousselle P, Ruggiero F (2012) In vivo evidence for a bridging role of a collagen V subtype at the epidermis-dermis interface. J Invest Dermatol 132:1841–1849

Bretaud S, Nauroy P, Malbouyres M, Ruggiero F (2019) Fishing for collagen function: about development, regeneration and disease. Semin Cell Dev Biol 89:100–108

Bulleid NJ, Dalley JA, Lees JF (1997) The C-propeptide domain of procollagen can be replaced with a transmembrane domain without affecting trimer formation or collagen triple helix folding during biosynthesis. EMBO J 16:6694–6701

Canty EG, Kadler KE (2005) Procollagen trafficking, processing and fibrillogenesis. J Cell Sci 118:1341–1353

Charvet B, Guiraud A, Malbouyres M, Zwolanek D, Guillon E, Bretaud S, Monnot C, Schulze J, Bader HL, Allard B, Koch M, Ruggiero F (2013) Knockdown of col22a1 gene in zebrafish induces a muscular dystrophy by disruption of the myotendinous junction. Development 140:4602–4613

Davis MN, Horne-Badovinac S, Naba A (2019) In-silico definition of the *Drosophila melanogaster* matrisome. Matrix Biol Plus 4:100015

Delbaere S, Dhooge T, Syx D, Petit F, Goemans N, Destrée A, Vanakker O, De Rycke R, Symoens S, Malfait F (2020) Novel defects in collagen XII and VI expand the mixed myopathy/Ehlers–Danlos syndrome spectrum and lead to variant-specific alterations in the extracellular matrix. Genet Med 22:112–123

Doege KJ, Fessler JH (1986) Folding of carboxyl domain and assembly of procollagen I. J Biol Chem 261:8924–8935

Eyre DR, Wu J, Woolley DE (1984) All three chains of 1α2α3α collagen from hyaline cartilage resist human collagenase. Biochem Biophys Res Commun 118:724–729

Fichard A, Kleman JP, Ruggiero F (1995) Another look at collagen V and XI molecules. Matrix Biol 14:515–531

Fitzgerald J, Bateman JF (2004) Why mice have lost genes for COL21A1, STK17A, GPR145 and AHRI: evidence for gene deletion at evolutionary breakpoints in the rodent lineage. Trends Genet 20:408–412

Fitzgerald J, Rich C, Zhou FH, Hansen U (2008) Three novel collagen VI chains, α4(VI), α5(VI), and α6(VI). J Biol Chem 283:20170–20180

Forlino A, Marini JC (2016) Osteogenesis imperfecta. Lancet 387:1657–1671

Gara SK, Grumati P, Urciuolo A, Bonaldo P, Kobbe B, Koch M, Paulsson M, Wagener R (2008) Three novel collagen VI chains with high homology to the α3 chain. J Biol Chem 283:10658–10670

Gebauer JM, Kobbe B, Paulsson M, Wagener R (2016) Structure, evolution and expression of collagen XXVIII: lessons from the zebrafish. Matrix Biol 49:106–119

Gonzaga-Jauregui C, Gamble CN, Yuan B, Penney S, Jhangiani S, Muzny DM, Gibbs RA, Lupski JR, Hecht JT (2015) Mutations in COL27A1 cause steel syndrome and suggest a founder mutation effect in the Puerto Rican population. Eur J Hum Genet 23:342–346

Gordon MK, Hahn R (2010) Collagens. Cell Tissue Res 339:247–257

Grimal S, Puech S, Wagener R, Ventéo S, Carroll P, Fichard-Carroll A (2010) Collagen XXVIII is a distinctive component of the peripheral nervous system nodes of Ranvier and surrounds nonmyelinating glial cells. Glia 58:1977–1987

Guillon E, Bretaud S, Ruggiero F (2016) Slow muscle precursors lay down a collagen XV matrix fingerprint to guide motor axon navigation. J Neurosci 36:2663–2676

Has C, Nyström A, Saeidian AH, Bruckner-Tuderman L, Uitto J (2018) Epidermolysis bullosa: molecular pathology of connective tissue components in the cutaneous basement membrane zone. Matrix Biol 71–72:313–329

Hicks D, Farsani GT, Laval S, Collins J, Sarkozy A, Martoni E, Shah A, Zou Y, Koch M, Bönnemann CG, Roberts M, Lochmüller H, Bushby K, Straub V (2014) Mutations in the collagen XII gene define a new form of extracellular matrix-related myopathy. Hum Mol Genet 7–10

Hynes RO, Naba A (2012) Overview of the matrisome-an inventory of extracellular matrix constituents and functions. Cold Spring Harb Perspect Biol 4:1–16

Izu Y, Sun M, Zwolanek D, Veit G, Williams V, Cha B, Jepsen KJ, Koch M, Birk DE (2011) Type XII collagen regulates osteoblast polarity and communication during bone formation. J Cell Biol 193:1115–1130

Jayadev R, Sherwood DR (2017) Basement membranes. Curr Biol 27:R207–R211

Kadler KE (2017) Fell Muir lecture: collagen fibril formation in vitro and in vivo. Int J Exp Pathol 98:4–16

Kirschner J, Hausser I, Zou Y, Schreiber G, Christen HJ, Brown SC, Anton-Lamprecht I, Muntoni F, Hanefeld F, Bönnemann CG (2005) Ullrich congenital muscular dystrophy: connective tissue abnormalities in the skin support overlap with Ehlers-Danlos syndromes. Am J Med Genet 132(A):296–301

Kuivaniemi H, Tromp G (2019) Type III collagen (COL3A1): gene and protein structure, tissue distribution, and associated diseases. Gene 707:151–171

Lamandé SR, Bateman JF (2018) Collagen VI disorders: insights on form and function in the extracellular matrix and beyond. Matrix Biol 71–72:348–367

Lees JF, Tasab M, Bulleid NJ (1997) Identification of the molecular recognition sequence which determines the type-specific assembly of procollagen. EMBO J 16:908–916

Makareeva E, Leikin S (2007) Procollagen triple helix assembly: an unconventional chaperone-assisted polding paradigm. PLoS One 2

Mäki JM, Kivirikko KI (2001) Cloning and characterization of a fourth human lysyl oxidase isoenzyme. Biochem J 355:381–387

Mäki JM, Tikkanen H, Kivirikko KI (2001) Cloning and characterization of a fifth human lysyl oxidase isoenzyme: the third member of the lysyl oxidase-related subfamily with four scavenger receptor cysteine-rich domains. Matrix Biol 20:493–496

Malfait F, Francomano C, Byers P, Belmont J, Berglund B, Black J, Bloom L, Bowen JM, Brady AF, Burrows NP, Castori M, Cohen H, Colombi M, Demirdas S, De Backer J, De Paepe A, Fournel-Gigleux S, Frank M, Ghali N, Giunta C, Grahame R, Hakim A, Jeunemaitre X, Johnson D, Juul-Kristensen B, Kapferer-Seebacher I, Kazkaz H, Kosho T, Lavallee ME, Levy H, Mendoza-Londono R, Pepin M, Pope FM, Reinstein E, Robert L, Rohrbach M, Sanders L, Sobey GJ, Van Damme T, Vandersteen A, van Mourik C, Voermans N, Wheeldon N, Zschocke J, Tinkle B (2017) The 2017 international classification of the Ehlers–Danlos syndromes. Am J Med Genet Part C Semin Med Genet 175:8–26

McCaughey J, Stephens DJ (2019) ER-to-Golgi transport: a sizeable problem. Trends Cell Biol 29:940–953

Mohamad Shah NS, Sulong S, Wan Sulaiman WA, Halim AS (2019) Two novel genes TOX3 and COL21A1 in large extended Malay families with nonsyndromic cleft lip and/or palate. Mol. Genet Genomic Med 7:1–10

Mok JW, Kim HS, Joo CK (2009) Q455V mutation in COL8A2 is associated with Fuchs' corneal dystrophy in Korean patients. Eye 23:895–903

Naba A, Clauser KR, Hoersch S, Liu H, Carr SA, Hynes RO (2012a) The Matrisome: in silico definition and in vivo characterization by proteomics of Normal and tumor extracellular matrices. Mol Cell Proteomics 11:M111.014647–M111.014647

Naba A, Hoersch S, Hynes RO (2012b) Towards definition of an ECM parts list: an advance on GO categories. Matrix Biol 31:371–372

Naba A, Clauser KR, Ding H, Whittaker CA, Carr SA, Hynes RO (2016) The extracellular matrix: tools and insights for the "omics" era. Matrix Biol 49:10–24

Nauroy P, Hughes S, Naba A, Ruggiero F (2018) The in-silico zebrafish matrisome: a new tool to study extracellular matrix gene and protein functions. Matrix Biol 65:5–13

Pagnon-Minot A, Malbouyres M, Haftek-Terreau Z, Kim HR, Sasaki T, Thisse C, Thisse B, Ingham PW, Ruggiero F, Le Guellec D (2008) Collagen XV, a novel factor in zebrafish notochord differentiation and muscle development. Dev Biol 316:21–35

Pölsler L, Schatz UA, Simma B, Zschocke J, Rudnik-Schöneborn S (2020) A Syrian patient with steel syndrome due to compound heterozygous COL27A1 mutations with colobomata of the eye. Am J Med Genet Part A 182:730–734

Randles M, Humphries MJ, Lennon R (2016) Proteomic definitions of basement membrane composition in health and disease. Matrix Biol 57-58:12

Rappu P, Salo AM, Myllyharju J, Heino J (2019) Role of prolyl hydroxylation in the molecular interactions of collagens. Essays Biochem 63:325–335

Ricard-Blum S (2011) The collagen family. Cold Spring Harb Perspect Biol 3:a004978

Ricard-Blum S, Ruggiero F (2005) The collagen superfamily: from the extracellular matrix to the cell membrane. Pathol Biol (Paris) 53:430–442

Sertié AL, Sossi V, Camargo AMA, Zatz M, Brahe C, Passos-Bueno MR (2000) Collagen XVIII, containing an endogenous inhibitor of angiogenesis and tumor growth, plays a critical role in the maintenance of retinal structure and in neural tube closure (Knobloch syndrome). Hum Mol Genet 9:2051–2058

Shinwari JMA, Khan A, Awad S, Shinwari Z, Alaiya A, Alanazi M, Tahir A, Poizat C, Al Tassan N (2015) Recessive mutations in COL25A1 are a cause of congenital cranial dysinnervation disorder. Am J Hum Genet 96:147–152

Teuscher AC, Jongsma E, Davis MN, Statzer C, Gebauer JM, Naba A, Ewald CY (2019) The in-silico characterization of the *Caenorhabditis elegans* matrisome and proposal of a novel collagen classification. Matrix Biol Plus 1:100001

Vallet SD, Ricard-Blum S (2019) Lysyl oxidases: from enzyme activity to extracellular matrix cross-links. Essays Biochem 63:349–364

Wells JM, Gaggar A, Blalock JE (2015) MMP generated matrikines. Matrix Biol 44–46:122–129

Winograd-Katz SE, Fässler R, Geiger B, Legate KR (2014) The integrin adhesome: from genes and proteins to human disease. Nat Rev Mol Cell Biol 15:273–288

Zaidel-Bar R (2013) Cadherin adhesome at a glance. J Cell Sci 126:373–378

Zaidel-Bar R, Itzkovitz S, Ma'ayan A, Iyengar R, Geiger B (2007) Functional atlas of the integrin adhesome. Nat Cell Biol 9:858–867

Zou Y, Zwolanek D, Izu Y, Gandhy S, Schreiber G, Brockmann K, Devoto M, Tian Z, Hu Y, Veit G, Meier M, Stetefeld J, Hicks D, Straub V, Voermans NC, Birk DE, Barton ER, Koch M, Bönnemann CG (2014) Recessive and dominant mutations in COL12A1 cause a novel EDS/myopathy overlap syndrome in humans and mice. Hum Mol Genet 23:1–14

Chapter 2
Procollagen Trafficking and its Implications in Osteogenesis Imperfecta

Shakib Omari, Elena Makareeva, and Sergey Leikin

Abstract Recent discoveries of skeletal dysplasias caused by mutations in COPII coat proteins responsible for a key checkpoint in the secretory pathway have renewed the interest of matrix biologists in procollagen trafficking. Osteogenesis imperfecta (OI) has long been considered a disorder related to malformations or malfunction of type I collagen and its procollagen precursor (type I collagenopathy). However, OI and chondrodysplasia features of patients with mutations in inner COPII coat proteins SEC24D and SEC23A have revealed that procollagen malformations and trafficking defects might cause similar pathology. The goal of this review is to bridge our knowledge of procollagen biosynthesis, secretory protein trafficking, degradative protein trafficking, and genetic defects in skeletal development, in order to understand the pathophysiology of these and related diseases. We argue that folding in the Endoplasmic Reticulum (ER), sorting by COPII-bound proteins at ER exit sites, transport by COPI carriers to Golgi, and autophagy of misfolded molecules are all intimately linked and crucial steps in procollagen homeostasis. Disruptions in these processes by mutations in key regulatory proteins underlie common collagenopathy-like skeletal pathologies in a wide variety of hereditary disorders ranging from OI to lysosomal storage diseases.

S. Omari
Eunice Kennedy Shriver National Institute of Child Health and Human Development, National Institutes of Health, Bethesda, MD, USA

Howard Hughes Medical Institute, Janelia Research Campus, Ashburn, VA, USA

E. Makareeva · S. Leikin (✉)
Eunice Kennedy Shriver National Institute of Child Health and Human Development, National Institutes of Health, Bethesda, MD, USA
e-mail: leikins@mail.nih.gov

F. Ruggiero (ed.), *The Collagen Superfamily and Collagenopathies*, Biology of Extracellular Matrix 8, https://doi.org/10.1007/978-3-030-67592-9_2

Abbreviations

ARC	arthrogryposis renal dysfunction and cholestasis syndrome
COP	coat protein complex I
COPII	coat protein complex II
ECM	extracellular matrix
EGAD	endosome and Golgi-associated degradation
ER	endoplasmic reticulum
ERAD	ER-associated degradation
ERES	ER exit site
ERGIC	ER-Golgi intermediate compartment
ISR	integrated cell stress response
MCDS	metaphyseal chondrodysplasia type Schmid
OI	osteogenesis imperfecta
TGN	trans-Golgi network
UPR	unfolded protein response

2.1 Introduction

Type I is the most abundant collagen, which is also by far the most abundant protein in all vertebrates. It is the main building block of fibers that provide structural scaffold for bones, skin, and other tissues and organs. Mutations in type I collagen and many other proteins that interact with or regulate type I collagen cause brittle bone disease characterized by bone fragility and deformities, which is scientifically known as osteogenesis imperfecta (OI). Disruptions in type I collagen homeostasis are frequently accompanied by abnormalities in tissues other than bones, resulting in secondary clinical features in OI or more complex genetic disorders that are clinically distinct from OI. Expression, structure, folding, and extracellular function of type I collagen are somewhat better understood. Readers interested in these processes and their dysregulation in OI and other diseases might find up to date information in other reviews (Bornstein and Sage 1989; Koide and Nagata 2005; Canty and Kadler 2005; Kadler et al. 2008; Marini et al. 2007, 2017; Bateman et al. 2009; Forlino et al. 2011; Eyre and Weis 2013; Arnold and Fertala 2013; Forlino and Marini 2016; Arseni et al. 2018; Hulmes 2019; Rossi et al. 2019; Shoulders and Raines 2009; Ishikawa and Bachinger 2013; Makareeva et al. 2011). In this chapter, we provide only a brief introduction and instead focus on less understood and reviewed processes of type I collagen quality control and trafficking in the cell, including secretory pathways, degradative pathways, and sorting of collagen precursors into these pathways. We discuss implications of dysregulation in these processes for OI and other disorders with overlapping bone phenotypes. Some of these disorders are type I collagenopathies, i.e., diseases caused by type I collagen mutations, malformation, and/or malfunction. Other ones have more complex pathophysiology involving multiple proteins and primary defects that are not directly related to collagen, yet their discussion is still useful in the context of collagenopathies.

2.2 Type I Collagen and Procollagen

Type I collagen is a heterotrimer of two α1(I) and one α2(I) chains wound together into ~300 nm long triple helix with short, non-helical terminal peptides (telopeptides) at each end. The triple helix consists of 338 consecutive Gly-Xaa-Yaa triplets within each chain. About half of the Xaa and Yaa positions are occupied by prolines. The obligatory Gly in every third position enables tight triple-helical conformation held together by interchain hydrogen bonding; the prolines further stabilize this conformation.

Type I collagen precursor (procollagen) is assembled from precursor proα1(I) and proα2(I) chains encoded by *COL1A1* and *COL1A2* genes, respectively. The chains are co-translationally translocated into the Endoplasmic Reticulum (ER), where they are posttranslationally modified and then folded together into trimeric procollagen molecules. The key posttranslational modifications are hydroxylation of most Yaa prolines and some Xaa prolines within the triple-helical regions into hydroxyprolines as well as hydroxylation and subsequent glycosylation of some lysines. Folded procollagen molecules are exported from the ER, trafficked through Golgi, secreted, and processed by proteolytic cleavage of N- and C-terminal propeptides into mature type I collagen molecules, which are then assembled into fibers (Koide and Nagata 2005; Canty and Kadler 2005; Kadler et al. 2008).

For more detailed up to date information on procollagen structure, synthesis, and folding, we refer the reader to dedicated reviews (Shoulders and Raines 2009; Ishikawa and Bachinger 2013; Makareeva et al. 2011; Lamande and Bateman 1999; Engel and Bachinger 2005), yet several unique features of procollagen briefly described below are important for understanding its quality control, sorting, and trafficking in the cell. (a) At normal body temperature (above 35 °C in humans), the triple helix is less thermodynamically stable than unfolded chains, favoring unfolding rather than folding of the native conformation (Leikina et al. 2002; Makareeva and Leikin 2007). (b) Procollagen folding in the ER is enabled by preferential binding of a collagen-specific chaperone HSP47 to the triple helix, which makes the native conformation more favorable (Makareeva and Leikin 2007; Koide et al. 2006). Cells lacking HSP47 can properly fold type I procollagen only below the normal body temperature (Fujii et al. 2019). (c) Preferential binding of HSP47 to the native triple helix results in its accrual rather than release during procollagen folding, indicating non-canonical procollagen quality control (Makareeva et al. 2011). This is opposite to canonical binding of ER chaperones to unfolded regions in protein chains, which prevents nonspecific interactions and signals that the protein is ready to be exported from the ER once the native conformation is achieved and the chaperones are released. (d) In contrast, HSP47 is released at reduced pH only after procollagen enters the secretory pathway, as discussed later in this chapter. The release of HSP47 makes procollagen thermally unstable; yet the triple helix unfolding is sufficiently slow to allow procollagen secretion, processing into collagen, and assembly into fibers (Leikina et al. 2002; Makareeva and Leikin 2007). In fibers, interactions between nearest neighbors prevent the unfolding of the whole molecule and enable collagen helices to withstand significant heating and mechanical loads (Miles and Ghelashvili 1999), while the unimpeded unfolding of small regions enhances fiber strength and elasticity (Privalov 1982; Privalov et al. 1979). Importantly, faster denaturation of structurally deficient (e.g., mutant) collagen molecules might prevent their incorporation into fibers (Makareeva et al. 2018; Mirigian et al. 2016).

2.3 Collagen Trafficking Disruptions in Heritable Bone Pathology

Another unique feature of type I collagen, which is important in the context of the present chapter, is the massive amount of procollagen molecules produced and secreted by osteoblasts, the cells responsible for bone synthesis. Not only is this the most abundant vertebrate protein, but it is also by far the most massively secreted protein, comparable only to type II collagen secretion by chondrocytes during cartilage development and in growth plates. Massive procollagen secretion presents a significant challenge for trafficking through the cell. Moreover, up to ~15% of newly synthesized procollagen molecules are misfolded and rerouted for intracellular degradation even in the absence of any pathology (Bienkowski et al. 1978). Together, massive synthesis and misfolding of procollagen burden the quality control, sorting, trafficking, and degradation machineries of the cell, which is likely why mutations affecting these machineries often have the same effects on bone as *COL1A1* and *COL1A2* mutations.

2.3.1 Osteogenesis Imperfecta

OI is a clinical rather than molecular/genetic diagnosis, yet most OI cases are caused by *COL1A1* and *COL1A2* mutations, and almost all deleterious mutations in these two genes cause OI (Marini et al. 2017; Rossi et al. 2019; Robinson and Rauch 2019). There are three possible bone pathology mechanisms for *COL1A1* and *COL1A2* mutations: (1) quantitative deficiency of type I collagen, e.g., "null allele" mutations that prevent translation of one *COL1A1* allele; (2) qualitative deficiency in assembly and function of extracellular collagen fibers, e.g., associated with structural defects in the collagen triple helix; and (3) deficiency in differentiation and function of osteoblasts, e.g., caused by the accumulation of misfolded mutant procollagen and the resulting disruption of the osteoblast ER. These mutation effects are interrelated and difficult to separate from each other. Indeed, deficient osteoblast function might result in lower collagen synthesis as well as abnormal assembly and function of collagen fibers even in the absence of any mutations. Similarly, abnormal composition and structure of extracellular matrix (ECM) that contains secreted mutant molecules might affect cell-matrix interactions and result in abnormal differentiation and function of osteoblasts. It is widely accepted that quantitative collagen deficiency causes only mild pathology, since most cases of mild OI are caused by null allele mutations in *COL1A1* (Forlino et al. 2011). However, relative contributions of ECM defects and osteoblast malfunction to severe OI are still being debated and might vary significantly for different mutations (Mirigian et al. 2016).

Recently discovered mutations in genes other than *COL1A1* and *COL1A2* that cause bone fragility disorders of OI spectrum might offer some clues (Marini et al. 2017; Besio et al. 2019a). These disorders include OI, Bruck syndrome (OI with contractures), osteoporosis-pseudoglioma syndrome (OI with eye involvement), Cole-Carpenter syndrome (OI with craniosynostosis), OI with calvarial doughnut lesions, etc. (Mortier et al. 2019). They are caused by mutations in a variety of proteins: (a) ER chaperones involved in procollagen folding (e.g., HSP47, P4HB/PDI, PPIB, and FKBP65); (b) regulators of ER and Golgi homeostasis (e.g., TMEM38B, CREB3L1, SGMS2, and S2P); (c) an actin-binding protein PLS3; (d) regulators of osteoblast differentiation and signaling (e.g., SP7, WNT1, and LRP5); and (e) potential regulators of osteoblasts (e.g., IFITM5 and PEDF). The common feature of all these mutations appears to be their known or likely effects on osteoblast function. These effects might be direct, e.g., disruption of osteoblast differentiation and signaling by SP7 and WNT1 deficiencies, respectively. They might also be indirect, e.g., disruption of osteoblast ER homeostasis by deficiencies in ER chaperones or possible disruption of osteoblast-ECM interaction by a gain of function mutation in the osteoblast plasma membrane protein IFITM5 in type V OI.

An interesting example is Cole-Carpenter syndrome 2 caused by mutations in SEC24D (Garbes et al. 2015; Moosa et al. 2016; Zhang et al. 2017), a subunit of the COPII coat regulating cargo export from the ER (Townley et al. 2008; Zanetti et al. 2011). The normal color of sclera and other clinical features of these patients are inconsistent with simple quantitative deficiency in procollagen I secretion. Overall, the clinical features are close to the Cole-Carpenter syndrome 1 caused by mutations in P4HB/PDI, the β subunit of prolyl-4-hydroxylases responsible for hydroxylation of Yaa prolines and therefore normal folding of the collagen triple helix (Balasubramanian et al. 2018; Rauch et al. 2015). These features of SEC24D mutations reinforce the idea that disruptions in cargo movement through the cell might be particularly detrimental for osteoblasts because of the massive burden of procollagen quality control, sorting, and trafficking. Like procollagen misfolding, SEC24D mutations might disrupt osteoblast function by causing excessive accumulation of procollagen in the ER.

2.3.2 Skeletal Pathology in Other Protein Trafficking Disorders

Bone, ligament, and tendon pathology were also reported for mutations in other proteins involved in cargo export from the ER, including COPII coat subunits SEC23A (Boyadjiev et al. 2006) and SEC31A (Halperin et al. 2019), sedlin/TRAPPC2 (Sacher et al. 2019), and TANGO1 (Lekszas et al. 2020). In these patients, the clinical features were inconsistent with OI, which is not surprising given the variable effects of different mutations on the protein function, alternative splicing isoforms, and variable expression and function of protein paralogs in different tissues. Even in the case of null mutations, the protein might be replaced by a paralog with variable success in different tissues, e.g., SEC23B was proposed to compensate for SEC23A (Khoriaty et al. 2018). The role of all these proteins and paralogs in the export of type I procollagen from the ER is still poorly understood.

Similarly, variable skeletal abnormalities suggestive of collagen homeostasis dysregulation were reported for mutations in multiple proteins involved in the secretory trafficking pathway downstream of ER export, including RIN2, ATP6V0A2, GORAB, COG4, COG7, dymeclin, RAB33B, S1P, S2P, and FGD1 (Mortier et al. 2019; Sacher et al. 2019; Vanakker et al. 2015; Dell'Angelica and Bonifacino 2019; Cutrona et al. 2018; Kondo et al. 2018; Ferreira et al. 2018; Bexiga and Simpson 2013; Lindert et al. 2016). An important example is skeletal dysplasia caused by heterozygous loss-of-function mutations in the δ subunit of COPI coat ARCN1 (Izumi et al. 2016). The pathology in these patients resembles Stickler syndrome caused by type II rather than type I collagen defects, suggesting that the mutation affects mostly chondrocytes, particularly in the growth plate. Apparently, chondrocytes cannot balance the loss of one *ARCN1* allele by upregulating the expression of the second allele. Notably, various chondrodysplasia symptoms were also described for patients with mutations in all aforementioned proteins that regulate secretory trafficking. For more information on cartilage disorders associated with type II collagen defects see Chap. 4 of this book.

A large fraction of patients with lysosomal storage diseases, particularly deficiencies in various lysosomal enzymes that degrade mucopolysaccharides, also have pronounced skeletal pathology (Clarke and Hollak 2015). A recent study of mucopolysaccharidoses in mice suggested that skeletal defects and type II collagen production by chondrocytes can be rescued upon restoring cargo delivery to lysosomes, even without altering deficient mucopolysaccharide degradation (Bartolomeo et al. 2017). Apparently mucopolysaccharidoses suppress cargo delivery by inducing lysosomal association and subsequent hyperactivation of mTORC1. Given that a significant fraction of newly synthesized procollagen is degraded by lysosomes (Bienkowski et al. 1978; Ripley and Bienkowski 1997), defects in degradative procollagen trafficking might have similar outcomes to defects in secretory trafficking.

2.4 Basic Trafficking Concepts for Secretory Proteins

2.4.1 Secretory Trafficking

To understand the ongoing debate on mechanisms and pathways of procollagen trafficking, it is important to begin from a brief summary of more extensively studied trafficking concepts for other proteins. The classical view of the secretory pathway (Bonifacino and Glick 2004) involves the following steps: (i) Secretory cargo is exported from the ER through ER exit sites (ERESs) and carried to an intermediate ER-Golgi compartment (ERGIC) in COPII-coated vesicles, from which it is transported to Golgi in COPI-coated vesicles. (ii) The cargo progresses through the Golgi stack from cis-Golgi to trans-Golgi network (TGN). (iii) Different cargoes are sorted at the TGN and delivered to the plasma membrane. The overall scheme is widely accepted, but there is less agreement on the details of each trafficking step, particularly in mammalian cells.

The organization and function of ERESs have been described in multiple reviews (Zanetti et al. 2011; Brandizzi and Barlowe 2013; Geva and Schuldiner 2014; Barlowe and Helenius 2016). Briefly, ERES formation is activated by GDP → GTP exchange at SAR1 GTPase by an ER-associated exchange factor SEC12. The activated SAR1-GTP attaches to the ERES membrane surface and recruits the first layer of the COPII coat, which is composed of SEC23-SEC24 heterodimers. Polymerization of $(SEC13)_2$-$(SEC31)_2$ tetramers on top of this inner coat produces an outer COPII coat. Protein cargo is loaded into ERES by cargo-specific receptors that bind to SEC24 in the inner COPII coat (Dancourt and Barlowe 2010). The COPII coat is disassembled by activation of the SAR1 GTPase activity. Recent studies suggest complex regulation of COPII coat formation and disassembly, cargo loading, and cargo export from ERES by SEC16 and a variety ERES-associated kinase complexes (Sprangers and Rabouille 2015; Centonze and Farhan 2019; Subramanian et al. 2019).

The COPII coat is clearly important for the export of secretory proteins from the ER (Dell'Angelica and Bonifacino 2019; Aridor 2018; McCaughey and Stephens 2018), although some protein cargoes can be delivered from the ER to Golgi even when COPII coat assembly is blocked (Mironov 2014). What is less clear is whether the COPII coat forms a cage around ERES that produces 60–100 nm COPII-coated transport vesicles, which is a model based on COPII cages reconstituted in vitro and similar structures observed in electron microscopy images of cultured cells (Zanetti et al. 2011; Venditti et al. 2014). Competing observations include TANGO1-COPII collar around the neck between ER and ERES (Raote and Malhotra 2019; Raote et al. 2018; Reynolds et al. 2019); finger-like ERES projections (Mironov et al. 2003) that might extend over 1 μm and recruit COPI coat (Weigel et al. 2019); and COPI rather than COPII transport intermediates moving along microtubules from the ER to Golgi (Presley et al. 1997; Scales et al. 1997).

One approach to reconciling these observations is the ever-malleable notion of ERGIC as a functionally distinct intermediate membrane compartment between the

ER and Golgi. Various ERGIC models have been proposed, e.g., sorting/redistribution facility that repackages cargo into COPI vesicles, dynamic system of transport intermediates formed by maturation/fusion of ERES-derived vesicles, tunnel-like ER-Golgi connections, etc. (Zanetti et al. 2011; Brandizzi and Barlowe 2013; Appenzeller-Herzog and Hauri 2006; Saraste and Marie 2018). However, neither immunofluorescence nor time lapse imaging of any marker or cargo protein shows ERGIC as a distinct compartment, in contrast to ER, ERES, and Golgi (Lippincott-Schwartz 2001; Lippincott-Schwartz et al. 2000). The commonly used ERGIC protein marker ERGIC53 is instead a lectin cargo receptor for COPII and COPI coats that rapidly cycles between ERES and Golgi, only reinforcing the idea of membrane structures referred to as ERGIC being just transient transport intermediates.

The COPI coat is believed to mediate transport intermediate scission from its origin and tethering to its destination, but its role in stabilizing transport intermediate structure is less clear (Cai et al. 2007; Faini et al. 2013; Hsu et al. 2009). Like COPII, COPI coat formation is activated by GDP → GTP exchange at a specific GTPase ARF1, which leads to the recruitment of COPI proteins. Unlike COPII, the seven subunit COPI coatomer is recruited from the cytosol to the membrane by ARF1-GTP *en bloc* (Bethune and Wieland 2018). Inhibition of GBF1-catalyzed GDP → GTP exchange at ARF1 by brefeldin A and other compounds stabilizes ARF1-GDP, causing the release of COPI coatomers to the cytosol (Hsu and Yang 2009). The same receptors are responsible for cargo recognition by COPII and COPI coats (Dancourt and Barlowe 2010). COPI coat structure is more dynamic and flexible than COPII, explaining why COPI coatomers might remain attached to transport intermediates and do not form cages in vitro (Faini et al. 2013). GTP hydrolysis by ARF1 is essential for COPI coatomer release, but its putative role in transport intermediate scission remains controversial (Faini et al. 2013; Bethune and Wieland 2018).

Concepts of secretory protein progression through the Golgi stack are continuously evolving as well (Pantazopoulou and Glick 2019; Mironov and Beznoussenko 2012; Beznoussenko et al. 2014). The classical view of this progression as protein movement through a quasi-stationary stack of Golgi cisternae has been supplemented by the ideas involving dynamic rearrangements of Golgi cisternae and cisternal maturation in time and space from the cis-Golgi to trans-Golgi network (TGN). Some soluble cargo, luminal, and membrane components diffuse through dynamic connections between the cisternae; some components remain in the maturing cisternae; and some are shuttled between the cisternae, returned to the ER by retrograde trafficking, or delivered to the endosomal compartment.

Protein sorting at the TGN is more complex and relatively well characterized only for some cargoes (Di Martino et al. 2019; Guo et al. 2014; Pakdel and von Blume 2018). It involves a separate pathway for transmembrane proteins. It also includes secretion through the endo-lysosomal compartment, regulated secretion through secretory storage granules, SPCA1 and CAB45-dependent sorting, and a poorly understood pathway for large ECM molecules briefly reviewed in Sect. 2.5.2.

2.4.2 Degradative Trafficking

Misfolded secretory proteins that do not attain their native conformation are believed to be exported from the ER to cytosol for proteasomal degradation (ER-associated degradation pathway, aka ERAD) or transported to lysosomes (autophagy pathway) (Fregno and Molinari 2019). Recently, a novel endosome and Golgi-associated degradation pathway (EGAD) was reported, in which membrane proteins are rerouted from Golgi to endosomes and then exported into the cytosol for proteasomal degradation (Schmidt et al. 2019).

ERAD is an integral part of unfolded protein response (UPR), and it is upregulated by activation of UPR receptors IRE1, ATF6, and PERK (Hwang and Qi 2018). ERAD substrate recognition in mammalian cells is not fully understood, but it is believed that misfolded proteins are targeted to ER membrane E3 ubiquitin ligase complexes (e.g., SEL1L-HRD1) for retrotranslocation to the cytosol, ubiquitination, and proteasomal degradation (Qi et al. 2017; Ruggiano et al. 2014). One proposed targeting mechanism is based on N-linked oligosaccharides attached to some polypeptide chains, which enable binding cycles of nascent polypeptide chains with lectin ER chaperones calnexin and calreticulin until the protein is folded. After futile cycling, the oligosaccharides are trimmed by ER mannosidases and recognized by lectin ERAD receptors OS-9 and XTP3-B (Roth and Zuber 2017). Another proposed mechanism is independent of N-linked oligosaccharides and based on retention of BiP/GRP78 and other ER chaperones by misfolded proteins, although it is still unclear how such misfolded proteins are distinguished from partially folded ones (Qi et al. 2017).

Autophagy is the process of cargo delivery to lysosomes for degradation and recycling, which has three main mechanisms: macroautophagy, microautophagy, and chaperone-mediated autophagy (Galluzzi et al. 2017). In macroautophagy, the cargo is surrounded and encapsulated by a specialized phagophore double membrane containing lipidated LC3 and/or GABARAP, creating an autophagosome that subsequently fuses with a lysosome or an endosome (Mercer et al. 2018; Papandreou and Tavernarakis 2017; Yu et al. 2018). In microautophagy, the cargo is directly surrounded by lysosomal or endosomal membrane and internalized (Li et al. 2012; Oku and Sakai 2018). In chaperone-mediated autophagy, a pentapeptide motif is recognized by a cytosolic chaperone HSC70, which facilitates translocation of cytosolic proteins targeted for lysosomal degradation through LAMP2A complexes in lysosomal membranes (Kaushik and Cuervo 2018). Interestingly, HSC70 can also facilitate selective macroautophagy and endosomal microautophagy of some cytosolic cargo (Tekirdag and Cuervo 2018).

Misfolded secretory proteins not captured by ERAD are commonly believed to be delivered to lysosomes by macroautophagy of ER regions where these proteins are accumulated (Hubner and Dikic 2020). Such selective macroautophagy of the ER is commonly referred as reticulophagy or as ER-phagy (Klionsky et al. 2016). Six mammalian receptors of ER-phagy in the ER membrane have been identified so far, FAM134B, RTN3L, CCPG1, SEC62, TEX264, and ATL3 (Hubner and Dikic

2020). All of them have a cytosolic LIR motif, which recruits the phagophore by binding to LC3 (or GABARAP), but the mechanisms of cargo recognition by these receptors are unknown. These receptors might also participate in other autophagic pathways of ER degradation (for consistency, here we restrict the usage of ER-phagy to its original definition given above). In particular, FAM134B or RTN3L mediates shedding and autophagy of ER vesicles containing misfolded cargo, while SEC62 mediates ER microautophagy (Chino and Mizushima 2020). Interestingly, FAM134B and RTN3L mediated ER fragmentation and autophagy appears to require SEC24C (Cui et al. 2019).

Although autophagy receptors in the ER facilitate ER-phagy, ER vesicle autophagy, and ER microautophagy, they are not required for these pathways. Phagophore or lysosomal membranes can also be recruited to any ER membrane protein with a ubiquitinated cytosolic tail by p62, which is a cytosolic protein with both ubiquitin-binding and LIR motifs (Chino and Mizushima 2020).

2.5 Procollagen Secretion

2.5.1 ER-Golgi Trafficking

Procollagen has long been a prototypical large cargo for secretory pathway studies, yet more recent interest has been sparked by the discovery of cranio-lenticulo-sutural dysplasia being caused by SEC23A mutations that impair procollagen trafficking (Boyadjiev et al. 2006; Fromme et al. 2007). Seemingly consistent with the classical paradigm of trafficking in 60–100 nm COPII vesicles, this finding renewed the interest in resolving whether ECM molecules over several hundred nanometers in size are packaged into much smaller COPII vesicles or exported from the ER in some other way (Jin et al. 2012; Kim et al. 2012).

The first proposed explanation was the formation of enlarged COPII vesicles (model 1, pathway 2a in Fig. 2.1). This model was initially based on the observation of increased collagen secretion by fetal human lung fibroblast cell line (IMR90) upon overexpression of E3 ubiquitin ligase Cullin-3 adapter KLHL12 involved in SEC31 monoubiquitination (Jin et al. 2012). The same study also reported reduced secretion and increased procollagen accumulation in the ER of human HT1080 fibrosarcoma cells by dominant negative Cullin-3 mutant. Subsequent studies revealed ~1 μm and even larger vesicle-like structures containing procollagen and KLHL12 (Gorur et al. 2017), which were also found to contain COPII coat proteins, TANGO1, and HSP47 (Yuan et al. 2018). Since ER to Golgi trafficking of procollagen was shown to be COPII dependent, these structures were interpreted as procollagen transport vesicles, suggesting that SEC31 ubiquitination mediated by KLHL12 might enable the enlargement of transport vesicles (Jin et al. 2012). However, neither of the two key characteristics of bona fide transport vesicles, which are directional movement along microtubules and cargo delivery to ERGIC or Golgi, were validated.

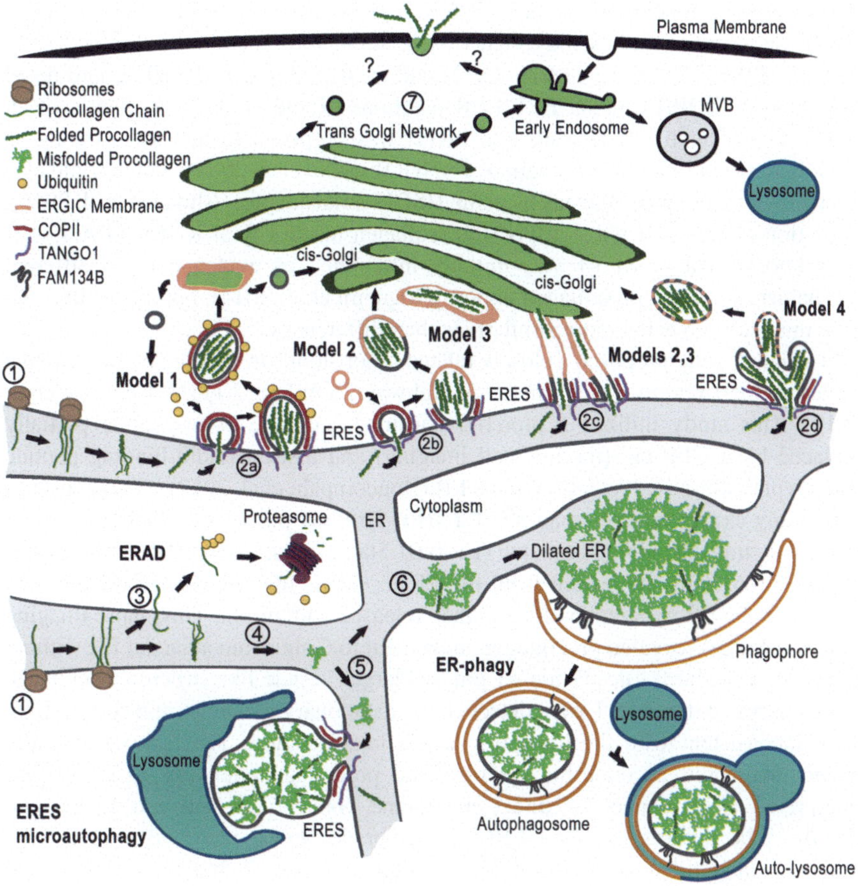

Fig. 2.1 Models of procollagen biosynthesis and trafficking in the cell. **(1)** Procollagen synthesis and folding in the ER lumen. **(2)** ERES → Golgi trafficking: ***(2a, model 1)*** *Large monoubiquitinated COPII vesicles*; ***(2b, models 2,3)*** *Large transport vesicles/intermediates derived from ERGIC membranes (model 2), which might mature into cis-Golgi cisternae without being transported through the cell (short-loop model 3).* ***(2c, models 2,3)*** *Direct ERES-cis-Golgi connections (tunnels);* ***(2d, model 4)*** *Morphologically diverse transport intermediates formed by ARF1/COPI at ERES that perform functions commonly attributed to ERGIC (might include ARF1/COPI-dependent direct ERES-cis-Golgi connections).* **(3)** Proteasomal degradation (ERAD). **(4)** Triple helix misfolding. **(5)** Lysosomal degradation via ERES microautophagy. **(6)** Procollagen aggregation in the ER lumen, the formation of dilated ER regions, and degradation of large aggregates by ER-phagy. **(7)** Secretory and degradative trafficking through Golgi, TGN, and endo-lysosomal pathway. Question marks indicate unknown or poorly characterized sorting/trafficking mechanisms, for which no well-defined models are available

A second explanation was the formation of large transport vesicles from ERGIC membranes that fuse with ERES (model 2, pathway 2b in Fig. 2.1). This model was initially based on the observation that ER-Golgi trafficking of type VII procollagen overexpressed in a human RDEB/FB/C7 cell line was dependent on SLY1 and

STX18 proteins mediating fusion of ERGIC membranes with the ER (Nogueira et al. 2014). Subsequent studies suggested ERGIC membrane recruitment by the TEER domain of the TANGO1 cytosolic tail bound to SEC23 and TANGO1-coordinated formation of COPII rings at ERES-ER junctions (Raote et al. 2018; Santos et al. 2015). Procollagen VII loading into ERES was proposed to be facilitated by its binding to the luminal SH3 domain of TANGO1 (Saito et al. 2009), which could also interact with other procollagens by using HSP47 as an adapter (Ishikawa et al. 2016). Retention of type VII but not type I procollagen in the ER of RDEB/FB/C7 cells after knockdown of SLY1 was initially interpreted as evidence for a different mechanism of type I procollagen export (Nogueira et al. 2014), but a later live cell imaging study revealed more similarities than differences.

The latter study proposed a closely related "short-loop" trafficking model with the same transport intermediates (model 3, pathway 2b in Fig. 2.1) (McCaughey et al. 2019). This study utilized a proα1(I) construct with the N-propeptide partially replaced by a GFP tag (for live cell imaging) and a streptavidin-binding peptide (for streptavidin-based retention in the ER (Boncompain et al. 2012)). This construct was stably expressed in human hTERT-RPE-1 retinal pigment epithelium cells or transiently transfected into human hTERT-BJ-5ta fibroblasts and IMR90 fibroblasts. To synchronize its movement from the ER to Golgi, fluorescent procollagen was trapped in the ER by streptavidin and then released with biotin. Time lapse imaging (15–30 s/frame) revealed procollagen movement to Golgi from adjacent ER without apparent vesicular traffic. While ~1 μm and larger vesicle-like structures similar to those described in model 1 were detected, their stationary dynamics and composition were inconsistent with the transport vesicle interpretation. The lack of vesicular traffic that could account for the observed procollagen delivery to Golgi was interpreted as evidence for direct membrane connections between ERES and ERGIC. The authors assumed ERGIC membranes to be recruited to ERES by TANGO1 and filled with procollagen similar to model 2, but then mature into proximal cis-Golgi without detectable movement after fission from ERES.

Faster time lapse imaging of fluorescently tagged procollagen in mouse MC3T3 osteoblasts (0.5–5 s/frame) revealed numerous rapidly moving procollagen puncta, which were generally <0.5 μm in size and appeared to travel along microtubules (Omari et al. 2018). In this study focused on procollagen autophagy, various fluorescent proteins were used to replace a part of proα2(I) N-propeptide. The experiments were performed within 24 h of transient transfection, to avoid disruption of cellular function. Some of the puncta could be traced leaving ERES and some entering Golgi, which was difficult to do even by super-resolution Airyscan microscopy at ~120 nm spatial resolution and 0.5 s/frame. Many puncta were moving so fast (~ 1 μm/s or faster) that they were at the detection limit even at 0.5 s/frame scanning speed. At 5 s/frame, most puncta became indistinguishable from stochastic fluctuations in the background fluorescence and many were likely missed even at 0.5 s/frame. These puncta were clearly transport intermediates delivering procollagen from the ER to Golgi, but none of them contained COPII coat proteins. Although vesicle-like structures containing procollagen and KLHL12 as well as procollagen and Cullin-3 were also observed, they were generally larger than the

ER-Golgi transport intermediates and exhibited only limited, stochastic rather than directional motion, in agreement with the study underlying model 3. All KLHL12 or Cullin-3 positive vesicle-like structures containing procollagen were also positive for LC3 and/or LAMP1, indicating that they were autophagic intermediates in the degradative pathway rather than transport intermediates in the secretory pathway.

Model 4 proposed COPI-dependent formation of procollagen transport intermediates at ERES (pathway 2d in Fig. 2.1) based on a follow-up study of type I procollagen trafficking in MC3T3 cells by Airyscan microscopy (Omari et al. 2020). In this model, procollagen carriers form by the maturation of distal ERES regions, which is accompanied by removal of the COPII coat, accumulation of ERGIC53, release of HSP47, and activation of ARF1 that recruits COPI coatomers. Accumulation of ERGIC53 could result from TANGO1-mediated ERES fusion with ERGIC53 membranes that are part of the dynamic pool of anterograde and retrograde ER-Golgi transport intermediates, as discussed in the preceding section. Since no vesicles have been observed merging or exiting any compartment between ERES and cis-Golgi, model 4 describes procollagen transport intermediates as dynamically maturing membrane structures that derive from ERES rather than ERGIC, even though they contain ERGIC53 and ARF1.

Another distinction of model 4 is the conclusion that HSP47 is released from folded procollagen at ERES before procollagen transport intermediate formation (Omari et al. 2020). Only several out of hundreds ER-Golgi procollagen transport intermediates traced by time lapse Airyscan microscopy were found to contain HSP47, consistent with HSP47 leakage rather than previously presumed (Satoh et al. 1996; Smith et al. 1995) co-trafficking with procollagen. True HSP47 and procollagen co-trafficking was observed only when the C-terminal RDEL motif in HSP47 was deleted or mutated to RNGL, or the cells were severely stressed (Omari et al. 2020). The study did not address whether reduced pH or other differences between the microenvironments in the ER lumen and distal ERES regions are responsible for the HSP47 release.

Perhaps the most important distinction of model 4 is the dependence of procollagen transport intermediate formation at ERES on ARF1 activation by GBF1 and therefore COPI coat recruitment. In collagen-producing cells, inhibition of the GBF1-catalyzed GDP $\rightarrow$ GTP exchange at ARF1 by brefeldin A causes accumulation of vesicle-like structures containing procollagen and HSP47 (Satoh et al. 1996), similar to inhibition of GTP hydrolysis by GTPγS at ARF1 and other GTPases (Smith et al. 1995). These structures were initially interpreted as transport vesicles that could not enter Golgi because of its dispersal by brefeldin A (Satoh et al. 1996) or inhibition of COPI coat dissociation by GTPγS (Smith et al. 1995). However, live cell imaging revealed a general propensity of procollagen and HSP47 to form large vesicle-like structures that are not transport vesicles (McCaughey et al. 2019; Omari et al. 2018). Super-resolution imaging of their movement, colocalization with organelle markers, and fluorescence recovery after photobleaching has shown that these structures are dilated regions of ER lumen, ERESs, or autophagic intermediates (Omari et al. 2020). Contrary to the initial assumption, brefeldin A has been shown to cause the disappearance of ER-Golgi procollagen transport intermediates by preventing their formation at ERESs. This inhibition of procollagen export causes the accumulation of dilated ER regions that have been misidentified as transport vesicles. Given the reports of COPI-coated tubular ERES projections (Weigel et al. 2019) and COPI rather than COPII coat at mammalian Golgi-destined transport vesicles (Presley et al. 1997; Scales et al. 1997), the requirement of ARF1 activation supports perhaps the simplest possible interpretation for ERGIC-like procollagen carriers. Distal ERES regions and transport intermediates emerging from them might perform all functions commonly attributed to ERGIC (Lippincott-Schwartz 2001; Lippincott-Schwartz et al. 2000), at least in the context of procollagen trafficking (Omari et al. 2020).

Overall, we leave it to the readers to decide for themselves whether these distinctions between models 2, 3, and 4 are substantive or mostly semantic. Each of these three models is consistent with all experimental data reported so far, unlike the idea of COPII-coated procollagen transport vesicles. Still, none of the models has been fully validated and many questions remain unanswered. Procollagens seem to follow a general trafficking pathway from the ER to Golgi. Their proposed ERES receptor, TANGO1 appears to recognize only few other large molecules (Saito and Maeda 2019). However, we do not know whether this is the only or just one out of many potential differences in the ERES trafficking machinery between procollagens and other proteins. The same G-protein and kinases appear to regulate ERES function and signaling for type I procollagen and VSVG, although VSVG is a rhabdovirus membrane protein that directly binds to SEC24 without an additional cargo receptor (Subramanian et al. 2019). Nonetheless, one cannot conclude yet that this is a general ERES regulation machinery. We do not fully understand the composition and function of such machinery in different cells under different conditions.

Mutation outcomes suggest that different ERES machinery components might have variable expression and roles in different cells as well as variable impact on trafficking of different secretory proteins. For instance, SEC24D mutations cause OI bone fragility with craniofacial involvement (Garbes et al. 2015; Moosa et al. 2016; Zhang et al. 2017). Mutations in SEC23A, which forms an inner COPII coat dimer with SEC24 paralogs, cause craniofacial deformities and short stature without bone fragility (Boyadjiev et al. 2006; Fromme et al. 2007). Mutations in TANGO1, which binds to SEC23A, cause craniofacial deformities and short stature without bone fragility as well, but they also cause diabetes, nephropathy, and mental retardation that do not accompany SEC23A mutations (Lekszas et al. 2020). Based on the symptoms, one might guess that SEC24D deficiency affects primarily type I and type II procollagen trafficking, SEC23A primarily type II procollagen trafficking, and TANGO1 affects trafficking of type II and IV procollagens as well as other proteins. Even more complex and variable effects of deficiencies in these proteins in mice and zebrafish add to the puzzle (Lu and Kim 2020; Malhotra and Erlmann 2015). More studies appear to be required to fully understand how all these components work together in different cells.

2.5.2 *Golgi Progression and TGN Sorting*

In Golgi cisternae, type I procollagen molecules appear to form side-by-side aggregates, likely due to the high concentration and reduced pH (Trelstad and Hayashi 1979; Leblond 1989). These immobile aggregates have been shown to progress from cis- to trans-Golgi by the maturation of the host cisternae in time and space rather than by moving from cisternae to cisternae (Beznoussenko et al. 2014; Bonfanti et al. 1998). Such cisternal maturation occurs via the dynamic exchange of much smaller Golgi proteins and other components between cisternae through direct connections and vesicular traffic (Pantazopoulou and Glick 2019). Aggregation makes procollagen modification by Golgi enzymes unlikely even within propeptide regions, but unfolded proα2(I) chains not incorporated into procollagen trimers have been described to undergo glycosylation and sialylation in Golgi (Ramachandran and Peterkofsky 1997).

In contrast to the straightforward progression through Golgi, sorting at TGN and delivery to the plasma membrane are perhaps the least understood steps in procollagen trafficking (Malhotra and Erlmann 2015). Deficiencies in some proteins have been shown to disrupt this process, but the sorting mechanism and nature of procollagen carriers remain unknown (pathway 7 in Fig. 2.1). These proteins include RAB GTPases, RAB3D, and RAB27B (Nabavi et al. 2012), RAB10 (Lerner et al. 2013; Banushi et al. 2016), RAB25 (Banushi et al. 2016), and RAB6A (Unlu et al. 2020); RIC1-RGP1 guanine exchange factor complex that activates RAB6A (Unlu et al. 2020); and VPS33B-VIPAR complex (Banushi et al. 2016). However, their multiple functions and general involvement in the endo-lysosomal pathway, which is crucial for many aspects of cellular function (Bonifacino and Neefjes 2017; Delevoye et al. 2019), complicates the interpretation of the reported disruptions in procollagen export from the TGN caused by deficiencies in these proteins. For instance, VPS33B and VIPAR mutations cause deficient type I procollagen secretion in cultured fibroblasts from patients with arthrogryposis, renal dysfunction and cholestasis syndrome (ARC) (Banushi et al. 2016), which is characterized not only by multiple joint contractures (arthrogryposis) but also recurring fractures (Gissen et al. 2006). However, it is unclear whether the primary cause of type I collagen matrix pathology in ARC patients and mouse models is related to procollagen trafficking, procollagen modification by LH3, or general disruption of the endo-lysosomal pathway in collagen-producing cells (Banushi et al. 2016).

2.6 Procollagen Degradation

Up to ~15% of newly synthesized type I procollagen chains were reported to be rerouted for intracellular degradation in confluent fibroblast cultures under normal conditions, likely due to misfolding (Berg et al. 1980; Barile et al. 1990). Disruption of folding by cis-4-hydroxyproline (Berg et al. 1980) or by proline underhydroxylation in rapidly proliferating cells (Bienkowski et al. 1978), was reported to increase the fraction of degraded procollagen up to ~30%. Procollagen misfolding or secretion pathology places an additional heavy burden on degradative trafficking pathways, which therefore might play a critical role both in normal collagen homeostasis and in disease. These pathways appear to be diverse, and even basal degradation of misfolded procollagen involves multiple routes to proteasomes and lysosomes (Ripley and Bienkowski 1997).

2.6.1 ERAD

Results of several studies suggest that unassociated proα1(I) but not proα2(I) chains or misfolded trimeric type I procollagen molecules follow the ERAD pathway from the ER to cytosol for proteasomal degradation (pathway 3 in Fig. 2.1) (Fitzgerald et al. 1999; Ishida et al. 2009; Lamande et al. 1995). Unassociated proα2(I) chains were instead found in Golgi and lysosomes and their degradation was affected by lysosomal but not proteasomal inhibitors, probably because these chains were rerouted to lysosomes from Golgi via the endo-lysosomal pathway (Gotkin et al. 2004). The difference cannot be explained by targeting based on oligosaccharide trimming after futile calnexin-calreticulin binding cycles, since both proα1(I) and proα2(I) chains undergo similar N-linked glycosylation within the C-propeptide region (Clark 1979). It is also unlikely to be related to only one proα2(I) vs. two or even three proα1(I) being incorporated into trimeric procollagen, since the same applies to proα2(VI) chains of type VI procollagen degraded by ERAD (Zamurs et al. 2015). The mechanism for endo-lysosomal rather than ERAD degradation of unassociated proα2(I) chains is unknown.

Misfolded trimeric procollagen molecules and their aggregates were proposed to be degraded by autophagy (Ishida and Nagata 2009), but ERAD of misfolded molecules with all three chains was also observed. Type I procollagen with underhydroxylated proline was found to be directed to ERAD by P4HB/PDI (Ko and Kay 2004). ERAD of trimeric type III procollagen was reported in cases of splicing defects (Thakker-Varia et al. 1995) and cortisol treatment (Mi et al. 2018). Both ERAD and lysosomal degradation were observed for different mutations in type X procollagen (Chan et al. 2001; Mullan et al. 2017). In other words, the mechanism of trimeric procollagen sorting between different degradation pathways is unknown as well.

2.6.2 Autophagy

Like other secretory proteins, procollagen can be directed to lysosomes from the ER, Golgi, or transport intermediates, but mechanistic studies have been focused mostly on autophagy of misfolded procollagen from the ER. Degradation of misfolded procollagen I by selective ER-phagy (pathway 6 in Fig. 2.1) was reported in a recent study of human osteosarcoma Saos2 cell line and mouse embryonic fibroblasts (Forrester et al. 2019). The ER-phagy receptor triggering phagophore recruitment and autophagosome formation appeared to be FAM134B, which has no luminal tail. Therefore accumulation of misfolded procollagen in the ER was proposed to be recognized by calnexin-FAM134B complexes (Forrester et al. 2019). Autophagosome formation and calnexin-FAM134B complexes were assayed after 6–12 h bafilomycin A1 treatment, once autophagic structures filled the entire cell. Since such treatment induces severe cell stress and significantly alters cell function, the role of this pathway in the physiological degradation of misfolded procollagen remains unclear.

A different pathway of direct, microautophagy-like capture of misfolded procollagen by lysosomes at ERES (pathway 5 in Fig. 2.1) was proposed in a live cell imaging study of MC3T3 mouse osteoblast cell line (Omari et al. 2018). Most misfolded procollagen molecules were observed to escape the quality control in ER lumen and enter the secretory pathway at ERES, causing ubiquitination of ERES surface and lysosome recruitment without a phagophore. This pathway was assayed by transient transfection of the cells with fluorescently tagged procollagen and fluorescent markers of various organelles. ERES microautophagy could degrade only exogenous procollagen as implied in (Fregno and Molinari 2019), but this interpretation would require an unlikely preexistence of a selective degradation pathway for molecules unknown to the cells. Moreover, a follow-up study confirmed ERES microautophagy of procollagen tagged not only by transient transfection with fluorescent constructs of either proα1(I) or proα2(I) but also by similar CRISPR/Cas modification of endogenous proα2(I) (Omari et al. 2020). The fluorescent tag could induce ERES microautophagy by enabling procollagen to escape the luminal quality control and enter ERES, but the minimal effect of knocking out ATG5 in osteoblasts in mice supported microautophagy being the primary pathway of misfolded procollagen removal from mouse osteoblast ER (Makareeva et al. 2019).

ATG5 is required for the conventional ER-phagy pathway described above because it is essential for LC3 and GABARAP lipidation, phagophore membrane formation, and phagophore recruitment by FAM134B and other ER-phagy receptors (Mizushima 2019). In mice, complete ATG5 knockout is perinatal lethal (Kuma et al. 2004). Nonetheless, neuron-specific transgenic expression of ATG5 leads to survival of ATG5 knockout mice for up to 6 months, indicating that conventional ER-phagy is dispensable in other tissues (Yoshii et al. 2016). In primary mouse osteoblasts, ATG5 knockout was confirmed to prevent the formation of LC3 and GABARAP membranes and therefore selective ER-phagy, but it did not have a significant effect on ERES microautophagy (Makareeva et al. 2019).

In general, cells are likely to utilize all defensive mechanisms at their disposal when challenged by significant procollagen misfolding and other adverse conditions. The benefits of ER-phagy, ER vesicle autophagy, and ER microautophagy might be (a) selective recognition of ER membrane cargo or ER membrane adapters of luminal cargo and (b) rapid ER fragmentation and recycling under severe cell stress (Wirth et al. 2019; Khaminets et al. 2015). The benefits of ERES microautophagy might be (a) degradation of cargo that escapes ER quality control; (b) removal of overfilled and damaged ERESs; and (c) secretory protein rerouting to degradation at the most natural check point (Omari et al. 2018). The latter feature might be particularly important for responding to less severe cell stress and nutritional deficiency. Therefore, we see these two misfolded procollagen autophagy pathways as complementary rather than competing; both might be utilized by a cell depending on its function as well as genetic and environmental factors.

Consistent with this idea, the osteoblast-specific ATG5 knockout has a minor effect on bones in a G610C mouse model of OI, while reduced ATG5 expression in all tissues causes 50% perinatal lethality due to lung failure (Makareeva et al. 2019). Indeed, pathology in these mice is caused by a Gly610 to Cys substitution in the triple-helical region of proα2(I) (Daley et al. 2010), which leads to procollagen misfolding, ER disruption, and cell malfunction (Mirigian et al. 2016). Only minor effects of the ATG5 knockout on bones and on procollagen autophagy in cultured cells point to microautophagy being the primary procollagen degradation pathway in osteoblasts. At the same time, dramatic disruption of lung collagen matrix in E18.5 embryos caused by ~ threefold nonspecific reduction in ATG5 expression points to procollagen ER-phagy being essential for embryonic lung fibroblast function. The different balance between the relative roles of these pathways might be related to the cell type, embryonic vs. postnatal development, and/or procollagen synthesis rate since no overt lung pathology was detected after weaning in the surviving mice with the G610C mutation.

2.6.3 Quality Control and Sorting

Clearly, misfolded procollagen is recognized and directed to ERES microautophagy, ER-phagy, ERAD, or other pathways we might not yet know, but our understanding of the cellular quality control and sorting machinery behind this rerouting is only hypothetical. Quality control in the ER is typically regulated by ER chaperones that target misfolded proteins to appropriate degradation pathways (Bateman et al. 2009; Hwang and Qi 2018; Ruggiano et al. 2014; Roth and Zuber 2017; Chino and Mizushima 2020; Adams et al. 2019). Two ER chaperones, calnexin, and HSP47 have been discussed as potential sensors of procollagen misfolding. As a lectin ER chaperone, calnexin binds to N-linked oligosaccharides in the C-propeptide, so that it might respond to C-propeptide misfolding by directing unassociated procollagen chains to ERAD, consistent with the general calnexin-calreticulin chaperone function in the ER (Roth and Zuber 2017). Nonetheless, calnexin clustering within the ER membrane caused by aggregation of misfolded procollagen in the ER lumen might also induce FAM134B clustering and selective ER-phagy (Forrester et al. 2019).

HSP47 must be crucial for quality control as a triple helix chaperone (triple helix folding is the most challenging step in procollagen biosynthesis and its misfolding is by far the most common cause of severe OI), yet this HSP47 function is not understood at all. The essence of the puzzle is that HSP47 preferentially binds and accumulates at properly folded rather than misfolded molecules, unlike all other ER chaperones. HSP47 is therefore unlikely to mediate recognition and retention of misfolded molecules in the ER lumen like the other chaperones do. Its putative function as an adapter of procollagen loading into ERES by TANGO1 (Ishikawa et al. 2016) is also inconsistent with the retention of misfolded procollagen in the ER lumen. Not only does HSP47 facilitate triple helix folding and procollagen loading into ERES, but it also appears to inhibit procollagen aggregation in the ER (Ishida et al. 2006). Taken together, these observations suggest that HSP47 should facilitate the release of procollagen from the ER to Golgi, but such release is instead enhanced by HSP47 deficiency (Christiansen et al. 2010).

This puzzle is, however, easily resolved by assuming that HSP47 performs procollagen quality control inside ERES rather than the ER lumen, unlike other ER chaperones. Indeed, the loss of misfolded procollagen rerouting to autophagy from ERES caused by HSP47 deficiency would explain both the increased procollagen accumulation in Golgi (Christiansen et al. 2010) and increased secretion of misfolded molecules (Christiansen et al. 2010; Nagai et al. 2000). This interpretation of HSP47 deficiency is consistent with procollagen aggregate accumulation in the ER lumen (Ishida et al. 2006) because of the less efficient procollagen loading into ERES and the loss of the inhibitory effect of HSP47 on aggregate formation in the ER lumen. It provides the mechanism behind the observed misfolded procollagen rerouting to autophagy from ERES (Omari et al. 2018; Makareeva et al. 2019) and explains HSP47 presence in autophagic structures formed at ERES but not in transport intermediates delivering procollagen from ERES to Golgi (Omari et al. 2020). It is also consistent with selective ER-phagy of misfolded procollagen aggregates from the ER lumen, when severe cell stress is accompanied by significant aggregate accumulation (Forrester et al. 2019). Moreover, different roles of HSP47 inside ERES and ER lumen might explain why its deficiency causes embryonic lethality in mice vs. moderate OI in humans despite similar effects on procollagen secretion. Other ER chaperones might compensate for HSP47 in the ER lumen in humans (Schwarze et al. 2019). The question this hypothesis does not answer is how HSP47 retains misfolded procollagen inside ERES and triggers ERES microautophagy. We hope that an answer to this question and the validity of the hypothesis will be provided by future studies.

A chaperone retaining procollagen in the ER lumen until the completion of triple helix folding might be P4HB/PDI. In addition to its function as a general ER protein disulfide isomerase, P4HB/PDI binds to unfolded regions of procollagen triple helix as a β subunit of prolyl 4-hydroxylases. Unlike HSP47, it conforms to the conventional protein folding paradigm of chaperone release prior to the export from ER. This hypothesis is consistent with the observation of P4HB/PDI-directed procollagen ERAD (Ko and Kay 2004), a more typical ER lumen degradation pathway for misfolded proteins that are not aggregated (Hwang and Qi 2018; Qi et al. 2017; Ruggiano et al. 2014). It explains OI bone fragility in patients with P4HB/PDI mutations (Balasubramanian et al. 2018; Rauch et al. 2015). It is also consistent with a very different, more complex outcome of a deficiency in P4HA1, a catalytic α subunit of prolyl 4-hydroxylase 1 (Zou et al. 2017). In the latter study, no ER retention of procollagen was observed despite significantly reduced proline 4-hydroxylation, perhaps because the chaperone function of P4HB/PDI was not affected. Unfortunately, effects of P4HB/PDI mutations on procollagen trafficking have not been investigated.

A molecule facilitating removal of misfolded procollagen from ERES by ERES microautophagy might be SEC24D. It is the only SEC24 paralog associated with skeletal dysplasia (see OMIM#607183–607,186, https://www.omim.org/). Mutations in SEC24D cause the same OI with craniosynostosis as mutations in P4HB/PDI, instead of very different disorders caused by mutations in proteins that dimerize with SEC24 paralogs (SEC23A and SEC23B). SEC24 paralogs are generally responsible for cargo receptor/adapter binding to COPII coat. A distinct phenotype of SEC24D mutations might therefore be related to a distinct function of SEC24D at ERES, e.g., as a COPII receptor for ERES microautophagy. Consistently, *Sec24d* expression is increased in the G610C mouse model of OI (our analysis of RNASeq data from (Jacobsen et al. 2014)); and *SEC24D* is a transcriptional target of CREB3L1, which is an ER membrane stress receptor involved in OI (Guillemyn et al. 2019; Keller et al. 2018). Although other explanations of similar clinical outcomes for SEC24D and P4HB/PDI mutations are possible, they are more difficult to reconcile with the ubiquitous expression of both these proteins.

Overall, this combined analysis of seemingly very different yet related observations points to a two-step procollagen quality control in the early secretory pathway. At the first step, unassociated proα1(I) chains and incompletely folded procollagen molecules are retained in the ER lumen by bound ER chaperones (likely P4HB/PDI, calnexin, and calreticulin, but perhaps other ones as well). At the second step, the misfolded procollagen that escapes the luminal quality control and enters ERES is rerouted to ERES microautophagy (perhaps by recruiting SEC24D). There is no direct evidence supporting this hypothesis, but it appears to be interesting and logical enough to merit future experimental verification.

2.7 Therapeutic Implications

Candidate pathways and regulatory mechanisms of secretory and degradative procollagen trafficking might not only explain pathology in OI and other skeletal dysplasias but also offer potential targets for intervention. Since ER homeostasis disruption by misfolded procollagen accumulation appears to be a common cause of pathology in many of these disorders, several approaches to reducing this accumulation are being explored already.

An attempt to treat OI by repurposing 4-phenyl butyric acid (4PBA), which is approved for the treatment of urea cycle disorders in humans, has produced encouraging results in zebrafish and cell culture studies (Gioia et al. 2017; Besio et al. 2019b). This compound is believed to improve ER homeostasis by increasing protein export through its interaction with the COPII coat, although it might have pleiotropic effects and the mechanism of its action is not fully understood (Ma et al. 2017). In cultured fibroblasts from OI patients, it reduces ER dilation and normalizes markers of ER stress by enhancing both secretory and degradative procollagen trafficking (Besio et al. 2019b). Preliminary results of 4PBA testing in the G610C mouse model of OI suggest some improvements in bone growth and trabecular architecture (Scheiber et al. 2019).

Testing of carbamazepine for the treatment of metaphyseal chondrodysplasia type Schmid (MCDS) associated with type X collagen misfolding has been even more successful, by significantly reducing skeletal pathology in a mouse model (Mullan et al. 2017). This drug, which is approved for use in humans as an anticonvulsant, enhances both proteasomal and lysosomal protein degradation (Vakifahmetoglu-Norberg et al. 2015; Hidvegi et al. 2010). The mouse model observations were sufficiently promising for initiating clinical trials in Europe. At least one study of carbamazepine in G610C mice is currently ongoing.

Another interesting idea is the inhibition of chronic integrated cell stress response (ISR) to procollagen misfolding by ISRIB, which also alleviated pathology in a mouse model of MCDS (Wang et al. 2018). ISRIB is an experimental drug, which suppresses ISR by targeting eIF2B and reversing the effects of eIF2α phosphorylation (Anand and Walter 2020). It appears to prevent ISR-associated arrest of hypertrophic chondrocyte differentiation in MCDS (Wang et al. 2018). It might similarly prevent delayed osteoblast maturation in OI, given that eIF2α phosphorylation is caused by type I procollagen misfolding as well (Mirigian et al. 2016). However, restoration of mutant procollagen synthesis by ISR suppression might not be as beneficial for osteoblasts as for hypertrophic chondrocytes; it might even be detrimental. Not only do osteoblasts produce much more collagen than hypertrophic chondrocytes, but their microenvironment, homeostasis, and interactions with other cells are different as well.

Attempts to selectively enhance autophagic degradation of misfolded procollagen in G610C mice have been less successful. A low protein diet treatment appeared to have the desired effect of improving osteoblast differentiation and function, but its side effect of stunting animal growth was too strong (Mertz et al. 2016). Activation of autophagy by mTOR inhibition with rapamycin resulted in some improvement in trabecular but not cortical bone and no net benefit for the animals (Bateman et al. 2019), unlike the reported beneficial effects of rapamycin in other proteinopathies (Vakifahmetoglu-Norberg et al. 2015; Rautou et al. 2010). A potential drawback in the treatment of bone pathology by autophagy enhancing drugs is the activation of bone resorption by osteoclasts, which utilize autophagy for degrading the bone matrix (Yin et al. 2019). To the best of our knowledge, a combination therapy with mTOR inhibitors and anti-osteoclast drugs has not been tried yet.

2.8 Concluding Remarks

Overall, significant progress has been made since the last comprehensive review dedicated to the implications of different procollagen trafficking aspects in ECM biology was published fifteen years ago (Canty and Kadler 2005). Many advancements can be attributed to the development of new techniques for visualizing subcellular compartments and structures in ever-increasing details. Modern focused ion beam scanning electron microscopy enables 3D visualization of the whole cell with ~5 nm spatial resolution (Kizilyaprak et al. 2019). Subcellular structures can now be precisely identified in fixed cells by correlative light and electron microscopy or just light microscopy with up to ~20–50 nm spatial resolution, utilizing immunofluorescence or transfected fluorescent markers (Sahl et al. 2017). Super-resolution time lapse imaging of fluorescently tagged proteins in live cells provides valuable dynamic information, which is particularly important for distinguishing subcellular structures that are similar in appearance yet different in function (McCaughey et al. 2019; Omari et al. 2018, 2020). Combined with biochemical and genetic approaches, all these techniques are providing tremendous (and often overwhelming) amounts of information.

On a cautionary note, each of these techniques and approaches has its drawbacks and the corresponding results might be misinterpreted. We described a few examples of similar observations being interpreted differently, depending on the experimental approach and investigator's perspective. Electron and super-resolution light microscopy of fixed cells provide the best spatial resolution, but the sample preparation procedures for these techniques might alter or distort membrane structures. Even without such artifacts, the structures observed in fixed cells might not be what they appear to be, e.g., we have described how vesicle-like regions of dilated ER have been mistaken for procollagen transport vesicles. Dynamic imaging of live cells identifies transport intermediates unequivocally from their origin, movement along microtubules, and destination, but a fluorescently tagged protein might be directed by the cell into a different trafficking pathway from the endogenous, untagged protein. Biochemical analysis of trafficking pathways requires cell-free model systems or separation of components after cell lysis, which necessarily creates artificial membrane structures. Drugs known to inhibit specific pathways usually have pleiotropic effects. Knockdowns and knockouts of genes might completely alter the studied pathway, even when they have no overt effects on cell appearance.

With all these caveats in mind, we have tried to provide a balanced analysis of different concepts reported in the literature and describe various hypotheses and models without representing them as established facts. As in any literature analysis, we have had to select, compare, evaluate, and express opinions, thereby injecting our own biases into this review. We, therefore, encourage the readers to make their own conclusions based on the presented evidence and ideas. The main take-home message of this review is that the studies of procollagen trafficking have only uncovered the tip of the iceberg. We are beginning to see possible pathways but can only speculate about the mechanisms regulating procollagen sorting and movement along these pathways. We do not know what lies beneath the surface and we might yet have to change the trafficking paradigms again, as it has happened before.

Acknowledgments This work was supported by funding from the Intramural Research Program of NICHD, NIH.

References

Adams BM, Oster ME, Hebert DN (2019) Protein quality control in the endoplasmic reticulum. Protein J 38(3):317–329

Anand AA, Walter P (2020) Structural insights into ISRIB, a memory-enhancing inhibitor of the integrated stress response. FEBS J 287(2):239–245

Appenzeller-Herzog C, Hauri HP (2006) The ER-Golgi intermediate compartment (ERGIC): in search of its identity and function. J Cell Sci 119(Pt 11):2173–2183

Aridor M (2018) COPII gets in shape: lessons derived from morphological aspects of early secretion. Traffic 19(11):823–839

Arnold WV, Fertala A (2013) Skeletal diseases caused by mutations that affect collagen structure and function. Int J Biochem Cell Biol 45(8):1556–1567

Arseni L, Lombardi A, Orioli D (2018) From structure to phenotype: impact of collagen alterations on human health. Int J Mol Sci 19(5):1407

Balasubramanian M, Padidela R, Pollitt RC, Bishop NJ, Mughal MZ, Offiah AC et al (2018) P4HB recurrent missense mutation causing Cole-carpenter syndrome. J Med Genet 55(3):158–165

Banushi B, Forneris F, Straatman-Iwanowska A, Strange A, Lyne AM, Rogerson C et al (2016) Regulation of post-Golgi LH3 trafficking is essential for collagen homeostasis. Nat Commun 7:12111

Barile FA, Guzowski DE, Ripley C, Siddiqi ZA, Bienkowski RS (1990) Ammonium chloride inhibits basal degradation of newly synthesized collagen in human fetal lung fibroblasts. Arch Biochem Biophys 276(1):125–131

Barlowe C, Helenius A (2016) Cargo capture and bulk flow in the early secretory pathway. Annu Rev Cell Dev Biol 32:197–222

Bartolomeo R, Cinque L, De Leonibus C, Forrester A, Salzano AC, Monfregola J et al (2017) mTORC1 hyperactivation arrests bone growth in lysosomal storage disorders by suppressing autophagy. J Clin Invest 127(10):3717–3729

Bateman JF, Boot-Handford RP, Lamande SR (2009) Genetic diseases of connective tissues: cellular and extracellular effects of ECM mutations. Nat Rev Genet 10(3):173–183

Bateman JF, Sampurno L, Maurizi A, Lamande SR, Sims NA, Cheng TL et al (2019) Effect of rapamycin on bone mass and strength in the alpha2(I)-G610C mouse model of osteogenesis imperfecta. J Cell Mol Med 23(3):1735–1745

Berg RA, Schwartz ML, Crystal RG (1980) Regulation of the production of secretory proteins: intracellular degradation of newly synthesized "defective" collagen. Proc Natl Acad Sci U S A 77(8):4746–4750

Besio R, Chow CW, Tonelli F, Marini JC, Forlino A (2019a) Bone biology: insights from osteogenesis imperfecta and related rare fragility syndromes. FEBS J 286(15):3033–3056

Besio R, Garibaldi N, Leoni L, Cipolla L, Sabbioneda S, Biggiogera M et al (2019b) Cellular stress due to impairment of collagen prolyl hydroxylation complex is rescued by the chaperone 4-phenylbutyrate. Dis Model Mech 12(6):dmm038521

Bethune J, Wieland FT (2018) Assembly of COPI and COPII vesicular coat proteins on membranes. Annu Rev Biophys 47:63–83

Bexiga MG, Simpson JC (2013) Human diseases associated with form and function of the Golgi complex. Int J Mol Sci 14(9):18670–18681

Beznoussenko GV, Parashuraman S, Rizzo R, Polishchuk R, Martella O, Di Giandomenico D et al (2014) Transport of soluble proteins through the Golgi occurs by diffusion via continuities across cisternae. elife 3:e02009

Bienkowski RS, Cowan MJ, McDonald JA, Crystal RG (1978) Degradation of newly synthesized collagen. J Biol Chem 253(12):4356–4363

Boncompain G, Divoux S, Gareil N, de Forges H, Lescure A, Latreche L et al (2012) Synchronization of secretory protein traffic in populations of cells. Nat Methods 9(5):493–498

Bonfanti L, Mironov AA Jr, Martinez-Menarguez JA, Martella O, Fusella A, Baldassarre M et al (1998) Procollagen traverses the Golgi stack without leaving the lumen of cisternae: evidence for cisternal maturation. Cell 95(7):993–1003

Bonifacino JS, Glick BS (2004) The mechanisms of vesicle budding and fusion. Cell 116 (2):153–166

Bonifacino JS, Neefjes J (2017) Moving and positioning the endolysosomal system. Curr Opin Cell Biol 47:1–8

Bornstein P, Sage H (1989) Regulation of collagen gene expression. Prog Nucleic Acid Res Mol Biol 37:67–106

Boyadjiev SA, Fromme JC, Ben J, Chong SS, Nauta C, Hur DJ et al (2006) Cranio-lenticulo-sutural dysplasia is caused by a SEC23A mutation leading to abnormal endoplasmic-reticulum-to-Golgi trafficking. Nat Genet 38(10):1192–1197

Brandizzi F, Barlowe C (2013) Organization of the ER-Golgi interface for membrane traffic control. Nat Rev Mol Cell Biol 14(6):382–392

Cai H, Reinisch K, Ferro-Novick S (2007) Coats, tethers, Rabs, and SNAREs work together to mediate the intracellular destination of a transport vesicle. Dev Cell 12(5):671–682

Canty EG, Kadler KE (2005) Procollagen trafficking, processing and fibrillogenesis. J Cell Sci 118 (Pt 7):1341–1353

Centonze FG, Farhan H (2019) Crosstalk of endoplasmic reticulum exit sites and cellular signaling. FEBS Lett 593(17):2280–2288

Chan D, Ho MS, Cheah KS (2001) Aberrant signal peptide cleavage of collagen X in Schmid metaphyseal chondrodysplasia. Implications for the molecular basis of the disease. J Biol Chem 276(11):7992–7997

Chino H, Mizushima N (2020) ER-Phagy: quality control and turnover of endoplasmic reticulum. Trends Cell Biol 30(5):384–398

Christiansen HE, Schwarze U, Pyott SM, AlSwaid A, Al Balwi M, Alrasheed S et al (2010) Homozygosity for a missense mutation in SERPINH1, which encodes the collagen chaperone protein HSP47, results in severe recessive osteogenesis imperfecta. Am J Hum Genet 86 (3):389–398

Clark CC (1979) The distribution and initial characterization of oligosaccharide units on the COOH-terminal propeptide extensions of the pro-alpha 1 and pro-alpha 2 chains of type I procollagen. J Biol Chem 254(21):10798–10802

Clarke LA, Hollak CE (2015) The clinical spectrum and pathophysiology of skeletal complications in lysosomal storage disorders. Best Pract Res Clin Endocrinol Metab 29(2):219–235

Cui Y, Parashar S, Zahoor M, Needham PG, Mari M, Zhu M et al (2019) A COPII subunit acts with an autophagy receptor to target endoplasmic reticulum for degradation. Science 365 (6448):53–60

Cutrona MB, Morgan NE, Simpson JC (2018) Heritable skeletal disorders arising from defects in processing and transport of type I Procollagen from the ER: perspectives on possible therapeutic approaches. In: UlloaAguirre A, Tao YX (eds) Targeting trafficking in drug development. Handbook of experimental pharmacology, vol 245. Springer, Berlin, pp 191–225

Daley E, Streeten EA, Sorkin JD, Kuznetsova N, Shapses SA, Carleton SM et al (2010) Variable bone fragility associated with an Amish COL1A2 variant and a knock-in mouse model. J Bone Miner Res 25(2):247–261

Dancourt J, Barlowe C (2010) Protein sorting receptors in the early secretory pathway. Annu Rev Biochem 79:777–802

Delevoye C, Marks MS, Raposo G (2019) Lysosome-related organelles as functional adaptations of the endolysosomal system. Curr Opin Cell Biol 59:147–158
Dell'Angelica EC, Bonifacino JS (2019) Coatopathies: genetic disorders of protein coats. Annu Rev Cell Dev Biol 35:131–168
Di Martino R, Sticco L, Luini A (2019) Regulation of cargo export and sorting at the trans-Golgi network. FEBS Lett 593(17):2306–2318
Engel J, Bachinger HP (2005) Structure, stability and folding of the collagen triple helix. In: Brinckmann J, Notbohm H, Muller PK (eds) Collagen: primer in structure, processing and assembly, Topics in current chemistry, vol 247. Springer, Heidelberg, pp 7–33
Eyre DR, Weis MA (2013) Bone collagen: new clues to its mineralization mechanism from recessive osteogenesis imperfecta. Calcif Tissue Int 93(4):338–347
Faini M, Beck R, Wieland FT, Briggs JA (2013) Vesicle coats: structure, function, and general principles of assembly. Trends Cell Biol 23(6):279–288
Ferreira CR, Xia ZJ, Clement A, Parry DA, Davids M, Taylan F et al (2018) A recurrent De novo heterozygous COG4 substitution leads to Saul-Wilson syndrome, disrupted vesicular trafficking, and altered proteoglycan glycosylation. Am J Hum Genet 103(4):553–567
Fitzgerald J, Lamande SR, Bateman JF (1999) Proteasomal degradation of unassembled mutant type I collagen pro-alpha1(I) chains. J Biol Chem 274(39):27392–27398
Forlino A, Marini JC (2016) Osteogenesis imperfecta. Lancet 387(10028):1657–1671
Forlino A, Cabral WA, Barnes AM, Marini JC (2011) New perspectives on osteogenesis imperfecta. Nat Rev Endocrinol 7(9):540–557
Forrester A, De Leonibus C, Grumati P, Fasana E, Piemontese M, Staiano L et al (2019) A selective ER-phagy exerts procollagen quality control via a Calnexin-FAM134B complex. EMBO J 38 (2):e99847
Fregno I, Molinari M (2019) Proteasomal and lysosomal clearance of faulty secretory proteins: ER-associated degradation (ERAD) and ER-to-lysosome-associated degradation (ERLAD) pathways. Crit Rev Biochem Mol Biol 54(2):153–163
Fromme JC, Ravazzola M, Hamamoto S, Al-Balwi M, Eyaid W, Boyadjiev SA et al (2007) The genetic basis of a craniofacial disease provides insight into COPII coat assembly. Dev Cell 13 (5):623–634
Fujii KK, Taga Y, Sakai T, Ito S, Hattori S, Nagata K et al (2019) Lowering the culture temperature corrects collagen abnormalities caused by HSP47 gene knockout. Sci Rep 9(1):17433
Galluzzi L, Baehrecke EH, Ballabio A, Boya P, Bravo-San Pedro JM, Cecconi F et al (2017) Molecular definitions of autophagy and related processes. EMBO J 36(13):1811–1836
Garbes L, Kim K, Riess A, Hoyer-Kuhn H, Beleggia F, Bevot A et al (2015) Mutations in SEC24D, encoding a component of the COPII machinery, cause a Syndromic form of Osteogenesis Imperfecta. Am J Hum Genet 96(3):432–439
Geva Y, Schuldiner M (2014) The back and forth of cargo exit from the endoplasmic reticulum. Curr Biol 24(3):R130–R136
Gioia R, Tonelli F, Ceppi I, Biggiogera M, Leikin S, Fisher S et al (2017) The chaperone activity of 4PBA ameliorates the skeletal phenotype of Chihuahua, a zebrafish model for dominant osteogenesis imperfecta. Hum Mol Genet 26(15):2897–2911
Gissen P, Tee L, Johnson CA, Genin E, Caliebe A, Chitayat D et al (2006) Clinical and molecular genetic features of ARC syndrome. Hum Genet 120(3):396–409
Gorur A, Yuan L, Kenny SJ, Baba S, Xu K, Schekman R (2017) COPII-coated membranes function as transport carriers of intracellular procollagen I. J Cell Biol 216(6):1745–1759
Gotkin MG, Ripley CR, Lamande SR, Bateman JF, Bienkowski RS (2004) Intracellular trafficking and degradation of unassociated proalpha2 chains of collagen type I. Exp Cell Res 296 (2):307–316
Guillemyn B, Kayserili H, Demuynck L, Sips P, De Paepe A, Syx D et al (2019) A homozygous pathogenic missense variant broadens the phenotypic and mutational spectrum of CREB3L1-related osteogenesis imperfecta. Hum Mol Genet 28(11):1801–1809

Guo Y, Sirkis DW, Schekman R (2014) Protein sorting at the trans-Golgi network. Annu Rev Cell Dev Biol 30:169–206
Halperin D, Kadir R, Perez Y, Drabkin M, Yogev Y, Wormser O et al (2019) SEC31A mutation affects ER homeostasis, causing a neurological syndrome. J Med Genet 56(3):139–148
Hidvegi T, Ewing M, Hale P, Dippold C, Beckett C, Kemp C et al (2010) An autophagy-enhancing drug promotes degradation of mutant alpha1-antitrypsin Z and reduces hepatic fibrosis. Science 329(5988):229–232
Hsu VW, Yang JS (2009) Mechanisms of COPI vesicle formation. FEBS Lett 583(23):3758–3763
Hsu VW, Lee SY, Yang JS (2009) The evolving understanding of COPI vesicle formation. Nat Rev Mol Cell Biol 10(5):360–364
Hubner CA, Dikic I (2020) ER-phagy and human diseases. Cell Death Differ 27(3):833–842
Hulmes DJS (2019) Roles of the procollagen C-propeptides in health and disease. Essays Biochem 63:313–323
Hwang J, Qi L (2018) Quality control in the endoplasmic reticulum: crosstalk between ERAD and UPR pathways. Trends Biochem Sci 43(8):593–605
Ishida Y, Nagata K (2009) Autophagy eliminates a specific species of misfolded procollagen and plays a protective role in cell survival against ER stress. Autophagy 5(8):1217–1219
Ishida Y, Kubota H, Yamamoto A, Kitamura A, Bachinger HP, Nagata K (2006) Type I collagen in Hsp47-null cells is aggregated in endoplasmic reticulum and deficient in N-propeptide processing and fibrillogenesis. Mol Biol Cell 17(5):2346–2355
Ishida Y, Yamamoto A, Kitamura A, Lamande SR, Yoshimori T, Bateman JF et al (2009) Autophagic elimination of misfolded procollagen aggregates in the endoplasmic reticulum as a means of cell protection. Mol Biol Cell 20(11):2744–2754
Ishikawa Y, Bachinger HP (2013) A molecular ensemble in the rER for procollagen maturation. Biochim Biophys Acta 1833(11):2479–2491
Ishikawa Y, Ito S, Nagata K, Sakai LY, Bachinger HP (2016) Intracellular mechanisms of molecular recognition and sorting for transport of large extracellular matrix molecules. Proc Natl Acad Sci U S A 113(41):E6036–E6E44
Izumi K, Brett M, Nishi E, Drunat S, Tan ES, Fujiki K et al (2016) ARCN1 mutations cause a recognizable craniofacial syndrome due to COPI-mediated transport defects. Am J Hum Genet 99(2):451–459
Jacobsen CM, Barber LA, Ayturk UM, Roberts HJ, Deal LE, Schwartz MA et al (2014) Targeting the LRP5 pathway improves bone properties in a mouse model of osteogenesis imperfecta. J Bone Miner Res 29(10):2297–2306
Jin L, Pahuja KB, Wickliffe KE, Gorur A, Baumgartel C, Schekman R et al (2012) Ubiquitin-dependent regulation of COPII coat size and function. Nature 482(7386):495–500
Kadler KE, Hill A, Canty-Laird EG (2008) Collagen fibrillogenesis: fibronectin, integrins, and minor collagens as organizers and nucleators. Curr Opin Cell Biol 20(5):495–501
Kaushik S, Cuervo AM (2018) The coming of age of chaperone-mediated autophagy. Nat Rev Mol Cell Biol 19(6):365–381
Keller RB, Tran TT, Pyott SM, Pepin MG, Savarirayan R, McGillivray G et al (2018) Monoallelic and biallelic CREB3L1 variant causes mild and severe osteogenesis imperfecta, respectively. Genet Med 20(4):411–419
Khaminets A, Heinrich T, Mari M, Grumati P, Huebner AK, Akutsu M et al (2015) Regulation of endoplasmic reticulum turnover by selective autophagy. Nature 522(7556):354–358
Khoriaty R, Hesketh GG, Bernard A, Weyand AC, Mellacheruvu D, Zhu G et al (2018) Functions of the COPII gene paralogs SEC23A and SEC23B are interchangeable in vivo. Proc Natl Acad Sci U S A 115(33):E7748–E7E57
Kim SD, Pahuja KB, Ravazzola M, Yoon J, Boyadjiev SA, Hammamoto S et al (2012) The [corrected] SEC23-SEC31 [corrected] interface plays critical role for export of procollagen from the endoplasmic reticulum. J Biol Chem 287(13):10134–10144
Kizilyaprak C, Stierhof YD, Humbel BM (2019) Volume microscopy in biology: FIB-SEM tomography. Tissue Cell 57:123–128

Klionsky DJ, Abdelmohsen K, Abe A, Abedin MJ, Abeliovich H, Acevedo Arozena A et al (2016) Guidelines for the use and interpretation of assays for monitoring autophagy (3rd edition). Autophagy 12(1):1–222
Ko MK, Kay EP (2004) PDI-mediated ER retention and proteasomal degradation of procollagen I in corneal endothelial cells. Exp Cell Res 295(1):25–35
Koide T, Nagata K (2005) Collagen biosynthesis. In: Brinckmann J, Notbohm H, Muller PK (eds) Collagen: primer in structure, processing and assembly, Topics in current chemistry-series, vol 247. Springer, Heidelberg, pp 85–114
Koide T, Nishikawa Y, Asada S, Yamazaki CM, Takahara Y, Homma DL et al (2006) Specific recognition of the collagen triple helix by chaperone HSP47. II. The HSP47-binding structural motif in collagens and related proteins. J Biol Chem 281(16):11177–11185
Kondo Y, Fu J, Wang H, Hoover C, McDaniel JM, Steet R et al (2018) Site-1 protease deficiency causes human skeletal dysplasia due to defective inter-organelle protein trafficking. JCI Insight 3(14):e121596
Kuma A, Hatano M, Matsui M, Yamamoto A, Nakaya H, Yoshimori T et al (2004) The role of autophagy during the early neonatal starvation period. Nature 432(7020):1032–1036
Lamande SR, Bateman JF (1999) Procollagen folding and assembly: the role of endoplasmic reticulum enzymes and molecular chaperones. Semin Cell Dev Biol 10(5):455–464
Lamande SR, Chessler SD, Golub SB, Byers PH, Chan D, Cole WG et al (1995) Endoplasmic reticulum-mediated quality control of type I collagen production by cells from osteogenesis imperfecta patients with mutations in the pro alpha 1 (I) chain carboxyl-terminal propeptide which impair subunit assembly. J Biol Chem 270(15):8642–8649
Leblond CP (1989) Synthesis and secretion of collagen by cells of connective tissue, bone, and dentin. Anat Rec 224(2):123–138
Leikina E, Mertts MV, Kuznetsova N, Leikin S (2002) Type I collagen is thermally unstable at body temperature. Proc Natl Acad Sci U S A 99(3):1314–1318
Lekszas C, Foresti O, Raote I, Liedtke D, Konig EM, Nanda I et al (2020) Biallelic TANGO1 mutations cause a novel syndromal disease due to hampered cellular collagen secretion. elife 9: e51319
Lerner DW, McCoy D, Isabella AJ, Mahowald AP, Gerlach GF, Chaudhry TA et al (2013) A Rab10-dependent mechanism for polarized basement membrane secretion during organ morphogenesis. Dev Cell 24(2):159–168
Li WW, Li J, Bao JK (2012) Microautophagy: lesser-known self-eating. Cell Mol Life Sci 69 (7):1125–1136
Lindert U, Cabral WA, Ausavarat S, Tongkobpetch S, Ludin K, Barnes AM et al (2016) MBTPS2 mutations cause defective regulated intramembrane proteolysis in X-linked osteogenesis imperfecta. Nat Commun 7:11920
Lippincott-Schwartz J (2001) The secretory membrane system studied in real-time. Robert Feulgen prize lecture, 2001. Histochem Cell Biol 116(2):97–107
Lippincott-Schwartz J, Roberts TH, Hirschberg K (2000) Secretory protein trafficking and organelle dynamics in living cells. Annu Rev Cell Dev Biol 16:557–589
Lu CL, Kim J (2020) Consequences of mutations in the genes of the ER export machinery COPII in vertebrates. Cell Stress Chaperones 25(2):199–209
Ma W, Goldberg E, Goldberg J (2017) ER retention is imposed by COPII protein sorting and attenuated by 4-phenylbutyrate. elife 6:e26624
Makareeva E, Leikin S (2007) Procollagen triple helix assembly: an unconventional chaperone-assisted folding paradigm. PLoS One 2(10):e1029
Makareeva E, Aviles NA, Leikin S (2011) Chaperoning osteogenesis: new protein-folding disease paradigms. Trends Cell Biol 21(3):168–176
Makareeva E, Sun G, Mirigian LS, Mertz EL, Vera JC, Espinoza NA et al (2018) Substitutions for arginine at position 780 in triple helical domain of the alpha1(I) chain alter folding of the type I procollagen molecule and cause osteogenesis imperfecta. PLoS One 13(7):e0200264

Makareeva E, Omari S, Roberts-Pilgrim AM, Gorrell L, Mertz E, Khoury B et al (2019) Therapeutic targeting of autophagy in Osteogenesis Imperfecta. J Bone Miner Res 32(Suppl 1):FRI-920. Available at https://www.asbmr.org//education/2019-abstracts

Malhotra V, Erlmann P (2015) The pathway of collagen secretion. Annu Rev Cell Dev Biol 31:109–124

Marini JC, Forlino A, Cabral WA, Barnes AM, San Antonio JD, Milgrom S et al (2007) Consortium for osteogenesis imperfecta mutations in the helical domain of type I collagen: regions rich in lethal mutations align with collagen binding sites for integrins and proteoglycans. Hum Mutat 28(3):209–221

Marini JC, Forlino A, Bachinger HP, Bishop NJ, Byers PH, Paepe A et al (2017) Osteogenesis imperfecta. Nat Rev Dis Primers 3:17052

McCaughey J, Stephens DJ (2018) COPII-dependent ER export in animal cells: adaptation and control for diverse cargo. Histochem Cell Biol 150(2):119–131

McCaughey J, Stevenson NL, Cross S, Stephens DJ (2019) ER-to-Golgi trafficking of procollagen in the absence of large carriers. J Cell Biol 218(3):929–948

Mercer TJ, Gubas A, Tooze SA (2018) A molecular perspective of mammalian autophagosome biogenesis. J Biol Chem 293(15):5386–5395

Mertz EL, Makareeva E, Mirigian LS, Koon KY, Perosky JE, Kozloff KM et al (2016) Makings of a brittle bone: unexpected lessons from a low protein diet study of a mouse OI model. Matrix Biol 52-54:29–42

Mi Y, Wang W, Lu J, Zhang C, Wang Y, Ying H et al (2018) Proteasome-mediated degradation of collagen III by cortisol in amnion fibroblasts. J Mol Endocrinol 60(2):45–54

Miles CA, Ghelashvili M (1999) Polymer-in-a-box mechanism for the thermal stabilization of collagen molecules in fibers. Biophys J 76(6):3243–3252

Mirigian LS, Makareeva E, Mertz EL, Omari S, Roberts-Pilgrim AM, Oestreich AK et al (2016) Osteoblast malfunction caused by cell stress response to Procollagen Misfolding in alpha2(I)-G610C mouse model of Osteogenesis Imperfecta. J Bone Miner Res 31(8):1608–1616

Mironov AA (2014) ER-Golgi transport could occur in the absence of COPII vesicles. Nat Rev Mol Cell Biol 15(3):1

Mironov AA, Beznoussenko GV (2012) The kiss-and-run model of intra-Golgi transport. Int J Mol Sci 13(6):6800–6819

Mironov AA, Mironov AA Jr, Beznoussenko GV, Trucco A, Lupetti P, Smith JD et al (2003) ER-to-Golgi carriers arise through direct en bloc protrusion and multistage maturation of specialized ER exit domains. Dev Cell 5(4):583–594

Mizushima N (2019) The ATG conjugation systems in autophagy. Curr Opin Cell Biol 63:1–10

Moosa S, Chung BHY, Tung JYL, Altmuller J, Thiele H, Nurnberg P et al (2016) Mutations in SEC24D cause autosomal recessive osteogenesis imperfecta. Clin Genet 89(4):517–519

Mortier GR, Cohn DH, Cormier-Daire V, Hall C, Krakow D, Mundlos S et al (2019) Nosology and classification of genetic skeletal disorders: 2019 revision. Am J Med Genet A 179 (12):2393–2419

Mullan LA, Mularczyk EJ, Kung LH, Forouhan M, Wragg JM, Goodacre R et al (2017) Increased intracellular proteolysis reduces disease severity in an ER stress-associated dwarfism. J Clin Invest 127(10):3861–3865

Nabavi N, Pustylnik S, Harrison RE (2012) Rab GTPase mediated procollagen trafficking in ascorbic acid stimulated osteoblasts. PLoS One 7(9):e46265

Nagai N, Hosokawa M, Itohara S, Adachi E, Matsushita T, Hosokawa N et al (2000) Embryonic lethality of molecular chaperone hsp47 knockout mice is associated with defects in collagen biosynthesis. J Cell Biol 150(6):1499–1506

Nogueira C, Erlmann P, Villeneuve J, Santos AJ, Martinez-Alonso E, Martinez-Menarguez JA et al (2014) SLY1 and Syntaxin 18 specify a distinct pathway for procollagen VII export from the endoplasmic reticulum. elife 3:e02784

Oku M, Sakai Y (2018) Three distinct types of microautophagy based on membrane dynamics and molecular machineries. BioEssays 40(6):e1800008

Omari S, Makareeva E, Roberts-Pilgrim A, Mirigian L, Jarnik M, Ott C et al (2018) Noncanonical autophagy at ER exit sites regulates procollagen turnover. Proc Natl Acad Sci U S A 115(43): E10099–E1E108

Omari S, Makareeva E, Gorrell L, Jarnik M, Lippincott-Schwartz J, Leikin S (2020) Mechanisms of procollagen and HSP47 sorting during ER-to-Golgi trafficking. Matrix Biol 93:79–94

Pakdel M, von Blume J (2018) Exploring new routes for secretory protein export from the trans-Golgi network. Mol Biol Cell 29(3):235–240

Pantazopoulou A, Glick BS (2019) A kinetic view of membrane traffic pathways can transcend the classical view of Golgi compartments. Front Cell Dev Biol 7:153

Papandreou ME, Tavernarakis N (2017) Autophagy and the endo/exosomal pathways in health and disease. Biotechnol J 12(1):1600175

Presley JF, Cole NB, Schroer TA, Hirschberg K, Zaal KJ, Lippincott-Schwartz J (1997) ER-to-Golgi transport visualized in living cells. Nature 389(6646):81–85

Privalov PL (1982) Stability of proteins. Proteins which do not present a single cooperative system. Adv Protein Chem 35:1–104

Privalov PL, Tiktopulo EI, Tischenko VM (1979) Stability and mobility of the collagen structure. J Mol Biol 127(2):203–216

Qi L, Tsai B, Arvan P (2017) New insights into the physiological role of endoplasmic reticulum-associated degradation. Trends Cell Biol 27(6):430–440

Ramachandran U, Peterkofsky B (1997) Aberrant O-glycosylation in the collagenous domain of pro alpha2(I) procollagen subunits synthesized by chemically transformed hamster fibroblasts. Arch Biochem Biophys 342(1):29–37

Raote I, Malhotra V (2019) Protein transport by vesicles and tunnels. J Cell Biol 218(3):737–739

Raote I, Ortega-Bellido M, Santos AJ, Foresti O, Zhang C, Garcia-Parajo MF et al (2018) TANGO1 builds a machine for collagen export by recruiting and spatially organizing COPII, tethers and membranes. elife 7:e32723

Rauch F, Fahiminiya S, Majewski J, Carrot-Zhang J, Boudko S, Glorieux F et al (2015) Cole-carpenter syndrome is caused by a heterozygous missense mutation in P4HB. Am J Hum Genet 96(3):425–431

Rautou PE, Mansouri A, Lebrec D, Durand F, Valla D, Moreau R (2010) Autophagy in liver diseases. J Hepatol 53(6):1123–1134

Reynolds HM, Zhang L, Tran DT, Ten Hagen KG (2019) Tango1 coordinates the formation of endoplasmic reticulum/Golgi docking sites to mediate secretory granule formation. J Biol Chem 294(51):19498–19510

Ripley CR, Bienkowski RS (1997) Localization of procollagen I in the lysosome/endosome system of human fibroblasts. Exp Cell Res 236(1):147–154

Robinson ME, Rauch F (2019) Mendelian bone fragility disorders. Bone 126:11–17

Rossi V, Lee B, Marom R (2019) Osteogenesis imperfecta: advancements in genetics and treatment. Curr Opin Pediatr 31(6):708–715

Roth J, Zuber C (2017) Quality control of glycoprotein folding and ERAD: the role of N-glycan handling, EDEM1 and OS-9. Histochem Cell Biol 147(2):269–284

Ruggiano A, Foresti O, Carvalho P (2014) Quality control: ER-associated degradation: protein quality control and beyond. J Cell Biol 204(6):869–879

Sacher M, Shahrzad N, Kamel H, Milev MP (2019) TRAPPopathies: an emerging set of disorders linked to variations in the genes encoding transport protein particle (TRAPP)-associated proteins. Traffic 20(1):5–26

Sahl SJ, Hell SW, Jakobs S (2017) Fluorescence nanoscopy in cell biology. Nat Rev Mol Cell Biol 18(11):685–701

Saito K, Maeda M (2019) Not just a cargo receptor for large cargoes; an emerging role of TANGO1 as an organizer of ER exit sites. J Biochem 166(2):115–119

Saito K, Chen M, Bard F, Chen S, Zhou H, Woodley D et al (2009) TANGO1 facilitates cargo loading at endoplasmic reticulum exit sites. Cell 136(5):891–902

Santos AJ, Raote I, Scarpa M, Brouwers N, Malhotra V (2015) TANGO1 recruits ERGIC membranes to the endoplasmic reticulum for procollagen export. elife 4:e10982
Saraste J, Marie M (2018) Intermediate compartment (IC): from pre-Golgi vacuoles to a semi-autonomous membrane system. Histochem Cell Biol 150(5):407–430
Satoh M, Hirayoshi K, Yokota S, Hosokawa N, Nagata K (1996) Intracellular interaction of collagen-specific stress protein HSP47 with newly synthesized procollagen. J Cell Biol 133 (2):469–483
Scales SJ, Pepperkok R, Kreis TE (1997) Visualization of ER-to-Golgi transport in living cells reveals a sequential mode of action for COPII and COPI. Cell 90(6):1137–1148
Scheiber A, Suzuki A, Enomoto-Iwamoto M, Iwamoto M, Leikin S, Otsuru S (2019) 4-phenylbutyrate (4PBA) ameliorates growth deficiency in G610C mouse model of Osteogenesis Imperfecta. J Bone Miner Res 32(Suppl 1):1135. Available at https://www.asbmr.org// education/2019-abstracts
Schmidt O, Weyer Y, Baumann V, Widerin MA, Eising S, Angelova M et al (2019) Endosome and Golgi-associated degradation (EGAD) of membrane proteins regulates sphingolipid metabolism. EMBO J 38(15):e101433
Schwarze U, Cundy T, Liu YJ, Hofman PL, Byers PH (2019) Compound heterozygosity for a frameshift mutation and an upstream deletion that reduces expression of SERPINH1 in siblings with a moderate form of osteogenesis imperfecta. Am J Med Genet A 179(8):1466–1475
Shoulders MD, Raines RT (2009) Collagen structure and stability. Annu Rev Biochem 78:929–958
Smith T, Ferreira LR, Hebert C, Norris K, Sauk JJ (1995) Hsp47 and cyclophilin B traverse the endoplasmic reticulum with procollagen into pre-Golgi intermediate vesicles. A role for Hsp47 and cyclophilin B in the export of procollagen from the endoplasmic reticulum. J Biol Chem 270 (31):18323–18328
Sprangers J, Rabouille C (2015) SEC16 in COPII coat dynamics at ER exit sites. Biochem Soc Trans 43(1):97–103
Subramanian A, Capalbo A, Iyengar NR, Rizzo R, di Campli A, Di Martino R et al (2019) Autoregulation of secretory flux by sensing and responding to the folded cargo protein load in the endoplasmic reticulum. Cell 176(6):1461–1476
Tekirdag K, Cuervo AM (2018) Chaperone-mediated autophagy and endosomal microautophagy: joint by a chaperone. J Biol Chem 293(15):5414–5424
Thakker-Varia S, Anderson DW, Kuivaniemi H, Tromp G, Shin HG, van der Rest M et al (1995) Aberrant splicing of the type III procollagen mRNA leads to intracellular degradation of the protein in a patient with Ehlers-Danlos type IV. Hum Mutat 6(2):116–125
Townley AK, Feng Y, Schmidt K, Carter DA, Porter R, Verkade P et al (2008) Efficient coupling of Sec23-Sec24 to Sec13-Sec31 drives COPII-dependent collagen secretion and is essential for normal craniofacial development. J Cell Sci 121(Pt 18):3025–3034
Trelstad RL, Hayashi K (1979) Tendon collagen fibrillogenesis: intracellular subassemblies and cell surface changes associated with fibril growth. Dev Biol 71(2):228–242
Unlu G, Qi X, Gamazon ER, Melville DB, Patel N, Rushing AR et al (2020) Phenome-based approach identifies RIC1-linked Mendelian syndrome through zebrafish models, biobank associations and clinical studies. Nat Med 26(1):98–109
Vakifahmetoglu-Norberg H, Xia HG, Yuan J (2015) Pharmacologic agents targeting autophagy. J Clin Invest 125(1):5–13
Vanakker O, Callewaert B, Malfait F, Coucke P (2015) The genetics of soft connective tissue disorders. In: Chakravarti A, Green E (eds) Annual review of genomics and human genetics, vol 16, pp 229–255
Venditti R, Wilson C, De Matteis MA (2014) Exiting the ER: what we know and what we don't. Trends Cell Biol 24(1):9–18
Wang C, Tan Z, Niu B, Tsang KY, Tai A, Chan WCW et al (2018) Inhibiting the integrated stress response pathway prevents aberrant chondrocyte differentiation thereby alleviating chondrodysplasia. elife 7:e37673

Weigel A, Chang C, Shtengel G, Hoffman D, Freeman M, Xu CS et al (2019) COPI and COPII cooperate at ER exit sites to support ER-to-Golgi protein trafficking revealed by 3D ultrastructure analyses and live-cell imaging. Mol Biol Cell M42:30. (abstract#)

Wirth M, Zhang W, Razi M, Nyoni L, Joshi D, O'Reilly N et al (2019) Molecular determinants regulating selective binding of autophagy adapters and receptors to ATG8 proteins. Nat Commun 10(1):2055

Yin X, Zhou C, Li J, Liu R, Shi B, Yuan Q et al (2019) Autophagy in bone homeostasis and the onset of osteoporosis. Bone Res 7:28

Yoshii SR, Kuma A, Akashi T, Hara T, Yamamoto A, Kurikawa Y et al (2016) Systemic analysis of Atg5-null mice rescued from neonatal lethality by transgenic ATG5 expression in neurons. Dev Cell 39(1):116–130

Yu L, Chen Y, Tooze SA (2018) Autophagy pathway: cellular and molecular mechanisms. Autophagy 14(2):207–215

Yuan L, Kenny SJ, Hemmati J, Xu K, Schekman R (2018) TANGO1 and SEC12 are copackaged with procollagen I to facilitate the generation of large COPII carriers. Proc Natl Acad Sci U S A 115(52):E12255–E12E64

Zamurs LK, Idoate MA, Hanssen E, Gomez-Ibanez A, Pastor P, Lamande SR (2015) Aberrant mitochondria in a Bethlem myopathy patient with a homozygous amino acid substitution that destabilizes the collagen VI alpha2(VI) chain. J Biol Chem 290(7):4272–4281

Zanetti G, Pahuja KB, Studer S, Shim S, Schekman R (2011) COPII and the regulation of protein sorting in mammals. Nat Cell Biol 14(1):20–28

Zhang H, Yue H, Wang C, Gu J, He J, Fu W et al (2017) Novel mutations in the SEC24D gene in Chinese families with autosomal recessive osteogenesis imperfecta. Osteoporos Int 28 (4):1473–1480

Zou Y, Donkervoort S, Salo AM, Foley AR, Barnes AM, Hu Y et al (2017) P4HA1 mutations cause a unique congenital disorder of connective tissue involving tendon, bone, muscle and the eye. Hum Mol Genet 26(12):2207–2217

Chapter 3
Collagens in the Physiopathology of the Ehlers–Danlos Syndromes

Fransiska Malfait, Robin Vroman, Marlies Colman, and Delfien Syx

Abstract The Ehlers–Danlos Syndromes (EDS) comprise a clinically and genetically heterogeneous group of complex hereditary disorders of connective tissue, with common features including joint hypermobility, soft and hyperextensible skin, abnormal wound healing, easy bruising, and signs of generalized connective tissue friability. Initial ultrastructural studies suggested that the abnormalities underlying EDS affected the collagen "wickerwork" of the connective tissue, and early biochemical and genetic studies identified defects in fibrillar types I, III, and V collagen, and in enzymes involved in their posttranslational modification, lysyl hydroxylase 1 and the procollagen amino-proteinase ADAMTS2. More recent discoveries have implicated a range of other, diverse extracellular matrix (ECM) molecules in the physiopathology of EDS, including the glycoprotein tenascin X, the FACIT type XII collagen, the intracellular chaperone and peptidylprolyl isomerase FKBP22, enzymes involved in glycosaminoglycan biosynthesis (D4ST1, DS-epi1, galactosyltransferase I and II), an intracellular zinc transporter ZIP13, (putative) transcription factors ZNF469 and PRDM5, factors involved in the classical complement pathway (C1r and C1s), and most recently, the ECM molecule AEBP1 that is involved in collagen polymerization. In this chapter, we give an overview of the different types of EDS and describe how the identification of their molecular underpinnings, and the study of pathophysiologic consequences of these defects in humans and in cellular and mouse models have provided key insights into the complex pathways of collagen fibrillogenesis and supramolecular organization of the collagen fibrils in the ECM.

F. Malfait (✉) · R. Vroman · M. Colman · D. Syx
Department of Biomolecular Medicine, Center for Medical Genetics, Ghent University and Ghent University Hospital, Ghent, Belgium
e-mail: Fransiska.Malfait@Ugent.be

F. Ruggiero (ed.), *The Collagen Superfamily and Collagenopathies*, Biology of Extracellular Matrix 8, https://doi.org/10.1007/978-3-030-67592-9_3

Abbreviations

AEBP1	Adipocyte enhancer-binding protein 1
ATCS	Adducted thumb–clubfoot syndrome
aEDS	Arthrochalasia type
BCS	Brittle cornea syndrome
C4ST	Chondroitin 4-*O*-sulfotransferase
C	Carboxy
CAH	Congenital adrenal hypoplasia
cEDS	Classical EDS
clEDS	Classical-like EDS
CS	Chondroitin sulfate
cvEDS	Cardiac-valvular EDS
D4ST1	Dermatan 4-*O*-sulfotransferase-1
dEDS	Dermatosparaxis type
DS	Dermatan sulfate
DS-epi	Dermatan sulfate epimerase
EDS	Ehlers–Danlos syndromes
ECM	Extracellular matrix
ER	Endoplasmic reticulum
FACIT	Fibril-associated collagens with interrupted triple helices
GAG	Glycosaminoglycan
GalNAc	*N*-acetylgalactosamine
GalT	Galactosyltransferase
GlcA	Glucuronic acid
hEDS	Hypermobile Ehlers–Danlos syndrome
HPs	Hydroxylysylpyridinolines
HS	Heparan sulfate
IdoA	Iduronic acid
kEDS	Kyphoscoliotic EDS
LH	Lysyl hydroxylase
LPs	Lysylpyridinolines
mcEDS	Musculocontractural EDS
mEDS	Myopathic EDS
N	Amino
NMD	Nonsense-mediated mRNA decay
OI	Osteogenesis imperfecta
pEDS	Periodontal EDS
PPIase	Peptidylprolyl isomerase
rER	Rough endoplasmic reticulum
RNAi	RNA interference
SEMD	Spondyloepimetaphyseal dysplasia
SDS-PAGE	Sodium dodecyl sulfate polyacrylamide gel electrophoresis
SLRP	Small leucine-rich proteoglycan
spEDS	Spondylodysplastic EDS

TNX	Tenascin X
TSP1	Thrombospondin type I repeats
TSPN	Thrombospondin N-terminal domain
vEDS	Vascular EDS
XylTs	Xylosyltransferases
ZIP13	Zrt/irt-like protein 13

3.1 The Ehlers–Danlos Syndromes and Collagen, an Intimate History

Throughout the ages, people have been fascinated by the Ehlers–Danlos Syndromes (EDS). The first report of this disorder dates from as far as the fourth century BC, when Hippocrates recorded that Nomads and Scythians "were unable to draw their bow because their shoulder joints were too lax" (Littré 1839). For many years, a fairground was not complete without an "Elastic Man," "India Rubber Man," or "Human Pretzel". Collectively enjoyed as curiosities, many affected individuals earned their livings by amazing their audiences with contortionist tricks or an astonishing ability to stretch their skin. The first (partial) clinical description is credited to Job Janszoon van Meek'ren (1611–1666), a Dutch surgeon who in 1657 observed a Spanish boy, Georg Albes, who had extreme extensibility of the skin (unusually, though, only at the right side of his body) (van Mee'ren 1668). Credit for the first comprehensive description (1892) of a syndrome showing skin laxity and fragility in combination with joint hypermobility goes to the Russian dermatologist, A. N. Chernogubow (1892). His report passed however largely unnoticed in Western Europe, likely because it was written in Russian. Two dermatologists, Edvard Ehlers and Henri-Alexandre Danlos separately described affected individuals in 1901 (Ehlers 1901) and 1908 (Danlos 1908), respectively, and between 1932 and 1936, the syndrome received its eponymous title, and as such became a scientifically respectable condition (Weber 1936). In 1936, Georg Sack, a German physician, described a patient with excessive friability of the arteries and termed the condition "status dysvascularis" (Sack 1936), later identified by Barabas as the arterial type of EDS (Barabas 1967).

In 1955, L.H. Jansen, a Dutch dermatologist, studied the skin of two patients with EDS by means of transmission electron microscopy. He concluded, "*We believe that the whole Ehlers-Danlos syndrome is based on the existence of (this) insufficient twining of the collagen elements in the dermis, the hypodermis and the joints,*" thus suggesting, for the first time, that the defect underlying EDS affects the collagen "wickerwork" (Jansen 1955). A year later Victor McKusick, who, in his hallmark work on heritable connective tissue disorders provided a first synthesis on the variable and multisystemic nature of EDS, commented on this: "*I am inclined to favor the view that the EDS is another heritable disorder of collagen* (…)" (McKusick 1956). Jackson and Bentley (1968) further elaborated on this hypothesis and suggested "*that the defect in EDS lies at a high level of organization of the collagen fibres, and that in this high-level binding the collagen fibrils interact with the various mucopolysaccharides and mucopolysaccharide-protein complexes of the 'ground substance'*" (Jackson and Bentley 1968).

For a long time, EDS was considered to be a single disease entity with broad variable expression, but in 1967 Barabas defined three types of EDS, classical, varicose and arterial, which he thought reflected distinct etiologies, rather than variable expression (Barabas 1967). By doing this, he started the classification of EDS. Not much later, Peter Beighton, in his hallmark clinical investigation of 100 individuals with EDS, confirmed Barabas' observations and expanded the classification to five types (gravis, mitis, hypermobile, vascular, and X-linked) (Beighton 1970).

Besides a few light and transmission electron microscopy studies that described structural abnormalities of the collagen fibrils and fibers in the dermis of EDS patients, further clues to the molecular basis of any of the EDS forms remained sparse. This changed with the introduction of biochemical studies on tissues obtained from humans and animals presenting EDS characteristics. In 1972 Pinnell and Krane and coworkers (Krane et al. 1972; Pinnell et al. 1972) performed studies on dermal collagen in two sisters presenting soft extensible skin, joint hypermobility, muscle hypotonia, scoliosis refractory to surgical intervention, and ocular globe fragility. They found that collagen extracted from biopsy specimens of these girls was normally soluble in non-denaturing solvents but had increased solubility in denaturing solvents and assumed that this was due to a defect in the intermolecular cross-linking of collagen in the skin. The most remarkable abnormality was a striking hydroxylysine deficiency in the skin of about 0.2 to 0.3 residues per 1000 amino acid residues (5–7% of normal). Collagen from lumbodorsal fascia was reduced to about 20%, and bone collagen to about 50% of normal. The activity of lysyl hydroxylase was shown to be reduced to ~10–15% of normal in the affected siblings, and to ~60% in their mother. Therefore, an autosomal recessive inheritance was assumed for the disorder. Lysyl hydroxylase deficiency represents the first congenital error of human collagen metabolism described at the biochemical level (Krane et al. 1972), and its identification established the paradigm of EDS as a collagen disorder.

Further clues to the key role of collagen abnormalities in the physiopathology of EDS came from studies in a cattle population of Central and High Belgium, that showed extreme fragility of the skin, and that was first described in 1967 as dermatosparaxis (Greek for "torn skin") (Hanset 1967). The dermis of these animals was shown to contain loosely packed, thin, and twisted ribbon-like collagen fibrils on cross-section that showed a typical "hieroglyphic" aspect (O'Hara et al. 1970). The condition showed an autosomal recessive inheritance pattern (Hanset 1974), and biochemical studies revealed that an amino (N)-terminal precursor peptide from the three α chains of type I collagen could not be processed in affected animals (Lenaers et al. 1971). These observations provided some of the first evidence for the existence of a soluble precursor form of collagen, "procollagen", and uncovered the existence of a "procollagen peptidase," that cleaves off these N-terminal propeptides (procollagen N-proteinase) (Lapiere et al. 1971). The skin tears in these animals were attributed to the incorporation of immature pN-collagen (the collagen molecules still containing the N-propeptide) into the fibrils, thereby distorting stacking of collagen molecules and sterically interfering with normal cross-linking (Pierard and Lapiere 1976). Dermatosparaxis was later also reported in sheep (Fjolstad and Helle 1974), cats (Counts et al. 1980; Holbrook et al. 1980), and dogs (Holbrook and Byers 1982).

Soon after the discovery of animal dermatosparaxis, three unrelated human individuals presenting with an EDS phenotype were reported to have an increased amount of pN-collagen in their skin and tendon in comparison with normal controls. The clinical manifestations of these patients were different from the dermatosparactic animals, in that they were characterized by stretchable, velvety skin with only minimal fragility and marked short stature in combination with severe joint hypermobility with multiple joint dislocations, especially bilateral hip dislocations (arthrochalasis multiplex congenita, later classified as EDS VII or EDS arthrochalasia type) (Lichtenstein et al. 1973). Initially also thought to be caused by a procollagen N-proteinase deficiency, biochemical studies on a skin biopsy of one of these patients however later showed the presence of equal amounts of pNα2(I) and mature α2(I) chains. The pNα2(I) chains were resistant to proteolytic cleavage by procollagen N-proteinase (in contrast to the pNα chains from dermatosparactic animals), and it was presumed that the patient carried a heterozygous mutation in or near the site at which procollagen N-proteinase cleaves the proα2(I) chain (Steinmann et al. 1980).

Finally, in 1975, Pope and coworkers reported biochemical studies on postmortem biopsy tissue specimens and dermal fibroblasts from a patient with the arterial form of EDS (later EDS IV or vascular EDS). They found a lack of type III collagen in the skin and aorta and in the proteins synthesized by the cultured fibroblasts and attributed the tissue weakness to this observation (Pope et al. 1975). Initially thought to be an autosomal recessive condition (based on the observation that some non-affected family members of the probands had lower levels of type III collagen) (Pope et al. 1977), genetic studies later provided proof of autosomal dominant inheritance (Tsipouras et al. 1986; Nicholls et al. 1988).

With the introduction of DNA sequencing techniques, the study of the molecular basis of heritable connective tissue disorders took its first steps into the genetic era, and in the second half of the 1980s the first mutations were identified in EDS patients. In 1988, Weil and coworkers identified in a patient with EDS type VII a heterozygous mutation in the splice donor site of intron 6 of the *COL1A2* gene, leading to loss of the 54 bp exon 6, thereby eliminating the procollagen N-proteinase cleavage site and the N-telopeptide cross-linking site (Weil et al. 1988). Not much later Superti-Furga et al. reported the first *COL3A1* mutation in a patient with EDS IV (Superti-Furga et al. 1988). These first genetic observations unequivocally linked EDS to defects in collagen.

3.2 General Clinical Presentation of the Ehlers–Danlos Syndromes

The cardinal features that are common to all EDS types, albeit to different degrees of severity, are joint hypermobility, skin hyperextensibility, and signs of soft connective tissue friability affecting the skin, ligaments, blood vessels, eyes, and internal organs.

One of the principal features of EDS is skin hyperextensibility (Fig. 3.1a), that is, the skin stretches easily but snaps back upon release (in contrast to cutis laxa, where the skin becomes inelastic and hangs in loose folds). The skin often has a smooth, soft, or velvety feel to it, and can be thin and translucent. It is fragile and tears easily, even following minor trauma, and so-called "cigarette paper scars" (widened and thin atrophic scars) (Fig. 3.1b), are frequently found, in particular at exposed areas and pressure points such as the forehead, elbows, knees, and shins. Easy bruising is common and may manifest itself as spontaneous or recurring hematomas. These may cause skin discoloration due to hemosiderin deposition, often referred to as "hemosiderotic" scars (Fig. 3.1c).

Joint hypermobility (Fig. 3.1e) is another cardinal sign. It is variable in severity and usually, but not always, generalized. Although sometimes regarded as an "asset" during childhood and adolescence, it can result in a serious burden over time and is often complicated by recurrent subluxations, dislocations, sprains, and chronic and debilitating joint pain that is poorly controlled by currently available treatments.

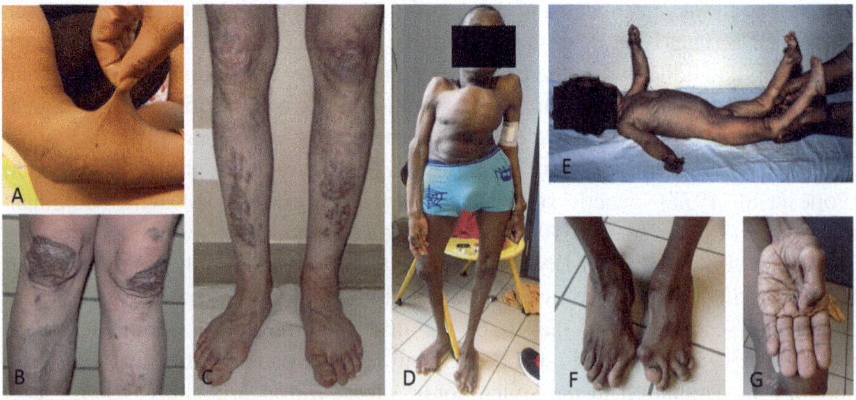

Fig. 3.1 Illustration of clinical manifestations of EDS. (**a**) Skin hyperextensibility in a patient with spondylodysplastic EDS (biallelic mutations *in B3GALT6*). (**b**) Widened atrophic scars in a patient with classical-like EDS type 2 (biallelic mutations in *AEBP1*). (**c**) Hemosiderotic deposition on the shins due to repetitive scarring and bruising in a patient with classical EDS (heterozygous mutation in *COL5A1*). (**d**) Short stature, severe kyphoscoliosis, pectus and foot deformities in a patient with spondylodysplastic EDS (biallelic mutations *in B3GALT6).* (**e**) Severe joint hypermobility with overextension of the knees in a child with Brittle Cornea Syndrome (biallelic mutations in *ZNF469*). (**f**) Foot deformities in a patient with spondylodysplastic EDS (biallelic mutations *in B3GALT6*). (**g**) Palmar wrinkling in a patient with spondylodysplastic EDS (biallelic mutations *in B3GALT6*)

Other frequently observed musculoskeletal features include congenital bilateral hip dislocation, spine deformities (scoliosis, kyphosis) (Fig. 3.1d), pectus deformities (pectus carinatum, pectus excavatum), club feet and other contractures, and deformities of the elbows, hands (Fig. 3.1g), knees, and feet (Fig. 3.1f). Muscle hypotonia is observed in a number of EDS types and, in combination with joint laxity, may cause floppy infant syndrome or a delay in motor development.

Signs of more generalized connective tissue weakness and fragility can be detected in varying degrees and may help to distinguish between the different EDS types (Table 3.1) (Malfait et al. 2017). Although rare, cardiovascular complications may have an important impact on the severity of the disorder, rupture of medium- and large-sized arteries is typical of vascular EDS but has been reported in a few other types as well (Table 3.1). Valvular defects and aortic root dilatation are rare and restricted to some of the rarer types of EDS. Gynecological and obstetrical complications like cervical insufficiency, premature rupture of membranes, vaginal tears and lacerations, organ prolapses (uterus, bladder, rectum), umbilical or hiatal hernia, and surgical complications, e.g., wound dehiscence and incisional hernia, can occur (Malfait et al. 2017).

3.3 Classification, Diagnostic Criteria, and Prevalence

Classification of the EDS types has been a dynamic process, and as the biochemical, and later, genetic bases of different EDS types were discovered, conditions were included or excluded accordingly. In 1986, an international group of experts convened at a dedicated workshop during the European Meeting for Human Genetics in Berlin, where they defined the International Nosology of Heritable Disorders of Connective Tissue (Beighton et al. 1988). Within this nosology, nine EDS types were recognized based on clinical presentation and mode of inheritance and they were defined by Roman numerals. With the elucidation of the biochemical and/or molecular basis of several EDS types, a simplified classification was published in 1998, known as the "Villefranche Nosology" (Beighton et al. 1998). This classification withheld six EDS types and assigned a descriptive name to each (classical, vascular, hypermobile, kyphoscoliotic, arthrochalasic, and dermatosparaxis types). In the meantime, most of these types were shown to be due to genetic defects in the genes encoding fibrillar collagens type I (arthrochalasis type), III (vascular type), or V (classical type), or their modifying enzymes lysyl hydroxylase 1 (kyphoscoliotic type) and procollagen I N-proteinase (dermatosparaxis type). Despite being the gold standard for diagnosis and classification of EDS for nearly two decades, with the ongoing identification of causative variants in new genes, the Villefranche Nosology became quite quickly outdated. In 2017, an international group of EDS experts (the International EDS Consortium) revised the EDS classification and defined 13 distinct clinical EDS types that result from pathogenic changes in 19 different genes (Malfait et al. 2017). Following the publication of the 2017 classification, another genetically distinct EDS type was described, thereby bringing the total number of

Table 3.1 Overview of the EDS types, their inheritance pattern, and the major and minor clinical criteria as defined by the 2017 International EDS classification

EDS type	Abbr.	IP	Major clinical criteria	Minor clinical criteria
Classical	cEDS	AD	Skin hyperextensibility with atrophic scarring Generalized joint hypermobility	Easy bruising Soft doughy skin Skin fragility (or traumatic splitting) Molluscoid pseudotumors (bluish-gray, spongy nodules, which are herniations of subcutaneous fat, seen over easily traumatized areas) Subcutaneous spheroids Hernia (or history thereof) Epicanthal folds Complications of joint hypermobility (e.g., sprain, (sub)luxation, pain, flexible flatfoot) Family history of a first-degree relative who meets criteria
Classical-like	clEDS	AR	Skin hyperextensibility with velvety skin texture and absence of atrophic scarring Generalized joint hypermobility Easy bruisable skin/spontaneous ecchymoses	Foot deformities Edema in legs in absence of cardiac failure Mild proximal and distal muscle weakness Axonal polyneuropathy Atrophy of muscle in hands and feet Acrogeric hands, mallet finger(s), clino- or brachydactyly Vaginal/uterine/rectal prolapse
Cardiac-valvular	cvEDS	AR	Severe progressive cardiac-valvular insufficiency Skin involvement Joint hypermobility (generalized or restricted to small joints)	Inguinal hernia Pectus deformity Joint dislocations Foot deformities: pes planus, pes planovalgus, hallux valgus
Vascular	vEDS	AD	Family history of vEDS with documented pathogenic variant in *COL3A1* Arterial rupture at a young age Spontaneous sigmoid colon perforation in the absence of	Bruising unrelated to identified trauma and/or in unusual sites such as cheeks and back Thin, translucent skin with increased venous visibility Characteristic facial features: large eyes, periorbital

			known colon disease Uterine rupture during third trimester of pregnancy Carotid-cavernous sinus fistula (in the absence of trauma)	pigmentation, small chin, sunken cheeks, thin nose and lips, lobeless ears Spontaneous pneumothorax Acrogeria Talipes equinovarus Congenital hip dislocation Small joint hypermobility Tendon and muscle rupture Gingival recession and gingival fragility Early-onset varicose veins
Hypermobile	hEDS	?	Generalized joint hypermobility Systemic manifestations of generalized connective tissue disorder Positive family history Musculoskeletal complaints Exclusion of other EDS types and other joint hypermobility-associated conditions (for a detailed description of clinical criteria, see (Malfait et al. 2017))	
Arthrochalasia	aEDS	AD	Congenital bilateral hip dislocation Severe generalized joint hypermobility with multiple dislocations Skin hyperextensibility	Muscle hypotonia Kyphoscoliosis Radiologically mild osteopenia Tissue fragility, including atrophic scars Easy bruising
Dermatosparaxis	dEDS	AR	Extreme skin fragility with congenital or postnatal tears Craniofacial features: large fontanel, puffy eyelids, excessive periorbital skin, downslanting palpebral fissures, blue sclerae, hypoplastic chin Progressively redundant, almost lax skin with excessive skin folds at wrists and ankles Increased palmar wrinkling Severe bruisability with risk of subcutaneous hematoma Umbilical hernia Postnatal growth retardation with short limbs Perinatal complications related to tissue fragility	Soft and doughy skin texture Skin hyperextensibility Atrophic scars Generalized joint hypermobility Complications of visceral fragility (e.g., rectal prolapse, bladder or diaphragm rupture) Delayed motor development Osteopenia Hirsutism Tooth abnormalities

(continued)

Table 3.1 (continued)

EDS type	Abbr.	IP	Major clinical criteria	Minor clinical criteria
				Refractive errors Strabismus
Kyphoscoliotic	kEDS	AR	Congenital muscle hypotonia Congenital or early-onset kyphoscoliosis Generalized joint hypermobility with (sub)luxations	Skin hyperextensibility Easy bruising Rupture/aneurysm of medium-sized artery Osteopenia/osteoporosis Blue sclerae Umbilical or inguinal hernia Pectus deformity Marfanoid habitus Talipes equinovarus Refractive errors *PLOD1* Skin fragility Microcornea Characteristic craniofacial features *FKBP14* Congenital hearing impairment Muscle atrophy Bladder diverticulae
Brittle cornea syndrome	BCS	AR	Thin cornea with/without rupture Early-onset progressive keratoconus and/or keratoglobus Blue sclerae	Enucleation or corneal scarring as a result of previous rupture Progressive loss of corneal stromal depth High myopia Retinal detachment

				Deafness (often mixed conductive and sensorineural) Hypercompliant tympanic membranes Developmental dysplasia of hip Hypotonia in infancy (usually mild) Scoliosis Arachnodactyly Hypermobility of distal joints Pes planus, hallux valgus Mild finger contractures Soft, velvety, and/or translucent skin
Musculocontractural	mcEDS	AR	Congenital multiple contractures (typically adduction/flexion contractures and talipes equinovarus) Characteristic craniofacial features: large fontanelle, short downslanting palpebral fissures, blue sclerae, hypertelorism, short nose with hypoplastic columella, low-set and rotated ears, long philtrum with thin upper lip vermillion, small mouth and hypoplastic chin Characteristic cutaneous features: skin hyperextensibility, easy bruising, skin fragility with atrophic scars increased palmar wrinkling	Recurrent/chronic dislocations Pectus deformities Spinal deformities Peculiar fingers Progressive talipes deformities Large subcutaneous hematomas Chronic constipation Colonic diverticulae Pneumo(hemo)thorax Nephrolithiasis/cystolithiasis Hydronephrosis Cryptorchidism in males Strabismus Refractive errors Glaucoma
Spondylodysplastic	spEDS	AR	Short stature (progressive in childhood) Muscle hypotonia (ranging from severe congenital to mild later-onset) Bowing of limbs	Skin hyperextensibility, soft and doughy, thin and translucent skin, Pes planus Delayed motor development Osteopenia Delayed cognitive development

(continued)

Table 3.1 (continued)

EDS type	Abbr.	IP	Major clinical criteria	Minor clinical criteria
				B4GALT7 Radioulnar synostosis Bilateral elbow contractures Single transverse palmar crease Characteristic craniofacial features Characteristic X-ray findings of skeletal dysplasia Clouded cornea *B3GALT6* Kyphoscoliosis (congenital or early-onset) Joint hypermobility (generalized or restricted to distal joints) Joint contractures (congenital or progressive) Peculiar fingers Characteristic craniofacial features Tooth discoloration, dysplastic teeth Characteristic X-ray findings of skeletal dysplasia Osteoporosis with spontaneous fractures Aortic aneurysm Lung hypoplasia, restrictive lung disease *SLC39A13* Protuberant eyes with bluish sclerae Hands with finely wrinkled palms Atrophy of thenar muscles and tapering fingers Hypermobility of distal joints Characteristic X-ray findings of skeletal dysplasia

Myopathic	mEDS	AD AR	Congenital muscle hypotonia and/or muscle atrophy Proximal joint contractures Hypermobility of distal joints	Soft, doughy skin Atrophic scarring Motor developmental delay Myopathy on muscle biopsy
Periodontal	pEDS	AD	Severe and intractable early-onset periodontitis Lack of attached gingiva Pretibial plaques Family history of first-degree relative who meets clinical criteria	Easy bruising Joint hypermobility, mostly distal Skin hyperextensibility and fragility, wide or atrophic scarring Increased infection rate Hernias Marfanoid facial features Acrogeria Prominent vasculature
Classical-like 2	clEDS2	AR	Skin hyperextensibility with atrophic scarring Generalized joint hypermobility Foot deformities Early-onset osteopenia	

A major criterion is presumed to have high diagnostic specificity because it is present in most individuals and presumably absent from the general population. It is characteristic of the disorder and may allow differentiation from other EDS types and/or other partially overlapping hereditary connective tissue disorders. A minor criterion is a sign of lesser diagnostic specificity, but its presence supports the diagnosis and often the combination of some minor criteria point even more strongly to the specific EDS diagnosis.
IP Inheritance pattern, *AD* autosomal dominant, *AR* autosomal recessive
Adapted from Malfait et al. (2017)

EDS-associated genes to 20 (Blackburn et al. 2018). The extended 2017 classification (Table 3.1) serves as a guide for the clinical diagnosis, molecular confirmation, genetic counseling and management of affected individuals and their family members. Both the Villefranche Nosology (Beighton et al. 1998) and the extended 2017 EDS classification (Malfait et al. 2017) (Table 3.1) defined major and minor clinical criteria. In the absence of genetic testing, these criteria were often the crucial parameters used to establish a specific diagnosis. With the increasing availability and the decreasing cost of next-generation sequencing-based genetic testing facilities, the presence of these clinical criteria should now open the door to diagnostic genetic testing. Since the genetic basis of hypermobile EDS (hEDS) remains unknown, this type is diagnosed solely based on clinical findings (Malfait et al. 2017).

Nearly two decades ago, the prevalence of EDS was estimated to be about 1 in 5000 individuals, apparently with no ethnic predisposition, but the basis of this estimate is not clear (Steinmann and Superti-Furga 2002). The prevalence of classical EDS (cEDS) and of vascular EDS (vEDS) has each been estimated at about 1:20,000 (Bowen et al. 2017) and 1:50,000–1:200,000 (Byers et al. 2017) globally. For the other EDS types with a known molecular basis, no prevalence estimates are available, but between ~5 and ~100 individuals per EDS type have been reported worldwide (Brady et al. 2017b). As cEDS and vEDS are considered the most common, molecularly solved EDS types, it is assumed that the prevalence of all other EDS subtypes is comparable to or below the cutoff of rare disease (defined in Europe as a condition affecting <5 individuals in 10,000, and in the United States as a condition affecting <200,000 people—www.rarediseaseday.org/article/what-is-a-rare-disease).

3.4 Defects in the Primary Structure, Processing, Folding, and Cross-Linking of Fibrillar Collagens

An overview of the different genes and proteins involved in the pathogenesis of the different EDS types is given in Table 3.2.

3.4.1 Collagen I Defects

Early biochemical studies had established that cultured fibroblasts of patients with the hallmarks of EDS arthrochalasia type (aEDS) (arthrochalasis multiplex congenita, EDS type VII) accumulated pN-collagen in the medium as a result of the heterozygous deletion of 24 or 18 amino acids, respectively, in either the proα1(I) (EDS VIIA) or the proα2(I) (EDS VIIB) chain (Steinmann et al. 1980). These amino acid sequences contain the procollagen N-proteinase cleavage site, the

Table 3.2 The Ehlers–Danlos syndromes and their molecular spectrum

Former nomenclature and other names	Villefranche nomenclature	New nomenclature	OMIM condition	Locus	Gene	OMIM gene	Protein	Mutation type
Defects in collagen primary structure and collagen processing								
Gravis/EDS I Mitis/EDS II	Classical type	Classical EDS (cEDS)	130000 130010	9q34.3 2q32.2	*COL5A1* *COL5A2*	120215 120190	Proα1(V) Proα2(V)	NS, FS, MS, SS, Gdel, Gdup FS, MS, SS, Indel, Gdel
		Classical EDS (cEDS)	/	17q21.33	*COL1A1*	120150	Proα1(I) p.(Arg312Cys)	MS
Arterial-Ecchymotic EDS EDS IV	Vascular type	Vascular EDS (vEDS)	130050	2q32.2	*COL3A1*	120180	Proα1(III)	NS, FS, MS, SS, Indel, Gdel
Arthrochalasis multiplex Congenita EDS VIIA EDS VIIB	Arthrochalasia type	Arthrochalasia EDS (aEDS)	130060 130060	17q21.33 7q21.3	*COL1A1* *COL1A2*	120150 120160	Proα1(I) Proα2(I)	SS†, Gdel† SS†, Gdel†
Human dermatosparaxis EDS VIIC	Dermatosparaxis type	Dermatosparaxis EDS (dEDS)	225410	5q35.3	*ADAMTS2*	604539	ADAMTS2	NS, FS, Gdel, initiating methionine
Cardiac-valvular EDS		Cardiac-valvular EDS (cvEDS)	225320	7q21.3	*COL1A2*	120160	proα2(I)	NS, FS, SS
Defects in collagen folding and collagen cross-linking								
Ocular-scoliotic EDS EDS VI EDS VIA	Kyphoscoliosis type	Kyphoscoliotic EDS (kEDS-*PLOD1*)	225400	1p36.22	*PLOD1*	153454	Lysyl hydroxylase 1	NS, FS, MS, SS, Indel,

(continued)

Table 3.2 (continued)

Former nomenclature and other names	Villefranche nomenclature	New nomenclature	OMIM condition	Locus	Gene	OMIM gene	Protein	Mutation type
								Gdel, Gdup
		Kyphoscoliotic EDS (kEDS-*FKBP14*)	614557	7p14.3	*FKBP14*	614505	FKBP22	NS, FS, SS, MS, Indel
Defects in structure and function of myomatrix, the interface between muscle and ECM								
		Classical-like EDS (clEDS)	606408	6p21.33-p21.32	*TNXB*	600985	Tenascin X	NS, FS, MS, Gdel
		Myopathic EDS (mEDS)	616471	6q13-q14	*COL12A1*	120320	Proα1(XII)	MS, SS, Gdel
Defects in glycosaminoglycan biosynthesis								
EDS Progeroid	EDS Progeroid type EDS Progeroid type 1	Spondylodysplastic EDS (spEDS-*B4GALT7*)	130070	5q35.3	*B4GALT7*	604327	Galactosyltransferase I β4GalT7	MS, FS, NS
EDS Progeroid type 2 β3GalT6-deficient EDS		Spondylodysplastic EDS (spEDS-*B3GALT6*)	615349	1p36.33	*B3GALT6*	615291	Galactosyltransferase II β3GalT6	NS, FS, MS, Indel, initiating methionine
Adducted thumb–clubfoot syndrome Dundar syndrome EDS Kosho type EDS Musculocontractural type D4ST1-deficient EDS	EDS VIB	Musculocontractural EDS (mcEDS-*CHST14*)	601776	15q15.1	*CHST14*	608429	Dermatan 4-*O*-sulfotransferase-1	NS, FS, MS
		Musculocontractural EDS (mcEDS-*DSE*)	615539	6q22.1	*DSE*	605942	Dermatan sulfate epimerase-1	NS, FS, MS
Defects in classical complement pathway								

EDSVIII	EDS periodontitis	Periodontal EDS (pEDS)	130080	12p13.31	*C1R* *C1S*	613785 120580	C1r C1s	MS, indel
Defects in intracellular processes								
Spondylocheirodysplastic EDS		Spondylodysplastic EDS (spEDS-*SLC39A13)*	612350	11p11.2	*SLC39A13*	608735	ZIP13	MS, indel
Brittle cornea syndrome	(EDSVIB)	Brittle cornea syndrome (BCS)	229200 614170	16q24 4q27	*ZNF469* *PRDM5*	612078 614161	ZNF469 PRDM5	NS, FS, MS, Gdel NS, FS, SS, MS, Gdel
Unresolved forms of EDS								
Hypermobile EDS III	Hypermobility type	Hypermobile EDS (hEDS)	130020	?	?		?	
Novel type of EDS (identified after 2017 classification)								
		Classical-like EDS type 2 (clEDS2) (provisional)	618000	7p13	*AEBP1*	602981	AEBP1 (ACLP)	NS, FS, SS, Indel

Adapted from Malfait et al. 2017

NS Nonsense, *FS* frameshift, *SS* splice site, *MS* missense, *Indel* small in-frame insertion/deletion, *Gdel* Genomic deletion (exon, multi-exon, gene level), *Gdup* Genomic duplication (exon, multi-exon, gene level) † causing entire or partial loss of exon 6

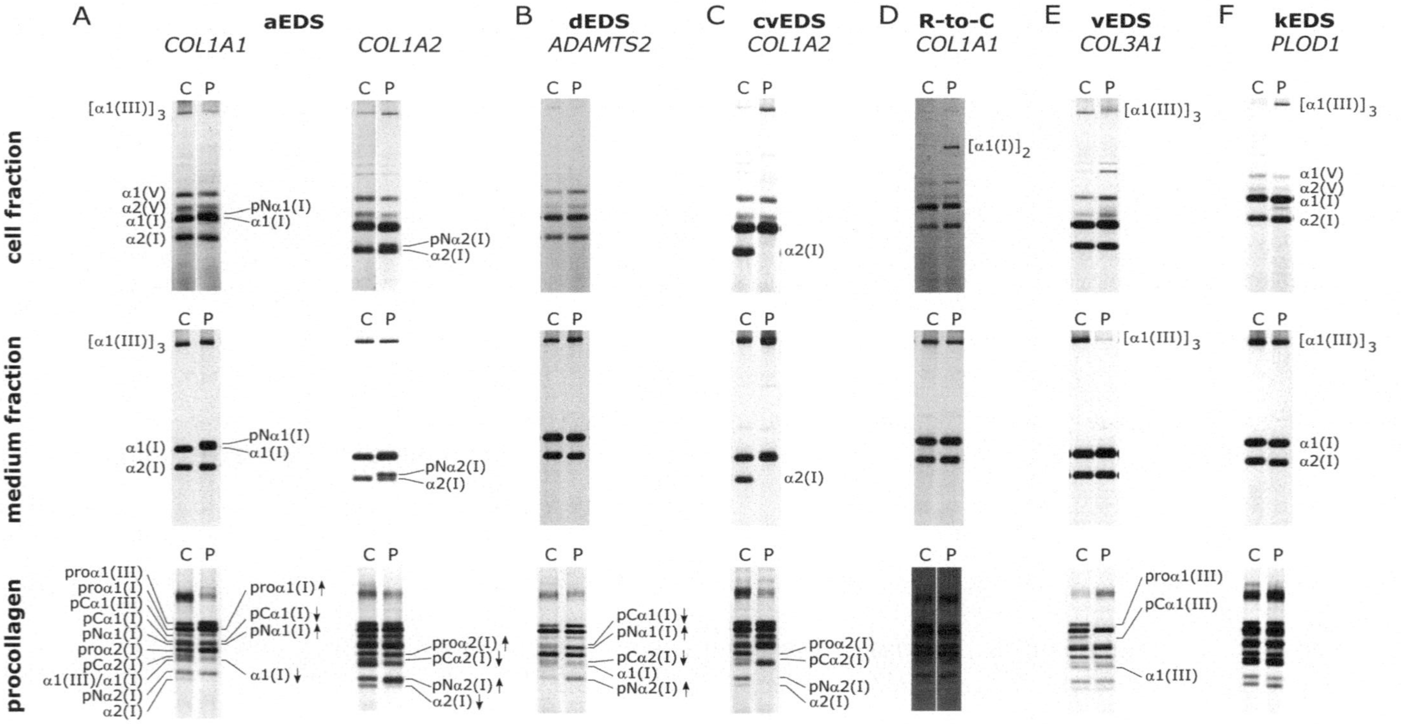

Fig. 3.2 Examples of (pro)collagen electrophoretic mobility patterns for different types of EDS. Metabolically labeled (pro)collagen chains are isolated from the cell and medium fraction of conditioned dermal fibroblast cultures and either partially digested with pepsin (cell and medium fraction) or not (procollagen) prior to SDS-PAGE. All individual intermediate and mature (pro)collagen chains are indicated on the left side of panel A as a reference. At the right side of the gels, (pro)collagen chains displaying a difference are highlighted. (**a**) Arthrochalasia EDS (aEDS) caused by a defect in either *COL1A1* (left panel) or *COL1A2* (right panel) results in the presence of additional mutant pNα1(I) and pNα2(I) chains, respectively, in the cell and medium fractions. (**b**) Dermatosparaxis EDS

(dEDS) due to biallelic *ADAMTS2* defects shows normal electrophoretic mobility for the α chains in the cell and medium fractions because propeptides are enzymatically removed during the sample preparation. The procollagen fraction shows a typical accumulation of the procollagen chains with a retained N-propeptide (pNα1(I) and pNα2(I)) and nearly complete absence of bands representing the pCα1(I) and pCα2(I) procollagen chains. (**c**) Cardiac-valvular EDS (cvEDS) is hallmarked by the complete absence of the α2(I) procollagen chain in cell and medium fractions and consequentially, procollagen analysis shows a lack of all intermediate and mature chains related to α2(I) (pro)collagen. (**d**) Arginine-to-cysteine (R-to-C) substitutions in the *COL1A1* gene (c.934C>T, p.(Arg312Cys)) result in abnormal disulfide-bonded α1(I) dimers in the cell layer, but the dimeric α1(I) chains are not visible in the medium fraction for this patient. (**e**) Vascular EDS (vEDS) caused by a heterozygous *COL3A1* mutation results in a severely reduced amount of the type III (pro)collagen homotrimer in both the cell and medium fraction as well as a diminished amount of the bands corresponding to the α1(III) procollagen chains. (**f**) Kyphoscoliotic EDS (kEDS) due to *PLOD1* mutations show a uniformly faster migration of the type I, III, and V (pro)collagen chains in both cell and medium fractions and on procollagen gels, demonstrating indirectly that lysyl residues are underhydroxylated and underglycosylated. *C* control, *P* patient

telopeptide lysine residue for cross-linking, a cleavage site for proteinases (including pepsin and α-chymotrypsin), and the first Gly-Xaa-Yaa triplet of the major triple-helical domain, and correspond exactly to the sequence encoding exon 6 of the *COL1A1* and the *COL1A2* gene, respectively (Chu et al. 1984; de Wet et al. 1987). After the identification of the first heterozygous mutation affecting the splice donor site of intron 6 of the *COL1A2* gene, molecular defects of other EDS VII patients were identified and turned out to be remarkably homogeneous in nature. Indeed, a number of alterations at the *COL1A1* and *COL1A2* acceptor and donor splice sites adjacent to exon 6 were reported, leading to deletion of part or all of the amino acids encoded by this exon (Byers et al. 1997). The observation that splice site alterations that result in the use of a cryptic acceptor site and loss of only the N-proteinase cleavage site (but preserving the lysine residue cross-linking site) produce the full EDS type VII phenotype suggested that the clinical picture is caused by the retention of the N-propeptide, rather than loss of the cross-linking lysine residue (Chiodo et al. 1992). Sodium dodecyl sulfate polyacrylamide gel electrophoresis (SDS–PAGE) analysis of pepsin-digested collagens produced by dermal fibroblasts from EDS VIIA and VIIB patients revealed the presence of pNα1(I)- and pNα2(I) chains, respectively, in addition to the normal α1(I) and α2(I) chains, and this was for a long time used as a diagnostic marker (Fig. 3.2a).

Since type I collagen consists of two α1(I) chains and one α2(I) chain, a *COL1A1* mutation will result in three quarters of type I collagen molecules containing one or two mutant pNα1(I) chains, whereas in the case of a *COL1A2* mutation, only half of the type I collagen molecules are affected. Because of this stoichiometric difference, *COL1A1*-associated aEDS (EDS VIIA) is assumed to be more severe compared to *COL1A2*-associated aEDS (EDS VIIB). At the ultrastructural level, incorporation of the pN-collagen molecules results in collagen fibrils with smaller diameters and irregular contours, which is more pronounced in *COL1A1* alterations (Byers et al. 1997). While for a long time it remained unclear whether such a genotype–phenotype correlation also held true at the clinical level, recent detailed clinical observations seem to confirm this, as patients with *COL1A1* mutations seem to be at the more severe end of the spectrum, especially in terms of musculoskeletal features (Klaassens et al. 2012; Liu et al. 2020).

Besides mutations leading to loss of the proα1(I) or the proα2(I) N-proteinase cleavage site, a number of mutations (both glycine substitutions and in-frame exon-skipping mutations) residing within the 85 most N-terminal triple helical amino acid residues of type I collagen, are associated with a distinct phenotype showing features of both EDS and osteogenesis imperfecta (OI) ("EDS/OI overlap" phenotype) (Cabral et al. 2005). This stretch of 85 amino acids functions as a stabilizing "anchor" for the N-terminal end of the triple helical domain of type I collagen. Alterations in this α1(I) N-anchor region were demonstrated to introduce a conformational change at the adjacent N-propeptide cleavage site, thereby leading to inefficient N-propeptide cleavage (Makareeva et al. 2006). Defects in the α2 (I) N-anchor region were also shown to delay type I procollagen N-propeptide processing (Malfait et al. 2013b). Thus, although the proper N-proteinase cleavage site remains intact, inefficiently processed collagen molecules will be included in the

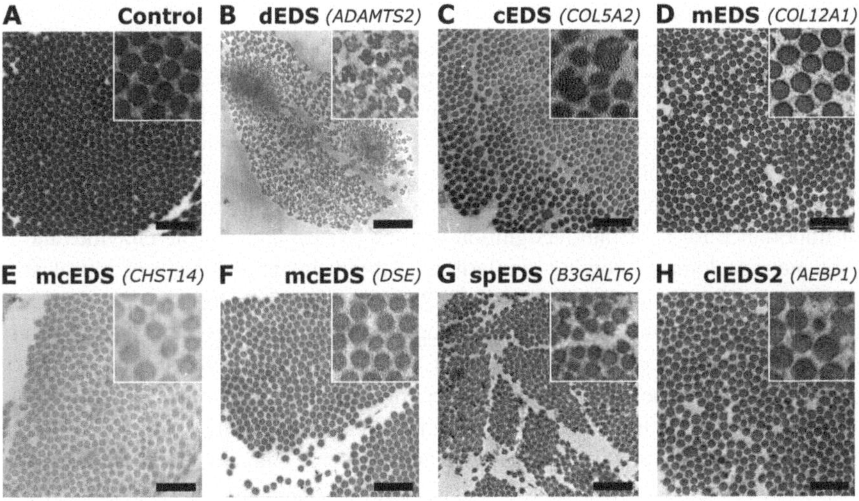

Fig. 3.3 Ultrastructural abnormalities of dermal collagen fibrils in different forms of EDS. (**a**) Control showing tightly packed collagen fibrils with uniform diameters. (**b**) Dermatosparaxis EDS (dEDS) due to biallelic mutations in *ADAMTS2*. Note the hieroglyphic aspect of the collagen fibrils. (**c**) Classical EDS (cEDS) due to a heterozygous mutation in *COL5A2*. Note the presence of some very large fibril diameters with irregular contours, called "collagen cauliflowers". (**d**) Myopathic EDS (mEDS) due to a heterozygous mutation in *COL12A1*. Note the increased interfibrillar spacing. (**e**) Musculocontractural EDS (mcEDS) due to biallelic mutations in *CHST14*. Note collagen fibrils with variable diameters, intermittent presence of small flower-like fibrils, and irregular interfibrillar spaces filled with granulofilamentous deposits. (**f**) Musculocontractural EDS (mcEDS) due to biallelic mutations in *DSE*. Note the grossly normal collagen fibril architecture but a mild increase in interfibrillar spacing. (**g**) Spondylodysplastic EDS (spEDS) due to biallelic mutations in *B3GALT6*. Note dispersed collagen fibrils with variable collagen fibril diameters, sporadic fibrils with very irregular contours, and granulofilamentous deposits between the collagen fibrils. (**h**) Classical-like EDS type 2 (clEDS2) due to biallelic mutations in *AEBP1*. Note the collagen fibrils with large and small collagen fibril diameters and irregular contours. Scale bars represent 500 nm

fibrils, resulting in irregularly contoured collagen fibrils with more variable diameters. Like for aEDS, the phenotype seems to be more severe when the defect is located in the α1(I) chain (Malfait et al. 2013b; Morlino et al. 2020).

Although dermatosparaxis represents the first collagen disorder that was characterized at the ultrastructural and biochemical level in the animal world, its human counterpart was not identified until 1992, with the description of three independent children with clinical and ultrastructural features reminiscent of those seen in animal dermatosparaxis (Fig. 3.3b) (Nusgens et al. 1992; Smith et al. 1992; Wertelecki et al. 1992). Skin extracts from one child were shown to contain collagen precursors with retained N-terminal extensions, and cultured fibroblasts from the three infants failed to cleave the N-terminal propeptides from the proα1(I) and proα2(I) chains. Incubation of collagen extracted from the patients' skin or extracts from normal cells, with purified procollagen N-proteinase or pepsin resulted in fully processed collagen chains, demonstrating that the defect resided in the enzyme, rather than in the substrate (Nusgens et al. 1992; Smith et al. 1992). Further studies by Colige and coworkers on bovine and human procollagen I N-proteinase provided evidence that this enzyme belongs to the metzincin superfamily of metalloproteinases (Colige et al. 1997). Currently known as ADAMTS2 (A disintegrin and metalloproteinase with thrombospondin motifs 2), it contains a Zn^{2+} binding catalytic site accounting for the procollagen processing activity, four thrombospondin type 1 repeats (TSP1) of importance for substrate recognition and the 3D structure of the enzyme, and a cysteine-rich domain showing similarities to the disintegrin domain of reprolysin. The first biallelic mutations in *ADAMTS2* were reported in 1999 in six unrelated human patients, as well as in the initially reported Belgian cattle strain suffering from dermatosparaxis (Colige et al. 1999). The total number of human EDS dermatosparaxis type (dEDS) patients reported to date remains sparse, with only 15 reported patients from 14 independent families (Van Damme et al. 2016). Eight human *ADAMTS2* mutations (splice site and nonsense mutations; exon and multi-exon deletions) have been reported, including an Ashkenazi pathogenic founder variant (c.673C>T) that originated in western Poland (Shi et al. 2017).

The surprising clinical differences observed between aEDS (mainly congenital bilateral hip dislocation and severe joint hypermobility) and dEDS (mainly excessive skin fragility) are still incompletely understood. There are several possible explanations that may contribute to these phenotypic outcomes. First, in contrast to aEDS (caused by heterozygous alterations affecting either the proα1(I) or the proα2(I) chain and where some normal molecules are still produced), variants in *ADAMTS2* result in a nonfunctional enzyme, affecting the cleavage and maturation of both the proα1(I) and proα2(I) chains (Fig. 3.2b), unless there is some residual enzyme activity and/or partial compensation by other enzymes (Colige et al. 1999, 2004). Second, a few other enzymes, mainly ADAMTS3 and ADAMTS14 (Fernandes et al. 2001; Colige et al. 2002; Le Goff et al. 2006), have been shown to display procollagen N-proteinase activity. Studies in dermatosparactic animals revealed that the proportion of pNα(I) chains to fully processed α(I) chains was much higher in the skin than in other tissues, such as tendon, which was further underscored by transmission electron microscopy analyses demonstrating much more severe collagen fibril abnormalities in the skin than in tendon (Cassidy et al. 1980). While ADAMTS2 is the major procollagen I N-proteinase in the dermis, in other tissues such as tendon, a larger proportion of type I procollagen might be cleaved by these other procollagen N-proteinases. Third, ADAMTS2 also shows N-endopeptidase activity for types II, III, and V procollagen, and perhaps other substrates (Bekhouche and Colige 2015; Bekhouche et al. 2016). Therefore, it is possible that insufficient cleavage of these substrates also contributes to the severity of this condition, for example, the extreme fragility and laxity of the skin (type V collagen) and the bruising and bleeding propensity (type III collagen) (Van Damme et al. 2016).

Defects in the primary structure type I collagen have also been identified in two other EDS forms. In 1987 and 1988, respectively, two unrelated Japanese patients with a classical EDS-like phenotype and cardiac-valvular defects were reported to have (near-) complete absence of the proα2(I) collagen chain (Fig. 3.2c) (Sasaki et al. 1987; Hata et al. 1988). At the time, they were not characterized at the molecular level. In 2001, Nicholls et al. reported a girl with an EDS/OI overlap phenotype who displayed a complete absence of the proα2(I) collagen chains. She was shown to harbor a homozygous mutation affecting the splice donor site in intron 46 of *COL1A2* introducing a premature termination codon (Nicholls et al. 2001). A few years later Schwarze et al. (2004) reported clinical, biochemical, and molecular data of two additional patients lacking proα2(I) collagen chains. Both were compound heterozygous for splice site mutations leading to the use of cryptic splice donor sites, and creating a downstream premature termination codon, and very unstable mRNA. Their studies on *COL1A2* mRNA splicing in cells from these patients versus controls provided some of the first evidence that the order of intron removal plays an important role in predicting the outcome of splice site mutations and that folding of the nascent mRNA can be a contributing factor. This autosomal recessive condition, currently known as "cardiac-valvular EDS" (cvEDS) is complicated by the development of polyvalvular disease, often requiring cardiac valve replacement in adulthood (Schwarze et al. 2004; Malfait et al. 2006). In the absence of α2(I) collagen molecules, $[\alpha 1(I)]_3$ homotrimers are formed, but the exact pathogenic mechanisms by which these homotrimers compromise ECM structure and homeostasis and give rise to these phenotypes remains elusive. Of note, total absence of the proα2(I) collagen chain, due to a homozygous frameshift mutation in *COL1A2*, was also documented in a boy with moderately severe OI (Nicholls et al. 1984). In this patient mRNA was stable, and mutant proα2(I) chains were synthesized but failed to associate with proα1(I) chains and were degraded (Deak et al. 1983; Pihlajaniemi et al. 1984). These observations suggest that, in the latter case, an unfolded protein response triggered by unstable mutant protein might contribute to the OI phenotype, whereas in the case of cvEDS, *COL1A2* mRNA instability and total absence of mutant proα2(I) chains have milder consequences on bone, but more severe effects on the soft connective tissues (Byers and Murray 2014).

The second group comprises *COL1A1* missense mutations leading to the substitution of an arginine residue by a cysteine residue in the proα1(I) triple helical domain. Nuytinck et al. reported two children with the clinical hallmarks of classical EDS, caused by a *COL1A1* c.934C>T, p.(Arg312Cys) mutation (Nuytinck et al. 2000). Our team identified the same mutation a few years later in an individual who had suffered a medium-sized arterial rupture in young adulthood (Malfait et al. 2007). The mutation has been reported a few times since in patients with a classical EDS phenotype, with or without vascular ruptures (Gaines et al. 2015; Colombi et al. 2017; Adham et al. 2019; Duong et al. 2019). Our team also identified two other arginine-to-cysteine substitutions (positions 574 and 1093) in the proα1(I) triple helical domain in two individuals with a vascular EDS-like propensity for rupture of median-sized arteries but without any other overt signs of EDS (or OI) (Malfait et al. 2007). These variants have not been reported in additional patients to date, and their causal effect remains unclear. Besides, arginine-to-cysteine substitutions at other positions in the proα1(I) triple helix domain have been associated with an EDS/OI overlap phenotype without arterial rupture (positions 1036 and 1066) (Cabral et al. 2007; Lund et al. 2008), and with Caffey disease (positions 918 and 1014) (Gensure et al. 2005) (Dhooge et al., submitted). The precise effects of these arginine-to-cysteine substitutions on the structure and secretion of type I collagen remain poorly understood. As a result of the presence of a cysteine residue in the triple helical domain, α1(I) dimers are produced (Fig. 3.2d). However, it is also possible that disulfide-bonding occurs with other proteins, either intracellularly, during transport through and from the rough endoplasmic reticulum (rER), or in the extracellular matrix (ECM). Together this may disturb secretion of collagen molecules, and/or disrupt normal physiological interactions. Furthermore, loss of an arginine residue may lead to local destabilization of the type I collagen molecules (Cabral et al. 2007; Malfait et al. 2007).

3.4.2 Collagen III Defects

Defects in type III collagen lead to vEDS. Patients with vEDS are at risk of life-threatening ruptures of tissues containing a high type III:I collagen ratio, such as medium- and large-sized arteries, the gastrointestinal tract, and the uterus. In contrast to most other EDS types, their skin is not hyperextensible, but thin and often translucent, with a prominent venous pattern (Byers et al. 2017). At the ultrastructural level, the dermis is indeed very thin (sometimes only one-fourth of the normal thickness), and collagen bundles in dermis and vessel walls are small, with fibril diameters either uniformly small or with a marked variation (Holbrook and Byers 1981). For many years, the diagnosis of vEDS relied on the demonstration of quantitative (reduced amounts) and/or qualitative (altered electrophoretic mobility, intracellular retention) abnormalities of type III procollagen, which were detected by means of SDS-PAGE of radioactively labeled collagens from cell and conditioned medium derived from dermal fibroblast cultures with a very high sensitivity (probably more than 95%) (Fig. 3.2e) (De Paepe and Malfait 2012). Nowadays, sequence analysis of the *COL3A1* gene, either single gene testing or gene panel testing (e.g., gene panel testing for familial thoracic aortic aneurysm syndrome), is the diagnostic gold standard in a patient suspected to have vEDS (Malfait et al. 2017). To date, >700 *COL3A1* mutations have been identified (https://www.le.ac.uk/genetics/collagen/). The majority (about two thirds) result in glycine substitutions in the Gly-Xaa-Yaa triplet repeats of the proα1(III) triple helix. A quarter of known variants comprise splice site variants leading to in-frame exon skips, and a small proportion results in short in-frame deletions or insertions. As a result of these heterozygous defects, equal amounts of normal and mutant α1(III) chains are synthesized. Because procollagen III is a homotrimer, seven-eighths of the homotrimers will contain either one, two, or three mutant chains and thus be abnormal. For some specific variants (mostly glycine substitutions disturbing the canonical triplets of the triple helix and exon-skipping variants at the carboxy (C)-terminal part of the triple helical domain) fibroblasts were demonstrated to be almost completely unable to secrete type III procollagen. The non-secreted type III procollagen is sequestered within the rER (which very often has a dilated aspect), where it is overmodified and very slowly degraded (Byers et al. 1981). The mechanisms by which these molecules are retained in the rER remain unclear. These mutations are associated with collagen fibril diameters that are considerably smaller than normal (65–80 nm vs. 93 ± 7.5, range 95–110 nm) (Hausser and Anton-Lamprecht 1994). In contrast, mutations close to the N-terminal end showed less signs of intracellular retention of abnormal collagen, suggesting rapid intracellular degradation of mutant protein molecules. The latter mutations were shown to result in collagen fibrils with a higher variability in diameter (85–120 nm) (Smith et al. 1997). In contrast to molecular findings in *COL1A1* and, as discussed later, in *COL5A1*, *COL3A1* null mutations that lead to loss of transcript stability (*COL3A1* haploinsufficiency) were not found for a long time. It remained unclear whether this was due to the fact that those were missed because they did not result in a phenotype, or that the phenotype differed from the usual vEDS phenotype with respect to the severity or the nature of the symptoms. Mice lacking one *Col3a1* allele, thereby mimicking *Col3a1* haploinsufficiency, did not show a vascular phenotype (Liu et al. 1997), but the late onset of signs occurring after the follow-up period of 18 months could be missed (Table 3.3). In 2001, Schwarze and coworkers reported the first human *COL3A1* haploinsufficiency mutations (Schwarze et al. 2001), and since then a small number of *COL3A1* mutations (about 5% of all identified *COL3A1* mutations) have been identified that introduce a premature termination codon and mRNA instability (Pepin et al. 2014). On SDS-PAGE analysis these mutations do not always lead to clear electrophoretic abnormalities, and hence might be missed. In terms of genotype–phenotype correlations, *COL3A1* haploinsufficiency is associated with a delayed onset of complications by two decades and a reduced penetrance, and complications seem to be limited to vascular events (Pepin et al. 2014; Byers et al. 2017). Other genotype–phenotype correlations recently emerged from studies in larger cohorts of vEDS patients (Pepin et al. 2014). Individuals harboring splice site variants leading to in-frame exon skips appear to have the lowest median survival, followed by patients with substitutions of a helical glycine residue by a bulky amino acid (arginine, aspartic acid, glutamic acid, valine), while milder phenotypes arise when glycine is substituted by a small residue (alanine, serine, cysteine). These different phenotypic outcomes and survival rates are potentially caused by a complex series of events triggered by the presence of abnormal type III collagen molecules, including effects of altered secretion and intracellular retention on cellular function and signaling as well as changes in the ECM itself. Missense variants affecting the proα1(III) C-propeptide, and substitutions altering the Xaa and Yaa positions of the triple helix can result in mild signs of vEDS and arterial fragility (Frank et al. 2015). The latter is exemplified by the recent identification of glutamic acid to lysine substitutions that lead to a skin phenotype more resembling cEDS (skin hyperextensibility, delayed wound healing), combined with gastrointestinal and vascular fragility (Ghali et al. 2019). A few individuals have been identified with biallelic *COL3A1* variants; they had a severe vEDS phenotype, associated with neuronal post-migrational disorder (polymicrogyria) (Plancke et al. 2009; Jorgensen et al. 2015; Horn et al. 2017; Vandervore et al. 2017).

Table 3.3 Mouse models for the Ehlers–Danlos syndromes

Model	Protein	Human EDS type	Effect on protein	Phenotypic characteristics	Effect on collagen	References
Col3a1^{tm1Jae}	Type III collagen	vEDS	KO	• 5% survival rate at weaning age • Reduced life span due to blood vessel rupture • Surviving adult mice are 15% smaller • Intestinal enlargement and rupture • Development of spontaneous skin wounds in 60% of surviving animals	• Skin and aorta: reduced number of collagen fibrils • Aorta: doubling of mean fibril diameter and reduced or absent collagen fibrils between vSMC and/or elastic fibers in aorta	Liu et al. (1997)
Col3a1$^{tm1Jae/J}$	Type III collagen	vEDS	HI	• Viable • Diminished tensile strength of aorta and colon, and increased compliance of colon wall	• Abdominal aorta: reduced collagen content • Aorta: reduced collagen type III content	Cooper et al. (2010)
Col3a1^{m1Lsmi}	Type III collagen	vEDS	In-frame deletion exon 33–39	• Sudden death due to dissection of the thoracic aorta in 28% (more frequent in males)	• Thoracic aorta: lower medial collagen content • Adventitia of thoracic aorta and skin: reduced number and variable collagen fibril diameters with oversized collagen fibrils	Smith et al. (2011), Dubacher et al. (2020)
Col3a1$^{Tg\text{-}G182S}$ (human p.(Gly183Cys) substitution, corresponds to the	Type III collagen	vEDS	Tg	• Viable • Development of spontaneous skin wounds in males, at 12 weeks necessitating euthanasia • Reduced tensile strength	• Skin: reduced total collagen content and abnormal collagen III:I ratio • Dermis: poor organization, with malformed and loosely packed collagen	D'hondt et al. (2018)

murine Gly182 residue)				of aorta • Reduced thickness of the adventitia of the aorta • Reduced tensile strength of skin	fibrils, reduced number of fibrils, variable fibril diameters with some extremely large fibrils • Aorta: abnormal distribution and morphology of vSMC with reduced contact with elastic lamina • Adventitia of aorta: variable diameter and irregular contours of collagen fibrils	
Col3a1$^{+/G183S}$ *(corresponds with murine G182S)*	Type III collagen	vEDS	KI	• Survival rate of 50% at 24 weeks with premature death due to rupture of the thoracic aorta • Mortality in male (60%) > female (25%) mice	• Aorta: lower collagen fibril density with heterogeneous diameters	Fontaine et al. (2015)
Col3a1$^{G209S/+}$	Type III collagen	vEDS	KI	• Sudden death due to aortic rupture (median lifespan 400 days)	• Aorta: reduced collagen content and more elastin breaks • Aortic wall: disruption of elastic lamellar units with reduced collagen fibrils between vSMC and elastic fibers, disarray of vSMC and thickened elastic fibers with a moth-eaten appearance • Aortic media: heterogeneous collagen fibril	Bowen et al. (2020)

(continued)

Table 3.3 (continued)

Model	Protein	Human EDS type	Effect on protein	Phenotypic characteristics	Effect on collagen	References
					diameter with a generally smaller size	
Col3a1$^{G938D/+}$	Type III collagen	vEDS	KI	• Sudden death due to aortic rupture (median life span 45 days) • Smaller aortas and body size • Decreased aortic wall thickness	• Aorta: reduced collagen content and more elastin breaks • Aortic wall: disruption of elastic lamellar units with reduced collagen fibrils between vSMC and elastic fibers, disarray of vSMC and thickened elastic fibers with a moth-eaten appearance • Aortic media: heterogeneous collagen fibril diameter with a generally smaller size	Bowen et al. (2020)
Col5a1$^{-/-}$	Type V collagen	–	KO	• Perinatal lethal • No blood-filled vessels at E10	• Lack of mesenchymal collagen fibrils at E10 • Small number of very large diameter fibrils with irregular contours in ectodermal basement membrane at E10	Wenstrup et al. (2004a)
Col5a1$^{+/-}$	Type V collagen	cEDS	HI	• Viable • Skin hyperextensibility • Reduced tensile strength of normal and wounded	• Collagen fiber disarray and reduced dermal connective tissue density • Dermis: reduced fibril	Wenstrup et al. (2006)

				skin • Decreased tendon stiffness • Decreased aortic stiffness and breaking strength	density with two subpopulations: relatively normal circular fibrils and very large, structurally aberrant fibrils	
Col5a2$^{pN/pN}$	Type V collagen	cEDS	In-frame deletion of exon 6 resulting in severely reduced proα2(V) levels	• 5% survival rate (most died within 2 days of unknown cause) • Soon after birth: progressive hunching of the back • Varying degree of spinal lordosis and kyphosis with normal morphology of vertebrae • Delayed growth rate of bone • Increased skin stretchability and severe fragility with multiple scars and bleeding lacerations	• Dermis: less tightly packed collagen fibrils with heterogenous diameters • Cornea: disorganized collagen fibrils	Andrikopoulos et al. (1995)
Col5a2$^{+/-}$	Type V collagen	cEDS?	HI	• Viable • Skin hyperextensibility and decreased tensile strength • Increased aortic compliance and decreased tensile strength	• Dermis: irregularities in fibril contours on longitudinal sections and a slight increase in the range of fibril diameters on cross-section	Park et al. (2015)
Adamts2$^{-/-}$	Procollagen *N*-proteinase	dEDS	KO	• Viable • Triangular face with shorter snout • Less dense hair with	• Partial processing of type I and II procollagen N-propeptides • Dermis: hieroglyphic	Li et al. (2001)

(continued)

Table 3.3 (continued)

Model	Protein	Human EDS type	Effect on protein	Phenotypic characteristics	Effect on collagen	References
				thinner hair follicles • Thin, soft, and very fragile skin • Males: sterile • Females: fertile	pattern of collagen fibrils in cross-section with reduced fibril diameter	
Plod1$^{-/-}$	Lysyl hydroxylase 1	kEDS	KO	• 15% dies of aortic rupture (most 1–4 months of age) • General muscle weakness and laxity • Gait abnormalities and difficulty in locomotion	• Dermis and aorta: larger variation in collagen fibril diameter, increased mean fibril diameter, and fibrils with irregular contours • Dermis and aorta: decreased lysyl hydroxylase activity, lower total hydroxylysine content • Tail tendon, cornea, lung, aorta, and femur: decreased HP content, increased LP content and dramatically increased LP/HP ratio and total amount of pyridinoline (HP + LP) cross-links was increased	Takaluoma et al. (2007)
Tnxb$^{-/-}$	Tenascin X	clEDS	KO	• Viable • Progressive skin hyperextensibility and reduced tensile strength • Mild muscle weakness	• Dermis: reduced collagen fibril density, but collagen fibrils normal in size and shape, slightly increased variability in fibril size • Tail and Achilles tendon: reduced collagen fibril density	Mao et al. (2002)

Col12a1$^{-/-}$	Type XII collagen	mEDS	KO	• Viable • Skeletal abnormalities: shorter, more slender long bones with decreased mechanical strength, altered vertebrae structure and kyphoscoliosis • Muscle weakness	• Muscle: collagen fibrils more diffusely localized throughout the endomysium (more in the middle of the endomysium)	Zou et al. (2014)
Chst14$^{-/-}$	Dermatan 4-*O* sulfotransferase-1	mcEDS-*CHST14*	KO	• Decreased survival rate • Smaller size • Altered vascular structure, and ischemic and/or necrotic-like changes in placenta • Abnormal structure of the basement membrane of capillaries in the placental villus • Reduced tensile strength of the skin	• Dermis: increased interfibrillar spaces with decreased collagen fibril density	Yoshizawa et al. (2018)
Dse$^{-/-}$	Dermatan sulfate epimerase-1	mcEDS-*DSE*	KO	• Viable • Smaller and reduced body weight • Kinked tail (not present after 4 weeks) • Reduced fertility of homozygous knockouts • Decreased mechanical strength of skin	• Dermis and hypodermis: shift toward thicker collagen fibrils • Tail tendon: milder shift toward thicker collagen fibrils	Maccarana et al. (2009)
Slc39a13-KO	ZIP13	spEDS-*SLC39A13*	KO	• Viable • Growth retardation • Decreased tensile strength of the dermis	• Dermis: Decreased amount of collagen • Dermis: Decreased number of collagen fibrils with	Fukada et al. (2008), Hirose et al. (2018)

(continued)

Table 3.3 (continued)

Model	Protein	Human EDS type	Effect on protein	Phenotypic characteristics	Effect on collagen	References
				• Abnormal teeth and craniofacial development • Kyphosis • Osteopenia • Abnormal cartilage development	decreased and variable fibril diameters • Cornea: Decreased fibril diameters and increased interfibrillar space	
Aebp1$^{-/-}$	Adipocyte enhancer-binding protein 1	clEDS2-*AEBP1*	KO	• Majority dies around birth and have anterior abdominal wall defect and a bent or looped tail • Nonhealing skin wounds and deficient wound healing in surviving mice	ND	Layne et al. (2001)
Dcn$^{-/-}$	Decorin	–	KO	• Viable • Skin: increased laxity and fragility • Reduced tensile strength and ductility of the skin	• Abdominal skin: dermal thinning and loose connective tissue • Dermis: abnormal organization of collagen fibrils with increased variability in diameter size (coexistence of thin collagen fibrils intermingled with large ones) and irregular outlines • Tail tendon: very irregular, ragged outline of collagen fibril cross-sections	Danielson et al. (1997)

Lum$^{-/-}$ (*Lum*$^{tm1Sc/tm1Sc}$)	Lumican	–	KO	• Viable • 10–20% of mice are smaller at birth • Loose and fragile skin with increased skin compliance and reduced tensile strength • Bilateral corneal clouding	• Dermis and cornea: presence of thicker, irregular shaped fibrils in cross-section with increased interfibrillar spacing	Chakravarti et al. (1998)
Dpt$^{-/-}$	Dermatopontin	–	KO	• Viable • Increased skin elasticity	• Dermis: reduced thickness and collagen content • Dermis: collagen fibrils with great variety in diameter and irregular contours • Cornea: increased fibril spacing within the posterior lamellae	Takeda et al. (2002), Cooper et al. (2006)
Mimecan$^{-/-}$	Mimecan/osteoglycin	–	KO	• Viable • Moderately reduced tensile strength in skin	• Dermis and cornea: thicker collagen fibrils and less orderly packed • Dorsal and tail skin: collagen fibrils with variable size and altered morphology	Tasheva et al. (2002)

HI haploinsuficiency, *KI* knockin, *KO* knockout, *Tg* transgenic, *ND* not determined, *vSMC* vascular smooth muscle cells

Type III collagen is the second most abundant fibrillar collagen and occurs in all soft tissues where it associates with type I collagen. It is mainly found in tissues displaying elastic properties such as the dermis, blood vessel wall, gastrointestinal tract and uterus, as well as in several other internal organs including the lungs, liver and spleen, where it represents about 10–30% of the total collagen content (Kadler et al. 2007). The exact role of type III collagen in the organization and biological properties of the ECM is not extensively studied, and hence not well understood. During different developmental stages, the type III:I collagen ratio was shown to be inversely proportional to collagen fibril diameter. Fibrils containing a high type III:I collagen ratio (like in skin, aorta, gut, and lung, which all exhibit a degree of elasticity) are thinner, whereas fibrils enriched in type I collagen often present with larger diameters on cross-section (Keene et al. 1987; Romanic et al. 1991). Because of this observation, it was hypothesized early on that type III collagen serves as a negative regulator of collagen fibril diameter (Keene et al. 1987; Romanic et al. 1991). Further evidence for a role for type III collagen in type I collagen fibrillogenesis came from different mouse models harboring *Col3a1* defects (Table 3.3). Mice completely lacking type III collagen in a *Col3a1* knockout model (*Col3a1*^{tm1Jae}) show a severe reduction of collagen fibril density both in skin and aorta, with a collagen fibril diameter in the latter that is severely increased compared to wild-type (Liu et al. 1997). Several murine models harboring structural *Col3a1* mutations (either substitutions of a glycine residue in the triple helix of the proα1(III) chain (D'hondt et al. 2018; Bowen et al. 2020) or a multi-exon deletion (Smith et al. 2011; Dubacher et al. 2020)) also show a reduced collagen fibril content in skin and/or aorta and collagen fibrils with a wide variation in diameter. Ultrastructural studies on postmortem tissues from a 28-year-old female with vEDS had previously also shown thin arterial walls with a markedly reduced collagen content, and decreased collagen fibril diameters in the media of all arteries and in the adventitia and intima of some, while they were increased in the vena cava (Crowther et al. 1991).

Transcriptome analysis on cultured dermal fibroblasts from three unrelated vEDS patients (two harboring a helical glycine substitution and one with an in-frame skip of exon 14 in the α1(III) chain) showed altered expression levels for several genes encoding ECM molecules and proteins involved in endoplasmic reticulum (ER) homeostasis (including *FKBP14*), possibly contributing to the retention of misfolded type III collagen in the rER (Chiarelli et al. 2018). Further studies are however necessary to examine the contribution of ER stress to the vEDS phenotype.

The observation that *COL3A1* haploinsufficiency results in a milder condition than a dominant negative mutation holds some promise for the development of (personalized) treatment strategies. One report studied the effect of allele-specific RNA interference (RNAi) in primary fibroblasts from a vEDS patient (harboring a p.(Gly252Val) substitution). This approach resulted in specific silencing (>90%) of the mutant allele, which was accompanied by a reduced unfolded protein response in the ER, and also restored the formation of proper collagen fibrils in vitro (Muller et al. 2012).

3.4.3 Collagen V Defects

The molecular basis of cEDS remained elusive for a long time. Transmission electron microscopy studies on skin biopsies from cEDS patients provided the first evidence of disturbances in the formation of heterotypic fibrils consisting of types I and V collagen (Vogel et al. 1979). These studies showed dramatic alterations in the dermal collagen fibril organization, with a large variation in fibril diameters, and the presence of aggregate formation or so-called "cauliflower" deformities, which have a highly irregular fibril shape and five to six times larger fibril diameters (Fig. 3.3c). However, defects in the genes encoding type I collagen were excluded by linkage studies in several cEDS pedigrees (Wordsworth et al. 1985, 1991; Sokolov et al. 1991). The involvement type V collagen in cEDS was first demonstrated in mice harboring a homozygous deletion in the *Col5a2* gene, so-called pN/pN mice, which showed clinical and ultrastructural characteristics reminiscent of human cEDS (Andrikopoulos et al. 1995) (Table 3.3). Soon thereafter Loughlin et al. found linkage with the *COL5A1* gene in one large and two smaller cEDS pedigrees (Loughlin et al. 1995). Eventually, the finding of a (9,X) chromosomal translocation, with a breakpoint in the *COL5A1* gene, in a female patient with hypomelanosis of Ito and cEDS, definitively linked type V collagen to cEDS (Toriello et al. 1996). Soon thereafter, several structural mutations in the *COL5A1* and the *COL5A2* genes were reported (Nicholls et al. 1996; Wenstrup et al. 1996; De Paepe et al. 1997; Burrows et al. 1998; Richards et al. 1998; Giunta and Steinmann 2000; Giunta et al. 2002; Takahara et al. 2002). Molecular analysis of *COL5A1* and *COL5A2* in two case series of cEDS patients identified mutations in only about half of the patients, the majority of which were shown to lead to *COL5A1* haploinsufficiency, caused by a nonfunctional *COL5A1* allele (Schwarze et al. 2000; Wenstrup et al. 2000; Malfait and De Paepe 2005). Based on more recent molecular studies in larger case series of cEDS patients, defects in either *COL5A1* or *COL5A2* are currently identified in about 90% of cEDS patients (Symoens et al. 2012; Ritelli et al. 2013). Approximately 75% of these defects reside in *COL5A1* and lead to haploinsufficiency. This haploinsufficiency is caused by nonsense variants, small out-of-frame genomic duplications/deletions and splicing errors, resulting in nonsense-mediated mRNA decay (NMD) or due to the deletion of one allele. Because type V procollagen molecules do not form with more than one proα2(V) chain, the reduced availability of proα1(V) chains leads to about half of the normal type V collagen amount. In contrast, proα1(V) chains can form functional homotrimers (Haralson et al. 1980, 1984; Fichard et al. 1997; Chanut-Delalande et al. 2001), so *COL5A2* heterozygous null variants appear to have no phenotype (no *COL5A2* null variants have been reported to date). Indeed, in the $Col5a2^{pN/pN}$ mice, the reduced levels of proα2 (V) chains were shown to result in the predominant formation of $[\alpha1(V)]_3$ homotrimers in skin but dermal collagen fibrils appeared more disorganized, less tightly packed, and heterogeneous in size with areas devoid of banded fibrils (Chanut-Delalande et al. 2004). Many of the other *COL5A1* variants (e.g., signal peptide mutations that compromise proper trafficking through the ER of the mutant proα1(V) collagen chains (Symoens et al. 2009), and defects that prevent the association of proα1(V) chains at their C-propeptide (Wenstrup et al. 1996; De Paepe et al. 1997)) can also lead to reduced secretion of type V procollagen. These observations confirm that "functional" haploinsufficiency of type V collagen plays a key role in cEDS pathogenesis (Symoens et al. 2012). In contrast to the landscape of molecular defects identified in *COL1A1, COL1A2, and COL3A1,* only a small number of missense variants causing glycine substitutions in the triple helix and splice site variants leading to single or multiple in-frame exon skips have been described in the *COL5A1* and *COL5A2* genes. The exact mode of action of these defects is yet to be completely elucidated, but most probably results from a combination of inefficient secretion and altered formation of heterotypic fibrils. To date, no clear genotype–phenotype correlations have been documented (Bowen et al. 2017). The phenotype associated with *COL5A2* missense and exon-skipping alterations is believed to be at the more severe end of the clinical spectrum, with an increased presentation of congenital hip dislocation, club feet, and severe scoliosis, but numbers are still too limited to draw any firm conclusions (Colman et al., unpublished). In about 10% of individuals with a clinical cEDS diagnosis, no type V collagen alterations are found (Symoens et al. 2012). This could be due to the fact that in those rare individuals the disorder is caused by causative variants in other, yet to be identified genes, and/or due to detection limits of the standard diagnostic procedures, thereby missing, for example, deep intronic variants affecting splicing or (intragenic) genomic rearrangements (Symoens et al. 2012).

Type V collagen comprises only 2–5% of the total amount of collagen in most tissues, such as the dermis, tendon, and bone, and thus represents a quantitatively minor fibrillar collagen when compared to types I and III collagen (Birk 2001). At least three molecular isoforms, $[\alpha1(V)]_2\alpha2(V)$, $[\alpha1(V)]_3$, and $\alpha1(V)\alpha2(V)\alpha3(V)$, have been recognized, of which the proα(V) chains are encoded by the *COL5A1, COL5A2*, and *COL5A3* genes (Sage and Bornstein 1979; Madri et al. 1982; Niyibizi et al. 1984; Broek et al. 1985; Birk et al. 1988; Fichard et al. 1995). No mutations have been found in *COL5A3*. The $[\alpha1(V)]_2\alpha2(V)$ heterotrimer is most often referred to as type V collagen and primarily assembles with type I collagen into heterotypic fibrils. Observations of an inverse correlation between type V:I collagen ratios and collagen fibril diameter in in vitro fibril assembly studies (Birk et al. 1990), cell cultures (Marchant et al. 1996; Wenstrup et al. 2004a), and in various tissues (Fichard et al. 1995) led to the hypothesis that type V collagen acts as a negative regulator of collagen fibril diameter. Later it was found that the latter function was mediated by the retained noncollagenous N-terminal propeptide (Birk et al. 1990; Linsenmayer et al. 1993; Niyibizi and Eyre 1993, 1994; Moradi-Ameli et al. 1994). This N-propeptide protrudes through the gap zone between adjoining type I collagen molecules (Birk et al. 1988; Linsenmayer et al. 1993; Marchant et al. 1996) and limits lateral fibril growth by sterical hindrance and charge interactions (Adachi and Hayashi 1986; Fichard et al. 1995). The vital importance of type V collagen as a nucleator of collagen fibril formation became evident from the study of cultured dermal fibroblasts from cEDS patients with *COL5A1* haploinsufficiency (Wenstrup et al. 2004b) and from the subsequent study of homozygous *Col5a1* knockout mice

(Wenstrup et al. 2004a) and heterozygous *Col5a1* haploinsufficient mice (Wenstrup et al. 2006). In cultured dermal fibroblasts from *COL5A1* haploinsufficient cEDS patients, the amount of collagen incorporated into fibrils was reduced by half, which was proportional to the decreased fibril number. The latter finding suggests a role for type V collagen in controlling type I collagen utilization during collagen fibril initiation, at least in some tissues (Wenstrup et al. 2004b). This was further underscored by the finding that *Col5a1*$^{-/-}$ mice virtually lack collagen fibrils in the mesenchyme at day 10 of embryogenesis and died of cardiovascular failure (Wenstrup et al. 2004a). *Col5a1*$^{+/-}$ mice on the other hand showed a cEDS-like phenotype and their dermis contained only about half the number of fibrils as their wild-type littermates, with the presence of a very large, heterogeneous fibril subpopulation with highly irregular fibril contours, similar to the dermal collagen "cauliflowers" described in human cEDS patients (Wenstrup et al. 2006) (Fig. 3.3c).

In vitro studies on fibroblasts from cEDS patients showed a disorganization of several ECM components (e.g., types III and type V collagen, fibronectin, and fibrillins), and of specific integrin receptors, as well as a reduced migration capacity (Zoppi et al. 2004). Transcriptome profiling of four cEDS fibroblasts (harboring either a p.(Gly1393Asp) substitution or a frameshift mutation in *COL5A1* or an in-frame deletion in *COL5A2*) further underscored alterations in ECM modeling and in wound healing. Furthermore, the expression of genes related to ER homeostasis and autophagy was dysregulated (Chiarelli et al. 2019). Additional studies are required however to broaden our understanding of how these processes contribute to the molecular pathogenesis of cEDS.

3.4.4 Defects in Collagen Folding and Cross-Linking

As discussed above, lysyl hydroxylase deficiency was the first human collagen disorder to be unraveled at the biochemical level (Krane et al. 1972). The disorder, initially called ocular-scoliotic type of EDS, later EDS VIA and now known as kyphoscoliotic EDS (kEDS), is clinically characterized by neonatal muscle hypotonia, progressive kyphoscoliosis, abnormal scarring and easy bruising, increased risk of fatal arterial ruptures and in some patients ocular fragility (Brady et al. 2017a). Lysyl hydroxylase is an ascorbate-dependent ER-resident homodimer that catalyzes the hydroxylation of lysine residues mostly, though not exclusively, in Xaa-Lys-Gly triplets of collagens as well as in proteins with collagen-like sequences (Kivirikko and Pihlajaniemi 1998). In 1992, lysyl hydroxylase (LH, now LH1) was cloned from human tissue, and the corresponding gene, *PLOD* (now *PLOD1*), was mapped to chromosome 1 (Hautala et al. 1992). A few years later, two isoenzymes, LH2 (Valtavaara et al. 1997) and LH3 (Passoja et al. 1998; Valtavaara et al. 1998), were identified. The cloning of LH1 and mapping of the *PLOD1* gene allowed to study the molecular defects underlying LH1 deficiency, and the first *PLOD1* mutations were identified in the early 1990s and included a homozygous nonsense mutation (Hyland et al. 1992), and a large homozygous duplication of 7 exons

(exons 10–16) (Hautala et al. 1993). Not much later, the same duplication was identified in an unrelated EDS VIA patient. It was shown to be mediated by a recombination of common Alu-Alu repeat elements which are dispersed throughout the human genome and thereby represents one of the very early illustrations of how repetitive elements can mediate the formation of deletions or duplications (Pousi et al. 1994). Subsequent molecular studies in other LH1-deficient patients showed that this 7-exon duplication accounts for approximately 30% of all *PLOD1* mutations (Brady et al. 2017a).

The formation of collagen cross-links starts with the conversion of specific lysine or hydroxylysine residues in the collagen telopeptides into the aldehydes allysine or hydroxy allysine, respectively, which is catalyzed by lysyl oxidase in the ECM (Kagan and Li 2003). Hydroxylation of the telopeptide lysyl residues is done by LH2 (Uzawa et al. 1999; van der Slot et al. 2003), whereas LH1 and LH3 hydroxylate the lysyl residues in the triple helical domain (Kivirikko and Myllyla 1982; Myllyla et al. 2007). The telopeptides are subsequently linked to the triple helix of an adjacent collagen molecule by the formation of difunctional immature cross-links in the characteristic staggered array of collagen molecules in a fibril (Fig. 3.5). If a hydroxyallysine residue is present in the telopeptide, the difunctional cross-links can mature into trifunctional non-reducible cross-links including lysylpyridinolines (LPs), when one lysine and two hydroxylysines are linked, and hydroxylysylpyridinolines (HPs) derived from three hydroxylysines (Yamauchi and Shiiba 2008). The trifunctional cross-links that are derived from the hydroxyallysine route are known as pyridinolines, which are particularly difficult to degrade (Takaluoma et al. 2007).

The hydroxylysine residues that are formed through the LH1 catalytic activity serve two important roles: they are necessary for the stability of the intermolecular cross-links that in turn provide tensile strength and mechanical stability to collagen fibrils, and they function as attachment sites for carbohydrate units, either the monosaccharide galactose or the disaccharide glucosylgalactose. Thus, deficiency of LH1 leads to underhydroxylation of triple helical lysyl residues as well as underglycosylation of hydroxylysyl residues (Rohrbach et al. 2011), which accounts for the faster electrophoretic mobility of the collagen α-chains during SDS-PAGE (Fig. 3.2f). In addition, patients with LH1 deficiency have a high urinary excretion of LP, resulting in a high LP:HP ratio (range 2–9, compared to a normal ratio of about 0.2, as measured by high-performance liquid chromatography), a finding that is highly sensitive and specific for EDS VIA (Rohrbach et al. 2011). The mechanical instability of the tissues that is observed in EDS VIA patients is attributed to impaired cross-link formation due to underhydroxylation of the lysyl residues. Nonetheless, the pathogenetic mechanisms underlying this disorder, e.g., the role of the underglycosylation due to the lack of hydroxylysine residues, are however not fully understood.

Biochemical data on patients with lysyl hydroxylase deficiency are mostly limited to the skin, where the residual hydroxylysine content was shown to be about 5%, whereas it was about 50 and 90% in the few tendon and cartilage samples that were analyzed, respectively (Steinmann and Superti-Furga 2002). Similar results were obtained for different types of collagen extracted from several tissues from a patient with EDS VIA: the extent of lysyl hydroxylation of type I collagen extracted from the skin was 0%, from bone 17%, from tendon 36%, and from lung 73%. Type III collagen extracted from skin and lung contained 0 and 50% of the normal hydroxylysine content, respectively, and types II, IV, and V collagen from, respectively, cartilage, kidney, and bone showed a nearly normal hydroxylation (Ihme et al. 1984). Complete inactivation of the *Plod1* gene in mice resulted in a phenotype reminiscent of human EDS VIA with muscle hypotonia, and soft and flaccid tissues; and about 15% of the *Plod1*$^{-/-}$ mice died of aortic rupture (Takaluoma et al. 2007) (Table 3.3). The total amounts of hydroxylysine residues were decreased in all murine *Plod1*$^{-/-}$ tissues, but the magnitudes of these showed marked variability between tissues, with the hydroxylysine amount in the *Plod1*$^{-/-}$ lung and femur being as high as 86% and 75% of that in the wild-type, respectively, while corresponding values in skin and tail tendon were only 22% and 24%. These data suggest that the two other LH isoenzymes can effectively hydroxylate triple helical lysine residues, and that the observed variation in hydroxylysine deficiency can probably be attributed to tissue-specific differences in the levels of the three isoenzymes (Takaluoma et al. 2007).

In 2012, Baumann et al. performed linkage analysis in a large Tyrolean pedigree displaying a phenotype that largely overlapped with EDS VIA, with muscle hypotonia, progressive kyphoscoliosis, joint hypermobility, skin hyperextensibility, myopathy, and also sensorineural hearing impairment, but with normal urinary pyridinoline excretion. They identified biallelic mutations in *FKBP14*, coding for the ER-resident mixed-function protein FKBP22 (Baumann et al. 2012; Boudko et al. 2014). Homozygous and compound heterozygous *FKBP14* mutations (including a few splice site variants leading to frameshifts, and one missense variant (p.(Met48Lys)) were subsequently identified in a series of additional patients with similar clinical features, and the phenotype was later expanded to also include arterial fragility (Murray et al. 2014; Dordoni et al. 2016; Giunta et al. 2018; Ishikawa et al. 2020). A c.362dupC (p.(Glu122Argfs*7)) mutation disrupting the reading frame thereby leading to mRNA instability accounts for approximately 70% of disease alleles and is linked to the same haplotype in all individuals that were analyzed to date, suggesting the presence of a founder effect (Murray et al. 2014).

FKBP22 was shown to preferentially bind to types III, VI, and X procollagen (Ishikawa and Bachinger 2014). As a molecular chaperone, FKBP22 exerts a quality control function on the folded type III collagen triple helix, and with its peptidylprolyl isomerase (PPIase) activity accelerates the formation of the type III collagen triple helix (Ishikawa and Bachinger 2014; Ishikawa et al. 2017). Thus, FKBP22 deficiency may lead to premature interaction and accumulation of collagen molecules in the ER, which could explain the ER dilatation observed in patients' fibroblasts (Baumann et al. 2012). A common mechanism linking kEDS due to LH1 deficiency and kEDS due to FKBP22 defects remains however elusive. Intriguingly, alterations in *PLOD2* (encoding LH2) and *FKBP10* (encoding FKBP65, another rER-resident PPIase) result in overlapping OI phenotypes, which could be attributed to the role of FKBP65 in LH2 dimerization, essential for LH2 function (Gjaltema et al. 2016). To date, no similar model has emerged for LH1 and FKBP22.

3.5 More than Collagen? Beyond the Fibrillar Collagens

3.5.1 Defects in Bridging Molecules

In 1997, Burch and coworkers (Burch et al. 1997) reported a new contiguous gene syndrome, encompassing the *TNXB* and *CYP21B* genes, in a 26-year-old patient combining signs of congenital adrenal hypoplasia (CAH) due to 21-hydroxylase deficiency, and a phenotype strongly similar to cEDS, with skin hyperextensibility, easy bruising and joint hypermobility, but without the typical atrophic scarring. Skin fibroblasts from the patient failed to produce tenascin X (TNX) protein in vitro or in vivo. The paternal allele was shown to carry a novel 30 kb deletion caused by a recombination event between the *TNXB* gene and its partially duplicated gene, *TNXA*, precluding TNX synthesis. Autosomal recessive inheritance was assumed (based on the total absence of *TNXB* mRNA and protein in the patient, among other reasons), but no second *TNXB* mutation was identified. The function of TNX was unknown at the time. Initially co-purified with the Fibril-Associated Collagens with Interrupted Triple Helices (FACIT) types XII and XIV collagens (Lethias et al. 1996), it was later shown to be expressed in various connective tissues such as dermis, tendon, and blood vessels (Lethias et al. 1996; Valcourt et al. 2015), and, unexpectedly, peripheral nerves (Matsumoto et al. 2002). It remained however unclear whether EDS due to TNX deficiency was inherited in a dominant or recessive fashion, and whether isolated TNX deficiency, without CAH, would also lead to EDS (Burch et al. 1997). These questions were answered in 2001, when Schalkwijk et al. (Schalkwijk et al. 2001) reported five unrelated EDS patients, also displaying skin hyperextensibility without atrophic scarring, easy bruising and joint hypermobility, who showed absence of serum TNX (measured by ELISA). Four of them harbored homozygous *TNXB* mutations, either the 30 kb deletion identified in the first index patient or truncating mutations; the fifth patient was heterozygous for the 30 kb deletion, but no second mutation could be identified. Since then, several other patients with biallelic *TNXB* mutations have been identified, and the phenotype was expanded to also include muscle weakness, axonal polyneuropathy, gastrointestinal rupture, and peripheral edema (Demirdas 2016; Brady et al. 2017b). The condition is currently termed classical-like EDS (clEDS).

In order to elucidate the function of TNX, Mao et al. inactivated *Tnxb* in mice (Mao et al. 2002). The skin of *Tnxb*$^{-/-}$ mice showed progressive hyperextensibility, similar to EDS patients, and reduced tensile strength was confirmed by biomechanical testing. In contrast to the irregular shapes of collagen fibrils observed in cEDS, ultrastructural analysis of skin from *Tnxb*$^{-/-}$ mice (and likewise, TNX-deficient human patients (Bristow et al. 2005)) revealed a regular collagen fibril morphology, but a markedly reduced collagen fibril density. In addition, cultured dermal fibroblasts obtained from *Tnxb*$^{-/-}$ mice failed to deposit type I collagen into the cell-associated matrix. Later, TNX was shown to have multiple direct and indirect interactions with several ECM components in vitro, including fibrillar types I, III, and V collagen, types XII and XIV collagen (Lethias et al. 1996; Lethias et al. 2006; Veit et al. 2006; Egging et al. 2007), type VI collagen (Minamitani et al. 2004), and the small leucine-rich proteoglycan (SLRP) decorin (Elefteriou et al. 2001).

Interestingly, a few mutations were recently found in *COL12A1*, which codes for type XII collagen, in individuals presenting a phenotype with features of both EDS and myopathy (resembling the type VI collagen associated Bethlem myopathy—see Chap. 6 from Paolo Bonaldo for details), hence the designation "myopathic EDS (mEDS)" (Hicks et al. 2014; Zou et al. 2014; Delbaere et al. 2020). Both heterozygous and homozygous variants (including a few heterozygous missense and in-frame exon skipping variants, and one homozygous splice site variant that creates a frameshift which introduces a premature termination codon) were found, most of them clustering to the thrombospondin N-terminal domain (TSPN) and the collagenous COL2 domain of the proα1(XII) chain (Delbaere et al. 2020). The patient with the biallelic *COL12A1* variants seemed to have a congenital and more severe disease, while the presentation in children with heterozygous variants seems subtler. Nevertheless, the number of patients with *COL12A1* mutations is still very limited and further studies are needed to confirm whether this correlation holds true.

Type XII collagen primarily co-localizes to type I collagen-rich tissues, including dermis, ligaments, and skeletal muscle (Keene et al. 1991; Chiquet et al. 2014), and it also interacts with several ECM molecules, including type I collagen (Koch et al. 1995), decorin (Font et al. 1996), and TNX (Veit et al. 2006). Similar to TNX deficiency, ultrastructural analysis of the dermis of an individual with mEDS showed a reduced collagen fibril density (Fig. 3.3d) (Delbaere et al. 2020). From these observations emerges a model in which TNX, type XII collagen, and their binding partners, such as decorin, are thought to function as flexible bridges between collagen fibrils and other noncollagenous ECM molecules. As such, these molecules have a regulatory role affecting the organization and mechanical properties of collagen fibrils in several tissues (Fig. 3.5) (Veit et al. 2006; Valcourt et al. 2015; Delbaere et al. 2020). Nonetheless, the precise structural and physiologic effects of these changes on the ECM and their correlation to phenotype remain poorly understood.

3.5.2 Defects in Glycosaminoglycan Biosynthesis

Although TNX-deficient EDS is often cited as the first identified EDS type to be caused by a defect in a noncollagenous or collagen-modifying protein, this is, strictly speaking, not correct. In the late 1970s to the early 1980s, Hernandez described a number of children with EDS features and signs of early aging (Hernandez et al. 1979, 1981, 1986). Kresse et al. (1987) described a child with comparable features and demonstrated that only half of the decorin core protein was converted into a mature, glycosaminoglycan (GAG) bearing proteoglycan in the patients' skin fibroblasts. Coined as a "progeroid type of EDS", Quentin et al. showed that the observed reduction in GAG substitution of decorin was the result of a marked deficiency (about 5% residual activity) of galactosyltransferase I, a key enzyme in the biosynthetic pathway of proteoglycans (Quentin et al. 1990). In 1999, Okajima et al. isolated cDNA from galactosyltransferase I, enabling them to discover the first biallelic missense mutations in the *B4GALT7* gene (coding for galactosyltransferase I) in the patient originally described by Kresse and collaborators (Okajima et al. 1999).

Proteoglycans comprise a heterogeneous group of molecules that are widely distributed in the ECM and on the surface of all animal cells and mediate versatile functions including cell–cell communication, cell–matrix interactions, cell growth and differentiation. They interact with multiple ECM components, including many of the collagens. Proteoglycans are composed of a core protein linked to one or more GAG side chain(s), such as heparan sulfate (HS), chondroitin sulfate (CS), and/or dermatan sulfate (DS). Central to the biosynthesis of these GAG chains is the formation of a common linker region by the sequential addition of four monosaccharides (xylose–galactose–galactose–glucuronic acid). Synthesis of this tetrasaccharide linker region is initiated by the addition of a xylose residue to a specific serine residue of the core protein, catalyzed by xylosyltransferase I and II (XylT-I/-II) encoded by the paralogues *XYLT1* and *XYLT2*, respectively (Gotting et al. 2000). Subsequently, two galactose residues are added by galactosyltransferase I (Almeida et al. 1999) (GalT-I or β4GalT7) encoded by *B4GALT7* and galactosyltransferase II (Bai et al. 2001) (GalT-II or β3GalT6) encoded by *B3GALT6,* respectively. Finally, the formation of the linker region is completed by the addition of a glucuronic acid, catalyzed by glucuronosyltransferase I (GlcAT-I) encoded by *B3GAT3* (Kitagawa et al. 1998) (Fig. 3.4). An important proteoglycan that has long been implicated in the physiopathology of EDS is decorin. It belongs to a special group of widely distributed proteoglycans, called SLRPs, that play pivotal roles during tissue assembly. Decorin primarily binds noncovalently to the surface of type I collagen (Scott 1991) and delays the collagen fibrillogenesis rate and degree in vitro (Vogel et al. 1984). This particular interaction is attributed to the core protein (Pogany and Vogel 1992), whereas the GAG chain of decorin maintains interfibrillar spacing by extending laterally from adjacent collagen fibrils (Scott 1988). The involvement of defects in decorin (and/or other SLRPs) in the physiopathology of EDS was further suggested based on studies in transgenic mice. Mice deficient for decorin core protein ($Dcn^{-/-}$) were shown to have an EDS-like skin fragility with markedly reduced tensile strength, and ultrastructural analysis of skin and tendon revealed aberrant collagen fibril morphology, with irregularly shaped collagen fibers (Danielson et al. 1997) (Table 3.3). This phenotypic resemblance prompted molecular analysis of the *DCN* gene in a series of cEDS patients without *COL5A1* or *COL5A2* mutations, but these efforts failed to detect any causal defects (Malfait et al. 2005). Also knockout mice for either lumican, dermatopontin or mimecan, other SLRPs, showed characteristics reminiscent of EDS, including fragility and laxity of

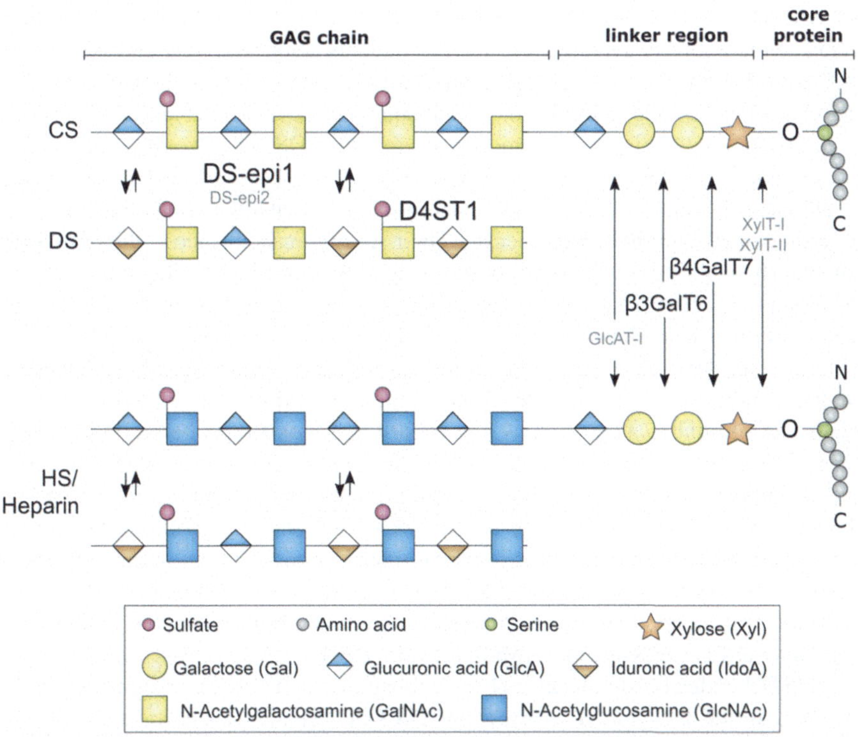

Fig. 3.4 Schematic representation of the biosynthetic pathway of CS/DS and HS/heparin proteoglycans. Proteoglycan core proteins are synthesized in the endoplasmic reticulum (ER) and are subsequently modified in the Golgi apparatus. Glycosaminoglycan (GAG) biosynthesis is initiated by the formation of a tetrasaccharide linker region consisting of four monosaccharides (Xyl-Gal-Gal-GlcA). The synthesis of this linker region is a stepwise process starting with the addition of a xylose to a specific serine residue in the core protein, catalyzed by xylosyltransferases I/II (XylT-I/-II). Subsequently, two galactose residues are added by galactosyltransferase I (β4GalT7) and II (β3GalT6), respectively, followed by the addition of a glucuronic acid (GlcA) by glucuronosyltransferase I (GlcAT-I). Following completion of the linker region, the addition of N-acetylglucosamine (GlcNAc) or N-acetylgalactosamine (GalNAc) initiates the formation of heparan sulfate (HS)/heparin or chondroitin/dermatan sulfate (CS/DS) backbones, respectively. The alternating addition of GlcA and GlcNAc or GalNAc then results in the assembly of the respective GAG chains. The GAG chains are further modified by epimerization and sulfation events. Epimerization of GlcA residues to iduronic acid (IdoA) is catalyzed by dermatan sulfate epimerases (DS-epi1 and DS-epi2). Subsequent 4-*O*-sulfation of GalNAc by dermatan 4-*O*-sulfotransferase-1 (D4ST1) prevents back-epimerization of the adjacent IdoA to GlcA and generates DS. Enzymes that are defective in EDS forms are depicted in bold

the skin, and ultrastructural alterations in the diameters of collagen fibrils (Chakravarti et al. 1998; Takeda et al. 2002; Tasheva et al. 2002) (Table 3.3).

The role of decorin in EDS pathogenesis became clearer when, in 2010, with the help of homozygosity mapping, homozygous mutations were identified in *CHST14* in a series of patients displaying an EDS VI-like phenotype, but with normal LH1 activity (coined EDS VIB by (Steinmann et al. 1975)). Miyake et al. identified *CHST14* mutations in a number of Japanese patients displaying joint and skin laxity and multisystemic fragility-related manifestations, combined with distinct craniofacial characteristics and multiple congenital contractures (Kosho et al. 2010; Miyake et al. 2010), recognizing their clinical overlap with EDS VIB (Kosho et al. 2005). They coined this condition "EDS Kosho type" (Miyake et al. 2014). Our team simultaneously found *CHST14* mutations in a few patients who had clinically also been diagnosed with EDS VIB (because their striking resemblance to the EDS VIB patients (two siblings of Pakistani origin) originally described by Steinmann et al.) (Malfait et al. 2010). This link between *CHST14*-associated EDS and EDS VIB was later confirmed when Janecke et al. identified *CHST14* mutations in those Pakistani siblings (Janecke et al. 2016). *CHST14* mutations had previously also been linked to another condition, called "adducted thumb–clubfoot syndrome" (ATCS) or "Dündar syndrome", initially categorized as a novel arthrogryposis type, based on the presence of congenital adducted thumbs and club feet, characteristic craniofacial features and a number of other anomalies, including signs of connective tissue fragility (Dundar et al. 2009). These three entities were later shown to form a phenotypic continuum, based on the fact that identical mutations were found in patients diagnosed with ATCS as in patients diagnosed with EDS VIB (Malfait et al. 2010). Therefore, they were merged into "musculocontractural EDS (mcEDS)" in the 2017 international EDS classification (Malfait et al. 2017).

CHST14 codes for dermatan 4-*O*-sulfotransferase-1 (D4ST1), an enzyme that is involved in the biosynthetic pathway of DS GAG chains (Pacheco et al. 2009a) (Fig. 3.4). Following completion of the linker region, this pathway starts with the formation of CS by the alternating addition of N-acetylgalactosamine (GalNAc) and glucuronic acid (GlcA) residues, which can be modified by several sulfotransferases (Trowbridge and Gallo 2002). DS synthesis necessitates the epimerization of GlcA toward iduronic acid (IdoA), which is catalyzed by DS epimerases-1 and -2 (DS-epi1 and DS-epi2 encoded by *DSE* and *DSEL*, respectively) (Maccarana et al. 2006; Pacheco et al. 2009b) (Fig. 3.4). Subsequently, D4ST1 is able to catalyze 4-*O*-sulfation of GalNAc, thereby preventing back-epimerization of the adjacent IdoA (Silbert and Sugumaran 2002; Kusche-Gullberg and Kjellen 2003). More recently, mutations in *DSE* were found in a few patients with a mcEDS phenotype (Muller et al. 2013; Syx et al. 2015; Lautrup et al. 2020). Both D4ST1 and DS-epi1 show a ubiquitous expression in human tissues such as skin, blood vessels, heart valves, tendon, and lungs (Thelin et al. 2013). Biochemical studies on skin fibroblasts of patients with *CHST14* or *DSE* mutations leading to defective activity of D4ST1 or DS-epi1, respectively, showed a severely compromised composition of the decorin GAG chain, which normally consists mainly of IdoA-containing DS moieties (Dundar et al. 2009; Malfait et al. 2010; Miyake et al. 2010; Muller et al. 2013; Syx et al. 2015). The lack of 4-*O*-sulfation of GalNAc due to deficiency of D4ST1 allows back-epimerization of the adjacent IdoA to GlcA. These GlcA residues are subsequently 4-*O*-sulfated by chondroitin 4-*O*-sulfotransferases (C4STs), resulting in excessive formation of CS and nearly complete absence of DS. In contrast, patients with DS-epi1 deficiency still possess a small amount of DS consisting of

IdoA-containing disaccharides, possibly due to residual DS-epi1 activity, or to partial compensation by DS-epi2 (Muller et al. 2013; Syx et al. 2015). This could explain the somewhat milder phenotype of patients with *DSE* mutations (Syx et al. 2015; Lautrup et al. 2020). The pathophysiological mechanisms of these enzyme deficiencies remain incompletely understood, but in contrast to normal skin, dermal collagen fibrils of patients with *CHST14* mutations are not regularly assembled and tightly packed (Fig. 3.3e) (Malfait et al. 2010; Hirose et al. 2019). In contrast, collagen fibril organization associated with *DSE* mutations is grossly normal (Fig. 3.3f), which can potentially be due to the small fraction of IdoA residues in dermal decorin (Syx et al. 2015). The high IdoA content of dermal decorin is important for collagen fibril organization, as IdoA residues allow flexibility of the GAG chain by switching between different conformations. The substitution of IdoA by the less flexible GlcA residues is therefore thought to contribute to the observed alterations in collagen fibril organization (Miyake et al. 2010; Thelin et al. 2013). Hirose et al. showed that the decorin GAG chains in the skin from patients harboring *CHST14* mutations are indeed linear, and project from the outer collagen fibril surface to adjacent collagen fibrils, whereas these GAG chains appeared curved in normal skin, and closely attached to collagen fibrils (Hirose et al. 2019). This structural alteration in the GAG chains could cause spatial disorganization of collagen networks, explaining the fragility-related symptoms in these patients. Besides decorin, D4ST1 and DS-epi1 deficiency likely compromise the function of several other DS-containing proteoglycans, including versican, biglycan, epiphycan, and syndecan-1, and disturb many interactions of CS/DS GAGs, thereby affecting many other processes and signaling cascades, including cell migration, wound healing, and coagulation. In addition, the differentiation and early development of many organs may be impaired by the imbalance between sulfated GAGs, which could also contribute to the complex multisystemic phenotype of these conditions.

The important role of correct GAG biosynthesis in the pathophysiology of EDS was further confirmed when biallelic mutations in *B3GALT6*, encoding galactosyltransferase II, which catalyzes the addition of the second galactose in the linker region (Fig. 3.4), were identified in a group of patients presenting with a complex pleiotropic connective tissue disorder that combined features of EDS (*i.e.*, skin hyperextensibility and joint hypermobility) with spondyloepimetaphyseal dysplasia (SEMD), bone fragility, progressive contractures, and muscle hypotonia (Malfait et al. 2013a; Nakajima et al. 2013). The phenotype largely overlaps with that caused by *B4GALT7* mutations, hence both conditions were included in the 2017 international EDS classification as spondylodysplastic EDS (spEDS). Enzyme activities of galactosyltransferase I or II were severely reduced in fibroblast cultures from spEDS-*B4GALT7* or spEDS-*B3GALT6* patients, respectively. As a consequence, HS and CS GAG chains were markedly reduced and/or shorter, and a partial or complete lack of the DS GAG chains on decorin was observed (Seidler et al. 2006; Gotte et al. 2008; Malfait et al. 2013a; Nakajima et al. 2013; Van Damme et al. 2018). Ultrastructural analyses revealed collagen fibrils with variable diameters and occasional fibrils with irregular contours, which are dispersed in the reticular dermis with granulofilamentous deposits between the collagen fibrils (Fig. 3.3g). The severe pleiotropic phenotypes associated with these enzyme deficiencies most likely arise due to an abnormal GAG configuration in decorin and other SLRPs, which compromises the interfibrillar spacing of collagen fibrils, as well as altered interactions with growth factors and other ligands that are normally mediated by HS and CS/DS proteoglycans thereby leading to altered cell signaling during development. Additional studies are still necessary to elucidate how disturbances in these processes contribute to the observed phenotypes.

3.5.3 Other Proteins

Over the past two decades, a number of additional rare variants of EDS were delineated through the combination of careful clinical phenotyping and the use of next-generation sequencing techniques. One of those EDS forms results from biallelic variants in the *SLC39A13* gene, encoding the homodimeric transmembrane Zrt/irt-like protein 13 (ZIP13) protein. ZIP13 belongs to the family of SLC39A/ZIP proteins regulating Zn^{2+} influx into the cytosol (Giunta et al. 2008). To date, only three homozygous pathogenic variants have been reported, including one 9 bp deletion, one missense, and one nonsense variant (Giunta et al. 2008; Dusanic et al. 2018). Mutations in *SLC39A13* were shown to result in generalized underhydroxylation of lysine and proline residues in collagen and aberrant collagen cross-linking in the ECM (Giunta et al. 2008). Several pathophysiological mechanisms were suggested, including Zn^{2+} overload in the ER and competition with Fe^{2+} for binding to lysyl hydroxylase and prolyl 4-hydroxylase (Giunta et al. 2008), trapping of Zn^{2+} in cytosolic vesicular stores ("zincosomes") leading to reduced Zn^{2+}availability in the ER and other cellular components and induction of ER stress (Jeong et al. 2012), and changes in the activation of BMP/TGFβ signaling regulated by the intracellular localization of Smad proteins in connective tissue-forming cells (Fukada et al. 2008). Recent findings in *Drosophila melanogaster* show that the fruit fly ZIP13 homolog functions as an Fe^{2+} exporter on the ER/Golgi membrane. Therefore, its absence might result in Fe^{2+}-depletion in the ER/Golgi compartment, which could result in the observed underhydroxylation of lysine and proline residues in collagen (Xiao et al. 2014). Initially called spondylocheirodysplastic EDS, this condition was merged with spEDS (spEDS-*SLC39A13*) in the 2017 EDS classification, because of its clinical overlap with spEDS-*B4GALT7* and spEDS-*B3GALT6* (Malfait et al. 2017). The mechanisms that link these conditions at the molecular level remain however unknown.

Brittle cornea syndrome (BCS) is a rare autosomal recessive condition, the hallmarks of which are thin and fragile corneas with an increased risk for spontaneous perforation, keratoglobus and keratoconus, and often also hearing loss (Ticho et al. 1980; Cameron 1993; Burkitt Wright et al. 2013). BCS was initially considered to be a form of EDS, because a subset of patients presented signs of generalized connective tissue fragility, including joint hypermobility, muscle hypotonia, kyphoscoliosis, and mild EDS-like skin abnormalities (Cameron 1993; Ogur et al. 1994; Macsai et al. 2000). At the time, these patients were also labeled as EDS VIB.

Part of the genetic basis of BCS was discovered in 2006. Through homozygosity mapping and genotyping, causal variants in *ZNF469* were identified (Abu et al. 2008). This gene encodes ZNF469, a protein containing five C2H2 zinc finger motifs and an arginine-rich region located between two amino-terminal proline-rich domains (Rohrbach et al. 2013). Little is known about ZNF469, but based on a limited homology of approximately 30% to the helical domains of certain collagens strongly expressed in the cornea (α1(I), α2(I), α1(IV) collagen), it was suggested that ZNF469 could act as (transcriptional) regulator in the synthesis or assembly of collagen fibrils (Abu et al. 2008), though this has not been shown in experimental studies. In 2011, a second BCS gene, *PRDM5* was discovered by combining autozygosity mapping with whole-exome sequencing (Burkitt Wright et al. 2011). The gene codes for PRDM5 (PR domain zinc finger protein 5), containing 16 C2H2 zinc finger motifs and one PR-SET domain. PRDM5 is a transcription factor and studies in pre-osteoblastic cells of mouse embryos demonstrated that PRDM5 is a regulator of the expression of multiple collagen and SLRP encoding genes (Galli et al. 2012). More specifically, PRDM5 binds to collagen genes and maintains RNA polymerase II activity thereby regulating collagen transcription and fibrillogenesis (Galli et al. 2012). Transcriptome analyses of fibroblasts from patients with *ZNF469* or *PRDM5* mutations showed a dysregulated expression of several genes involved in the development and maintenance of the ECM, including downregulation of *COL4A1*, *COL11A1,* and *HAPLN1* (the latter encoding the hyaluronan and proteoglycan link protein 1) (Burkitt Wright et al. 2011). Moreover, studies on fibroblasts from patients with defects in *PRDM5* or *ZNF469* showed an abnormal deposition of type I collagen, type III collagen and fibronectin in the ECM and an altered distribution of α2β1 and α5β1 integrins on fibroblast surfaces (Burkitt Wright et al. 2011, 2013). Together, these observations suggest that ZNF469 and PRDM5 may play a regulatory role in the organization of the ECM both in the eye and in other connective tissues.

Periodontal EDS (pEDS) was first described in 1977 and is characterized by aggressive periodontal disease and often premature tooth loss, with mild joint hypermobility and pretibial plaques (Stewart et al. 1977). Linkage analysis in three families revealed a disease locus of 5 Mb at chromosome 12p13 (Rahman et al. 2003), and exome sequencing demonstrated the presence of heterozygous missense variants or in-frame insertion/deletions in *C1R* and *C1S*, contiguous genes located within the linked region (Kapferer-Seebacher et al. 2016). These genes encode the C1r and C1s subunits of the first factor (C1) of the classical complement pathway. A complete C1 molecule is formed by combining two C1r and two C1s subunits, which assemble into a heterotetramer, with six C1q subunits (Cooper 1985; Arlaud et al. 1987, 1998, 2001). During this assembly, the N-terminal collagenous domain of C1q binds to the CUB domains of C1r and C1s (Bally et al. 2009). The evolutionary conserved CUB domains mediate protein–protein interactions and are present in several ECM proteins, such as the C-proteinase for type I procollagen, BMP1, and its enhancer PCPE1 (Bork and Beckmann 1993). It is the CUB domain that allows

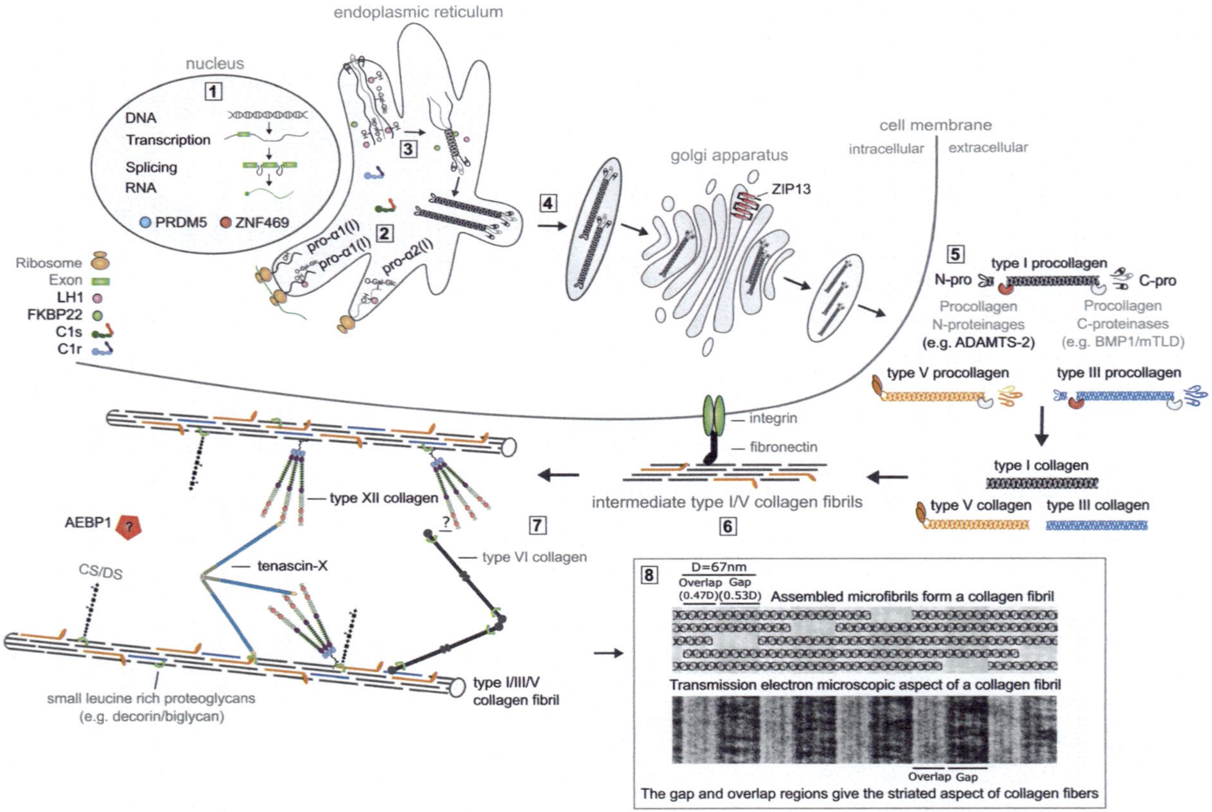

Fig. 3.5 Schematic illustration of collagen fibrillogenesis in the context of the Ehlers–Danlos syndromes. (1) The biosynthetic pathway of fibrillar types I, III, and V collagen, and all other molecules involved in the pathogenesis of EDS starts with transcription in the nucleus of the respective genes, followed by

translation in the endoplasmic reticulum (ER). (2) In the ER, nascent proα chains undergo extensive posttranslational modification by prolyl and lysyl hydroxylases, including lysyl hydroxylase 1 (LH1) (Steinmann and Superti-Furga 2002). The resulting hydroxylysine residues serve important functions in the formation of intermolecular cross-links and as attachment sites for carbohydrate units. (3) Triple helix formation is initiated in the ER by the association of the C-propeptide domains of three proα chains which then fold in a zipper-like fashion toward the N-terminal end. (4) After folding, posttranslational modification ceases, and the procollagen molecules are transported from the ER to the Golgi apparatus, during which they begin to aggregate laterally in secretory vesicles and are eventually secreted into the extracellular environment. (5) During transport and/or in the ECM, the N- and C-propeptides are enzymatically removed (for procollagen type I, this is done by ADAMTS and BMP1/tolloid-like proteinases, respectively). (6) Following the cleavage of the N- and C-propeptides, the resulting mature collagen molecules assemble into striated fibrils. This process requires the assistance of several proteins, which can be categorized as organizers, nucleators, and regulators (Kadler et al. 2008). At the plasma membrane, fibronectin and integrins function as "organizers" of fibril assembly. "Nucleators", such as type V collagen, initiate the assembly of immature fibrils at the cell surface. The triple helical domain of type V collagen is embedded within the heterotypic collagen fibrils, while its partially cleaved N-propeptide protrudes at the fibril surface and controls fibrillogenesis by sterical hindrance (Birk 2001). (7) The intermediate fibrils are then deposited into the ECM where they are stabilized by the interaction with "regulators", including decorin, tenascin X, and type XII collagen, which affect the rate of assembly, size, and structure of the collagen fibrils. (8) Increasing numbers of covalent cross-links are formed between the intermediate collagen fibrils which results in stabilization of the mature collagen fibrils. These fibrils display a quarter-staggered arrangement with a characteristic 67 nm axial periodicity (D-periodicity) (Kadler et al. 1996) in which gap and overlap zones can be distinguished. The gap zone is present between the N- and C-termini of adjacent molecules, whereas complete molecular overlap is observed in the overlap zone. On transmission electron microscopy, this is seen as a characteristic pattern of alternating light and dark bands. The resulting collagen fibrils are indeterminate in length and have diameters ranging from 12 to >500 nm, depending on the developmental stage and tissue (Canty and Kadler 2005)

PCPE1, BMP1, and C1s to bind to the triple helix of collagen and/or the propeptides (Steiglitz et al. 2002; Vadon-Le Goff et al. 2011). These findings suggest that some characteristics of pEDS can be caused by the abnormal interaction of C1r and C1s with components of the ECM (Kapferer-Seebacher et al. 2016). Pathogenetic *C1R* variants lead to a gain-of-function, resulting in constitutive intracellular activation of C1s and C1r serine proteases, which could result in cleavage of C4 and local activation of the complement cascade (Grobner et al. 2019). The unexpected discovery that defects in components of the complement pathway are associated with an EDS type has opened possibilities to understand the interplay between the immune system and connective tissues.

After the publication of the 2017 EDS classification, an autosomal recessive EDS type was described including skin hyperextensibility with atrophic scarring, generalized joint hypermobility, foot deformities, and early-onset osteopenia. Whole-exome sequencing revealed the causal involvement of biallelic variants in the *AEBP1* gene (Blackburn et al. 2018). *AEBP1* encodes the ECM-associated adipocyte enhancer-binding protein 1 (AEBP1; also known as aortic carboxypeptidase-like protein, ACLP), which is expressed in tissues with a high collagen content (Layne et al. 2001; Ith et al. 2005). This protein binds to fibrillar types I, III, and V collagen, and assists in type I collagen polymerization (Blackburn et al. 2018). Ultrastructural studies of the reticular dermis of two patients with biallelic *AEBP1* mutations showed collagen fibrils with variable diameters and irregular contours, underscoring a role for AEBP1 in collagen fibril formation (Syx et al. 2019) (Fig. 3.3h). The involvement of AEBP1 in connective tissue development and homeostasis had previously also been demonstrated in *Aebp1* knockout mice ($Aebp1^{-/-}$). While the majority of these mice died perinatally due to gastroschisis, surviving $Aebp1^{-/-}$ mice developed spontaneous skin lesions and delayed wound healing, which is correlated with reduced dermal fibroblast proliferation (Layne et al. 2001). The presence of these wound healing defects correlates with the phenotype of human patients and suggests the involvement of AEBP1 during wound repair. In addition, AEBP1 was also shown to be implicated in the transition of fibroblasts to myofibroblasts mediated by the activation of TGFβ receptors (Schissel et al. 2009; Tumelty et al. 2014), and in the development and homeostasis of bone through activation of the canonical Wnt-signaling pathway mediated by frizzled 8 and LRP6 (Teratani et al. 2018). The exact mechanisms that give rise to the observed EDS phenotype are currently unknown. Because of the clinical resemblance to cEDS, this EDS variant is currently coined classical-like EDS type 2 (clEDS2-AEBP1) (Ritelli et al. 2019).

3.6 Concluding Remarks

An exciting series of genetic discoveries have provided major advances in our understanding of the molecular basis of EDS, and in hindsight, Jansen's concept of "defects in the collagen wickerwork" and Jackson and Bentley' words—"*that the defect in EDS lies at a high level of organization of the collagen fibres, and that in this high-level binding the collagen fibrils interact with the various mucopolysaccharides and mucopolysaccharide-protein complexes of the 'ground substance'"* (Jackson and Bentley 1968)—turned out to be very accurate. The elucidation of the biochemical and molecular defects of the different types of EDS and the study of their consequences, in human tissues, in in vitro models, and in mouse models, have been intimately intertwined with our understanding and appreciation of the complexity of collagen fibril formation, both intracellularly and in the ECM (illustrated in Fig. 3.5).

Nonetheless, for most EDS types, the exact pathways from the gene alteration to the phenotype remain elusive. To truly understand the pathophysiologic consequences of these molecular defects and appreciate how to influence their broad phenotypic effects, more in-depth studies are needed to investigate these pathways and the nodes for modification. Such studies include, among others, the performance of transcriptome and proteome studies, the assessment of signaling influences as a result of aberrant production or processing of the molecules, and the study of known and new cellular and animal models that mimic the genetic defects and pathophysiological mechanisms of patients with different EDS types. Despite the availability of several mouse models for EDS, the study of these models for understanding the pathophysiology of EDS is still in its infancy, and additional models (murine, but also other animal models, such as zebrafish models), harboring different genetic defects and different types of genetic variants, need to be generated. These models are and will be valuable resources that can uncover novel insights in pathogenic mechanisms, identify reliable biomarkers, and pinpoint relevant cellular processes or signaling pathways that can serve as targets for the development (personalized) therapies.

Acknowledgments This work was supported by the Research Foundation Flanders (FWO), Belgium (11B7921N to MC, 12Q5920N to DS, 1842318N to FM and 3G041519 to FM) and a Methusalem Grant from Ghent University (grant number BOFMET2015000401). The authors declare no conflicts of interest.

References

Abu AFM, Marek D, Pras E, Nir U, Reznik-Wolf H, Pras E (2008) Deleterious mutations in the zinc-finger 469 gene cause brittle cornea syndrome. Am J Hum Genet 82:1217–1222

Adachi E, Hayashi T (1986) In vitro formation of hybrid fibrils of type V collagen and type I collagen. Limited growth of type I collagen into thick fibrils by type V collagen. Connect Tissue Res 14:257–266

Adham S, Dupuis-Girod S, Charpentier E, Mazzella JM, Jeunemaitre X, Legrand A (2019) Classical Ehlers-Danlos syndrome with a propensity to arterial events: a new report on a French family with a COL1A1 p.(Arg312Cys) variant. Clin Genet 97:357.

Almeida R, Levery SB, Mandel U, Kresse H, Schwientek T, Bennett EP, Clausen H (1999) Cloning and expression of a proteoglycan UDP-galactose:beta-xylose beta1,4-galactosyltransferase I. A

seventh member of the human beta4-galactosyltransferase gene family. J Biol Chem 274:26165–26171.
Andrikopoulos K, Liu X, Keene DR, Jaenisch R, Ramirez F (1995) Targeted mutation in the col5a2 gene reveals a regulatory role for type V collagen during matrix assembly. Nat Genet 9:31–36.
Arlaud GJ, Colomb MG, Gagnon J (1987) A functional model of the human C1 complex emergence of a functional model. Immunol Today 8:106–111.
Arlaud GJ, Rossi V, Thielens NM, Gaboriaud C, Bersch B, Hernandez JF (1998) Structural and functional studies on C1r and C1s: new insights into the mechanisms involved in C1 activity and assembly. Immunobiology 199:303–316.
Arlaud GJ, Gaboriaud C, Thielens NM, Rossi V, Bersch B, Hernandez JF, Fontecilla-Camps JC (2001) Structural biology of C1: dissection of a complex molecular machinery. Immunol Rev 180:136–145.
Bai X, Zhou D, Brown JR, Crawford BE, Hennet T, Esko JD (2001) Biosynthesis of the linkage region of glycosaminoglycans: cloning and activity of galactosyltransferase II, the sixth member of the beta 1,3-galactosyltransferase family (beta 3GalT6). J Biol Chem 276:48189–48195.
Bally I, Rossi V, Lunardi T, Thielens NM, Gaboriaud C, Arlaud GJ (2009) Identification of the C1q-binding sites of human C1r and C1s: a refined three-dimensional model of the C1 complex of complement. J Biol Chem 284:19340–19348.
Barabas AP (1967) Heterogeneity of the Ehlers-Danlos syndrome: description of three clinical types and a hypothesis to explain the basic defect(s). Br Med J 2:612–613.
Baumann M et al (2012) Mutations in FKBP14 cause a variant of Ehlers-Danlos syndrome with progressive kyphoscoliosis, myopathy, and hearing loss. Am J Hum Genet 90:201–216.
Beighton P (1970) Ehlers-Danlos syndrome. Ann Rheum Dis 29:332–333
Beighton P et al (1988) International nosology of heritable disorders of connective tissue, Berlin, 1986. Am J Med Genet 29:581–594.
Beighton P, De Paepe A, Steinmann B, Tsipouras P, Wenstrup RJ (1998) Ehlers-Danlos syndromes: revised nosology, Villefranche, 1997. Ehlers-Danlos National Foundation (USA) and Ehlers-Danlos Support Group (UK). Am J Med Genet 77:31–37
Bekhouche M, Colige A (2015) The procollagen N-proteinases ADAMTS2, 3 and 14 in pathophysiology. Matrix Biol 44–46:46–53.
Bekhouche M et al (2016) Determination of the substrate repertoire of ADAMTS2, 3, and 14 significantly broadens their functions and identifies extracellular matrix organization and TGF-beta signaling as primary targets. FASEB J 30:1741–1756.
Birk DE (2001) Type V collagen: heterotypic type I/V collagen interactions in the regulation of fibril assembly. Micron (Oxford, England : 1993) 32:223–237.
Birk DE, Fitch JM, Babiarz JP, Linsenmayer TF (1988) Collagen type I and type V are present in the same fibril in the avian corneal stroma. J Cell Biol 106:999–1008.
Birk DE, Fitch JM, Babiarz JP, Doane KJ, Linsenmayer TF (1990) Collagen fibrillogenesis in vitro: interaction of types I and V collagen regulates fibril diameter. J Cell Sci 95(Pt 4):649–657
Blackburn PR et al (2018) Bi-allelic alterations in AEBP1 Lead to defective collagen assembly and connective tissue structure resulting in a variant of Ehlers-Danlos syndrome. Am J Hum Genet 102:696–705.
Bork P, Beckmann G (1993) The CUB domain. A widespread module in developmentally regulated proteins. J Mol Biol 231:539–545.
Boudko SP, Ishikawa Y, Nix J, Chapman MS, Bachinger HP (2014) Structure of human peptidyl-prolyl cis-trans isomerase FKBP22 containing two EF-hand motifs. Protein Sci 23:67–75.
Bowen JM, Sobey GJ, Burrows NP, Colombi M, Lavallee ME, Malfait F, Francomano CA (2017) Ehlers-Danlos syndrome, classical type. Am J Med Genet C Semin Med Genet 175:27–39.
Bowen CJ et al (2020) Targetable cellular signaling events mediate vascular pathology in vascular Ehlers-Danlos syndrome. J Clin Invest 130:686–698.
Brady AF et al (2017a) The Ehlers-Danlos syndromes, rare types. Am J Med Genet Part C Semin Med Genet 175:70–115.
Brady AF et al (2017b) The Ehlers-Danlos syndromes, rare types. Am J Med Genet C 175:70–115.

Bristow J, Carey W, Egging D, Schalkwijk J (2005) Tenascin-X, collagen, elastin, and the Ehlers-Danlos syndrome. Am J Med Genet C Semin Med Genet 139C:24–30.

Broek DL, Madri J, Eikenberry EF, Brodsky B (1985) Characterization of the tissue form of type V collagen from chick bone. J Biol Chem 260:555–562

Burch GHGY, Liu W, Dettman RW, Curry CJ, Smith L, Miller WL, Bristow J (1997) Tenascin-X deficiency is associated with Ehlers-Danlos syndrome. Nat Genet 17:104–108

Burkitt Wright EM et al (2011) Mutations in PRDM5 in brittle cornea syndrome identify a pathway regulating extracellular matrix development and maintenance. Am J Hum Genet 88:767–777.

Burkitt Wright EMPL, Spencer HL, Clayton-Smith J, Au L, Munier FL, Smithson S, Suri M, Rohrbach M, Manson FD, Black GC (2013) Brittle cornea syndrome: recognition, molecular diagnosis and management. Orphanet J Rare Dis 8:6

Burrows NP, Nicholls AC, Richards AJ, Luccarini C, Harrison JB, Yates JR, Pope FM (1998) A point mutation in an intronic branch site results in aberrant splicing of COL5A1 and in Ehlers-Danlos syndrome type II in two British families. Am J Hum Genet 63:390–398.

Byers PH, Murray ML (2014) Ehlers-Danlos syndrome: a showcase of conditions that lead to understanding matrix biology. Matrix Biol 33:10–15.

Byers PH, Holbrook KA, Barsh GS, Smith LT, Bornstein P (1981) Altered secretion of type III procollagen in a form of type IV Ehlers-Danlos syndrome. Biochemical studies in cultured fibroblasts. Lab Investig 44:336–341

Byers PH et al (1997) Ehlers-Danlos syndrome type VIIA and VIIB result from splice-junction mutations or genomic deletions that involve exon 6 in the COL1A1 and COL1A2 genes of type I collagen. Am J Med Genet 72:94–105.

Byers PH et al (2017) Diagnosis, natural history, and management in vascular Ehlers-Danlos syndrome. Am J Med Genet Part C Semin Med Genet 175:40–47.

Cabral WA et al (2005) Mutations near amino end of alpha1(I) collagen cause combined osteogenesis imperfecta/Ehlers-Danlos syndrome by interference with N-propeptide processing. J Biol Chem 280:19259–19269.

Cabral WA et al (2007) Y-position cysteine substitution in type I collagen (alpha1(I) R888C/p. R1066C) is associated with osteogenesis imperfecta/Ehlers-Danlos syndrome phenotype. Hum Mutat 28:396–405.

Cameron (1993) Corneal abnormalities in Ehlers-Danlos syndrome type VI. Cornea 12:54–59.

Canty EG, Kadler KE (2005) Procollagen trafficking, processing and fibrillogenesis. J Cell Sci 118:1341–1353.

Cassidy K, Eikenberry EF, Olsen B, Brodsky B (1980) X-ray diffraction investigations of collagen fibril structure in dermatosparactic lamb tissues. Lab Investig 43:542–546

Chakravarti S, Magnuson T, Lass JH, Jepsen KJ, LaMantia C, Carroll H (1998) Lumican regulates collagen fibril assembly: skin fragility and corneal opacity in the absence of lumican. J Cell Biol 141:1277–1286.

Chanut-Delalande H, Fichard A, Bernocco S, Garrone R, Hulmes DJ, Ruggiero F (2001) Control of heterotypic fibril formation by collagen V is determined by chain stoichiometry. J Biol Chem 276:24352–24359.

Chanut-Delalande H, Bonod-Bidaud C, Cogne S, Malbouyres M, Ramirez F, Fichard A, Ruggiero F (2004) Development of a functional skin matrix requires deposition of collagen V heterotrimers. Mol Cell Biol 24:6049–6057.

Chernogubow AN (1892) Uber einen Fall von Cutis Laxa Jahresber. Ges Med 27:562

Chiarelli N, Carini G, Zoppi N, Ritelli M, Colombi M (2018) Transcriptome analysis of skin fibroblasts with dominant negative COL3A1 mutations provides molecular insights into the etiopathology of vascular Ehlers-Danlos syndrome. PLoS One 13:e0191220.

Chiarelli N, Carini G, Zoppi N, Ritelli M, Colombi M (2019) Molecular insights in the pathogenesis of classical Ehlers-Danlos syndrome from transcriptome-wide expression profiling of patients' skin fibroblasts. PLoS One 14:e0211647.

Chiodo AA, Hockey A, Cole WG (1992) A base substitution at the splice acceptor site of intron 5 of the COL1A2 gene activates a cryptic splice site within exon 6 and generates abnormal type I procollagen in a patient with Ehlers-Danlos syndrome type VII. J Biol Chem 267:6361–6369
Chiquet M, Birk DE, Bonnemann CG, Koch M (2014) Collagen XII: protecting bone and muscle integrity by organizing collagen fibrils. Int J Biochem Cell Biol 53:51–54.
Chu ML et al (1984) Human pro alpha 1(I) collagen gene structure reveals evolutionary conservation of a pattern of introns and exons. Nature 310:337–340.
Colige A VI, Thiry M, Lambert CA, Van Beeumen J, Li SW, Prockop DJ, Lapiere CM, Nusgens BV (2002) Cloning and characterization of ADAMTS-14, a novel ADAMTS displaying high homology with ADAMTS-2 and ADAMTS-3. J Biol Chem 277:5756–5766
Colige A, Li SW, Sieron AL, Nusgens BV, Prockop DJ, Lapiere CM (1997) cDNA cloning and expression of bovine procollagen I N-proteinase: a new member of the superfamily of zinc-metalloproteinases with binding sites for cells and other matrix components. Proc Natl Acad Sci USA 94:2374–2379
Colige ASA, Li SW, Schwarze U, Petty E, Wertelecki W, Wilcox W, Krakow D, Cohn DH, Reardon W, Byers PH, Lapière CM, Prockop DJ, Nusgens BV (1999) Human Ehlers-Danlos syndrome type VII C and bovine dermatosparaxis are caused by mutations in the procollagen I N-proteinase gene. Am J Hum Genet 65:308–317
Colige A et al (2004) Novel types of mutation responsible for the dermatosparactic type of Ehlers-Danlos syndrome (type VIIC) and common polymorphisms in the ADAMTS2 gene. J Invest Dermatol 123:656–663.
Colombi M, Dordoni C, Venturini M, Zanca A, Calzavara-Pinton P, Ritelli M (2017) Delineation of Ehlers-Danlos syndrome phenotype due to the c.934C>T, p.(Arg312Cys) mutation in COL1A1: report on a three-generation family without cardiovascular events, and literature review. Am J Med Genet A 173:524–530.
Cooper NR (1985) The classical complement pathway: activation and regulation of the first complement component. Adv Immunol 37:151–216.
Cooper LJ et al (2006) The role of dermatopontin in the stromal organization of the cornea. Invest Ophthalmol Vis Sci 47:3303–3310.
Cooper TK et al (2010) The haploinsufficient Col3a1 mouse as a model for vascular Ehlers-Danlos syndrome. Vet Pathol 47:1028–1039.
Counts DF, Byers PH, Holbrook KA, Hegreberg GA (1980) Dermatosparaxis in a Himalayan cat: I. biochemical studies of dermal collagen. J Invest Dermatol 74:96–99
Crowther MA, Lach B, Dunmore PJ, Roach MR (1991) Vascular collagen fibril morphology in type IV Ehlers-Danlos syndrome. Connect Tissue Res 25:209–217.
Danielson KG, Baribault H, Holmes DF, Graham H, Kadler KE, Iozzo RV (1997) Targeted disruption of decorin leads to abnormal collagen fibril morphology and skin fragility. J Cell Biol 136:729–743.
Danlos M (1908) Un cas de cutis laxa avec tumeurs par contusion chronique des coudes et des genoux (xanthome juvénile pseudo-diabétique de MM. Hallopeau et Macé de Lépinay). Bull Soc Fr Dermatol Syphiligr 19:70–72
De Paepe A, Malfait F (2012) The Ehlers-Danlos syndrome, a disorder with many faces. Clin Genet 82:1–11.
De Paepe A, Nuytinck L, Hausser I, Anton-Lamprecht I, Naeyaert JM (1997) Mutations in the COL5A1 gene are causal in the Ehlers-Danlos syndromes I and II. Am J Hum Genet 60:547–554
de Wet W, Bernard M, Benson-Chanda V, Chu ML, Dickson L, Weil D, Ramirez F (1987) Organization of the human pro-alpha 2(I) collagen gene. J Biol Chem 262:16032–16036
Deak SB, Nicholls A, Pope FM, Prockop DJ (1983) The molecular defect in a nonlethal variant of osteogenesis imperfecta. Synthesis of pro-alpha 2(I) chains which are not incorporated into trimers of type I procollagen. J Biol Chem 258:15192–15197

Delbaere S et al (2020) Novel defects in collagen XII and VI expand the mixed myopathy/Ehlers-Danlos syndrome spectrum and lead to variant-specific alterations in the extracellular matrix Genet Med 22:112–123

Demirdas SDE, Robert L, Kempers M, van Beek D, Micha D, van Engelen BG, Hamel B, Schalkwijk J, Loeys B, Maugeri A, Voermans NC (2016) Recognizing the tenascin-X deficient type of Ehlers-Danlos syndrome: a cross-sectional study in 17 patients. Clin Genet 91:411.

D'hondt S et al (2018) Type III collagen affects dermal and vascular collagen fibrillogenesis and tissue integrity in a mutant Col3a1 transgenic mouse model. Matrix Biol 70:72–83.

Dordoni C, Ciaccio C, Venturini M, Calzavara-Pinton P, Ritelli M, Colombi M (2016) Further delineation of FKBP14-related Ehlers-Danlos syndrome: a patient with early vascular complications and non-progressive kyphoscoliosis, and literature review. Am J Med Genet A 170:2031–2038.

Dubacher N et al (2020) Celiprolol but not losartan improves the biomechanical integrity of the aorta in a mouse model of vascular Ehlers-Danlos syndrome. Cardiovasc Res 116:457–465.

Dundar M et al (2009) Loss of dermatan-4-sulfotransferase 1 function results in adducted thumb-clubfoot syndrome. Am J Hum Genet 85:873–882.

Duong J et al (2019) A family with classical Ehlers-Danlos syndrome (cEDS), mild bone fragility and without vascular complications, Caused by the p.Arg312Cys mutation in COL1A1. Eur J Med Genet 103730.

Dusanic M et al (2018) Novel nonsense mutation in SLC39A13 initially presenting as myopathy: case report and review of the literature. Mol Syndromol 9:100–109.

Egging D, van den Berkmortel F, Taylor G, Bristow J, Schalkwijk J (2007) Interactions of human tenascin-X domains with dermal extracellular matrix molecules. Arch Dermatol Res 298:389–396.

Ehlers E (1901) Cutis laxa, neigung zu haemorrhagien in der haut, lockerung meherer artikulationen. Dermatol Z 8:173–174

Elefteriou F, Exposito JY, Garrone R, Lethias C (2001) Binding of tenascin-X to decorin. FEBS Lett 495:44–47.

Fernandes RJ, Hirohata S, Engle JM, Colige A, Cohn DH, Eyre DR, Apte SS (2001) Procollagen II amino propeptide processing by ADAMTS-3. Insights on dermatosparaxis. J Biol Chem 276:31502–31509.

Fichard A, Kleman JP, Ruggiero F (1995) Another look at collagen V and XI molecules. Matrix Biol 14:515–531.

Fichard A, Tillet E, Delacoux F, Garrone R, Ruggiero F (1997) Human recombinant alpha1 (V) collagen chain. Homotrimeric assembly and subsequent processing. J Biol Chem 272:30083–30087.

Fjolstad M, Helle O (1974) A hereditary dysplasia of collagen tissues in sheep. J Pathol 112:183–188.

Font B, Eichenberger D, Rosenberg LM, van der Rest M (1996) Characterization of the interactions of type XII collagen with two small proteoglycans from fetal bovine tendon, decorin and fibromodulin. Matrix Biol 15:341–348

Fontaine E, Faugeroux J, Beugnon C, Verpont M-C, Nematalla H, Bruneval P, Hadchouel J, Jeunemaitre X (2015) Caractérisation d'un modèle murin du Syndrome d'Ehlers-Danlos vasculaire. J Mal Vasc 40:119

Frank M et al (2015) The type of variants at the COL3A1 gene associates with the phenotype and severity of vascular Ehlers-Danlos syndrome. Eur J Hum Genet 23:1657–1664

Fukada T et al (2008) The zinc transporter SLC39A13/ZIP13 is required for connective tissue development; its involvement in BMP/TGF-beta signaling pathways. PLoS One 3:e3642.

Gaines R, Tinkle BT, Halandras PM, Al-Nouri O, Crisostomo P, Cho JS (2015) Spontaneous ruptured dissection of the right common iliac artery in a patient with classic Ehlers-Danlos syndrome phenotype. Ann Vasc Surg 29:595.e511–595.e594.

Galli GG et al (2012) Prdm5 regulates collagen gene transcription by association with RNA polymerase II in developing bone. PLoS Genet 8:e1002711.

Gensure RC et al (2005) A novel COL1A1 mutation in infantile cortical hyperostosis (Caffey disease) expands the spectrum of collagen-related disorders. J Clin Invest 115:1250–1257.

Ghali N et al (2019) Atypical COL3A1 variants (glutamic acid to lysine) cause vascular Ehlers-Danlos syndrome with a consistent phenotype of tissue fragility and skin hyperextensibility. Genet Med 21:2081–2091.

Giunta C, Steinmann B (2000) Compound heterozygosity for a disease-causing G1489E [corrected] and disease-modifying G530S substitution in COL5A1 of a patient with the classical type of Ehlers-Danlos syndrome: an explanation of intrafamilial variability? Am J Med Genet 90:72–79.

Giunta C, Nuytinck L, Raghunath M, Hausser I, De Paepe A, Steinmann B (2002) Homozygous Gly530Ser substitution in COL5A1 causes mild classical Ehlers-Danlos syndrome. Am J Med Genet 109:284–290.

Giunta C et al (2008) Spondylocheiro dysplastic form of the Ehlers-Danlos syndrome--an autosomal-recessive entity caused by mutations in the zinc transporter gene SLC39A13. Am J Hum Genet 82:1290–1305.

Giunta C et al (2018) A cohort of 17 patients with kyphoscoliotic Ehlers-Danlos syndrome caused by biallelic mutations in FKBP14: expansion of the clinical and mutational spectrum and description of the natural history. Genet Med 20:42–54.

Gjaltema RA, van der Stoel MM, Boersema M, Bank RA (2016) Disentangling mechanisms involved in collagen pyridinoline cross-linking: the immunophilin FKBP65 is critical for dimerization of lysyl hydroxylase 2. Proc Natl Acad Sci USA 113:7142–7147.

Gotte M, Spillmann D, Yip GW, Versteeg E, Echtermeyer FG, van Kuppevelt TH, Kiesel L (2008) Changes in heparan sulfate are associated with delayed wound repair, altered cell migration, adhesion and contractility in the galactosyltransferase I (beta4GalT-7) deficient form of Ehlers-Danlos syndrome. Hum Mol Genet 17:996–1009.

Gotting C, Kuhn J, Zahn R, Brinkmann T, Kleesiek K (2000) Molecular cloning and expression of human UDP-d-xylose:proteoglycan core protein beta-d-xylosyltransferase and its first isoform XT-II. J Mol Biol 304:517–528.

Grobner R et al (2019) C1R mutations trigger constitutive complement 1 activation in periodontal Ehlers-Danlos syndrome. Front Immunol 10:2537.

Hanset RAM (1967) Dermatosparaxis (peau dechirée) chez le veau: un defaut général du tissu conjonctif, de nature héréditaire. Ann Méd Vétérinaire 7:451–470

Hanset RLC (1974) Inheritance of dermatosparaxis in the calf. A genetic defect of connective tissues. J Hered 65:356–358

Haralson MA, Mitchell WM, Rhodes RK, Kresina TF, Gay R, Miller EJ (1980) Chinese hamster lung cells synthesize and confine to the cellular domain a collagen composed solely of B chains. Proc Natl Acad Sci USA 77:5206–5210.

Haralson MA, Mitchell WM, Rhodes RK, Miller EJ (1984) Evidence that the collagen in the culture medium of Chinese hamster lung cells contains components related at the primary structural level to the alpha1(V) collagen chain. Arch Biochem Biophys 229:509–518.

Hata R, Kurata S, Shinkai H (1988) Existence of malfunctioning pro alpha2(I) collagen genes in a patient with a pro alpha 2(I)-chain-defective variant of Ehlers-Danlos syndrome. Eur J Biochem 174:231–237

Hausser I, Anton-Lamprecht I (1994) Differential ultrastructural aberrations of collagen fibrils in Ehlers-Danlos syndrome types I-IV as a means of diagnostics and classification. Hum Genet 93:394–407.

Hautala T, Byers MG, Eddy RL, Shows TB, Kivirikko KI, Myllyla R (1992) Cloning of human lysyl hydroxylase: complete cDNA-derived amino acid sequence and assignment of the gene (PLOD) to chromosome 1p36.3–p36.2. Genomics 13:62–69.

Hautala T, Heikkinen J, Kivirikko KI, Myllyla R (1993) A large duplication in the gene for lysyl hydroxylase accounts for the type VI variant of Ehlers-Danlos syndrome in two siblings. Genomics 15:399–404.

Hernandez A, Aguirre-Negrete MG, Ramirez-Soltero S, Gonzalez-Mendoza A, Martinez y Martinez R, Velazquez-Cabrera A, Cantu JM (1979) A distinct variant of the Ehlers-Danlos syndrome. Clin Genet 16:335–339

Hernandez A, Aguirre-Negrete MG, Liparoli JC, Cantu JM (1981) Third case of a distinct variant of the Ehlers-Danlos syndrome (EDS). Clin Genet 20:222–224

Hernandez A et al (1986) Ehlers-Danlos features with progeroid facies and mild mental retardation. Further delineation of the syndrome. Clin Genet 30:456–461

Hicks D et al (2014) Mutations in the collagen XII gene define a new form of extracellular matrix-related myopathy. Hum Mol Genet 23:2353–2363.

Hirose T, Suzuki I, Takahashi N, Fukada T, Tangkawattana P, Takehana K (2018) Morphometric analysis of cornea in the Slc39a13/Zip13-knockout mice. J Vet Med Sci 80:814–818.

Hirose T et al (2019) Structural alteration of glycosaminoglycan side chains and spatial disorganization of collagen networks in the skin of patients with mcEDS-CHST14. Biochim Biophys Gen Subj 1863:623–631.

Holbrook KA, Byers PH (1981) Ultrastructural characteristics of the skin in a form of the Ehlers-Danlos syndrome type IV. Storage in the rough endoplasmic reticulum. Lab Investig 44:342–350

Holbrook KA, Byers PH (1982) Structural abnormalities in the dermal collagen and elastic matrix from the skin of patients with inherited connective tissue disorders. J Invest Dermatol 79(Suppl 1):7s–16s

Holbrook KA, Byers PH, Counts DF, Hegreberg GA (1980) Dermatosparaxis in a Himalayan cat: II. Ultrastructural studies of dermal collagen. J Invest Dermatol 74:100–104

Horn D et al (2017) Biallelic COL3A1 mutations result in a clinical spectrum of specific structural brain anomalies and connective tissue abnormalities. Am J Med Genet A 173:2534–2538.

Hyland J, Ala-Kokko L, Royce P, Steinmann B, Kivirikko KI, Myllyla R (1992) A homozygous stop codon in the lysyl hydroxylase gene in two siblings with Ehlers-Danlos syndrome type VI. Nat Genet 2:228–231.

Ihme A et al (1984) Ehlers-Danlos syndrome type VI: collagen type specificity of defective lysyl hydroxylation in various tissues. J Invest Dermatol 83:161–165.

Ishikawa Y, Bachinger HP (2014) A substrate preference for the rough endoplasmic reticulum resident protein FKBP22 during collagen biosynthesis. J Biol Chem 289:18189–18201.

Ishikawa Y, Mizuno K, Bachinger HP (2017) Ziploc-ing the structure 2.0: endoplasmic reticulum-resident peptidyl prolyl isomerases show different activities toward hydroxyproline. J Biol Chem 292:9273–9282.

Ishikawa Y et al (2020) The novel missense mutation Met48Lys in FKBP22 changes its structure and functions. Sci Rep 10:497.

Ith B, Wei J, Yet SF, Perrella MA, Layne MD (2005) Aortic carboxypeptidase-like protein is expressed in collagen-rich tissues during mouse embryonic development. Gene Expression Patt 5:533–537.

Jackson DS, Bentley JP (1968) Collagen-glycosaminoglycan interactions. In: Gould BS (ed) Treatise on collagen, vol 2., Part 1. Academic, New York, pp 189–214

Janecke AR et al (2016) The phenotype of the musculocontractural type of Ehlers-Danlos syndrome due to CHST14 mutations. Am J Med Genet A 170A:103–115.

Jansen LH (1955) The structure of the connective tissue, an explanation of the symptoms of the Ehlers-Danlos syndrome. Dermatologica 110:108–120.

Jeong J et al (2012) Promotion of vesicular zinc efflux by ZIP13 and its implications for spondylocheiro dysplastic Ehlers-Danlos syndrome. Proc Natl Acad Sci USA 109:E3530–E3538.

Jorgensen A et al (2015) Vascular Ehlers-Danlos syndrome in siblings with biallelic COL3A1 sequence variants and marked clinical variability in the extended family. Eur J Hum Genet 23:796–802.

Kadler KE, Holmes DF, Trotter JA, Chapman JA (1996) Collagen fibril formation. Biochem J 316 (Pt 1):1–11.

Kadler KE, Baldock C, Bella J, Boot-Handford RP (2007) Collagens at a glance. J Cell Sci 120:1955–1958.

Kadler KE, Hill A, Canty-Laird EG (2008) Collagen fibrillogenesis: fibronectin, integrins, and minor collagens as organizers and nucleators. Curr Opin Cell Biol 20:495–501.

Kagan HM, Li W (2003) Lysyl oxidase: properties, specificity, and biological roles inside and outside of the cell. J Cell Biochem 88:660–672.

Kapferer-Seebacher I et al (2016) Periodontal Ehlers-Danlos syndrome is caused by mutations in C1R and C1S, which encode subcomponents C1r and C1s of complement. Am J Hum Genet 99:1005–1014.

Keene DR, Sakai LY, Bachinger HP, Burgeson RE (1987) Type III collagen can be present on banded collagen fibrils regardless of fibril diameter. J Cell Biol 105:2393–2402.

Keene DR, Lunstrum GP, Morris NP, Stoddard DW, Burgeson RE (1991) Two type XII-like collagens localize to the surface of banded collagen fibrils. J Cell Biol 113:971–978.

Kitagawa H et al (1998) Molecular cloning and expression of glucuronyltransferase I involved in the biosynthesis of the glycosaminoglycan-protein linkage region of proteoglycans. J Biol Chem 273:6615–6618.

Kivirikko KI, Myllyla R (1982) Posttranslational enzymes in the biosynthesis of collagen: intracellular enzymes. Methods Enzymol 82(Pt A):245–304.

Kivirikko KI, Pihlajaniemi T (1998) Collagen hydroxylases and the protein disulfide isomerase subunit of prolyl 4-hydroxylases. Adv Enzymol Relat Areas Mol Biol 72:325–398.

Klaassens M et al (2012) Ehlers-Danlos arthrochalasia type (VIIA-B)—expanding the phenotype: from prenatal life through adulthood. Clin Genet 82:121–130.

Koch M, Bohrmann B, Matthison M, Hagios C, Trueb B, Chiquet M (1995) Large and small splice variants of collagen XII: differential expression and ligand binding. J Cell Biol 130:1005–1014.

Kosho T, Takahashi J, Ohashi H, Nishimura G, Kato H, Fukushima Y (2005) Ehlers-Danlos syndrome type VIB with characteristic facies, decreased curvatures of the spinal column, and joint contractures in two unrelated girls. Am J Med Genet A 138A:282–287.

Kosho T et al (2010) A new Ehlers-Danlos syndrome with craniofacial characteristics, multiple congenital contractures, progressive joint and skin laxity, and multisystem fragility-related manifestations. Am J Med Genet A 152a:1333–1346.

Krane SM, Pinnell SR, Erbe RW (1972) Lysyl-protocollagen hydroxylase deficiency in fibroblasts from siblings with hydroxylysine-deficient collagen. Proc Natl Acad Sci USA 69:2899–2903

Kresse H, Rosthoj S, Quentin E, Hollmann J, Glossl J, Okada S, Tonnesen T (1987) Glycosaminoglycan-free small proteoglycan core protein is secreted by fibroblasts from a patient with a syndrome resembling progeroid. Am J Hum Genet 41:436–453

Kusche-Gullberg M, Kjellen L (2003) Sulfotransferases in glycosaminoglycan biosynthesis. Curr Opin Struct Biol 13:605–611.

Lapiere CM, Lenaers A, Kohn LD (1971) Procollagen peptidase: an enzyme excising the coordination peptides of procollagen. Proc Natl Acad Sci USA 68:3054–3058.

Lautrup CK et al (2020) Delineation of musculocontractural Ehlers-Danlos syndrome caused by dermatan sulfate epimerase deficiency. Mol Genet Genom Med 8:e1197.

Layne MD, Yet SF, Maemura K, Hsieh CM, Bernfield M, Perrella MA, Lee ME (2001) Impaired abdominal wall development and deficient wound healing in mice lacking aortic carboxypeptidase-like protein. Mol Cell Biol 21:5256–5261.

Le Goff C, Somerville RP, Kesteloot F, Powell K, Birk DE, Colige AC, Apte SS (2006) Regulation of procollagen amino-propeptide processing during mouse embryogenesis by specialization of homologous ADAMTS proteases: insights on collagen biosynthesis and dermatosparaxis. Development 133:1587–1596.

Lenaers A, Ansay M, Nusgens BV, Lapiere CM (1971) Collagen made of extended -chains, procollagen, in genetically-defective dermatosparaxic calves. Eur J Biochem 23:533–543

Lethias C, Descollonges Y, Boutillon MM, Garrone R (1996) Flexilin: a new extracellular matrix glycoprotein localized on collagen fibrils. Matrix Biol 15:11–19.

Lethias C, Carisey A, Comte J, Cluzel C, Exposito JY (2006) A model of tenascin-X integration within the collagenous network. FEBS Lett 580:6281–6285.

Li SW et al (2001) Transgenic mice with inactive alleles for procollagen N-proteinase (ADAMTS-2) develop fragile skin and male sterility. Biochem J 355:271–278.

Lichtenstein JRMG, Kohn LD, Byers PH, McKusick VA (1973) Defect in conversion of procollagen to collagen in a form of Ehlers–Danlos syndrome. Science 182:298–300

Linsenmayer TF, Gibney E, Igoe F, Gordon MK, Fitch JM, Fessler LI, Birk DE (1993) Type V collagen: molecular structure and fibrillar organization of the chicken alpha 1(V) NH2-terminal domain, a putative regulator of corneal fibrillogenesis. J Cell Biol 121:1181–1189.

Littré E (1839) Oeuvres Complètes d'Hippocrate, vol 2. Paris

Liu X, Wu H, Byrne M, Krane S, Jaenisch R (1997) Type III collagen is crucial for collagen I fibrillogenesis and for normal cardiovascular development. Proc Natl Acad Sci USA 94:1852–1856.

Liu YA et al (2020) Pathologic skull fracture in a near-term neonate with arthrochalasia type Ehlers-Danlos syndrome: a case report. Fetal Pediatr Pathol 1–6.

Loughlin J, Irven C, Hardwick LJ, Butcher S, Walsh S, Wordsworth P, Sykes B (1995) Linkage of the gene that encodes the alpha 1 chain of type V collagen (COL5A1) to type II Ehlers-Danlos syndrome (EDS II). Hum Mol Genet 4:1649–1651.

Lund A, Joensen F, Christensen E, Duno M, Skovby F, Schwartz M (2008) A novel arginine-to-cysteine substitution in the triple helical region of the alpha1(I) collagen chain in a family with an osteogenesis imperfecta/Ehlers-Danlos phenotype. Clin Genet 73:97–101.

Maccarana M et al (2006) Biosynthesis of dermatan sulfate: chondroitin-glucuronate C5-epimerase is identical to SART2. J Biol Chem 281:11560–11568.

Maccarana M, Kalamajski S, Kongsgaard M, Magnusson SP, Oldberg A, Malmstrom A (2009) Dermatan sulfate epimerase 1-deficient mice have reduced content and changed distribution of iduronic acids in dermatan sulfate and an altered collagen structure in skin. Mol Cell Biol 29:5517–5528.

Macsai MS, Lemley HL, Schwartz T (2000) Management of oculus fragilis in Ehlers-Danlos type VI. Cornea 19:104–107.

Madri JA, Foellmer HG, Furthmayr H (1982) Type V collagens of the human placenta: trimer alpha-chain composition, ultrastructural morphology and peptide analysis. Coll Relat Res 2:19–29.

Makareeva E, Cabral WA, Marini JC, Leikin S (2006) Molecular mechanism of alpha 1(I)-osteogenesis imperfecta/Ehlers-Danlos syndrome: unfolding of an N-anchor domain at the N-terminal end of the type I collagen triple helix. J Biol Chem 281:6463–6470.

Malfait F, De Paepe A (2005) Molecular genetics in classic Ehlers-Danlos syndrome. Am J Med Genet C Semin Med Genet 139C:17–23.

Malfait F, Coucke P, Symoens S, Loeys B, Nuytinck L, De Paepe A (2005) The molecular basis of classic Ehlers-Danlos syndrome: a comprehensive study of biochemical and molecular findings in 48 unrelated patients. Hum Mutat 25:28–37.

Malfait F, Symoens S, Coucke P, Nunes L, De Almeida S, De Paepe A (2006) Total absence of the alpha2(I) chain of collagen type I causes a rare form of Ehlers-Danlos syndrome with hypermobility and propensity to cardiac valvular problems J Med Genet 43:e36

Malfait F et al (2007) Three arginine to cysteine substitutions in the pro-alpha (I)-collagen chain cause Ehlers-Danlos syndrome with a propensity to arterial rupture in early adulthood. Hum Mutat 28:387–395.

Malfait F et al (2010) Musculocontractural Ehlers-Danlos syndrome (former EDS type VIB) and adducted thumb clubfoot syndrome (ATCS) represent a single clinical entity caused by mutations in the dermatan-4-sulfotransferase 1 encoding CHST14 gene. Hum Mutat 31:1233–1239.

Malfait F et al (2013a) Defective initiation of glycosaminoglycan synthesis due to B3GALT6 mutations causes a pleiotropic Ehlers-Danlos-syndrome-like connective tissue disorder. Am J Hum Genet 92:935–945.

Malfait F et al (2013b) Helical mutations in type I collagen that affect the processing of the amino-propeptide result in an osteogenesis imperfecta/Ehlers-Danlos syndrome overlap syndrome. Orphanet J Rare Dis 8:78.
Malfait F et al (2017) The 2017 international classification of the Ehlers-Danlos syndromes. Am J Med Genet Part C Semin Med Genet 175:8–26.
Mao JR et al (2002) Tenascin-X deficiency mimics Ehlers-Danlos syndrome in mice through alteration of collagen deposition. Nat Genet 30:421–425.
Marchant JK, Hahn RA, Linsenmayer TF, Birk DE (1996) Reduction of type V collagen using a dominant-negative strategy alters the regulation of fibrillogenesis and results in the loss of corneal-specific fibril morphology. J Cell Biol 135:1415–1426.
Matsumoto K, Sawa H, Sato M, Orba Y, Nagashima K, Ariga H (2002) Distribution of extracellular matrix tenascin-X in sciatic nerves. Acta Neuropathol 104:448–454.
McKusick VA (1956) Heritable disorders of connective tissue. IV. The Ehlers-Danlos syndrome. J Chronic Dis 3:2–24.
Minamitani T, Ariga H, Matsumoto K (2004) Deficiency of tenascin-X causes a decrease in the level of expression of type VI collagen. Exp Cell Res 297:49–60.
Miyake N et al (2010) Loss-of-function mutations of CHST14 in a new type of Ehlers-Danlos syndrome. Hum Mutat 31:966–974.
Miyake N, Kosho T, Matsumoto N (2014) Ehlers-Danlos syndrome associated with glycosaminoglycan abnormalities. Adv Exp Med Biol 802:145–159.
Moradi-Ameli M et al (1994) Diversity in the processing events at the N-terminus of type-V collagen. Eur J Biochem 221:987–995.
Morlino S et al (2020) COL1-related overlap disorder: a novel connective tissue disorder incorporating the osteogenesis imperfecta/Ehlers-Danlos syndrome overlap. Clin Genet 97:396–406.
Muller GA, Hansen U, Xu Z, Griswold B, Talan MI, McDonnell NB, Briest W (2012) Allele-specific siRNA knockdown as a personalized treatment strategy for vascular Ehlers-Danlos syndrome in human fibroblasts. FASEB J 26:668–677.
Muller T et al (2013) Loss of dermatan sulfate epimerase (DSE) function results in musculocontractural Ehlers-Danlos syndrome. Hum Mol Genet 22:3761–3772.
Murray ML, Yang M, Fauth C, Byers PH (2014) FKBP14-related Ehlers-Danlos syndrome: expansion of the phenotype to include vascular complications. Am J Med Genet A 164A:1750–1755.
Myllyla R et al (2007) Expanding the lysyl hydroxylase toolbox: new insights into the localization and activities of lysyl hydroxylase 3 (LH3). J Cell Physiol 212:323–329.
Nakajima M et al (2013) Mutations in B3GALT6, which encodes a glycosaminoglycan linker region enzyme, cause a spectrum of skeletal and connective tissue disorders. Am J Hum Genet 92:927–934.
Nicholls AC et al (1984) The clinical features of homozygous alpha 2(I) collagen deficient osteogenesis imperfecta. J Med Genet 21:257–262
Nicholls AC, De Paepe A, Narcisi P, Dalgleish R, De Keyser F, Matton M, Pope FM (1988) Linkage of a polymorphic marker for the type III collagen gene (COL3A1) to atypical autosomal dominant Ehlers-Danlos syndrome type IV in a large Belgian pedigree. Hum Genet 78:276–281
Nicholls AC, Oliver JE, McCarron S, Harrison JB, Greenspan DS, Pope FM (1996) An exon skipping mutation of a type V collagen gene (COL5A1) in Ehlers-Danlos syndrome. J Med Genet 33:940–946.
Nicholls AC, Valler D, Wallis S, Pope FM (2001) Homozygosity for a splice site mutation of the COL1A2 gene yields a non-functional pro(alpha)2(I) chain and an EDS/OI clinical phenotype. J Med Genet 38:132–136
Niyibizi C, Eyre DR (1993) Structural analysis of the extension peptides on matrix forms of type V collagen in fetal calf bone and skin. Biochim Biophys Acta 1203:304–309.
Niyibizi C, Eyre DR (1994) Structural characteristics of cross-linking sites in type V collagen of bone. Chain specificities and heterotypic links to type I collagen. Eur J Biochem 224:943–950.

Niyibizi C, Fietzek PP, van der Rest M (1984) Human placenta type V collagens. Evidence for the existence of an alpha 1(V) alpha 2(V) alpha 3(V) collagen molecule. J Biol Chem 259:14170–14174
Nusgens BV, Verellen-Dumoulin C, Hermanns-Le T, De Paepe A, Nuytinck L, Pierard GE, Lapiere CM (1992) Evidence for a relationship between Ehlers-Danlos type VII C in humans and bovine dermatosparaxis. Nat Genet 1:214–217.
Nuytinck L, Freund M, Lagae L, Pierard GE, Hermanns-Le T, De Paepe A (2000) Classical Ehlers-Danlos syndrome caused by a mutation in type I collagen. Am J Hum Genet 66:1398–1402.
O'Hara PJ, Read WK, Romane WM, Bridges CH (1970) A collagenous tissue dysplasia of calves. Lab Investig 23:307–314
Ogur G, Baykan N, De Paepe A, Steinmann B, Quatacker J, Kuseyri F, Yuksel-Apak M (1994) Clinical, ultrastructural and biochemical studies in two sibs with Ehlers-Danlos syndrome type VI-B-like features. Clin Genet 46:417–422.
Okajima T, Fukumoto S, Furukawa K, Urano T (1999) Molecular basis for the progeroid variant of Ehlers-Danlos syndrome. Identification and characterization of two mutations in galactosyltransferase I gene. J Biol Chem 274:28841–28844
Pacheco B, Maccarana M, Malmstrom A (2009a) Dermatan 4-O-sulfotransferase 1 is pivotal in the formation of iduronic acid blocks in dermatan sulfate. Glycobiology 19:1197–1203.
Pacheco B, Malmstrom A, Maccarana M (2009b) Two dermatan sulfate epimerases form iduronic acid domains in dermatan sulfate. J Biol Chem 284:9788–9795.
Park AC et al (2015) Homozygosity and heterozygosity for null Col5a2 alleles produce embryonic lethality and a novel classic Ehlers-Danlos syndrome-related phenotype. Am J Pathol 185:2000–2011.
Passoja K, Rautavuoma K, Ala-Kokko L, Kosonen T, Kivirikko KI (1998) Cloning and characterization of a third human lysyl hydroxylase isoform. Proc Natl Acad Sci USA 95:10482–10486.
Pepin MG, Schwarze U, Rice KM, Liu M, Leistritz D, Byers PH (2014) Survival is affected by mutation type and molecular mechanism in vascular Ehlers-Danlos syndrome (EDS type IV). Genet Med 16:881–888.
Pierard GE, Lapiere M (1976) Skin in dermatosparaxis. Dermal microarchitecture and biomechanical properties. J Investig Dermatol 66:2–7.
Pihlajaniemi T, Dickson LA, Pope FM, Korhonen VR, Nicholls A, Prockop DJ, Myers JC (1984) Osteogenesis imperfecta: cloning of a pro-alpha 2(I) collagen gene with a frameshift mutation. J Biol Chem 259:12941–12944
Pinnell SR, Krane SM, Kenzora JE, Glimcher MJ (1972) A heritable disorder of connective tissue. Hydroxylysine-deficient collagen disease. N Engl J Med 286:1013–1020.
Plancke A, Holder-Espinasse M, Rigau V, Manouvrier S, Claustres M, Khau Van Kien P (2009) Homozygosity for a null allele of COL3A1 results in recessive Ehlers-Danlos syndrome. Eur J Hum Genet 17:1411–1416.
Pogany G, Vogel KG (1992) The interaction of decorin core protein fragments with type I collagen. Biochem Biophys Res Commun 189:165–172.
Pope FM, Martin GR, Lichtenstein JR, Penttinen R, Gerson B, Rowe DW, McKusick VA (1975) Patients with Ehlers-Danlos syndrome type IV lack type III collagen. Proc Natl Acad Sci USA 72:1314–1316.
Pope FM, Martin GR, McKusick VA (1977) Inheritance of Ehlers-Danlos type IV syndrome. J Med Genet 14:200–204.
Pousi B, Hautala T, Heikkinen J, Pajunen L, Kivirikko KI, Myllyla R (1994) Alu-Alu recombination results in a duplication of seven exons in the lysyl hydroxylase gene in a patient with the type VI variant of Ehlers-Danlos syndrome. Am J Hum Genet 55:899–906
Quentin E, Gladen A, Roden L, Kresse H (1990) A genetic defect in the biosynthesis of dermatan sulfate proteoglycan: galactosyltransferase I deficiency in fibroblasts from a patient with a progeroid syndrome. Proc Natl Acad Sci USA 87:1342–1346

Rahman N et al (2003) Ehlers-Danlos syndrome with severe early-onset periodontal disease (EDS-VIII) is a distinct, heterogeneous disorder with one predisposition gene at chromosome 12p13. Am J Hum Genet 73:198–204.
Richards AJ, Martin S, Nicholls AC, Harrison JB, Pope FM, Burrows NP (1998) A single base mutation in COL5A2 causes Ehlers-Danlos syndrome type II. J Med Genet 35:846–848.
Ritelli M, Cinquina V, Venturini M, Pezzaioli L, Formenti AM, Chiarelli N, Colombi M (2019) Expanding the clinical and mutational spectrum of recessive AEBP1-related classical-like Ehlers-Danlos syndrome. Genes 10.
Ritelli M et al (2013) Clinical and molecular characterization of 40 patients with classic Ehlers-Danlos syndrome: identification of 18 COL5A1 and 2 COL5A2 novel mutations. Orphanet J Rare Dis 8:58.
Rohrbach MSH, Porter LF, Burkitt-Wright EM, Bürer C, Janecke A, Bakshi M, Sillence D, Al-Hussain H, Baumgartner M, Steinmann B, Black GC, Manson FD, Giunta C (2013) ZNF469 frequently mutated in the brittle cornea syndrome (BCS) is a single exon gene possibly regulating the expression of several extracellular matrix components. Mol Genet Metab 109:289–295
Rohrbach M et al (2011) Phenotypic variability of the kyphoscoliotic type of Ehlers-Danlos syndrome (EDS VIA): clinical, molecular and biochemical delineation. Orphanet J Rare Dis 6:46.
Romanic AM, Adachi E, Kadler KE, Hojima Y, Prockop DJ (1991) Copolymerization of pNcollagen III and collagen I. pNcollagen III decreases the rate of incorporation of collagen I into fibrils, the amount of collagen I incorporated, and the diameter of the fibrils formed. J Biol Chem 266:12703–12709
Sack G (1936) Status dysvascularis, ein Fall von "Cutis laxa" (Vater und Sohn). Muench Med Wochenschr 15:259–260
Sage H, Bornstein P (1979) Characterization of a novel collagen chain in human placenta and its relation to AB collagen. Biochemistry 18:3815–3822.
Sasaki T, Arai K, Ono M, Yamaguchi T, Furuta S, Nagai Y (1987) Ehlers-Danlos syndrome. A variant characterized by the deficiency of pro alpha 2 chain of type I procollagen. Arch Dermatol 123:76–79
Schalkwijk J et al (2001) A recessive form of the Ehlers-Danlos syndrome caused by tenascin-X deficiency. N Engl J Med 345:1167–1175.
Schissel SL, Dunsmore SE, Liu X, Shine RW, Perrella MA, Layne MD (2009) Aortic carboxypeptidase-like protein is expressed in fibrotic human lung and its absence protects against bleomycin-induced lung fibrosis. Am J Pathol 174:818–828.
Schwarze U, Atkinson M, Hoffman GG, Greenspan DS, Byers PH (2000) Null alleles of the COL5A1 gene of type V collagen are a cause of the classical forms of Ehlers-Danlos syndrome (types I and II). Am J Hum Genet 66:1757–1765.
Schwarze U, Hata R, McKusick VA, Shinkai H, Hoyme HE, Pyeritz RE, Byers PH (2004) Rare autosomal recessive cardiac valvular form of Ehlers-Danlos syndrome results from mutations in the COL1A2 gene that activate the nonsense-mediated RNA decay pathway. Am J Hum Genet 74:917–930.
Schwarze U et al (2001) Haploinsufficiency for one COL3A1 allele of type III procollagen results in a phenotype similar to the vascular form of Ehlers-Danlos syndrome, Ehlers-Danlos syndrome type IV. Am J Hum Genet 69:989–1001.
Scott JE (1988) Proteoglycan-fibrillar collagen interactions. Biochem J 252:313–323.
Scott PG (1991) Physical studies on the protein core of skin dermatan sulphate proteoglycan II (decorin). Biochem Soc Trans 19:377S.
Seidler DG et al (2006) Defective glycosylation of decorin and biglycan, altered collagen structure, and abnormal phenotype of the skin fibroblasts of an Ehlers-Danlos syndrome patient carrying the novel Arg270Cys substitution in galactosyltransferase I (beta4GalT-7). J Mol Med (Berl) 84:583–594.

Shi L et al (2017) Comprehensive population screening in the Ashkenazi Jewish population for recurrent disease-causing variants. Clin Genet 91:599–604.

Silbert JE, Sugumaran G (2002) Biosynthesis of chondroitin/dermatan sulfate. IUBMB Life 54:177–186.

Smith LB et al (2011) Haploinsufficiency of the murine Col3a1 locus causes aortic dissection: a novel model of the vascular type of Ehlers-Danlos syndrome. Cardiovasc Res 90:182–190.

Smith LT, Schwarze U, Goldstein J, Byers PH (1997) Mutations in the COL3A1 gene result in the Ehlers-Danlos syndrome type IV and alterations in the size and distribution of the major collagen fibrils of the dermis. J Invest Dermatol 108:241–247.

Smith LTWW, Milstone LM, Petty EM, Seashore MR, Braverman IM, Jenkins TG, Byers PH (1992) Human dermatosparaxis: a form of Ehlers-Danlos syndrome that results from failure to remove the amino-terminal propeptide of type I procollagen. Am J Hum Genet 51:235–244

Sokolov BP, Prytkov AN, Tromp G, Knowlton RG, Prockop DJ (1991) Exclusion of COL1A1, COL1A2, and COL3A1 genes as candidate genes for Ehlers-Danlos syndrome type I in one large family. Hum Genet 88:125–129.

Steiglitz BM, Keene DR, Greenspan DS (2002) PCOLCE2 encodes a functional procollagen C-proteinase enhancer (PCPE2) that is a collagen-binding protein differing in distribution of expression and post-translational modification from the previously described PCPE1. J Biol Chem 277:49820–49830.

Steinmann BGR, Vogel A, Grant ME, Harwood R, Sear CH (1975) Ehlers-Danlos syndrome in two siblings with deficient lysyl hydroxylase activity in cultured skin fibroblasts but only mild hydroxylysine deficit in skin. Helv Paediatr Acta 30:255–274

Steinmann BRP, Superti-Furga A (2002) The Ehlers–Danlos syndrome. Connective tissue and its heritable disorders, 2nd edn. Wiley-Liss, New York

Steinmann BTL, Peltonen L, Martin GR, McKusick VA, Prockop DJ (1980) Evidence for a structural mutation of procollagen type I in a patient with the Ehlers-Danlos syndrome type VII. J Biol Chem 255:8887–8893

Stewart RE, Hollister DW, Rimoin DL (1977) A new variant of Ehlers-Danlos syndrome: an autosomal dominant disorder of fragile skin, abnormal scarring, and generalized periodontitis. Birth Defects 13:85–93

Superti-Furga A, Gugler E, Gitzelmann R, Steinmann B (1988) Ehlers-Danlos syndrome type IV: a multi-exon deletion in one of the two COL3A1 alleles affecting structure, stability, and processing of type III procollagen. J Biol Chem 263:6226–6232

Symoens S et al (2009) COL5A1 signal peptide mutations interfere with protein secretion and cause classic Ehlers-Danlos syndrome. Hum Mutat 30:E395–E403.

Symoens S et al (2012) Comprehensive molecular analysis demonstrates type V collagen mutations in over 90% of patients with classic EDS and allows to refine diagnostic criteria. Hum Mutat 33:1485–1493.

Syx D et al (2019) Bi-allelic AEBP1 mutations in two patients with Ehlers-Danlos syndrome. Hum Mol Genet 28:1853–1864.

Syx D et al (2015) Genetic heterogeneity and clinical variability in musculocontractural Ehlers-Danlos syndrome caused by impaired dermatan sulfate biosynthesis. Hum Mutat 36:535–547.

Takahara K et al (2002) Order of intron removal influences multiple splice outcomes, including a two-exon skip, in a COL5A1 acceptor-site mutation that results in abnormal pro-alpha1(V) N-propeptides and Ehlers-Danlos syndrome type I. Am J Hum Genet 71:451–465.

Takaluoma K et al (2007) Tissue-specific changes in the hydroxylysine content and cross-links of collagens and alterations in fibril morphology in lysyl hydroxylase 1 knock-out mice. J Biol Chem 282:6588–6596.

Takeda U et al (2002) Targeted disruption of dermatopontin causes abnormal collagen fibrillogenesis. J Invest Dermatol 119:678–683.

Tasheva ES et al (2002) Mimecan/osteoglycin-deficient mice have collagen fibril abnormalities. Mol Vis 8:407–415

Teratani T et al (2018) Aortic carboxypeptidase-like protein, a WNT ligand, exacerbates nonalcoholic steatohepatitis. J Clin Investig 128:1581–1596.

Thelin MA et al (2013) Biological functions of iduronic acid in chondroitin/dermatan sulfate. FEBS J 280:2431–2446.

Ticho U, Ivry M, Merin S (1980) Brittle cornea, blue sclera, and red hair syndrome (the brittle cornea syndrome). Br J Ophthalmol 64:175–177.

Toriello HV, Glover TW, Takahara K, Byers PH, Miller DE, Higgins JV, Greenspan DS (1996) A translocation interrupts the COL5A1 gene in a patient with Ehlers-Danlos syndrome and hypomelanosis of Ito. Nat Genet 13:361–365.

Trowbridge JM, Gallo RL (2002) Dermatan sulfate: new functions from an old glycosaminoglycan. Glycobiology 12:117R–125R.

Tsipouras P et al (1986) Ehlers-Danlos syndrome type IV: cosegregation of the phenotype to a COL3A1 allele of type III procollagen. Hum Genet 74:41–46.

Tumelty KE, Smith BD, Nugent MA, Layne MD (2014) Aortic carboxypeptidase-like protein (ACLP) enhances lung myofibroblast differentiation through transforming growth factor beta receptor-dependent and -independent pathways. J Biol Chem 289:2526–2536.

Uzawa K, Grzesik WJ, Nishiura T, Kuznetsov SA, Robey PG, Brenner DA, Yamauchi M (1999) Differential expression of human lysyl hydroxylase genes, lysine hydroxylation, and cross-linking of type I collagen during osteoblastic differentiation in vitro. J Bone Miner Res 14:1272–1280.

Vadon-Le Goff S et al (2011) Procollagen C-proteinase enhancer stimulates procollagen processing by binding to the C-propeptide region only. J Biol Chem 286:38932–38938.

Valcourt U, Alcaraz LB, Exposito JY, Lethias C, Bartholin L (2015) Tenascin-X: beyond the architectural function. Cell Adhes Migr 9:154–165.

Valtavaara M, Papponen H, Pirttila AM, Hiltunen K, Helander H, Myllyla R (1997) Cloning and characterization of a novel human lysyl hydroxylase isoform highly expressed in pancreas and muscle. J Biol Chem 272:6831–6834.

Valtavaara M, Szpirer C, Szpirer J, Myllyla R (1998) Primary structure, tissue distribution, and chromosomal localization of a novel isoform of lysyl hydroxylase (lysyl hydroxylase 3). J Biol Chem 273:12881–12886.

Van Damme T et al (2016) Expanding the clinical and mutational spectrum of the Ehlers-Danlos syndrome, dermatosparaxis type. Genet Med 18:882–891.

Van Damme T et al (2018) Biallelic B3GALT6 mutations cause spondylodysplastic Ehlers-Danlos syndrome. Hum Mol Genet 27:3475–3487.

van der Slot AJ et al (2003) Identification of PLOD2 as telopeptide lysyl hydroxylase, an important enzyme in fibrosis. J Biol Chem 278:40967–40972.

van Mee'ren J (1668) Een Rekkelÿke Spanjert Heel- en Geneeskonstige Aanmerkingen Van Job van Meek'ren; in sijn leven Heelmeester der Stadt, Admiraliteyt en 't Gasthuys binnen Amster-dam Met koopere Plaaten verçiert T'Amsterdam, by Casparus Cooelijn, op 't Water, in de Waarheyt 170–172

Vandervore L et al (2017) Bi-allelic variants in COL3A1 encoding the ligand to GPR56 are associated with cobblestone-like cortical malformation, white matter changes and cerebellar cysts. J Med Genet 54:432–440.

Veit G, Hansen U, Keene DR, Bruckner P, Chiquet-Ehrismann R, Chiquet M, Koch M (2006) Collagen XII interacts with avian tenascin-X through its NC3 domain. J Biol Chem 281:27461–27470.

Vogel A, Holbrook KA, Steinmann B, Gitzelmann R, Byers PH (1979) Abnormal collagen fibril structure in the gravis form (type I) of Ehlers-Danlos syndrome. Lab Investig 40:201–206

Vogel KG, Paulsson M, Heinegard D (1984) Specific inhibition of type I and type II collagen fibrillogenesis by the small proteoglycan of tendon. Biochem J 223:587–597.

Weber FP (1936) Ehlers-Danlos syndrome. Proc R Soc Med 30:30–31

Weil D, Bernard M, Combates N, Wirtz MK, Hollister DW, Steinmann B, Ramirez F (1988) Identification of a mutation that causes exon skipping during collagen pre-mRNA splicing in an Ehlers-Danlos syndrome variant. J Biol Chem 263:8561–8564

Wenstrup RJ, Florer JB, Brunskill EW, Bell SM, Chervoneva I, Birk DE (2004a) Type V collagen controls the initiation of collagen fibril assembly. J Biol Chem 279:53331–53337.

Wenstrup RJ, Florer JB, Cole WG, Willing MC, Birk DE (2004b) Reduced type I collagen utilization: a pathogenic mechanism in COL5A1 haplo-insufficient Ehlers-Danlos syndrome. J Cell Biochem 92:113–124.

Wenstrup RJ et al (2006) Murine model of the Ehlers-Danlos syndrome. col5a1 haploinsufficiency disrupts collagen fibril assembly at multiple stages. J Biol Chem 281:12888–12895.

Wenstrup RJ et al (2000) COL5A1 haploinsufficiency is a common molecular mechanism underlying the classical form of EDS. Am J Hum Genet 66:1766–1776.

Wenstrup RJ, Langland GT, Willing MC, D'Souza VN, Cole WG (1996) A splice-junction mutation in the region of COL5A1 that codes for the carboxyl propeptide of pro alpha 1 (V) chains results in the gravis form of the Ehlers-Danlos syndrome (type I). Hum Mol Genet 5:1733–1736.

Wertelecki W, Smith LT, Byers P (1992) Initial observations of human dermatosparaxis: Ehlers-Danlos syndrome type VIIC. J Pediatr 121:558–564

Wordsworth BP, Ogilvie DJ, Sykes BC (1991) Segregation analysis of the structural genes of the major fibrillar collagens provides further evidence of molecular heterogeneity in type II Ehlers-Danlos syndrome. Br J Rheumatol 30:173–177.

Wordsworth P, Ogilvie D, Smith R, Sykes B (1985) Exclusion of the alpha 1(II) collagen structural gene as the mutant locus in type II Ehlers-Danlos syndrome Annals of the rheumatic diseases 44:431–433.

Xiao G, Wan Z, Fan Q, Tang X, Zhou B (2014) The metal transporter ZIP13 supplies iron into the secretory pathway in *Drosophila melanogaster*. eLife 3:e03191.

Yamauchi M, Shiiba M (2008) Lysine hydroxylation and cross-linking of collagen. Methods Mol Biol (Clifton, NJ) 446:95–108.

Yoshizawa T et al (2018) Vascular abnormalities in the placenta of Chst14−/− fetuses: implications in the pathophysiology of perinatal lethality of the murine model and vascular lesions in human CHST14/D4ST1 deficiency. Glycobiology 28:80–89.

Zoppi N, Gardella R, De Paepe A, Barlati S, Colombi M (2004) Human fibroblasts with mutations in COL5A1 and COL3A1 genes do not organize collagens and fibronectin in the extracellular matrix, down-regulate alpha2beta1 integrin, and recruit alphavbeta3 Instead of alpha5beta1 integrin. J Biol Chem 279:18157–18168.

Zou Y et al (2014) Recessive and dominant mutations in COL12A1 cause a novel EDS/myopathy overlap syndrome in humans and mice. Hum Mol Genet 23:2339–2352.

Chapter 4
Cartilage Collagens and Associated Disorders

Uwe Hansen

Abstract The tissue-specific extracellular matrix is important for normal development and tissue function, and therefore, mutations in genes responsible for ECM components cause a variety of serious inherited connective tissue disorders. In articular cartilage, the collagens are indispensably connected with the characteristic strength of the tissue. Cartilage disorders involve primarily alterations of collagen II, IX, and XI and the cartilage oligomeric protein (COMP). These diseases include a variety of clinical phenotypes from common osteoarthrosis to different types of mostly inherited chondrodysplasias. More than 100 distinct disorders of chondrodysplasias are described with different subclasses and disproportionate stature, short limbs, dwarfism, premature osteoarthrosis, and eye complications are typical findings for most of these disorders. The typical mutations in collagen genes are null mutations that result in the loss of protein and a reduction of total quantity of collagen protein in tissue. In contrast, small deletions or substitutions of bases can lead to the synthesis of mutated α-chains that are able to form a triple-helix. The altered molecule is secreted and results in a compromised supramolecular assembly with altered ECM suprastructures. In general, these human diseases are difficult to treat, especially when the pathological processes start before birth affecting the complete skeleton.

Abbreviations

CNBr	Cyanogen bromid
COMP	Cartilage oligomeric protein
DFNA13	Deafness, autosomal dominant, type 13 (DFNA-13)
ECM	Extracellular matrix
EDM2	Epiphyseal dysplasia, multiple, 2
ER stress	Endoplasmic reticulum stress

U. Hansen (✉)
University Hospital Münster, Institute for Musculoskeletal Medicine, Münster, Germany
e-mail: uhansen@uni-muenster.de

F. Ruggiero (ed.), *The Collagen Superfamily and Collagenopathies*, Biology of Extracellular Matrix 8, https://doi.org/10.1007/978-3-030-67592-9_4

FGF	Fibroblast growth factor
MED	Multiple epiphyseal dysplasia
NMD	Nonsense-mediated decay
OA	Osteoarthritis
OI	Osteogenesis imperfecta
PTC	Premature termination codon
SED	Spondyloepiphyseal Dysplasia
SEMD	Metaphyseal chondrodysplasia, Schmid type
UPR	Unfolded protein response
VEGF	Vascular endothelial growth factor
VNTR	Variable number tandem repeats
Wnt	Acronym of homologous wingless (wg) and Int-1

4.1 Introduction

The tissue-specific extracellular matrix (ECM) is important for normal development and tissue function, and therefore, mutations in genes responsible for ECM components cause a variety of serious inherited connective tissue disorders. In general, the ECM is composed of a complex combination of different macromolecules that could be divided into four major groups: collagens, proteoglycans, glycoproteins, and elastin. The superfamily of collagens includes up to now 28 different collagen types that assemble into a variety of different matrix suprastructures like fibrils or network-like structures. The different members of the collagen superfamily display an impressive heterogeneity in structure, function, and tissue distribution (Ricard-Blum and Ruggiero 2005). The combination of diverse matrix components provides tissue-specific features ranging from the rigid bone to the elasticity of skin and ligaments, and from the long lifespan of articular cartilage to the more transient state of growth plate cartilage (Bateman et al. 2009). All collagens contain also non-collagenous domains with important functions often distinct from the collagenous domains. The collagen synthesis is a complex process including eight post-translational enzymes (of interest as pharmaceutical targets for the treatment of diseases like fibrosis in different tissues and organs). The structural role of the ECM is well known for many years but many other important functions of the ECM are still not systematically analyzed. The collagens represent up to 80–90% of the total protein content in cartilage, bone, tendon, and skin forming highly organized suprastructures in a self-assembly process like the characteristic D-periodic banded collagen fibrils or network-like structures. The quantitatively major components of collagen fibrils in tissues like bone, skin, tendons, or cartilage are the fibrillar collagens I, II, and III. These collagens in combination with further minor collagens like collagen V, IX, XI, XII, and XIV are always expressed in a tissue-specific manner, and thereby, forming tissue-specific matrix suprastructures.

The ECM is not only an important scaffold for cells and tissues providing structural integrity but has also an important impact on processes like cell proliferation, migration, differentiation, signalling modulation, and remodelling (Arseni et al. 2018). The ECM contributes in different manners to these developmental steps where it plays a double role, both as functional as well as structural supporting element. The ability of cells to migrate is crucial for embryogenesis as well as maintenance of multicellular organisms. Moreover, many recent studies have shown that the ECM can sequester growth factors, chemokines, and cytokines, like Wnts, VEGFs, and FGFs. The ECM serves as a reservoir for growth factors, and thereby, controls their bioavailability. Therefore, the ECM plays an important role as a reservoir for different kinds of signalling molecules. Finally, in order to maintain the structural integrity, the ECM needs a constant and strictly regulated remodelling in a precise and orchestrated manner for tissue homeostasis and developmental processes. The proper balance between ECM degradation and synthesis is crucial for correct tissue integrity.

Mutations in some of these structural genes result in abnormalities in different tissues (Prockop and Kivirikko 1995; Olsen 1995; Lamande and Bateman 2020). In general, three major alterations could lead to changes in tissue organization resulting in collagen-related disorders: Defects in collagen structure, defects in processing enzymes, and changes in gene expression (Fig. 4.1). Since collagens are present throughout the entire body, mutations resulting in changes with respect to quality or quantity of collagenous structures or activity of modifying enzymes can affect any tissue or organ. Moreover, due to the fact that numerous collagens are expressed in different tissues or organs, alterations may result in overlapping phenotypes (Arseni et al. 2018). The important role of collagens for diseases is obvious by the more than 1000 mutations identified in different collagen genes (and related ECM genes) resulting in mild-to-severe human diseases or a predisposition for certain common diseases with a wide spectrum of clinical phenotypes that affect different tissues and organs such as cartilage, bone, skin, kidney, muscle, cornea, and vessels (Myllyharju and Kivirikko 2001; Lamande and Bateman 2020).

The result of alterations in genes coding for collagens or ECM components, in general, can be classified in different manners. One classification based on the affected tissue or organ, i.e. skeletal and cartilage abnormalities, skin alterations, hearing loss and visual defects, muscle weakness, and small vessel anomalies and kidney diseases, respectively (Arseni et al. 2018). Other criteria for a basic characterization of the phenotypes of matrix disorders are based on the biological characteristics of the ECM and the specific properties of their components. The fact that most if not all matrix molecules are oligomers consisting of more than one polypeptide chain, incorporation of one or more mutated polypeptides will cause abnormal folding and assembly of the molecule. This is especially true for collagens. Therefore, mutations that result in the synthesis of structurally modified polypeptide chains produce in general more severe phenotypes than heterozygous null alleles (Bruckner-Tuderman and Bruckner 1998). Two different possibilities are likely: The substitution of amino acids may result in an interruption of the protein folding process and lead to the generation of unfolded and abnormal proteins. This leads

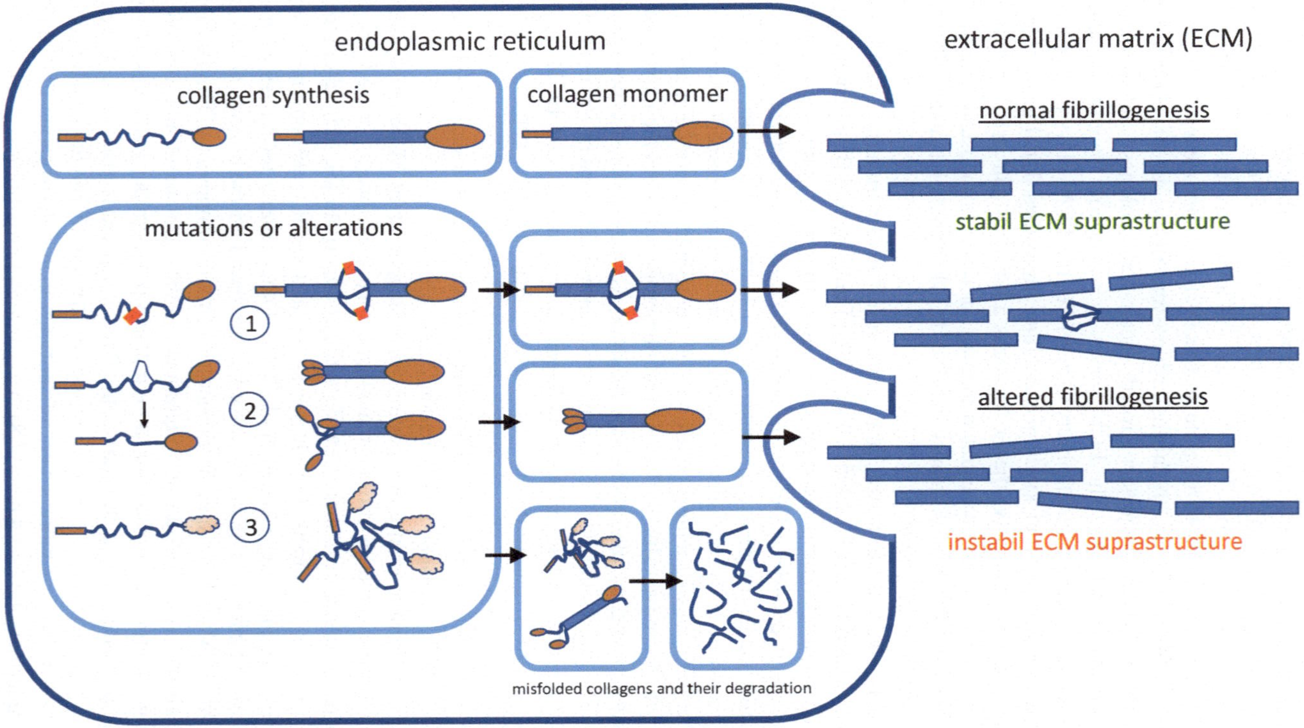

Fig. 4.1 Schematic representation of the outcome. *Legend*: Schematic representation of the outcome of point mutations or deletions in collagen-related genes on the synthesis of collagen chains, collagen folding, and the formation of collagen fibrils in the extracellular matrix. (1) Point mutation; (2) Deletion; (3) Mutation localized in the trimerization region of the triplehelix. Modified from Bonod-Bidaud and Ruggiero (2013)

to an intracellular accumulation and degradation of mutated and normal polypeptide chains present in the same trimeric protein by default mechanisms (called as protein suicide) (Prockop and Kivirikko 1995; Bateman et al. 2009). On the other hand, an amino acid substitution may not result in a complete unfolding and intracellular degradation but in an incorporation of mutated polypeptide chains into the trimeric protein. This has a massive effect on the molecular structure and the ability to incorporate into matrix suprastructures resulting in the formation of abnormal matrix suprastructures. This effect is called dominant negative. The negative effect on fibril formation was demonstrated by investigating in vitro collagen fibril formation. Collagen molecules with altered amino acid sequences copolymerize into collagen fibrils together with the normal molecules but the presence of the altered molecules leads to a delay in the fibril formation, a reduction of the total amount of collagen molecules incorporated into the fibrils, and most importantly, massive alterations in the structure of the resulting collagen fibril (Kadler et al. 1991).

The typical alterations in collagen genes are null mutations resulting in the absence of the protein (i.e. translation of α-chains that cannot form a triple-helix resulting in intracellular degradation) and a reduction of total quantity of collagen in tissue. In contrast, small deletions or substitution of nucleotide bases can result in the synthesis of mutated α-chains that are able to form a triple-helix. Therefore, altered molecules are secreted and lead to a compromised supramolecular assembly resulting in an altered ECM suprastructure. Thus, mutations in collagen genes result in an insufficient matrix assembly and organization that in turn can affect cell functions. However, dominant-negative mutations can be more severe than null mutations, especially in the case of large supramolecular assemblies, like collagen structures. However, an increasing number of studies demonstrated that the synthesis of huge amounts of mutated collagen molecules can result in endoplasmic reticulum stress (ER stress) with consequences ranging from cell recovery to death (Bonod-Bidaud and Ruggiero 2013; for details, see Chap. 2 Sergey Leikin and co-authors).

However, the correlation between the position of a point mutation and the severity of the resulting disorder is not clear. Nevertheless, based on the direction of the propagation of the collagen triple-helix synthesis, mutations located in the position responsible for the amino terminus of the fibrillar triple-helix result in mild phenotypes whereas mutations located in the region responsible for the carboxy terminus of the fibrillar triple-helix are often lethal. The substitution of glycine in the Gly-Xaa-Yaa repeats and the neighbouring amino acid sequences could result in different biochemical as well as clinical consequences. The resulting effects involve: (a) a slower triple-helix formation and over-glycosylation (Raghunath et al. 1994); (b) alterations in the processing of procollagen (Lightfoot et al. 1994); (c) intracellular retention of unfolded and abnormal proteins, leading in ER stress; and (d) formation of unstable trimeric molecules, resulting in a disrupted fibril formation.

The replacement of glycine residues by other amino acids disrupts the triple-helix and leads to structural changes in rod-like structures with rigid kinks or flexible hinges (Vogel et al. 1988; Bachmann et al. 2005; Shoulders and Raines 2009) causing major disruptions in helix formation, delayed propagation of the triple-helix at the position of mutation. The delay in helix formation is the reason for additional post-translational modifications resulting in an increasing hydroxylation and glycosylation. This is probably the underlying reason for mild or severe connective tissue disorders (Myllyharju and Kivirikko 2004; Malfait and De Paepe 2009). The glycine residue is important for the proper formation of the triple-helix, because glycine is the smallest amino acid that only fits into the centre of the triple-helix. Most of the dominant-negative mutations in collagen genes are the result of replacement of one of the glycines in the collagenous domains of the α-chains with larger amino acids. Examples for connective tissue disorders with an underlying glycine substitution are osteogenesis imperfecta (OI) (mutations in collagen I) (for details, see Chap. 2 Sergey Leikin and co-authors), chondrodysplasias (mutations in collagen II), some types of Ehlers-Danlos syndrome (mutations in collagen III causing vascular Ehlers–Danlos syndrome) (for details, see Chap. 3 Fransiska Malfait and co-authors) or the Alport syndrome (Bruckner-Tuderman and Bruckner 1998; Bateman et al. 2009; for details see Chap. 5 Tom van Agtmal and co-authors). The clinical outcomes depend strongly on the structural function of the helix in the respective collagen and the function of the collagen type in the ECM architecture of the respective tissue and organ (Bateman et al. 2009). The severity of the disease depends strongly on the kind of substitution. The critical result of the impaired collagen folding is the reduced secretion of trimers containing mutant collagen chains. As just one mutant chain in a trimer will impair helix formation, heterogeneous mutations have a dominant-negative effect (Bateman et al. 2009).

Mutations resulting in the introduction of premature termination codons (PTCs) are the most frequent genetic causes for a reduced synthesis of gene products. The presence of a PTC triggers an mRNA surveillance process and nonsense-mediated decay (NMD), whereby aberrant mRNAs are rapidly degraded and distinguished from the normal mRNAs (Bateman et al. 2009). Examples of mutations in structural ECM genes resulting in NMD and haploinsufficiency are collagen II (*COL2A1)* in the Stickler syndrome and collagen X (*COL10A1*) in metaphyseal chondrodysplasias, Schmidt type, which are both cartilage diseases. Although NMD can be 100% efficient, is it more common that the abundance of PTC-containing transcripts is reduced to 5–25% of the normal allele product resulting in dominant-negative or gain-of-function effects (Bateman et al. 2009).

Mutations with an impact on the triple-helix formation are the most prevalent group of collagen mutations. The glycine substitutions interrupting the essential Gly-Xaa-Yaa repeat sequence that leads to a misfolded and abnormal triple-helix are most common. Collagen molecules containing altered α-chains are retained in the ER to a large extent and consequently only poorly secreted. In addition to reduced rates of synthesis and secretion or disturbed interactions in the ECM, misfolded ECM proteins including collagens induce ER stress and trigger the unfolded protein response (UPR) as a cause for initiation and progression of many collagen-related disorders (Bateman et al. 2009; Boot-Handford and Briggs 2010; for details, see Chap. 2 Sergey Leikin and co-authors).

4.2 Defects in Structural Proteins of Cartilage

Cartilage is a semi-rigid but flexible connective tissue found at several sites of the body. It is possible to distinguish between three types of cartilage mainly based on biomechanical and histological criteria: hyaline, elastic, and fibrous cartilage. The most common type is the hyaline cartilage with the articular cartilage that forms the smooth and lubricated gliding surface of joints such as the hip or knee (Bruckner 2006). The cellular component of cartilage, the chondrocytes, accounts for only a very small volume. The chondrocytes are embedded in a dense extracellular matrix without direct cell–cell contacts. In contrast to most other tissues, no blood vessels, nerves, or lymphatics are present in articular cartilage. The chondrocytes contribute to the different zones of articular cartilage: the superficial zone, the middle zone, the deep zone, and the calcified zone (Fig. 4.2). Within each zone, three structurally and compositionally different regions could be distinguished: the pericellular region, the territorial region, and the interterritorial region. The pericellular matrix is a thin layer adjacent to the chondrocytes, which contains mainly proteoglycans. The territorial layer surrounds the pericellular matrix as a basket-like structure. This network consists of thin collagen fibrils with a weak cross-banding pattern and a diameter of approximately 20 nm. The third zone is the largest of the three matrix regions and is called as interterritorial region. This zone is characterized by bundles of large collagen fibrils with an obvious longitudinal cross-banding pattern. This fibril population is more restricted and found in regions more remote from the chondrocytes (Fig. 4.3a, b). In articular cartilage, collagens accounts for two-thirds of the dry weight and are indispensably connected with the characteristic strength of

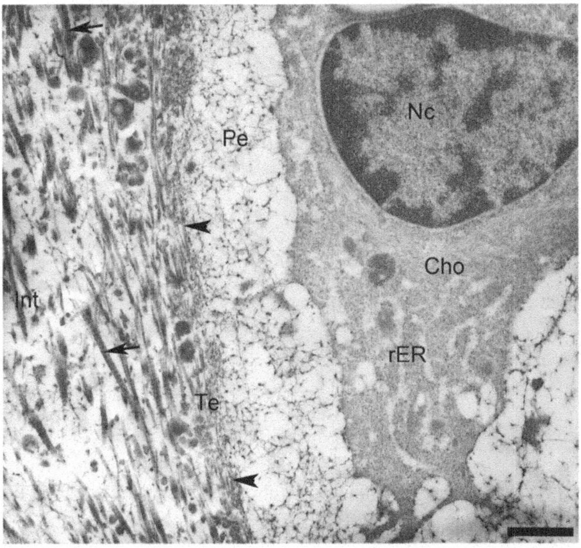

Fig. 4.2 Electron micrograph—cartilage with chondrocyte. *Legend*: Electron micrograph of human articular cartilage of the knee showing the typical organisation of collagen fibrils in two distinct major populations. Thin fibrils with an uniform diameter of approximately 20 nm (arrow heads) and thick and clearly banded collagen fibrils (arrows). Scale bar: 200 nm

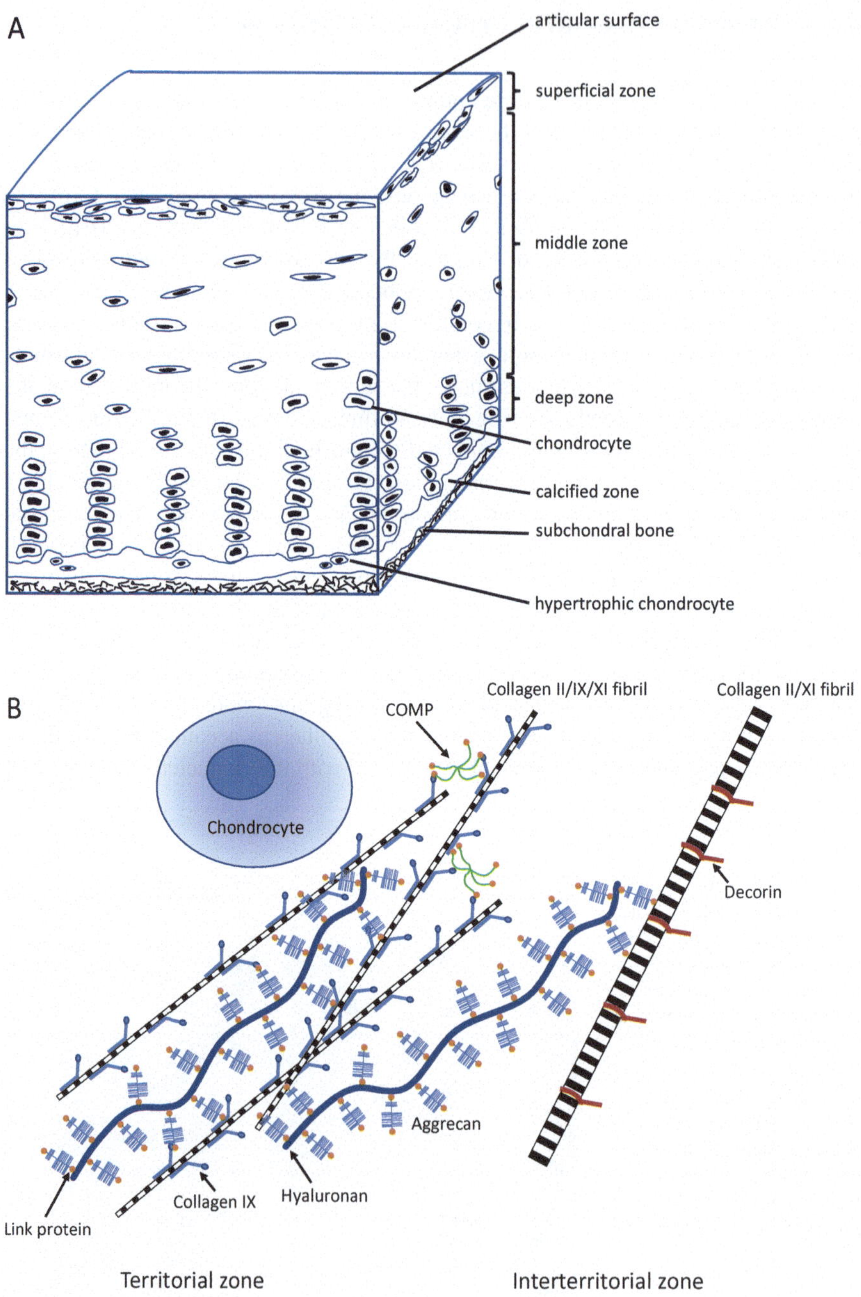

Fig. 4.3 (**a**) Schematic of the organization. (**b**) Schematic—territorial and interterritorial. *Legend*: (**a**) Schematic showing the organization of articular cartilage in the typical zones: superficial zone, middle zone, deep zone, calcified zone, and the following subchondral bone. Modified from Sophia Fox et al. (2009). (**b**) Extracellular matrix of articular cartilage. The major collagenous component of articular cartilage is collagen II that forms uniform thin fibrils together with collagens IX and XI

the tissue. At the molecular level, the major functions of articular cartilage are maintained by two suprastructural compartments, the fibrillar collagen network with the characteristic D-periodically banded collagen fibrils and the extrafibrillar network with the proteoglycan aggrecan as a major component. In addition, some fibrils are also associated with small leucine-rich proteoglycans like decorin, biglycan, or fibromodulin. Moreover, matrilins, the cartilage oligomeric matrix protein (COMP), and other ECM proteins that are neither collagens nor proteoglycans are also present (Fig. 4.3b). Collagen II is the quantitatively major component in cartilage representing 90% of total collagen in cartilage. It is present in all types of hyaline cartilage. The cartilage fibrils are good examples for the concept that matrix suprastructures are macromolecular composites/alloys. In tissue, collagen II occurs always as macromolecular composites. In the territorial matrix, the collagenous components always include collagen II, IX, and XI (Mendler et al. 1989) and to a small amount, collagen II, XI, and XVI (Kassner et al. 2003). The ability to buffer against forces exerted at the articular surface of bones is provided by hyaline cartilage. The complex network of collagen fibrils can resist tensile strength and the highly negatively charged proteoglycan network has a high capacity to absorb water resulting in an ECM with a high capability to absorb the compressing forces directed against the articular surface. Interestingly, after the formation of the collagenous network during development, it seems that articular chondrocytes have a limited capacity to remodel the typical collagen architecture if mature cartilage is injured (Eyre 2002).

Cartilage disorders are associated with abnormalities of collagen II, IX, and XI and also COMP or matrilin 3 (MATN3). These diseases display a wide variety of clinical outcomes from the common osteoarthrosis to several types of chondrodysplasias, most of them are inherited (Table 4.1). The most common cartilage disease is osteoarthrosis. In contrast, chondrodysplasias are rare disorders affecting cartilage and/or the growth plate of long bones. Till now, more than 450 genetic skeletal disorders are recognized and at least 200 of them have primarily an effect on cartilage with a broad clinical spectrum ranging from lethal conditions in utero to later onsets of the diseases involving joints and spine (Bateman 2001; Bonafe et al. 2015; Lamande and Bateman 2020). The mutations causing the majority of chondrodysplasias can be divided into genes responsible for local regulation of cartilage growth, genes responsible for cartilage metabolic pathways, and genes encoding cartilage structural proteins (Bateman 2001; Krakow 2015) leading to distinct subclasses of chondrodysplasias such as achondrogenesis, hypochondrogenesis, spondyloepimetaphyseal dysplasia, and the Stickler syndrome

Fig. 4.3 (continued) typical for the territorial matrix and clearly banded thick fibrils together with collagen XI typical for the interterritorial matrix. Proteoglycans are the major non-collagenous component. The most abundant is aggrecan that interacts with hyaluronan to form large proteoglycan aggregates occupying the interfibrillar space important for the capability to resist compressive loads. In addition, some fibrils are decorated with molecules like, e.g. decorin, a member of the family of small leucine-rich proteoglycans (SLRPs) or the non-collagenous ECM protein COMP

Table 4.1 Mutations in cartilage collagens and the related hereditary disorders

Gene	Protein	Chromosome	Major site of expression	Disease	Position of mutation
COL2A1	Collagen II	12q13-q14	Cartilage, vitreous humor	Spondyloepiphyseal dysplasia; Achondrogenesis, type II; Hypochondrogenesis; Kniest dysplasia; Stickler syndrome, type I; osteoarthritis with mild dysplasia; Wagner syndrome	Most mutations located in the triple-helix; glycine replacement predominates in collagen II collagenopathies; mutations in the N- or C-terminal propeptide region are rare; Stickler syndrome: Variants resulting in premature termination of translation (decreased synthesis of collagen II)
COL9A1 *COL9A2* *COL9A3*	Collagen IX	6q12-q141p32 20q13.3	Cartilage, vitreous humor	Multiple epiphyseal dysplasia	Splicing mutations are restricted to specific exons encoding an equivalent region of the COL3 domain in all three alpha (IX) chains
COL10A1	Collagen X	6q21-q22	Hypertrophic cartilage	Metaphyseal chondrodysplasia, Schmid type	Except two mutations (cleavage of signal peptide affected) all mutations found in the NC1 domain
COL11A1 *COL11A2*	Collagen XI	1p22 6p21.2	Bone, cartilage, vitreous humor, placenta	Stickler syndrome; Marshall syndrome Otospondylomegaepiphyseal dysplasia; Sensorineural deafness; Stickler syndrome; Weissenbach-Zweymüller syndrome	Disruption of Gly-Xaa-Yaa collagen sequence: Splice site variants; missense variants and in-frame deletions In-frame deletions and variants resulting in aberrant splicing and exon skipping (non-ocular Stickler syndrome)

(Schwartz and Domowicz 2002; Krakow 2015). The majority of these syndromes are rare, but autosomal or X-linked inheritance has been identified with both dominant and recessive inheritance patterns (Vikkula et al. 1994). Disproportionate stature, dwarfism, short limbs, premature osteoarthrosis, and eye involvement are typical clinical findings for most of these disorders. Myopia, retinal detachment, and cataract are frequently found, especially in Stickler syndrome and Wagner syndrome. The hyaline cartilage is in most of these disorders severely affected. Generally, the severity of the diseases as a result of mutations in the *COL2A1* gene varies from perinatal lethality to mild disease and is correlated with the type of mutation, i.e. haploinsufficiency as a result of a PTC will cause a milder phenotype than glycine substitutions in the coding region of the triple-helix that disrupts the formation of the collagen II triple-helix. Alterations in genes of minor collagens generally cause milder forms of chondrodysplasias. Osteoarthritis (OA) is very common in chondrodysplasias, and alterations in cartilage collagen genes have been identified in some families with early-onset OA. OA is one of the most common age-related chronic disorders of articular cartilage, joints, and bone. Multiple gene variations associated with an increased risk for OA were found in genetic studies. Moreover, further studies substantiate the notion that mutations in genes for structural components of articular cartilage cause the rare forms of highly penetrant inherited diseases that are associated with early-onset OA, whereas the more common age-related forms of the same disease are associated with genetic risk factors in form of frequent population polymorphisms (Reginato and Olsen 2002).

4.3 Diseases by Collagen Types

4.3.1 Collagen II

The major collagen of cartilage is collagen II and consists of three identical polypeptide α1 chains encoded by the *COL2A1* gene, which is located on chromosome 12. The *COL2A1* gene consists of 54 separated exons and it is only transcribed in a few tissues such as the hyaline cartilage, the nucleus pulposus, and the vitreous body of the eye. Many polymorphisms have been described in *COL2A1*, including a VNTR marker at the 3′ end of its encoding region (Vikkula et al. 1994; Barat-Houari et al. 2016). Mutations in the *COL2A1* gene result in a large spectrum of clinical phenotypes varying from lethal-to-mild arthropathy. Due to the fact that collagen II is found in cartilage, the intervertebral disc, and the vitreous humour of the eye, also ocular problems such as myopia and retinal detachment and vertebral abnormalities are quite common in these dysplasias. These disorders include achondrogenesis type II, or hypochondrogenesis. The most serious variant of achondrogenesis has a striking micromelia, a hydrophobic appearance, and death in utero or in the early hours of life. The characteristics of hypochondrogenesis are short stature and hydrops at birth with a flat face, widely spaced eyes, and small rib cage, which leads to death within hours or weeks from respiratory failure. The skeleton is less severely affected but the vertebral bodies are ossified and the tubular bones are longer. Single amino acid substitutions for a glycine residue, which are all clustered near the carboxy terminal end were found in both types of these diseases. Hypochondrogenesis, however, is a type II collagenopathy as a result of a heterozygous glycine substitution interrupting the crucial Gly-Xaa-Yaa repeat of the triple-helical domain of α1 chain. The mutations lead to a defective ECM of the cartilage with a reduced amount of collagen II with increased levels of post-translational modifications. The marked decrease of collagen II in the diseased cartilage ECM is accompanied by a predominant increase of the fibrillar collagens I and III and of the minor collagen XI. Due to the posttranslational over modification, even the low number of collagen II molecules were intracellularly degraded. However, bones were formed leading to the assumption that cartilage formation and bone development can happen in the absence of collagen II. Achondrogenesis type II, and hypochondrogenesis were previously regarded as separate disorders, but due to several overlaps, these disorders could be better regarded as one disorder representing a spectrum with a large phenotypic variability (Deng et al. 2016).

In all mutations of collagen II, only one of the two *COL2A1* alleles is affected and localized in the triple-helical domain. It is characteristic for collagenous polypeptide chains that every third amino acid is glycine which is an absolute requirement for a correct folding of the triple-helix. In collagen II, about half of the mutations resulting in osteoarthrosis and chondrodysplasias are substitutions resulting in the replacement of one of the glycine residues along the polypeptide chain by another amino acid. Therefore, all identified glycine substitutions have altered different glycine residues along the triple-helical domain with the consequence that the majority of the affected patients and families have an individual mutation. The substitution in the *COL2A1* gene is leading to at least five clinical different phenotypes: the Wagner syndrome, SEMD (spondyloepimetaphyseal dysplasia), SED (spondyloepiphyseal dysplasia), hypochondrogenesis, and achondrogenesis. The abnormalities vary from lethal perinatal types, to those of intermediate severity involving SED congenita and Kniest dysplasia, to the mildest variants involving SED tarda with early osteoarthritis. Moreover, nonsense mutations were also found in the *COL2A1* gene. These PTPs result in a shortened polypeptide. This type of mutation is found in the Stickler syndrome. In this disease, both hyaline cartilage of the joints leading to premature osteoarthrosis and severe alterations in the eye resulting in myopia and retinal detachment, are affected. In addition to the glycine mutations and the nonsense mutations several further mutations have been found in the *COL2A1* gene. Further analysis revealed three arginine to cysteine conversions, splicing errors, and a deletion and duplication mutation resulting in premature osteoarthrosis, the Kniest syndrome and SED.

The Stickler syndrome, also called hereditary arthro-ophthalmopathy, is a dominant inherited group of multisystem connective disorders characterized by clinical symptoms affecting the eye and cartilage structures. The ocular phenotypes include myopia and vitreoretinal degeneration with the risk of retinal detachment, cataract, and glaucoma. In cartilage, the symptoms comprise premature joint degeneration and joint hypermobility, cleft palate, orofacial abnormalities, and deafness (Baitner et al. 2000; Stickler et al. 2001). The Stickler syndrome can be divided into six distinguishable types (Stickler syndrome type I–V and Stickler syndrome atypical) primarily based on the genetic alterations. Alterations in one of the six genes (*COL2A1, COL11A1, COL11A2, COL9A1, COL9A2,* and *COL9A3*) have been identified in association with Stickler syndrome (Robin et al. 2017). However, the most affected gene involved in the Stickler syndrome is the *COL2A1* gene. Most mutations in the *COL2A1* cause a PTC resulting in haploinsufficiency with a decreased amount of normal collagen II (Baitner et al. 2000). The phenotype of patients with Stickler syndrome is quite different, and even patients with alterations in the same gene show different phenotypes. Due to locus and allelic heterogeneity, the phenotypic expression is quite variable within and among families. The inter- and intrafamilial variation of the disease is probably the result of locus and allelic heterogeneity. Nevertheless, it is possible to make a few general genotype–phenotype correlations. However, there are a few families with characteristics of Stickler syndrome which are not associated with any of these six loci but alterations in other genes may also cause the disorder (Robin et al. 2017).

The Kniest syndrome is characterized by a short stature, a cleft palate, and flat facial features as well as ocular involvement, similar to Stickler syndrome and Wagner syndrome. Mutations in the collagen II gene were also described in some patients with early-onset in familial osteoarthritis (Ritvaniemi et al. 1995). In five families, the early-onset osteoarthritis with very mild chondrodysplasia is caused by a Y position arginine-519 to cysteine substitution in the $\alpha 1$(II) chain. However, some other identified *COL2A1* mutations with a similar phenotype are likely to cause only about 2% of cases of early-onset of osteoarthritis. The variants of SED are characterized by different single nucleotide substitutions throughout the molecule as well as deletions and insertions. Patients with Kniest dysplasia have exon skipping mutations mostly located around the amino terminal end of the molecule. The Kniest dysplasia is a severe form of SED. The resting cartilage is characterized by an abnormal morphology with large chondrocytes embedded in a loosely woven ECM with numerous empty spaces, which led to the term “Swiss cheese cartilage syndrome”. In young children, the growth plate is hypercellular with large chondrocytes and little ECM in between. Since vascular penetration is irregular, broad, short, irregular spicules of calcified cartilage and bone are formed. With increasing age, there is some effort visible for column formation but the cartilage is still hypercellular. The epiphyseal growth plate zone is disorganized with little evidence of proliferation and hypertrophic cell arrangement in columns. The proliferative zone is abnormally short. In addition, ultrastructural analysis revealed that the chondrocytes have dilated cisternae of the rough endoplasmic reticulum and cytoplasmic inclusions. In the hypertrophic zone, irregular columns consisting of several enlarged hypertrophic cells with cytoplasmic inclusions are found. Moreover, the growth plate is characterized by scattered foci of fibrillated matrix with mucoid degeneration and larger acellular areas (Gilbert-Barnes et al. 1996; Barat-Houari et al. 2016; Deng et al. 2016).

The majority of the disorders other than the Stickler syndrome are felt to disrupt the formation of the triple-helix of collagen II with the result that abnormally deformed molecules are degraded prematurely or incorporated in an imperfect manner into the extracellular fibrillar network. Rimoin (1996) identified a direct correlation between the ratio of collagens I and II in cartilage and the clinical severity in type II collagenopathy skeletal dysplasias. In the most severe and lethal achondrogenesis II disorder only collagen I is present in cartilage, whereas under normal conditions, collagen I is not present in cartilage. In the slightly less severe hypochondrogenesis, both collagen I and post-translational overmodified collagen II are present, whereas in SED type I collagen I is not present and collagen II is present in the normal and overmodified form.

Moreover, several studies in which collagen II was analyzed from patients with chondrodysplasias have shown that in most cases the cyanogen bromide (CNBr) peptides from collagen II α1 chains exhibit a different electrophoretic mobility in comparison to healthy samples (Vikkula et al. 1994). In addition, in some of the most severely affected patients with chondrodysplasias, collagen II is completely missing in cartilage. Interestingly, the severity of the clinical outcome corresponds with the distance of the alteration to the carboxy terminus. Moreover, it is also presumed that changes in the mobility of CNBr fragments could be the result of an overmodification of polypeptides caused by a mutation in one of the α-chains resulting in a delay of the triple-helix folding (Murray and Rimoin 1988; Murray et al. 1989). However, it is important to mention that not all articular cartilage-related syndromes are associated with collagen II.

4.3.2 Collagen IX

Collagen IX makes up 1–5% of total collagen in adult articular cartilage and 10% of that in fetal cartilage (Luo et al. 2017; Eyre et al. 2006). It belongs to the group of fibril-associated collagens with interrupted helices (FACIT) and plays an important role in cartilage fibril formation and as a bridging structure between cartilage collagen fibrils and the extrafibrillar matrix. Collagen IX is together with collagen XI crucial for the maintenance of the cartilage matrix and the formation of the collagen meshwork. It is found as a fibrillar component of hyaline cartilage, intervertebral disc, vitreous humor, and the inner ear. The collagen IX molecule is a heterotrimer consisting of three genetically distinct chains, α1(IX)–α3(IX), and possesses three collagenous domains, COL1–COL3, flanked by four non-collagenous domains, NC1–NC4. In addition, collagen IX is also a proteoglycan as the NC3 domain of the α2(IX) chain is covalently attached to a glycosaminoglycan side chain.

Mutations in the gene encoding for collagen IX have been identified in the multiple epiphyseal dysplasias (MED), a clinical and genetically heterogeneous group of autosomal dominant diseases characterized by short stature and early onset of osteoarthrosis. All identified mutations result in an in-frame deletion of sequences encoding for the most N-terminal collagen part encoding by *COL9A1*, *COL9A2*, and *COL9A3* genes. Important to mention, this disease is also caused by alterations in COMP and matrilin-3, although these disorders are so far phenotypically comparable to the variants of collagen IX (Jackson et al. 2010). Histopathologic changes are similar in variants of the disease such as Jansen (the most severe) or Schmid (the mildest) (Wasylenko et al. 1980), and McKusick (cartilage-hair disease variant). Typical findings are an uneven staining of the cartilage, a fibrillar appearance of the ECM, and the larger size of the chondrocytes. Moreover, instead of linear columns of chondrocytes clusters of proliferating and hypertrophic cells surrounded by dense collagen staining are present. The septae are wider than normal and composed of dense fibrous material. In addition, the metaphyseal vascular invasion is irregular, and cells of the hypertrophic cartilage often persist in the metaphysis. Tongue-like extensions of cartilage into the metaphysis as a result of radiolucent linear metaphyseal streaking are often found. Furthermore, analysis by electron microscopy often revealed that chondrocytes at all levels contain rough dilated endoplasmic reticulum (Cooper and Ponseti 1973; Maynard et al. 1981).

The link between MED and the locus of *COL9A2* was first characterized by Briggs and co-authors (1994), and later confirmed by the identification of a mutation in the *COL9A2* gene in a different family (Muragaki et al. 1996a, b). The reason for the mutation is a skipping of exon 3 during the spicing of the primary mRNA transcript. Moreover, there is also a report of a large family with MED caused by mutations in the *COL9A2* gene in which the disease is always affecting the knee joints and often other more peripheral upper and lower extremity joints but without any problem with the shoulder or hip joint. This disorder is currently referred to as MED type 2 (or EDM 2). Moreover, mutations in the α1 and α3 chains of collagen IX (i.e. *COL9A1* and *COL9A3*, respectively) have also been identified in families with MED. Paassilta et al. (1999) and Bonnemann et al. (2000) described families with MED and mutations in the COL9A3 gene also with invariable knee involvement. In one family with MED, a mutation in intron 8 of *COL9A1* is causing three different splicing defects (Czarny-Ratajczak et al. 2001). The altered splice variant of the α1(IX) mRNA lacked sequences of either exon 8 or 10 or both. Moreover, there is an association of gene polymorphisms in *COL9A2* and *COL9A3* in intervertebral diseases (Annunen et al. 1999; Paassilta et al. 2001). In general, the deficiency of collagen IX results in changes in the supramolecular assembly of cartilage collagens. In addition, many further mutations in the COMP gene lead to the same disease (Briggs et al. 1995; Hecht et al. 1995). Moreover, two mutations in collagen IX genes have been identified in patients with intervertebral disc disease. One mutation in the *COL9A2* gene leads to a change of a codon for an X position glutamine to one for tryptophan and is responsible for a dominantly inherited disc disease in several families. The other mutation is a mutation in the *COL9A3* gene which results in a conversion of a codon for a Y position arginine to one for tryptophan. The amino acid tryptophan is rarely found in the primary sequences of collagen triple-helices as tryptophan poorly fits in the densely packed collagen triple-helix.

4.3.3 Collagen XI

Collagen XI is a minor component of cartilage collagen fibrils and assembles together with collagen II and IX to prototypic cartilage collagen fibrils. It plays an important role in the fibril diameter control (Blaschke et al. 2000). It accounts for 3–10% of total collagen in adult cartilage and fetal cartilage, respectively. Moreover, collagen XI is also found in the ear, the vitreous humor, and the intervertebral disc (Luo et al. 2017; Eyre 2002). It is a heterotrimer composed of α1(XI), α2(XI), and α3 (XI) chains. The gene coding for the α1 chain is located on chromosome 1 and that of the α2 chain on chromosome 6 whereas the α3 chain of collagen XI is located on chromosome 12. Moreover, the α3 chain of collagen XI is identical with the α1 chain of collagen II.

A common clinical phenotype of all disorders as a result of alterations in the collagen XI genes are facial anomalies, cleft palate, and hearing defects, whereas ocular defects are only present in some disorders (Spranger 1998). Mutations in the gene coding for the α1(XI) chain, *COL11A1*, are the reason for dominantly inherited chondrodysplasias, the Stickler and Marshall syndromes (Richards et al. 1996; Griffith et al. 1998; Annunen et al. 1999) as well as the non-syndromic Robin sequence (Melkoniemi et al. 2003). Mutations in the *COL11A1* gene have been identified in three families with Stickler syndrome. All identified alterations are different, one mutation is a heterozygous single base change resulting a glycine to valine substitution in the triple-helical domain of the α1(XI) chain, the other is an exon skipping mutation, and the third one is a multi-exon deletion (Richards et al. 1996; Martin et al. 1999). Moreover, phenotype overlapping Stickler and Marshall syndromes, are caused by glycine substitutions and in-frame deletions (Annunen et al. 1999). Important to note that the differentiation between the phenotypes of mild chondrodysplasias such as Stickler syndrome, non-ocular Stickler syndrome, and Marshall syndrome is sometimes difficult. The Marshall syndrome is a rare, autosomal dominant chondrodysplasia. The clinical phenotypes are different as patients with Marshall syndrome have more often a short stature, deafness, more abnormalities in cranial ossification, more pronounced dysmorphic features, and less frequent retinal detachment (Annunen et al. 1999). In patients with Marshall syndrome, several mutations have been found in the *COL11A1* gene, resulting in the skipping of a 54-bp exon in the region coding for the C-terminal half of the α1(XI) molecule (Griffith et al. 1998; Annunen et al. 1999). Mutations in the *COL11A2* gene have been associated with a Stickler-like syndrome associated with mild spondyloepiphyseal dysplasia, osteoarthritis, and sensorineural hearing loss and no eye involvement. The common characteristic caused by mutations in *COL11A2* is the absence of the eye involvement typically caused by mutations in the collagen XI genes, *COL2A1* and *COL11A1*, because this chain is not expressed in the eye. The α2(V) chain substitutes the α2(XI) chain in collagen XI in the vitreous body (Mayne et al. 1993). Further *COL11A2* alterations have been found in otospondylomegaepiphyseal dysplasia and Weissenbacher–Zweymüller syndromes, diseases with overlapping phenotypes with the Stickler and Marshall syndromes, and patients with non-syndromic hearing loss (Myllyharju and Kivirikko 2001). Mutations in the *COL2A1* gene coding for α3(XI) chain, cause a variety of chondrodysplasias. However, patients lacking α2(XI) have skeletal deformities characteristic of Stickler syndrome but without ocular complications. This is plausible because of the different compositions of vitreous fibrils that contain collagen XI hybrid molecules with a chain composition of $[\alpha1(XI)]_2\alpha2(V)$ instead of the predominant α1(XI)α2(XI)α3(XI) heterotrimer found in cartilage (Vikkula et al. 1995). The non-syndromic hearing loss is from the genetic point of view very heterogeneous. The autosomal dominant form is associated with more than 40 genes and the autosomal recessive form is associated with 30 genes (Bitner-Glindzicz 2002). However, certain types of mutations in the *COL11A2* gene are the reason for the autosomal dominant non-syndromic hearing loss. In correlation with this phenotype and according to the nomenclature, the *COL11A2* locus is called DFNA13. The mutations causing the disease are heterogenous missense mutations in the triple-helix region (McGuirt et al. 1999).

4.3.4 Collagen X

Collagen X is an $[\alpha1(X)]_3$ homotrimer and normally restricted to the thin layer of calcified cartilage at the interface between articular cartilage and bone. It is a short-chain collagen that forms hexagonal networks. The human α1(X) collagen chain has a short collagenous domain flanked by the globular NC1 and NC2 domains at the C- and N- terminal ends. The expression of collagen X is limited to the hypertrophic zone of the growth plate during endochondral bone formation and the basal calcified zone of articular cartilage. Moreover, collagen X is linked to the onset of cartilage calcification and extracellular matrix remodeling (Wilson et al. 2005; Luo et al. 2017). The biological function of collagen X includes the maintaining of tissue stiffness, the regulation of chondrocyte metabolism and the interaction with hypertrophic chondrocytes (Luckman et al. 2003). In addition, it is also known that collagen X triggers the process of calcification, the normal distribution of matrix vesicles and proteoglycans within the growth plate.

Mutations in the *Col10A1* gene disrupt growth plate function and cause Schmid metaphyseal chondrodysplasia (SMED). It is an autosomal dominant disease of the osseous skeleton associated with phenotypes like short stature, growth plate abnormalities, and a waddling gait. Except from two mutations which affect signal peptide cleavage, all other known mutations are located in the NC1 domain. Due to missing variations in the resulting phenotype of the affected patients, the identified mutations vary from amino acid substitutions to nonsense mutations and deletions resulting in PTCs. SMED mutations are mostly located in the NC1 domain which is important for trimer formation, and later, the extracellular assembly. The subsequent disruption of triple-helix formation results in a reduction of collagen X molecules by half as collagen X chains failed to enter into molecule formation (Chan and Jacenko 1998; Wilson et al. 2005). Moreover, recent studies suggest that *COL10A1* mutations in the

C-terminal NC-1 domain cause misfolding and ER retention and an induction of the canonical cellular UPR pathway and that the clinical phenotype of SMED is a direct result of the pathological UPR signalling (Rajpar et al. 2009; Lamande and Bateman 2020).

4.4 Outlook: Therapy

Inherited skeletal disorders are a very heterogeneous and complex group of rare diseases including phenotypes with a spectrum in severity from relatively mild to severe and lethal forms. In general, these human diseases are difficult to treat, especially when the pathological process starts before birth affecting the complete skeleton. Moreover, it is now more clear that alterations of the skeleton have a close relationship with physiological conditions of many other tissues and organs in the body. Therefore, alterations in structural genes may have serious effects on other organs including the peripheral nervous system, brain, bone marrow, immune system, pancreas, kidney, heart, muscle, and tendon. The development of new treatments or the identification of effective biomarkers is very limited for these inherited disorders, especially due to the clinical and genetic variability. However, ER stress has been associated with different genetic diseases and chronic conditions such as, e.g. skeletal dysplasias that make ER stress an attractive target. The increasing knowledge suggests that the primary genetic alteration may be less crucial than the response of the cell to the expression of an altered gene product (comprehensive review by Briggs et al. 2015; Marzin and Cormier-Daire 2020).

References

Annunen S, Paassilta P, Lohiniva J, Perala M, Pihlajamaa T, Karppinen J et al (1999) An allele of COL9A2 associated with intervertebral disc disease. Science 285(5426):409–412

Arseni L, Lombardi A, Orioli D (2018) From structure to phenotype: impact of collagen alterations on human health. Int J Mol Sci 19(5)

Bachmann A, Kiefhaber T, Boudko S, Engel J, Bachinger HP (2005) Collagen triple-helix formation in all-trans chains proceeds by a nucleation/growth mechanism with a purely entropic barrier. Proc Natl Acad Sci USA 102(39):13897–13902

Baitner AC, Maurer SG, Gruen MB, Di Cesare PE (2000) The genetic basis of the osteochondrodysplasias. J Pediatr Orthop 20(5):594–605

Barat-Houari M, Sarrabay G, Gatinois V, Fabre A, Dumont B, Genevieve D et al (2016) Mutation update for COL2A1 gene variants associated with type II Collagenopathies. Hum Mutat 37 (1):7–15

Bateman JF (2001) The molecular genetics of inherited cartilage disease. Osteoarthritis Cartilage 9 (Suppl A)):S141–S149

Bateman JF, Boot-Handford RP, Lamande SR (2009) Genetic diseases of connective tissues: cellular and extracellular effects of ECM mutations. Nat Rev Genet 10(3):173–183

Bitner-Glindzicz M (2002) Hereditary deafness and phenotyping in humans. Br Med Bull 63:73–94

Blaschke UK, Eikenberry EF, Hulmes DJ, Galla HJ, Bruckner P (2000) Collagen XI nucleates self-assembly and limits lateral growth of cartilage fibrils. J Biol Chem 275(14):10370–10378
Bonafe L, Cormier-Daire V, Hall C, Lachman R, Mortier G, Mundlos S et al (2015) Nosology and classification of genetic skeletal disorders: 2015 revision. Am J Med Genet A 167A (12):2869–2892
Bonod-Bidaud C, Ruggiero F (2013) Inherited connective tissue disorders of collagens: lessons from targeted mutagenesis. In: Figurski D (ed) Genetic manipulation of dna and protein-examples from current research, InTech, pp 253–270
Bonnemann CG, Cox GF, Shapiro F, Wu JJ, Feener CA, Thompson TG et al (2000) A mutation in the alpha 3 chain of type IX collagen causes autosomal dominant multiple epiphyseal dysplasia with mild myopathy. Proc Natl Acad Sci USA 97(3):1212–1217
Boot-Handford RP, Briggs MD (2010) The unfolded protein response and its relevance to connective tissue diseases. Cell Tissue Res 339(1):197–211
Briggs MD, Choi H, Warman ML, Loughlin JA, Wordsworth P, Sykes BC et al (1994) Genetic mapping of a locus for multiple epiphyseal dysplasia (EDM2) to a region of chromosome 1 containing a type IX collagen gene. Am J Hum Genet 55(4):678–684
Briggs MD, Hoffman SM, King LM, Olsen AS, Mohrenweiser H, Leroy JG et al (1995) Pseudoachondroplasia and multiple epiphyseal dysplasia due to mutations in the cartilage oligomeric matrix protein gene. Nat Genet 10(3):330–336
Briggs MD, Bell PA, Wright MJ, Pirog KA (2015) New therapeutic targets in rare genetic skeletal diseases. Expert Opin Orphan Drugs 3(10):1137–1154
Bruckner P (2006) Supramolecular structure of cartilage matrix. In: Seibel MJ, Robins SP, Bilezikian JP (eds) Dynamics of bone and cartilage metabolism. Elsevier, London, pp 407–420
Bruckner-Tuderman L, Bruckner P (1998) Genetic diseases of the extracellular matrix: more than just connective tissue disorders. J Mol Med (Berl) 76(3–4):226–237
Chan D, Jacenko O (1998) Phenotypic and biochemical consequences of collagen X mutations in mice and humans. Matrix Biol 17(3):169–184
Cooper RR, Ponseti IV (1973) Metaphyseal dysotosis: description of an ultrastructural defect in the epiphyseal plate chondrocytes. J Bone Joint Surg Am 55(3):485–495
Czarny-Ratajczak M, Lohiniva J, Rogala P, Kozlowski K, Perala M, Carter L et al (2001) A mutation in COL9A1 causes multiple epiphyseal dysplasia: further evidence for locus heterogeneity. Am J Hum Genet 69(5):969–980
Deng H, Huang X, Yuan L (2016) Molecular genetics of the COL2A1-related disorders. Mutat Res Rev Mutat Res 768:1–13
Eyre D (2002) Collagen of articular cartilage. Arthritis Res 4(1):30–35
Eyre DR, Weis MA, Wu JJ (2006) Articular cartilage collagen: an irreplaceable framework? Eur Cell Mater 12:57–63
Gilbert-Barnes E, Langer LO Jr, Opitz JM, Laxova R, Sotelo-Arila C (1996) Kniest dysplasia: radiologic, histopathological, and scanning electronmicroscopic findings. Am J Med Genet 63 (1):34–45
Griffith AJ, Sprunger LK, Sirko-Osadsa DA, Tiller GE, Meisler MH, Warman ML (1998) Marshall syndrome associated with a splicing defect at the COL11A1 locus. Am J Hum Genet 62 (4):816–823
Hecht JT, Nelson LD, Crowder E, Wang Y, Elder FFB et al (1995) Mutations in Cartilage Oligomeric Matrix Protein (Comp) Cause Pseudoachondroplasia. Am J Hum Genet 57(4):242
Jackson GC, Marcus-Soekarman D, Stolte-Dijkstra I, Verrips A, Taylor JA, Briggs MD (2010) Type IX collagen gene mutations can result in multiple epiphyseal dysplasia that is associated with osteochondritis dissecans and a mild myopathy. Am J Med Genet A 152a(4):863–869
Kadler KE, Torre-Blanco A, Adachi E, Vogel BE, Hojima Y, Prockop DJ (1991) A type I collagen with substitution of a cysteine for glycine-748 in the alpha 1(I) chain copolymerizes with normal type I collagen and can generate fractal like structures. Biochemistry 30(20):5081–5088

Kassner A, Hansen U, Miosge N, Reinhardt DP, Aigner T, Bruckner-Tuderman L et al (2003) Discrete integration of collagen XVI into tissue-specific collagen fibrils or beaded microfibrils. Matrix Biol 22(2):131–143
Krakow D (2015) Skeletal dysplasias. Clin Perinatol 42(2):301–319. viii
Lamande SR, Bateman JF (2020) Genetic disorders of the extracellular matrix. Anat Rec (Hoboken) 303(6):1527–1542
Lightfoot SJ, Atkinson MS, Murphy G, Byers PH, Kadler KE (1994) Substitution of serine for glycine 883 in the triple-helix of the pro alpha 1 (I) chain of type I procollagen produces osteogenesis imperfecta type IV and introduces a structural change in the triple-helix that does not alter cleavage of the molecule by procollagen N-proteinase. J Biol Chem 269 (48):30352–30357
Luckman SP, Rees E, Kwan AP (2003) Partial characterization of cell-type X collagen interactions. Biochem J 372(Pt 2):485–493
Luo YY, Sinkeviciute D, He Y, Karsdal M, Henrotin Y, Mobasheri A et al (2017) The minor collagens in articular cartilage. Protein Cell 8(8):560–572
Malfait F, De Paepe A (2009) Bleeding in the heritable connective tissue disorders: mechanisms, diagnosis and treatment. Blood Rev 23(5):191–197
Martin S, Richards AJ, Yates JR, Scott JD, Pope M, Snead MP (1999) Stickler syndrome: further mutations in COL11A1 and evidence for additional locus heterogeneity. Eur J Hum Genet 7 (7):807–814
Marzin P, Cormier-Daire V (2020) New perspectives on the treatment of skeletal dysplasia. Ther Adv Endocrinol Metab 11:2042018820904016
Maynard JA, Ippolito EG, Ponseti IV, Mickelson MR (1981) Histochemistry and ultrastructure of the growth plate in metaphyseal dysostosis: further observations on the structure of the cartilage matrix. J Pediatr Orthop 1(2):161–169
Mayne R, Brewton RG, Mayne PM, Baker JR (1993) Isolation and characterization of the chains of type V/type XI collagen present in bovine vitreous. J Biol Chem 268(13):9381–9386
McGuirt WT, Prasad SD, Griffith AJ, Kunst HP, Green GE, Shpargel KB et al (1999) Mutations in COL11A2 cause non-syndromic hearing loss (DFNA13). Nat Genet 23(4):413–419
Mendler M, Eich-Bender SG, Vaughan L, Winterhalter KH, Bruckner P (1989) Cartilage contains mixed fibrils of collagen types II, IX, and XI. J Cell Biol 108(1):191–197
Melkoniemi M, Koillinen H, Mannikko M, Warman ML et al (2003) Collagen XI sequence variations in nonsyndromic cleft palate, Robin sequence and micrognathia. Eur J Hum 11 (3):265–270
Muragaki Y, Mariman EC, van Beersum SE, Perala M, van Mourik JB, Warman ML et al (1996a) A mutation in COL9A2 causes multiple epiphyseal dysplasia (EDM2). Ann NY Acad Sci 785:303–306
Muragaki Y, Mariman EC, van Beersum SE, Perala M, van Mourik JB, Warman ML et al (1996b) A mutation in the gene encoding the alpha 2 chain of the fibril-associated collagen IX, COL9A2, causes multiple epiphyseal dysplasia (EDM2). Nat Genet 12(1):103–105
Murray LW, Rimoin DL (1988) Abnormal type II collagen in the spondyloepiphyseal dysplasias. Pathol Immunopathol Res 7(1–2):99–103
Murray LW, Bautista J, James PL, Rimoin DL (1989) Type II collagen defects in the chondrodysplasias. I. Spondyloepiphyseal dysplasias. Am J Hum Genet 45(1):5–15
Myllyharju J, Kivirikko KI (2001) Collagens and collagen-related diseases. Ann Med 33(1):7–21
Myllyharju J, Kivirikko KI (2004) Collagens, modifying enzymes and their mutations in humans, flies and worms. Trends Genet 20(1):33–43
Olsen BR (1995) New insights into the function of collagens from genetic analysis. Curr Opin Cell Biol 7(5):720–727
Paassilta P, Lohiniva J, Annunen S, Bonaventure J, Le Merrer M, Pai L et al (1999) COL9A3: a third locus for multiple epiphyseal dysplasia. Am J Hum Genet 64(4):1036–1044
Paassilta P, Lohiniva J, Goring HH, Perala M, Raina SS, Karppinen J et al (2001) Identification of a novel common genetic risk factor for lumbar disk disease. JAMA 285(14):1843–1849

Prockop DJ, Kivirikko KI (1995) Collagens: molecular biology, diseases, and potentials for therapy. Annu Rev Biochem 64:403–434

Raghunath M, Bruckner P, Steinmann B (1994) Delayed triple-helix formation of mutant collagen from patients with osteogenesis imperfecta. J Mol Biol 236(3):940–949

Rajpar MH, McDermott B, Kung L, Eardley R, Knowles L, Heeran M et al (2009) Targeted induction of endoplasmic reticulum stress induces cartilage pathology. PLoS Genet 5(10): e1000691

Reginato AM, Olsen BR (2002) The role of structural genes in the pathogenesis of osteoarthritic disorders. Arthritis Res 4(6):337–345

Ricard-Blum S, Ruggiero F (2005) The collagen superfamily: from the extracellular matrix to the cell membrane. Pathol Biol (Paris) 53(7):430–442

Richards AJ, Yates JR, Williams R, Payne SJ, Pope FM, Scott JD et al (1996) A family with Stickler syndrome type 2 has a mutation in the COL11A1 gene resulting in the substitution of glycine 97 by valine in alpha 1 (XI) collagen. Hum Mol Genet 5(9):1339–1343

Rimoin DL (1996) Molecular defects in the chondrodysplasias. Am J Med Genet 63(1):106–110

Ritvaniemi P, Korkko J, Bonaventure J, Vikkula M, Hyland J, Paassilta P et al (1995) Identification of COL2A1 gene mutations in patients with chondrodysplasias and familial osteoarthritis. Arthritis Rheum 38(7):999–1004

Schwartz NB, Domowicz M (2002) Chondrodysplasias due to proteoglycan defects. J Glycobiol 12 (4):57r–68r

Robin NH, Moran RT, Ala-Kokko L (2017) Stickler syndrome. In: Adam MP, Ardinger HH, Pagon RA, Wallace SE, Bean LJH, Stephens K et al (eds) Gene reviews((R)). Seattle

Shoulders MD, Raines RT (2009) Collagen structure and stability. Annu Rev Biochem 78:929–958

Sophia Fox AJ, Bedi A, Rodeo SA (2009) The basic science of articular cartilage: structure, composition, and function. Sports Health 1(6):461–468

Spranger J (1998) The type XI collagenopathies. Pediatr Radiol 28(10):745–750

Stickler GB, Hughes W, Houchin P (2001) Clinical features of hereditary progressive arthro-ophthalmopathy (Stickler syndrome): a survey. Genet Med 3(3):192–196

Vikkula M, Metsaranta M, Ala-Kokko L (1994) Type II collagen mutations in rare and common cartilage diseases Annals of medicine. Ann Med 28(2):107–114

Vikkula M, Mariman EC, Lui VC, Zhidkova NI, Tiller GE, Goldring MB et al (1995) Autosomal dominant and recessive osteochondrodysplasias associated with the COL11A2 locus. Cell 80 (3):431–437

Vogel H, Nilsson L, Rigler R, Voges KP, Jung G (1988) Structural fluctuations of a helical polypeptide traversing a lipid bilayer. Proc Natl Acad Sci USA 85(14):5067–5071

Wasylenko MJ, Wedge JH, Houston CS (1980) Metaphyseal chondrodysplasia, Schmid type. A defect of ultrastructural metabolism: case report. J Bone Joint Surg Am 62(4):660–663

Wilson R, Freddi S, Chan D, Cheah KS, Bateman JF (2005) Misfolding of collagen X chains harboring Schmid metaphyseal chondrodysplasia mutations results in aberrant disulfide bond formation, intracellular retention, and activation of the unfolded protein response. J Biol Chem 280(16):15544–15552

Chapter 5
Collagen IV-Related Diseases and Therapies

Afshan Dean and Tom Van Agtmael

Abstract While traditionally the function of the extracellular matrix and the basement membrane was considered to be providing structural support, it is now clear that this only covers one aspect of its multiple functions. This is also illustrated by our growing knowledge of the role of collagen IV, a major component of basement membranes, in development, health, and disease. With the extracellular matrix and collagen IV increasingly being recognised as key players in a growing number of diseases from stroke and vascular defects to kidney disease, deafness, and eye abnormalities, it is paramount that we increase our fundamental understanding of these complex molecules ranging from their biosynthesis to their role in human disease. Recently, exciting progress has been made in delineating the mechanisms by which mutations in collagen IV cause disease, and these are being exploited to develop mechanism-based treatments. Yet many important questions remain that need addressing to develop treatments for diseases associated with collagen IV.

Abbreviations

AS	Alport syndrome
ADAS	Autosomal dominant Alport syndrome
ARAS	Autosomal recessive Alport syndrome
Arg	Arginine
Asp	Aspartic acid
BM	Basement membrane
bp	Base pairs
CSVD	Cerebral small vessel disease
EC	Endothelial cell
ECM	Extracellular matrix
ER	Endoplasmic reticulum

A. Dean · T. Van Agtmael (✉)
Institute of Cardiovascular and Medical Sciences, University of Glasgow, Glasgow, UK
e-mail: tom.vanagtmael@glasgow.ac.uk

F. Ruggiero (ed.), *The Collagen Superfamily and Collagenopathies*, Biology of Extracellular Matrix 8, https://doi.org/10.1007/978-3-030-67592-9_5

FDA	Food and Drug Administration
GBM	Glomerular basement membrane
GP	Goodpasture syndrome
HANAC	Hereditary angiopathy, nephropathy, aneurysm and cramps
IAC	Integrin adhesion complexes
ICH	Intracerebral haemorrhage
kB	Kilo base pairs
kDa	Kilodalton
miR	miRNA
nm	Nanometre
P3H	Prolyl3-hydroxylase
P4H	Prolyl4-hydroxylase
PADMAL	Pontine autosomal dominant microangiopathy and leukoencephalopathy
PDI	Protein disulphide isomerase
TBMN	Thin basement membrane nephropathy
TGFβ1	Transforming growth factor beta 1
VR3	Variable region 3
XLAS	X-linked Alport syndrome
4PBA	Sodium 4-phenylbutyrate

5.1 Introduction

The vertebrate genome encodes 28 types of collagen, which is the most abundant protein of the human body, of which type IV collagen is evolutionarily the most ancient (Aouacheria et al. 2006; Fidler et al. 2017, 2018). Collagen IV is a major protein component found almost exclusively in the basement membrane (BM), constituting up to 50% of its protein mass, although recent evidence from *Drosophila* has proposed a non-BM mediated role in adipocyte adhesion (Dai et al. 2017).

The BM is a specialised sheet-like structure in the extracellular matrix (ECM), present in almost all tissues. BMs underlie epithelial and endothelial cell layers, surround adipocytes (Haraida et al. 1996; Noro et al. 2013), neurons (Nguyen et al. 2013), skeletal muscle fibres (Sanes 2003), cardiomyocytes (Walker and Spinale 1999), and smooth muscle cells (Hedin et al. 1999), and are present in the synaptic cleft of neuromuscular junctions (Fox et al. 2007; Ricard-Blum 2011). BMs have several important functions, including compartmentalising tissues, providing structural support, and modulating cell behaviour, signalling, and tissue repair (Vracko 1974; Wang et al. 2008; Yurchenco 2011). For a more in-depth review on BM function, we direct the reader to (Jayadev and Sherwood 2017; Pozzi et al. 2017). Major BM components are laminin, nidogen, the heparan sulphate proteoglycan perlecan, and collagen of which collagen IV is the major type (Van Agtmael and Bruckner-Tuderman 2010). In vertebrates, laminin is required for initial BM assembly and collagen IV recruitment (Smyth et al. 1999; Poschl et al. 2004) although recent elegant data from *Caenorhabditis elegans* showed this may not apply to all BMs (Jayadev et al. 2019). Collagen IV is necessary for BM maintenance and nidogens act as a bridge between laminin and collagen IV (Van Agtmael and Bruckner-Tuderman 2010; Poschl et al. 2004; Dai et al. 2018). Perlecan is the predominant proteoglycan and its deposition is dependent on collagen IV in at least some BMs (Pastor-Pareja and Xu 2011).

Each BM has its own composition and combined with the interactions of these BM components with cells and other secreted proteins (e.g., growth factors), allows BMs to carry out their different functions. Thus, each BM is uniquely tailored. The importance of BM and collagen IV for biology and physiology is illustrated by the diseases associated with alterations in the BM and collagen IV. In general, these are rare severe multisystemic disorders for which treatments are urgently needed. However, it is also becoming clear that the BM and collagen IV play an important role in common forms of disease in the general population (Yamada et al. 2008; Tarasov et al. 2009; O'Donnell et al. 2011; Schunkert et al. 2011; Rannikmae et al. 2015, 2017; Traylor et al. 2016; Chung et al. 2019).

Here we will provide an overview of collagen IV biology and focus primarily on Mendelian disorders due to collagen IV mutations. Collagen IV also has been analysed in a wide spectrum of pathologies including fibrosis, diabetes, cardiomyopathy to name but a few. While some of these will be mentioned, these are not the focus of this review.

5.2 Collagen IV

In 1966, Dr. Nicholas Kefalides described collagen IV in vertebrates while studying extracted glomerular basement membranes from dogs as a glycoprotein with abnormal levels of hydroxyproline and hydroxylysine (Kefalides 1966). Dr. Nicholas Kefalides and others subsequently identified the end of collagen IV protomers are not cleaved in contrast to fibrillary collagens (Kefalides 1973; Minor et al. 1976). Since then collagen IV has remained the topic of intense research ranging from its fundamental biochemistry, gene and protein structure function, as well as its expanding role in pathophysiology and disease.

The transition from single cell organisms to multicellular tissues necessitated a cellular microenvironment (Fidler et al. 2017; Hynes 2012), which was provided by the ECM. The spongin variant of collagen IV was a primordial component (Exposito et al. 1991) and enabled ECM assembly and genesis of multicellular tissues (Aouacheria et al. 2006; Fidler et al. 2017, 2018). Ctenophora represents one of the earliest branching extant animal phyla in which a BM and collagen IV have been identified, and their features are comparable between non-bilaterian and bilateria animal phyla (Fidler et al. 2017; Ryan et al. 2013; Moroz et al. 2014; Draper et al. 2019). In contrast, related unicellular groups do not encode collagen IV (Fidler et al. 2017). Remarkably, ctenophores express up to twenty collagen IV genes (Fidler et al. 2017), compared to six in vertebrates, illustrating conservation and adaptation dating back >0.5 billion years (Boute et al. 1996; Nauroy et al. 2018).

5.3 Gene Organisation and Regulation

In vertebrates three pairs of paralogous genes (*COL4A1, COL4A2, COL4A3, COL4A4, COL4A5*, and *COL4A6*), encode the collagen IV alpha chains α1(IV)–α 6(IV) (Fig. 5.1).

The genes are arranged in a head-to-head configuration separated by a communal promoter and transcribed in opposite directions (Poschl et al. 1988; Hostikka et al. 1990; Mariyama et al. 1994; Sugimoto et al. 1994; Zhou et al. 1994a; Momota et al. 1998). *COL4A1* is paired with *COL4A2, COL4A3* with *COL4A4*, and *COL4A5* with *COL4A6* (Fig. 5.1). The alpha chains interact to generate three collagen IV protomers, α1α1α2(IV), α3α4α5(IV), and α5α5α6(IV), that form the different collagen IV networks in the BM.

This genomic organisation is the result of three independent duplication events in evolution following the duplication and inversion of the ancestral *COL4A1* gene to give rise to *COL4A2* (Zhou et al. 1994a). The next duplication resulted in the *COL4A3/COL4A4* pair with the final duplication generating the more closely related *COL4A5/COL4A6* (Zhou et al. 1994a). Based on intron–exon structure and protein sequences the α-chains can be divided into the α1(IV) group containing α1(IV), α3 (IV), and α5(IV), and the α2(IV) group with α2(IV), α4(IV) α6(IV) (Netzer et al. 1998).

5.3.1 COL4A1 *and* COL4A2

The *COL4A1* and *COL4A2* genes consist of 52 and 48 exons, respectively, and are located in a ~370 kB locus on human chromosomal region 13q34 and mouse chromosome 8. Throughout the 1980s and 1990s experiments using plasmid expression systems provided the first insights into the regulation of *COL4A1* and *COL4A2* transcription. *COL4A1* and *COL4A2* are separated by a mere 127 nucleotides, in which there is a bidirectional promoter (Fig. 5.1) (Poschl et al. 1988; Soininen et al. 1988). The promoter has a palindromic sequence with an A/T intense region 30 base pairs (bp) upstream from the transcription start site and accordingly lacks the usual TATA box (Poschl et al. 1988; Soininen et al. 1988; Heikkila et al. 1993). The promoter contains three key elements, found also in other BM proteins: a CT box, a CCAAT box, and GC box, that bind the transcription factors CTCBF, CCAAT binding protein, and Sp1, respectively (Poschl et al. 1988; Soininen et al. 1988; Fischer et al. 1993; Schmidt et al. 1993) (Fig. 5.1). The bidirectional promoter can be considered as a small region of two overlapping gene specific promoters (Schmidt et al. 1993) that are not transcriptionally active (Poschl et al. 1988). Transcription

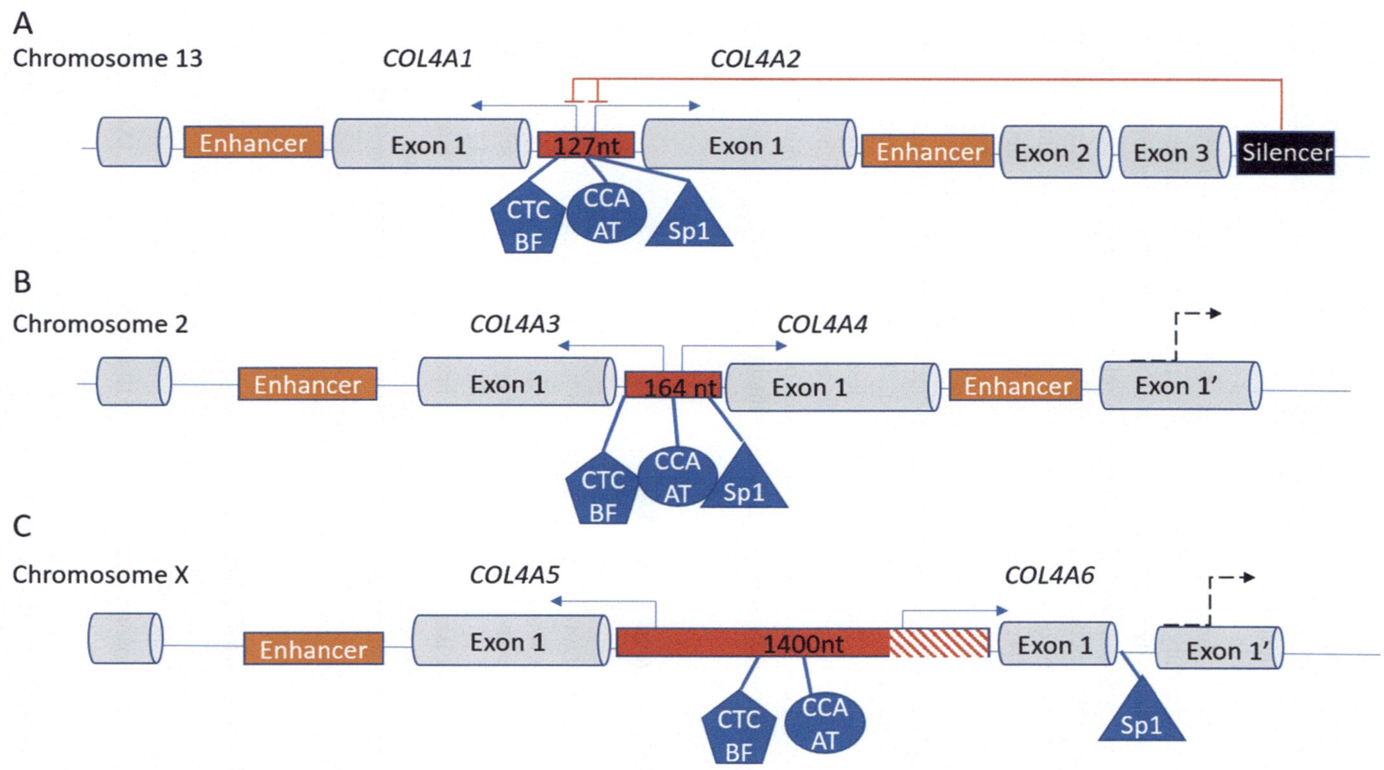

Fig. 5.1 Genome organisation for *COL4A1/COL4A2*, *COL4A3/COL4A4* and *COL4A5/COL4A6*. Each pair of collagen IV genes contain a bi-directional promoter (length in nucleotides provided) and *cis*-acting elements. The *cis*-acting elements (GC box, the CAAT motif and a CTC box) bind specifically to transcription factor SP1, CAAT binding factor (CCAAT) and CTCBF. (**a**) *COL4A1/COL4A2* on chromosome 13. A silencer influences transcription of both *COL4A1* and *COL4A2*. (**b**) *COL4A3/COL4A4* on chromosome 2 contains motifs similar to *COL4A1/COL4A2* present in the bidirectional 167 nucleotide promoter. *COL4A4* has two alternative first exons. (**c**) *COL4A5/COL4A6* have a larger bidirectional promoter. *COL4A6* has potential alternative promoter elements depicted by the red diagonal lines and an alternative first exon

depends on gene specific enhancers in introns 1 of *COL4A1* and *COL4A2* that promote binding of CTCBF, CCAAT binding protein and Sp1 (Poschl et al. 1988; Fischer et al. 1993; Schmidt et al. 1993) in the short promoter (Schmidt et al. 1993). These *cis*-acting enhancers act cooperatively and competitively (Pollner et al. 1997), perhaps in part, by causing steric hindrance, enabling gene-specific transcription. There is also a silencer in intron 3 of *COL4A2* that affects transcription of both genes (Pollner et al. 1997; Haniel et al. 1995) (Fig. 5.1). These *cis*- and *trans*-acting elements enable a tightly controlled regulation of transcription with different levels of transcriptional activity of *COL4A1* versus *COL4A2*. This correlates with a ~2:1 ratio of *COL4A1* versus *COL4A2* mRNA levels reflecting α1α1α2(IV) composition (Pollner et al. 1997; Schmidt et al. 1992). Collagen IV is also transcriptionally regulated by members of the Smad family (including Smad1 and Smad2) and DDR1 (Abe et al. 2004; Jiang et al. 2010; Chiusa et al. 2019), but their binding sites remain poorly defined.

COL4A1/COL4A2 expression is also regulated post-transcriptionally including by binding of micro-RNAs (miR) to the 3′UTR of *COL4A1* including miR-29 (Takahashi et al. 2012; Verdura et al. 2016). The miR-29 family consists of miR-29a, b, and c. In vitro *COL4A1* mRNA levels were regulated by miR-29c (Griffiths et al. 2019), which is also active in endothelial cells (Licholai et al. 2016). However, our knowledge regarding the post-transcriptional regulation of collagen IV expression by noncoding RNAs remains poor.

5.3.2 COL4A3 *and* COL4A4

The *COL4A3* and *COL4A4* genes are located on human chromosomal region 2q36, and in mice on chromosome 1 (Turner et al. 1992; Kamagata et al. 1992). They are located in a ~314 kB locus, are transcribed in the opposite direction and separated by a 164 bp bidirectional promoter (Fig. 5.1). Similar to *COL4A1/COL4A2,* the promoter contains a CpG island, GC boxes, CTC boxes, and a CCAAT box, and no TATA box (Fig. 5.1). *COL4A3* and *COL4A4* compromise of 52 and 48 exons respectively but two alternative first exons have been described for *COL4A4*, which can give rise to two *COL4A4* isoforms (Momota et al. 1998).

5.3.3 COL4A5 *and* COL4A6

The *COL4A5* and *COL4A6* genes consist of 51 and 46 exons, respectively, and are found on the X chromosome within a single ~510 kB locus in humans (Xq22) and mice (Zhou et al. 1994a; Oohashi et al. 1994; Zhang et al. 1996). Their promoter is more complex and larger (~1400 bp) than for *COL4A1/COL4A2* and *COL4A3/COL4A4*, as *COL4A6* has two promoters (Fig. 5.1). This enables a tissue-specific *COL4A6* expression including divergent expression from *COL4A5* (Ninomiya et al.

1995; Sund et al. 2005). The two promoters lead to usage of two transcription start sites generating two alternative transcripts with different first *COL4A6* exons, and two α6(IV) isoforms with different signal peptides (Sugimoto et al. 1994; Sund et al. 2005; Segal et al. 2001). As for the other gene pairs, the promoter has characteristics of those for housekeeping genes including the absence of the TATA box and presence of CCAAT, CTC, and GC boxes (Fig. 5.1).

It should, however, be noted, that in part due to the complex organisation of the different loci and the large size of the introns, our knowledge of the regulation of gene expression at the transcriptional level remains incomplete.

5.4 Collagen IV Protein Domain Structure and Biosynthesis

Seminal rotary shadowing electron microscopy experiments revealed that collagen IV proteins have a rod-like structure with a length of ~400 nm in which a central triple helical collagen domain of around ~330 nm is book-ended by two globular domains that, in contrast to fibrillar collagens, remain attached in the mature protein (Timpl et al. 1981) (Fig. 5.2). Collagen IV α-chains have similar domain structures and 50–70% homology at the amino acid level. Each α-chain, a rope-like structure, consists of a short N-terminal ~120 amino acid 7S domain, a central collagenous triple helical domain (~1400 residues), and a C-terminal globular non-collagenous (NC) 1 domain (~230 residues) (Fig. 5.2) (Timpl et al. 1981; Risteli et al. 1980; Raija et al. 1987; Vandenberg et al. 1991; Hudson et al. 2003; Parkin et al. 2011). The collagen domain, the defining feature of all collagens, is made up of a 3 amino acid repeat (Gly-Xaa-Yaa) in which every third residue is a glycine, and Xaa and Yaa can be any residue. Each chain also has functional subdomains and potential binding sites for interacting proteins (Parkin et al. 2011). The NC1 domains contain a cryptic peptide that can be released upon proteolytic cleavage. Following the initial identification of the antiangiogenic fragment tumstatin from α3(IV) (Maeshima et al. 2000), it became apparent that similar anti-angiogenic fragments (24–28 kDa) can be released from the other α-chains (Mundel and Kalluri 2007): *arresten from* α1 (IV), canstatin from α2(IV), tetrastatin from α4(IV), pentastatin from α5(IV), and hexastatin from α6(IV) (Mundel and Kalluri 2007; Petitclerc et al. 2000; Colorado et al. 2000; He et al. 2003; Boosani and Sudhakar 2006; Koskimaki et al. 2010; Brassart-Pasco et al. 2012).

In contrast to archetypical fibrillar collagens, the collagen domain contains 21–26 interruptions of the Gly-Xaa-Yaa triplet repeat (Mariyama et al. 1994; Zhou et al. 1994a; Parkin et al. 2011 #202; Brazel et al. 1987, 1988; Hostikka and Tryggvason 1988; Zhou et al. 1992, 1994b). It has been proposed that all collagens have evolved from a single 54 bp ancestral exon and that alternative splicing leading to incorporation of intronic sequences has generated these interruptions to the Gly-Xaa-Yaa repeat (Yamada et al. 1980; Buttice et al. 1990). The interruptions vary in length from 1 to 24 nucleotides and while their sequence is relatively poorly conserved between α-chains, their positions are highly conserved (Leinonen et al. 1994). These interruptions provide flexibility to collagen IV to form a network, while also enabling proteolytic cleavage of the triple helix or serve as cell-binding sites (Vandenberg et al. 1991). The interruptions have also been proposed to strengthen lateral association between chains by enabling intra-chain cross-linking (Yurchenco and Furthmayr 1984; Yurchenco and Ruben 1987, 1988).

As with all secreted proteins, during translation the nascent polypeptide is translocated into the endoplasmic reticulum (ER) *via* transient signal peptides. In the ER, three collagen IV α-chains interact to form a triple helical molecule called a protomer. Six different chains can theoretically form 56 different combinations of triple helices, however, only three heterotrimers exist: α1α1α2(IV), α3α4α5(IV), and α5α5α6(IV). The formation of protomers begins at the C-terminal NC1 domain and in a zip-like fashion the α-chains wrap around each other to form a triple helix through to the N-terminal 7S domain. Following protomer folding and secretion in ECM, protomers self-assemble into a three-dimensional network (Fig. 5.2).

5.4.1 NC1 Domains and Initiation of Protomer Folding

The collagen IV NC1 domain is a globular ~13 nm domain (Timpl et al. 1981), ~229 amino acid residues for α1(IV) and α2(IV), that is highly conserved in evolution (~95% identity) and between α-chains (Mariyama et al. 1994; Oberbaumer et al. 1985; Pihlajaniemi et al. 1985; Schwarz-Magdolen et al. 1986). The first stage of protomer formation is the interaction between NC1 domains of three α-chains (Soder and Poschl 2004). The 12 cysteine residues within the NC1 domains from intra- and inter-chain disulphide bonds made by protein disulphide isomerase (PDI). The intrachain bonds form a knot to give a compact NC1 domain structure similar to a four leaf clover (Siebold et al. 1988). The interchain bisulphide bonds are important for alignment of the three α-chains before triple helix formation. Recognition motifs within the NC1 domain of an α-chain establish protomer composition by selecting which α-chain it will bind to. This includes domain swapping interactions whereby each domain extends a β-hairpin motif into a corresponding docking site located in the variable region 3 (VR3) of the NC1 domain of the adjacent α-chain (Khoshnoodi et al. 2006a). Non-covalent interactions occur between the VR3 and the β-hairpin that determine chain specificity. For example, the α2(IV) chain β-hairpin loop binds to the α1(IV) VR3 to initiate α1α1α2(IV) formation (Sundaramoorthy et al. 2002; Than et al. 2002) whereby the reduced affinity of α2(IV) for α2(IV), as compared to α1(IV), drives the 2:1 α-chain composition (Khoshnoodi et al. 2006a). For α3α4α5 (IV) and α5α5α6(IV) this is driven by α4(IV) and α6(IV), respectively (Khoshnoodi et al. 2006a, b).

The NC1 domain also prevents inappropriate intracellular cross-linking/network formation and initiates extracellular network assembly. This is related to the difference in chloride (Cl^-) concentration between inside the cell (~12 mM) and the ECM (~100 mM) (Ziomber et al. 2008). The low intracellular Cl^- concentration leads to the formation of an intrachain salt bridge between residue Arg76 and Asp78 (and to a

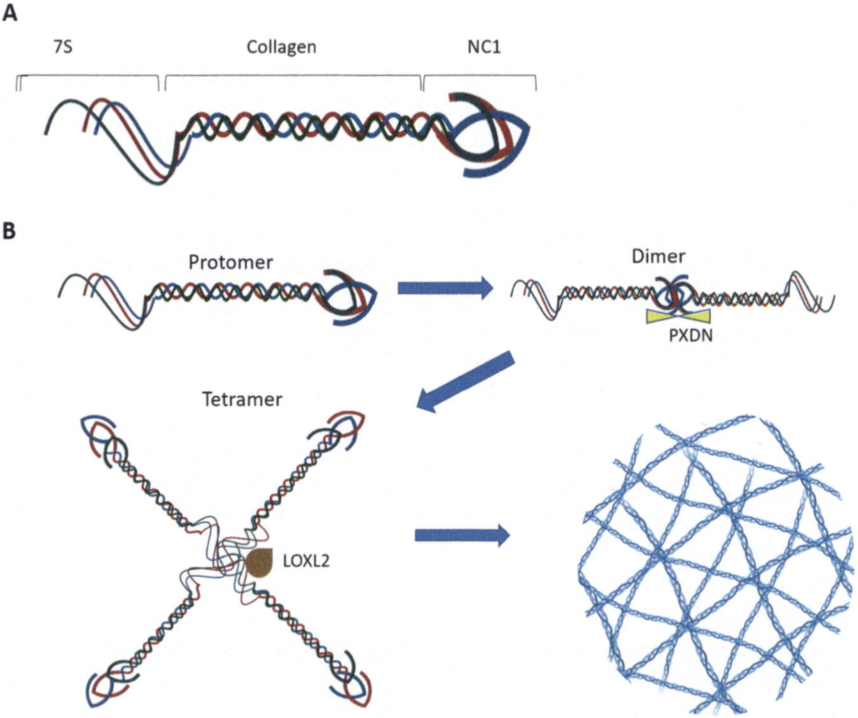

Fig. 5.2 (**a**) Protein domain structure of collagen IV protomer includes a N-terminal 7S domain, a central triple helical collagen domain and a C-terminal NC1 domain. (**b**) Collagen network formation in the ECM includes interaction of two protomers *via* their NC1 domains to form dimers. Peroxidasin mediates the sulfilimine bond formation between the NC1 trimers. Four protomers interacts via their 7S domains to form tetramers and cross linking is mediated by LOXL2. Combined this results in a strong but flexible mesh network in the ECM (**d**)

lesser extent Glu40) that blocks protomer oligomerisation (Cummings et al. 2016). Once exposed to the higher extracellular Cl^- concentration the salt bridge destabilises to enable initiation of network formation in the ECM (Cummings et al. 2016).

5.4.2 Post-translational Modification of Collagen IV α-chains

The accurate intracellular assembly and folding of collagens require a large set of proteins and much of what we know has been obtained through analysis of fibrillar collagens (see also Chap. 2 by Sergey Leikin and co-authors). Over the years, excellent reviews (Myllyharju and Kivirikko 2004; Makareeva et al. 2011; Forlino and Marini 2016) have covered this topic in depth and we will provide a brief collagen IV orientated overview. It should be noted that much of the folding and secretion of collagen IV remains unclear, and this represents a knowledge gap that needs to be addressed to increase our fundamental understanding of biology and diseases in which collagen IV is implicated.

Multiple enzymes, co-factors, and chaperones are implicated in the successful assembly and secretion of collagen IV. Following the interaction between NC1 domains, triple helix formation occurs that is dependent on the Gly-Xaa-Yaa repeat. This protein folding requires post-translational modification of the single unfolded α-chains (Fig. 5.3). Within the Gly-Xaa-Yaa repeat, Xaa and Yaa can be any residue but most often are proline (~28%) and hydroxy-proline (38%), respectively (Shoulders and Raines 2009). Proline can be hydroxylated at the fourth or third carbon of the proline ring by the enzymes prolyl4-hydroxylase (P4H) and prolyl3-hydroxylase (P3H) respectively (Fig. 5.3) (Shoulders and Raines 2009; Kivirikko and Pihlajaniemi 1998; Myllyharju 2008; Tiainen et al. 2008). This hydroxylation provides thermal stability to the triple helix by promoting electrostatic interactions (Shoulders and Raines 2009).

Vertebrates have three isoforms of P4H and P3H (P3H1-P3H3). P4H is composed of two subunits (α and β) whereby PDI makes up the β subunit (Myllyharju 2008) (Fig. 5.3). The importance of proline hydroxylation has become clear through phenotypes of model organisms deficient for the hydroxylases *C. elegans* deficient for *P4H* are embryonic lethal with defects similar to worms with mutations in *Emb-9* (*COL4A1* ortholog) (Friedman et al. 2000), while in mice heterozygous *P4ha1* deficiency caused embryonic lethality and defective collagen IV assembly (Holster et al. 2007). Interestingly, mutations in the catalytic domain of P4H1 cause a connective tissue/myopathy disorder affecting tendon, bone, muscle, and eye with reduced collagen IV staining in muscle BMs (Zou et al. 2017).

P3H acts on proline in the Xaa position of the collagen repeat (Gly-Xaa-HyP) after prolyl-4-hydroxylation has occurred. Isoform 1 of P3H (P3H1) together with cyclophilin B and cartilage-associate protein (CRTAP) forms a protein complex with chaperone activity (Ishikawa et al. 2009), and mutations in these proteins cause recessive osteogenesis imperfecta (Forlino and Marini 2016). Prolyl-3-hydroxylation causes a slight decrease in thermal stability (Ishikawa et al. 2009) and is involved in mediating interactions with other ECM components including nidogen in the BM (Pokidysheva et al. 2014; Mizuno et al. 2004; Montgomery et al. 2018). The expression pattern of P3H2 overlaps to a large extent with that of collagen IV (Tiainen et al. 2008) and P3H2 deficient mice develop a myopia-like phenotype with altered collagen IV hydroxylation in the lens capsule (Hudson et al. 2015). The absence of overt phenotypes can be explained by a functional redundancy (Montgomery et al. 2018), possibly with P3H3, which has a similar expression pattern (Vranka and Stadler 2009). For a detailed review on proline hydroxylases, we refer the reader to (Rappu et al. 2019).

Lysine residues that are at the Yaa position of the collagen triplet can also undergo hydroxylation by lysyl hydroxylases to provide binding sites for intermolecular collagen crosslinks, influencing the biomechanical properties of the ECM, and subsequent carbohydrate attachment (Fig. 5.3) (Myllyharju and Kivirikko

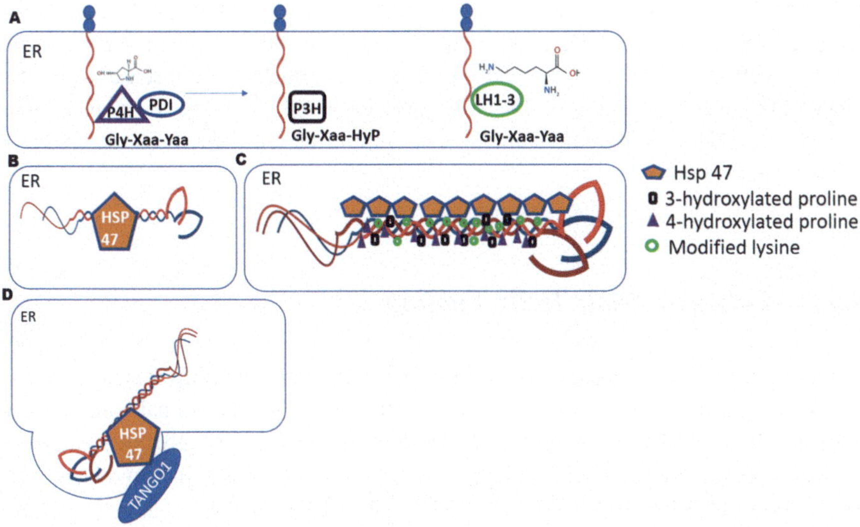

Fig. 5.3 Collagen IV protomer modification. (**a**) Co-translational and post-translational modification in the ER of single α-chains includes proline hydroxylation at the fourth or third carbon of the proline ring by the enzymes prolyl4-hydroxylase (P4H) and prolyl3-hydroxylase (P3H) respectively. Lysine residues at the Yaa position of the collagen triplet can also undergo hydroxylation by lysyl hydroxylases (LH1-3). (**b**) HSP47 binds and stabilises collagen IV protomers, in the ER and during transit to the *cis*-Golgi. (**c**) Schematic diagram of modified Collagen IV protomer. (**d**) HSP47 can act as anchor between TANGO1 to aid secretion of collagen IV from ER

2004). Mammals have three isoforms of lysyl hydroxylases, LH1-LH3, which are encoded by the genes *PLOD1-3* (procollogen-lysine 1, 2-oxoglutarate 5-dioxygenase). The importance of lysine hydroxylation is underscored by the embryonic lethality around 9.5-day postcoitum, and aberrant BM and collagen IV secretion in mice deficient for Lh3 (Rautavuoma et al. 2004). Defects in collagen processing and secretion also occur in lysyl hydroxylase-deficient worms and cause contraction-induced muscle defects similar to that observed with collagen IV mutations (Norman and Moerman 2000). In contrast to LH1 and LH2, LH3 is a multifunctional enzyme that also possesses galactosyl- (GT) and glycosyltransferase (GGT) activities responsible for modification of hydroxylysine to galactosylhydroxylysyl or glucosylgalactosylhydroxylysyl, which is fundamental for collagen IV secretion and BM structure. Very elegant mouse studies determined that the lethality of Lh3 deficient embryos was due to lack of GT and GGT activities, and not LH activity, highlighting the critical importance of the glycosyltransferase activities (Ruotsalainen et al. 2006).

The clinical importance of galactosylation of hydroxylysine has also become apparent by the identification of mutations in *COLGALT1*, which encodes collagen β (1-O) galactosyltransferase 1, in patients with cerebral small vessel disease (Miyatake et al. 2018). Intriguingly, the clinical characteristics correspond with

COL4A1 syndrome and functional analysis in cell lines supported that the mutations affect α1(IV) processing (Miyatake et al. 2018). While this requires validation in vascular cell types and in vivo, and we need to establish if the disease is due to processing defects of collagen IV alone (as compared to collagen IV and other collagens), it suggests mutations in collagen modifying enzymes may contribute to *COL4A1* syndrome.

5.4.3 Collagen Triple Helix Folding

Collagen triple helix formation commences following binding between NC1 domains. The α-chains wrap around each other in C to N-terminal direction to generate the central triple helical collagen domain whereby the glycine residue is located in the centre of the helix. Areas that have formed a triple helix will cease to be subject to post-translational hydroxylation and glycosylation.

For triple helix formation to occur all proline residues require to be in *trans* conformation but proline occurs in nascent collagen polypeptide in a *cis* or *trans* conformation. Peptidylprolyl isomerases (PPIases) mediate this conformational change and mutations in the PPIases Cyclophilin B, FK506 binding protein, and parvulins (Perrucci et al. 2015) cause the classical collagen disorder osteogenesis imperfecta (Forlino and Marini 2016), illustrating the importance of PPI for collagen folding.

Several chaperones are also involved in collagen folding. This includes PDI, which has chaperone activity in addition to its other roles, and Heat Shock Protein 47 (HSP47). HSP47 is a collagen-specific chaperone that binds and stabilises collagen IV protomers, amongst other collagens, in the ER and during their transit to the *cis*-Golgi (Ito and Nagata 2019) (see also Chap. 2 by Sergey Leikin and co-authors). The decrease in pH in the *cis*-Golgi causes HSP47 to dissociate from the collagen and for it to be recycled back to the ER. Mutations in the gene encoding HSP47, *SERPINH1*, cause osteogenesis imperfecta (Forlino and Marini 2016) while in mice and cells deficiency of HSP47 caused accumulation of collagen IV in the ER due to a lower secretion rate and absence in the BM (Nagai et al. 2000; Matsuoka et al. 2004). The quality of the secreted collagen was also affected as it has reduced thermal stability and increased sensitivity to protease digestion, supporting a critical stabilising role for HSP47 (Fujii et al. 2019). The mouse phenotype (including lethality at 11.5-day postcoitum) was very similar to mice deficient for α1α1α2 (IV) (Poschl et al. 2004). These data elegantly underscore the importance of the HSP47 for collagen IV processing and secretion, and the “quality” of secreted collagen (Fig. 5.3).

5.4.4 *Transport from the Intracellular to the Extracellular Space*

The regulation of the secretion of large cargos such as collagens has received increased attention over the past decade with very elegant and paradigm-shifting papers (Saito et al. 2009; Omari et al. 2018; McCaughey et al. 2019; Chang et al. 2020). This is a very active and fast-moving area of collagen and cell biology research. Much of this research has centred on collagen I and collagen VII although effects on collagen IV secretion have been documented (Malhotra and Erlmann 2015). For a more detailed overview, we refer the reader to Chap. 2 by Sergey Leikin and co-authors in this book and (McCaughey and Stephens 2019).

Collagen IV translocates from the rough ER *via* the ERGIC (ER Golgi Intermediate Compartment) to the Golgi for secretion to the ECM. Smaller proteins are secreted via coat protein complex II (COPII) secretory vesicles that are generated at ER exit sites (ERES) and have a diametre of 60–90 nm (McCaughey and Stephens 2019). However, collagens are extraordinarily large semi-rigid proteins (e.g. collagen IV ~400 nm) too big for normal COPII vesicles. Thus, their transport necessitates specialised transport vesicles to accommodate these large cargos. An important role for Transport and Golgi organisation 1 protein (TANGO1, also called MIA3) has been identified (Saito et al. 2009), as its deletion in mice leads to intracellular accumulation of collagen VII and other collagens including type IV. TANGO1 can recruit collagens via its luminal SH3 domain, and in parallel its cytoplasmic domain recruits other proteins to form extended COPII vesicles (Saito et al. 2009; McCaughey and Stephens 2019; Ishikawa et al. 2016). Recent evidence revealed a role for HSP47 in this process by acting as an anchor molecule between the SH3 domain of TANGO1 and collagens, allowing packaging of collagen molecules in the growing carrier (Fig. 5.3). The cytoplasmic domains of TANGO1 interact with Sec23A/Sec24C, the outer molecules of COPII vesicles. This brings collagen into contact with the COPII vesicle to allow them to grow sufficiently to accommodate the cargo and subsequently transport the cargo from the ERES to the ERGIC vesicles (McCaughey and Stephens 2019). Recent data, however, identified an alternative but not mutually exclusive mechanism that does not involve large carriers, but whereby collagen accumulates at the ERES in a TANGO1-COPII dependent manner and then flows more directly between ER and GOLGI (McCaughey and Stephens 2019). This is made possible through the proposed fusion of membranes of the ERES and ERGIC adopting a tunnel-like structure (McCaughey et al. 2019; McCaughey and Stephens 2019; Raote and Malhotra 2019).

5.4.5 Extracellular Network Formation

Following secretion the NC1 domains of two protomers associate end-to-end forming an NC1 hexamer, and the 7S domains of four protomers form a 7S tetramer, giving rise to a lattice style collagen IV network (Fig. 5.2).

The increased extracellular chloride concentration in the ECM "breaks" the Arg76-Asp78 intra-chain salt bridge in the NC1 domains, and Arg76 is re-orientated towards an opposing NC1 trimer. Six salt bridges are formed across two NC1 trimers between Arg76 residue, and Glu175 and Asn187 from the other protomers (Donald et al. 2011; Brown et al. 2017). Additionally, each bound chloride ion directly engages six electrostatic interactions per hexamer (Cummings et al. 2016). Combined, this allows for two protomers to bind, and the structural integrity of the NC1 hexamers is further strengthened by a sulfilimine covalent bond between methionine 93 and hydroxylysine 211 of opposing NC1 trimers. This sulfilimine bond is catalysed by peroxidasin (Bhave et al. 2012) and is fundamental to the structural integrity and tensile force of the collagen IV scaffold as illustrated by distorted BM, tissue morphology, and lethality in *Drosophila* due to loss sulfilimine bonds (Brown et al. 2017; McCall et al. 2014).

Following collagen IV dimer formation, four protomers bind through their N-terminal 7S domains at an angle to generate the lattice configuration of the network (Fig. 5.2). The 7S domains, which consists of a lysine and cysteine-rich region and a triple helical region, interact in a parallel and antiparallel fashion (Siebold et al. 1988). Disulphide bonds between the 5 cysteine residues as well as lysine–hydroxylysine crosslinking ensures resistance against proteolytic cleavage by enzymes such as collagenase, and provides mechanical stability. The lysine–hydroxylysine crosslinking is performed by lysyl oxidase-like-2 (LOXL2) (Brown et al. 2017; Anazco et al. 2016), which has also been implicated in angiogenesis (Bignon et al. 2011).

The site of collagen IV production can be distant from the site of incorporation into the BM and/or cells that do not reside on the BM also contribute to the BM by secretion of its components (Matsubayashi et al. 2017). Consequently, a mechanism that can delay extracellular assembly and maintain collagen IV folding might be employed. Secreted Protein Acidic and Rich in Cysteine (SPARC) is a matricellular glycoprotein that binds collagens (Hohenester et al. 1997; Bradshaw 2009) and can function as an extracellular chaperone to ensure proper spatial folding of collagen IV before incorporation into the BM (Pastor-Pareja and Xu 2011; Isabella and Horne-Badovinac 2015; Chioran et al. 2017). This is supported by data from *Drosophila* and *C. elegans* whereby *SPARC* knockdown and deficiency, respectively, led to local inappropriate accumulation of collagen IV rather than diffusion to distal sites (Pastor-Pareja and Xu 2011; Shahab et al. 2015; Morrissey et al. 2016). These studies suggest SPARC is important for allowing collagen IV to diffuse from sites of production and assemble on the surfaces of distal tissue. Recent studies have however suggested that this mechanism may only apply to maintenance of existing BM rather than for de novo BM assembly (Matsubayashi et al. 2017).

5.5 Collagen IV Receptors

Besides providing structural support, collagen IV also plays a role in cell biological processes including cell differentiation, survival, and migration, mediated by the interactions of collagen IV with cells. Collagen IV binds multiple cell types such as endothelial cells, hepatocytes, platelets, keratinocytes, pancreatic cells, mesangial, as well as tumour cells (Murray et al. 1979; Rubin et al. 1981; Santoro 1986; Herbst et al. 1988; Dedhar et al. 1993; Setty et al. 1998; Kaido et al. 2004). The collagen receptors can be divided into integrin and non-integrin receptors.

5.5.1 Integrin Receptors

Integrin receptors provide critical mechanosensing functions and convert the spatio-temporal information from the ECM into cellular signalling through the formation of integrin adhesion complexes (IAC) (Humphries et al. 2019). These transmembrane glycoproteins exist as noncovalent heterodimers made up of α and β subunits (Campbell and Humphries 2011). There are 18 α and 8 β subunits that combine to form 24 receptors with different binding and expression patterns. The integrin protein family can be divided into RGD-binding integrins and collagen-binding integrins (Zeltz and Gullberg 2016). The β1 subgroup of integrins are the main collagen receptors, particularly α1β1 and α2β1, and although these can both bind collagen IV, α1β1 does so with a greater affinity (Kern et al. 1993). Decreased expression of either of these receptors led to a decrease in adhesion and migration (α1β1), and adhesion and morphogenesis (α2β1) (Keely et al. 1995; Gardner et al. 1996). β1 integrin has also been implicated in recruiting collagen IV to some BMs in *C. elegans* (Jayadev et al. 2019). Other collagen IV binding integrins include α3β1, α10β1, and α11β1, but their functional significance and role remain unclear (Elices et al. 1991; Tulla et al. 2001; Tiger et al. 2001). Altered integrin signalling has been put forward as a potential mechanism in Alport syndrome (Cosgrove and Liu 2017), and for some mutations in HANAC syndrome that are located close to CB3 domain of α1(IV), a major integrin binding site (Vandenberg et al. 1991; Knight et al. 2000; Plaisier et al. 2007, 2010; Guiraud et al. 2017).

5.5.2 Non-integrin Receptors

The discoidin domain receptors (DDR1 and DDR2) are type 1 transmembrane tyrosine kinase proteins that are activated by the triple helix found in some collagens (Leitinger and Hohenester 2007). DDR1 can be activated by most collagens, such as collagen I–IV and VIII, whilst DDR2 by collagen I, II, III, and V but not IV (Itoh 2015). Several lines of evidence have highlighted a role for DDR1 in collagen IV biology. DDR1 knockout mice develop renal defects including thickening of the glomerular BM (GBM) and proteinuria, which also occur in Alport syndrome due to collagen IV mutations (Funk et al. 2018a). Interestingly, recent evidence indicates a novel unexpected role for DDR1 in mediating collagen IV transcription. In fibrosis, following collagen binding, DDR1 is translocated to the nucleus, where it binds chromatin and drives collagen IV expression (Chiusa et al. 2019). This not only reveals a novel role for DDR1 but also indicates an additional layer of regulation and complexity relevant to collagen disease that requires further investigation.

Besides integrins and DDRs, collagen IV can also bind and activate a G-protein-coupled receptor, GPR126, which is essential for Schwann cell myelination, ear canal formation, and heart development (Paavola et al. 2014). Finally, collagen IV can bind via its 3-hydroxylated proline residue to glycoprotein VI (Pokidysheva et al. 2014; Montgomery et al. 2018), an important platelet collagen receptor, and absence of this binding has been associated with increased platelet aggregation in mice (Pokidysheva et al. 2014).

5.6 Collagen IV Expression Patterns

During vertebrate development α1α1α2(IV) is ubiquitously expressed. In mice absence of α1α1α2(IV) due to deletion of exons 1 of *Col4a1* and *Col4a2*, cause embryonic lethality around 10.5–11.5 days postcoitum due to placental defects, and embryos displayed growth retardation and haemorrhaging (Poschl et al. 2004). The critical role of α1α1α2(IV) is also nicely illustrated by *Col4a2* mutant embryos homozygous for a deletion in exon 18. Embryos survived organogenesis but died by 14.5 days postcoitum and defects occurred in almost every organ (Reissig et al. 2019). This included cardiovascular, cardiac, skeletal, lung and endocrine organ defects. The later embryonic lethality of *Col4a1/Col4a2* knockout embryos can be explained by residual *Col4a2* expression (Poschl et al. 2004; Reissig et al. 2019), presumably through alternative splicing.

During embryonic development, in some BMs α1α1α2(IV) is replaced by α3α4α5(IV) or α5α5α6(IV). For example, replacement by α3α4α5(IV) occurs in the GBM (Kalluri et al. 1997). This developmental switch can also give rise to a mixed network such as in Bowman's capsule of the kidney and the lens capsule (Kelley et al. 2002). Consequently, in adults α1α1α2(IV) is present in BMs of all tissues, whereas α3α4α5(IV) occurs in the glomerular and tubular BM of the kidney, the alveolar BM of the lung, as well as BM in testis and inner ear cochlea (Mariyama et al. 1994). The α5α5α6(IV) protomer is expressed in the BM of the bronchial epithelium, the oesophagus, smooth muscle cells, skin, and Bowman's capsule of the kidney and the synovia (Ninomiya et al. 1995; Poduval et al. 2007). The tissues that are affected in patients and mice with mutations in collagen IV reflect the expression patterns of the different networks. This means that invariably mutations cause complex multi-systemic syndromes.

5.7 Collagen IV Pathologies

Mutations affecting every α-chain have been described and the first mutation affecting collagen IV was described in patients with Alport Syndrome (AS). Tables 5.1 and 5.2 provide an overview of the some of the animal models for collagen IV diseases and some of the clinical features of the Mendelian diseases, respectively.

5.7.1 *COL4A1 Syndrome and Mutations in* COL4A1 *and* COL4A2

The identification of the first mutations in *COL4A1/COL4A2* in patients was based on analysis of *Col4a1/Col4a2* mutant mouse models that were generated in large ENU mutagenesis screens (Gould et al. 2005; Van Agtmael et al. 2005; Favor et al. 2007) (Table 5.1). This represents a case whereby the mouse models led to the description of the human disease, which is underscoring the power of these models to determine disease mechanisms.

COL4A1/COL4A2 mutations cause COL4A1 syndrome, a rare dominant multi-systemic disorder that causes a wide range of abnormalities affecting the brain, eye, kidney, vasculature, and muscle (Table 5.2). However, COL4A1 syndrome remains poorly characterised clinically. The vast majority of mutations are missense mutations affecting the glycine residues, although missense mutations have been described in each of the protein domains and affecting the glycine, Xaa, and Yaa residues of the triplet (Meuwissen et al. 2015). From mice studies, a general genotype–phenotype correlation has been established whereby mutations affecting the Xaa or Yaa residues cause milder disease, and that more C-terminal glycine mutations are associated with more severe intracerebral haemorrhaging (Van Agtmael et al. 2005; Jeanne et al. 2015). From patients and mice it has become apparent that *COL4A2* mutations tend to cause a milder disease (Favor et al. 2007; Murray et al. 2014; Jeanne and Gould 2017). This can be explained by α1α1α2 (IV) composition whereby a *COL4A2* mutation can affect 50% of the promoter compared to 75% with a *COL4A1* mutation. Nonsense mutations have also been described (Verbeek et al. 2012; Lemmens et al. 2013) as well as mutations in the 3′UTR (Verdura et al. 2016) and copy number mutations (Saskin et al. 2018; Renard et al. 2014). It has also become apparent that the de novo mutation rate in *COL4A1/COL4A2* is high (~40%) (Meuwissen et al. 2015), which can give rise to apparent sporadic cases. Intriguingly very recently the first case has been reported of a recessive form of COL4A1 disease (Yaramis et al. 2020).

We will provide a brief overview of some of the major clinical features of COL4A1 syndrome (Table 5.2) but for a more in-depth overview of the clinical features, we refer the reader to (Meuwissen et al. 2015; Zagaglia et al. 2018).

5.7.1.1 Cerebrovascular and Cerebral Defects in COL4A1 Syndrome

Cerebrovascular disease is the predominant feature of COL4A1 syndrome (OMIM #614519,#175780, #614483) but patients can display a large variability in clinical presentation and severity due to the large amount of variable expressivity and penetrance of the disease (Table 5.2) (Meuwissen et al. 2015; Murray et al. 2014). The different clinical characteristics of the same mutation can occur within and between families (Meuwissen et al. 2015; Murray et al. 2014; Jeanne and Gould 2017).

A key feature of COL4A1 syndrome is recurrent intracerebral haemorrhage (ICH) but the brain and cerebrovascular features also cover porencephaly (development of cyst or cavities due to prenatal or neonatal ICH), white matter defects, and cortical defects (Table 5.2) (Meuwissen et al. 2015; Tonduti et al. 2012; Yoneda et al. 2013; Vitale et al. 2019). The cortical defects are associated with signs of white matter vascular insult (Zagaglia et al. 2018), supporting the vascular involvement in disease pathogenesis. Neuronal migration defects also occur in *Col4a1* mutant mice (Labelle-Dumais et al. 2011) and cause lissencephaly (Moon and Wynshaw-Boris 2013), which can occur in COL4A1 syndrome (Meuwissen et al. 2015; Zagaglia et al. 2018). The consequences of these defects are varied and severe with (infantile) hemiparesis, seizures, developmental delay, mental retardation, hydrocephalus, cerebral palsy, and epilepsy (Zagaglia et al. 2018; Shah et al. 2012).

Many of the cerebral clinical features of COL4A1 syndrome are considered to be due to the cerebral small vessel disease (CSVD) caused by these mutations (Meuwissen et al. 2015). CSVD is a disorder of the small penetrating capillaries and arterioles and in the general population accounts for ~30% of stroke and is the leading cause of vascular cognitive decline (vascular dementia) (Wardlaw et al. 2013, 2019).

5.7.1.2 PADMAL and Altered Collagen IV Levels

While the majority (~60%) of mutations are missense mutations affecting the collagen triplet (Kinoshita et al. 2020), nonsense mutations including splice site mutations and premature termination codon mutations that lead to reduced collagen IV levels, have also been identified in COL4A1 syndrome associated with haemorrhaging (Verbeek et al. 2012; Lemmens et al. 2013). In contrast, mutations affecting the miR-29 binding site of the 3′ UTR of *COL4A1* that increase *COL4A1* mRNA levels cause PADMAL (Pontine autosomal dominant microangiopathy and leukoencephalopathy, also known as multi-infarct dementia Swedish type; OMIM #618564). PADMAL is characterised by recurrent ischemic strokes causing lacunar infarcts, leukoencephalopathy, progressive dementia, motor impairment, and spinal cord defects (Verdura et al. 2016; Siitonen et al. 2017; Zhao et al. 2019a). The different clinical features including ischemic stroke indicate PADMAL can be considered a clinical sub-entity within COL4A1 syndrome. Intriguingly, duplication

Table 5.1 Frequently used animal models of collagen-related genetic diseases. Due to space limitation, only the original reference describing the model has been provided

BM component	Affected gene	Animal model	Disease phenotype (or human equivalent)	References
Collagen IV	*Col4a1*	Mouse missense mutations	COL4A1 syndrome HANAC syndrome	Gould et al. (2005), Van Agtmael et al. (2005), Favor et al. (2007), Chen et al. (2016)
	Col4a1/ Col4a2 double null	Mouse	Embryonically lethal, growth retardation vascular defects	Poschl et al. (2004)
	Col4a1 (Cg25c) and *Col4a2 (Vkg)*	Drosophila missense and loss of function mutation	Intestinal defects, myopathy	Kelemen-Valkony et al. (2012)
	emb-9; let-2 (Cola4a1, Col4a2)	*C. elegans* missense mutations	Embryonic lethal	Gupta et al. (1997)
	Col4a2	Mouse missense mutations	COL4A1 syndrome	Favor et al. (2007)
		Exon 18 deletion	Embryonic lethality Multiple organ defects	Reissig et al. (2019)
	Col4a3	Mouse knockout and missense mutation	ARAS and ADAS	Pieri et al. (2014), Cosgrove et al. (1996), Miner and Sanes (1996)
	Col4a3 and *Col4a4* double null	Mouse	Juvenile form of AS	Lu et al. (1999)
	Col4a4	Mouse missense mutation	ARAS	Korstanje et al. (2014)
	Col4a5	Mouse knockout and nonsense mutation	X-linked Alport Syndrome	Rheault et al. (2004), Hashikami et al. (2019)
	Col4a5	Zebrafish in-frame deletion	Defective retinal axon guiding	Xiao and Baier (2007)
	Col4a6	Zebrafish in-frame deletion	Defective axon guiding, cerebellar granule cells defects	Takeuchi et al. (2015)

and triplication of a genomic region harbouring *COL4A1* and *COL4A2* have been observed in patients with CSVD including lacunar infarcts (Saskin et al. 2018; Colin

Table 5.2 Overview of some of the clinical features of the Mendelian diseases due to mutations in collagen IV

Disease	Tissue	Feature	References
COL4A1 syndrome	Cerebrovascular	Intracerebral haemorrhage Cerebral small vessel disease	Meuwissen et al. (2015), Zagaglia et al. (2018)
		Porencephaly	
	Brain white matter defect	White matter hyperintensities leukoencephalopathy	Meuwissen et al. (2015), Zagaglia et al. (2018)
	Brain cortical defects	Lissencephaly, hydranencephaly Schizencephaly Polymicrogyria focal cortical dysplasia nodular heterotopia	Meuwissen et al. (2015), Zagaglia et al. (2018)
	Eye	Arterial Tortuosity Anterior segment dysgenesis Retinal haemorrhage glaucoma High myopia Micro-ophthalmia	Meuwissen et al. (2015), Coupry et al. (2010)
	Muscle	Increased CK Muscle cramps Myopathy Walker Warburg Syndrome Muscle Eye Brain disease	Meuwissen et al. (2015), Tonduti et al. (2012), Labelle-Dumais et al. (2011)
	Heart	Supraventricular arrhythmia Congenital heart defect, common arterial trunk	Plaisier et al. (2007), Meuwissen et al. (2015), McMahon et al. (2015)
	Kidney	Congenital anomalies of the kidney and urinary tract Urinary retention	Meuwissen et al. (2015), Kitzler et al. (2019), Rouaud et al. (2010)
		Glomerulopathy Cystic kidney disease	
		Haematuria Tubular defect	
	PADMAL	Lacunar infarct Ischaemic stroke Leukoencephalopathy	Verdura et al. (2016), Siitonen et al. (2017)
	HANAC	CSVD, intracranial aneurysms, retinal tortuosity, ASD, muscle cramps, Raynaud phenomenon renal cysts, haematuria	Plaisier et al. (2007), Meuwissen et al. (2015), Alamowitch et al. (2009)
Alport syndrome			
	Kidney	Proteinuria, haematuria, end stage kidney disease	Hudson et al. (2003), Nozu et al. (2019)
	Eye	Posterior cataract, lenticonus Dot fleck retinopathy	Hudson et al. (2003), Nozu et al. (2019)
	Ear	Sensorineural deafness	Hudson et al. (2003), Nozu et al. (2019)
Alport syndrome—diffuse leiomyomatosis		Alport syndrome Smooth muscle cell tumour	Hudson et al. (2003)

et al. 2014). These cerebrovascular defects occur due to alteration in α1α1α2 (IV) composition and level, with reduced and increased levels being pathogenic.

5.7.1.3 Ocular Disease

Besides the vascular BMs, the eye contains BMs in the anterior and posterior structures, and the lens (Table 5.2). In mice and patients, mutations in *COL4A1/ COL4A2* affect anterior and/or posterior ocular structures, and can lead to anterior segment dysgenesis (ASD) and microphthalmia (OMIM #180000, #175780) (Van Agtmael et al. 2005; Sibon et al. 2007; Gould et al. 2007; Coupry et al. 2010; Deml et al. 2014; Kuo et al. 2014) (Tables 5.1 and 5.2). The anterior defects include congenital cataracts, iridocorneal adhesions, ciliary body defects, and buphthalmos (Van Agtmael et al. 2005; Sibon et al. 2007; Gould et al. 2007; Coupry et al. 2010; Kuo et al. 2014) some of which affecting the ocular drainage structures, causing impaired vision and glaucoma (Gould and John 2002).

In addition to these anterior defects, arterial retinal tortuosity occurs in mouse models and patients whereby the arteries become tortuous and prone to bleeding (Plaisier et al. 2007; Van Agtmael et al. 2005; Gould et al. 2007; Coupry et al. 2010; Alavi et al. 2016), potentially causing retinal detachment and (temporary) vision loss. The retinal vascular defects reflect aspects of the cerebrovascular disease and have been proposed as a potential biomarker for cerebrovascular disease severity (Jeanne et al. 2015; Ratelade et al. 2018). However, not all patients display ocular defects and there is a large variability in the clinical presentation (Meuwissen et al. 2015; Rodahl et al. 2013). This variability reflects the importance of genetic background and modifiers to the ocular disease, which has been illustrated by *Col4a1* mutant mice on different genetic backgrounds (Gould et al. 2007; Mao et al. 2017).

5.7.1.4 Neuromuscular Disease

Up to one-third of COL4A1 syndrome patients present with muscle pathology and myopathy (Jeanne and Gould 2017), which also occurs in mice models as demonstrated by centralised nuclei, reduced grip strength and elevated serum creatine kinase levels (Plaisier et al. 2007; Guiraud et al. 2017; Labelle-Dumais et al. 2011, 2019) (Table 5.2). The muscle phenotype can be part of congenital muscular dystrophy (Labelle-Dumais et al. 2011); a spectrum of childhood diseases encompassing muscular, ocular, and cerebral malformations whereby patients develop muscle weakness, hypotonia, and in some cases severe myopathy, which can be fatal. *COL4A1* mutations have been detected in dominant forms of muscle–eye–brain disease (MEB; OMIM #175780) and Walker Warburg Syndrome (WWS). In contrast to the more common recessive diseases forms, α1α1α2(IV) associated WWS/MEB is independent of glycosylation of dystroglycan (Labelle-Dumais et al. 2011). In mice, the myopathy was associated with myelination and conductivity defects in the peripheral nerves (Labelle-Dumais et al. 2019), but vascular defects were also associated with the myopathy (Guiraud et al. 2017). In addition, the neuromuscular junction plays a key role in muscle contraction, and α1α1α2(IV) is present in the synaptic cleft, and plays a role in clustering of synaptic vesicles and nerve terminal maturation (Fox et al. 2007). Finally, in *C. elegans* mutations in *Col4a1* and *Col4a2* orthologs (*emb-9* and *let-2*, respectively) result in muscle fibre rupture (Gupta et al. 1997), while in *Drosophila*, mutations in *Col4a1* or *Col4a2* orthologs (*Cg25c* and *Viking*, respectively) causes myopathy (Kelemen-Valkony et al. 2012) with defective muscle attachment due to altered collagen IV-integrin signalling (Kiss et al. 2019). Combined, these data indicate an intricate and complex mechanism of the myopathy with vascular, muscular, and neuronal contributions.

5.7.1.5 HANAC and Kidney Defects

In the kidney, α1α1α2(IV) is expressed in all BMs during development but is replaced by α3α4α5(IV) in the GBM, and forms a mixed network with α5α5α6 (IV) in Bowman's capsule (Hudson et al. 2003). In patients, much of the renal component of COL4A1 syndrome has been characterised in HANAC syndrome (hereditary angiopathy, nephropathy, aneurysm, and cramps; OMIM #611773) which, similar to PADMAL, has been proposed as a clinical sub-entity of COL4A1 syndrome (Plaisier et al. 2007, 2010). HANAC syndrome patients can develop cerebrovascular, eye, kidney, and muscular defects (Table 5.2), but compared to non-HANAC COL4A1 syndrome patients, the brain is relatively mildly affected in HANAC syndrome (Plaisier et al. 2007, 2010; Alamowitch et al. 2009). HANAC is due to mutations in exons 24 and 25 of *COL4A1* in or near the CB3 region of α1(IV), an important domain for integrin signalling (Plaisier et al. 2007, 2010), supporting that location of mutations in the α-chain affects clinical outcome. This is also supported by the mouse model of HANAC syndrome, harbouring the G498V mutation, which is the only *Col4a1* mutant mouse model that is viable when homozygous mutant (Guiraud et al. 2017; Ratelade et al. 2018; Chen et al. 2016).

In mice, *Col4a1* mutations cause a progressive kidney pathology with early-onset defects to Bowman's capsule that are associated with MMP activation and activation of the parietal epithelial cells and tubular dysfunction (Van Agtmael et al. 2005; Chen et al. 2016; Jones et al. 2016, 2019) that could reflect hydronephrosis. The glomeruli are prone to cyst formation, while subtle defects in the GBM have also been observed (Jones et al. 2016, 2019). As these defects occur in multiple mouse models, this indicates they are not specific to HANAC. *COL4A1* mutant mice also develop apparent dysmorphia of the papilla/atrophy of the medulla (Chen et al. 2016; Jones et al. 2016) and the recent identification of *COL4A1* mutations in patients with vesicoureteral reflux (VUR), in which urine flows retrograde, from the bladder into the ureters/kidneys (Kitzler et al. 2019), provides a potential mechanism to this apparent hydronephrosis. These patients were diagnosed with CAKUT (Congenital anomalies of the kidney and urinary tract) (Kitzler et al. 2019), extending the spectrum of features of COL4A1 syndrome and highlighting the importance of α1α1α2(IV) for kidney morphogenesis as well as a developmental origin to some of these defects.

5.7.1.6 Additional Clinical Features in COL4A1 Syndrome

In contrast to the vascular, eye, kidney, and muscle defects, only a few reports have highlighted a cardiac component to COL4A1 syndrome. Arrhythmia has been noted in HANAC syndrome and patients who harbour a deletion of a locus containing *COL4A1* and *COL4A2* develop congenital cardiac defects (Plaisier et al. 2007; McMahon et al. 2015; Wang et al. 2017a). Support for severely reduced α1α1α2 (IV) levels causing cardiac defects is also supported by mouse studies (Reissig et al. 2019).

The broad expression patterns of α1α1α2(IV) coupled with defects in almost every tissue in mice deficiency for α2(IV) (Reissig et al. 2019), and the variable expressivity and penetrance of COL4A1 syndrome (Meuwissen et al. 2015), indicates that mutations may lead to defects in tissues that have so far remained poor or uncharacterised. For example, BM defects have been observed in the skin of a patient with a *COL4A2* mutation and in oesophagus of *Col4a1* mutant mice, but no overt disease was described in this particular family and mouse, respectively (Van Agtmael et al. 2005; Murray et al. 2014). It is therefore likely that additional clinical features of COL4A1 syndrome will be uncovered, which will be important for molecular diagnosis and management of patients.

5.7.1.7 The Role of α1α1α2(IV) in Common Disease

The identification of de novo mutations in apparently sporadic cases (Meuwissen et al. 2015; Meuwissen et al. 2011) suggested that variants in α1α1α2(IV) could occur and contribute to common forms of disease. Analysis of ~100 patients with sporadic ICH identified rare pathogenic mutations in *COL4A1* and *COL4A2* (Jeanne et al. 2012; Weng et al. 2012), indicating very rare missense mutations in α1α1α2 (IV) contribute to sporadic forms of adult ICH and haemorrhagic stroke. In addition, large-scale genetic analysis uncovered that common variants in *COL4A1-COL4A2* are risk factors for ICH (Rannikmae et al. 2015; Chung et al. 2019; Malik et al. 2018) and white matter hyperintensities (Traylor et al. 2017) in the general population. The proportion of ICH patients in which collagen IV plays a role as well as the identity and mechanisms of the variants remains unknown and requires investigation of larger cohorts. First inroads into the frequency of pathogenic variants in *COL4A1* in the general population have been made by Paré and colleagues who determined that within ~100,000 samples from the Genome Aggregation Database, the prevalence varies across different ethnicities and was highest in African/African-American at 0.3% (Grami et al. 2020).

In addition to stroke, *COL4A1* and *COL4A2* have also been associated with vascular calcification, vascular stiffness, coronary artery disease, and myocardial infarct in the general population (Yamada et al. 2008; O'Donnell et al. 2011; Schunkert et al. 2011). In the case of myocardial infarct and coronary artery disease, the intronic variant in *COL4A2* reduced expression. The lower *COL4A2* levels were associated with smooth muscle cell apoptosis and the presence of more unstable atherosclerotic plaques, which are known to increase the risk of myocardial infarct (Yang et al. 2016). These data clearly demonstrate that these genes and the basement membranes are important determinants of vascular health and physiology.

5.7.2 *Mutations and Diseases of α3α4α5(IV) and α5α5α6(IV)*

5.7.2.1 Alport Syndrome

Mutations in *COL4A3, COL4A4, and COL4A5* affect the α3α4α5(IV) network and cause Alport syndrome (AS) (OMIM # 301050, # 203780, # 104200) (Table 5.2). This multi-systemic disease leads to progressive glomerulonephritis accompanied by sensorineural hearing loss and ocular pathology (Hudson et al. 2003; Funk et al. 2018a). AS patients display altered thickness of the GBM leading to progressive splitting, and patients develop haematuria, proteinuria, and eventually end-stage kidney disease. Ocular pathology in AS patients includes a thin fragile lens capsule termed anterior lenticonus as well as temporal retinal thinning and dot and fleck retinopathy (Savige et al. 2015). The ocular phenotypes can be used for disease diagnosis as their presence correlates with renal failure before the age of 30 years (Savige et al. 2015).

AS can occur as X-linked Alport syndrome (XLAS) caused by *COL4A5* mutations which affect both α3α4α5(IV) and α5α5α6(IV) networks. Mutations in *COL4A3* or *COL4A4* cause autosomal recessive AS (ARAS) and autosomal dominant AS (ADAS) and affect only the α3α4α5(IV) network (Hudson et al. 2003). The distribution of these cases is as follows: 80% XLAS, 15% ARAS, and 5% ADAS, with a combined prevalence of AS of 1–9/100,000 (Hudson et al. 2003). The inheritance patterns affect disease severity. XLAS is more severe than autosomal AS, reflecting the effects on two collagen IV networks. For the autosomal forms, ARAS is more severe than ADAS whereby early-onset end stage renal disease occurs in ARAS, and ADAS patients may not progress to end stage renal disease (Hudson et al. 2003). As expected for X-linked disorders, XLAS causes a more severe disease in males than in females, who can be considered heterozygous carriers. In females the X chromosome inactivation pattern, which causes a characteristic mosaic expression pattern of α3α4α6(IV) (Rheault 2012), influences disease severity and results in a wide range of clinical features.

AS can be caused by deletion, insertion, splice site, and missense mutations, and the type of mutation impacts disease severity and age of onset (Gubler 2008). Nonsense mutations are associated with a more severe disease and earlier age of onset, juvenile in the case of *COL4A5* mutations, compared with missense mutations. This establishes that absence or reduced levels of α3α4α5(IV) are a driving force of disease (Hudson et al. 2003). The majority of missense mutations affect the glycine residue in the Gly-Xaa-Yaa repeat (Chew and Lennon 2018). The position of the glycine mutation in the collagen domain impacts on disease severity with C-terminal mutations being associated with a more severe disease (Gubler 2008). ARAS patients can harbour compound heterozygous or homozygous *COL4A3* or *COL4A4* mutations (Hudson et al. 2003; Gubler 2008). More recently it has become clear that AS can also be due to digenic inheritance whereby the two *COL4A3-COL4A5* mutations are located on different or same chromosomes (Mencarelli et al. 2015; Fallerini et al. 2017; Zhao et al. 2019b). Furthermore, molecular analysis of ADAS families also provided intriguing data that digenic inheritance may involve non-collagen IV genes. In this case a *COL4A4* mutation, which was insufficient to cause disease by itself in some pedigree members, was co-inherited with a hypomorphic mutation in *LAMA5* (laminin alpha 5 chain) (Daga et al. 2019), which plays a key role in maintaining the glomerular filtration barrier. The second mutation may also not be limited to other BM components as a hypomorphic mutation in *NPHS2* (encoding the podocin protein in which mutations cause congenital nephrotic syndrome) was also found to be co-inherited with a heterozygous *COL4A4* mutation (Bouchireb et al. 2014). This led the authors to propose a model whereby digenic or even oligogenic inheritance with other BM components and filtration barrier components should be considered for cases whereby the pathogenic effects of the collagen IV mutation is milder (Daga et al. 2019).

Until recently autosomal dominant thin basement membrane nephropathy (TBMN, also known as familial benign haematuria; OMIM # 141200) was considered a separate entity. However, 40% of TBMN patients carry heterozygous *COL4A3* or *COL4A4* mutations and the clinical features of haematuria and BM thinning also occur in carriers in families with AS (Heidet et al. 2003). Furthermore, a similar fraction of ADAS (24%) and TBMN patients (20%) develop end stage renal disease (Voskarides et al. 2007; Marcocci et al. 2009), and heterozygous *Col4a3*$^{+/-}$ mice, which exhibit TBMN, develop chronic renal failure and have a reduced life expectancy (Beirowski et al. 2006). In light of these findings, the criteria for AS were reconsidered and TBMN is now considered part of AS (Kashtan et al. 2018).

During development the GBM consists of α1α1α2(IV) and this is gradually replaced by α3α4α5(IV). The GBM of AS patients contains no α3α4α5(IV) but the embryonic α1α1α2(IV) chain persists, which is less cross-linked and more susceptible to proteolytic degradation and damage due to increased urine flow, leading to GBM damage (Kalluri et al. 1997). This disease mechanism is supported by the structural defects in mature GBM of X-linked Alport's syndrome patients that contains α1α1α2(IV), and filtration defects in *Col4a3*$^{-/-}$ mice in areas of the GBM that are structurally normal (Abrahamson et al. 2007). These differences in network characteristics are also illustrated by the lack of improved outcomes in ARAS patients who have mosaic α5α5α6(IV) expression (Murata et al. 2016). However, there do appear to be species differences as in the *Col4a3*$^{-/-}$ AS mouse model induced α5α5α6(IV) expression reduced disease severity (Kang et al. 2006). These species differences will complicate translation of findings between mice and patients, which will need to be considered for disease mechanism-based treatment strategies.

5.7.2.2 X-linked Alport Syndrome with Diffuse Leiomyomatosis

X-linked Alport syndrome may occur combined with diffuse leiomyomatosis (OMIM #308940), which is characterised by benign smooth muscle tumours leading to oesophageal dysfunction and genital leiomyomas (Hudson et al. 2003). The mutations are deletions that affect *COL4A5* and *COL4A6* leading to the original hypothesis that the absence of α3α4α5(IV) and α5α5α6(IV) causes the disease (Zhou et al. 1993). However, this was challenged by deletions extending beyond exon 3 of *COL4A6* that do not cause diffuse leiomyomatosis (Heidet et al. 1995), and by the identification of a deletion in *COL4A5* that caused AS-diffuse leiomyomatosis (Sá et al. 2013). However, in the latter study *COL4A6* expression was not assessed. Moreover, recent analysis of deletion breakpoints within a substantial cohort of AS-diffuse leiomyomatosis patients does support that inactivation of *COL4A5* and *COL4A6* is causative (Nozu et al. 2017) and disease could be due to effects on regulatory elements (Nozu et al. 2017).

5.7.2.3 Goodpasture Syndrome

Goodpasture syndrome (GP, OMIM # 233450) is an autoimmune disorder in which autoantibodies attack the GBM leading to glomerulonephritis and pulmonary haemorrhage. GP is a rare disorder (1–2 cases per million per year) (Bolton 1996; Pedchenko et al. 2018), however, it can be fatal without medical intervention, whereby patients quickly progress to end stage renal failure. A combination treatment of plasma exchange and immunosuppression improves renal outcome and survival time (Henderson and Salama 2018). In GP, autoantibodies target epitopes within the NC1 domain of α3(IV) and α5(IV) in α3α4α5(IV) in the GBM (Heidet et al. 2003). These epitopes are normally hidden by the NC1 hexamer structure, and thus require to be unmasked. Several mechanisms have been put forward including reactive oxygen species, hexamer structure, and stability (Kalluri et al. 2000; Borza et al. 2005). While the exact aetiology remains to be determined, studies suggest a conformational change in α3α4α5(IV) results from perturbation of sulfilimine crosslinking of NC1 through the enzyme peroxidasin, or excess phosphorylation of the NC1 domain by increased expression of goodpasture binding protein (Bhave et al. 2012; Pedchenko et al. 2018; Raya et al. 2000). This would expose the cryptic epitopes only accessible to the autoantibodies.

Autoimmunity can also be a major complication for AS patients with nonsense mutations that have undergone a renal transplant. The presence of "foreign" collagen IV chains post-transplantation can cause anti-GBM nephritis. For ARAS the epitopes are also in the NC1 domain of α3(IV) and α4(IV) (Wang et al. 2005) whilst for X-linked AS the epitope is in α5(IV) (Brainwood et al. 1998; Kang et al. 2007).

However, unlike in GP, these epitopes are accessible in the NC1 hexamer (Kang et al. 2007).

Recent genetic data has significantly increased our understanding of the role of α3α4α5(IV) and α5α5α6(IV) in biology and disease. For α3α4α5(IV), mutations in *COL4A3 and COL4A4* have been associated with kidney disorders including diabetic kidney disease (Guan et al. 2016; Wang et al. 2018), focal segmental sclerosis (Voskarides et al. 2007), and steroid-resistant nephrotic syndrome (Bullich et al. 2015). This has been complemented in the identification of altered axonogenesis in zebrafish due to *COL4A5/COL4A6* mutations (Takeuchi et al. 2015) (Table 5.1). Excitingly missense mutations in *COL4A6* that affect a glycine residue within the collagen domain have been recently identified as a cause for non-syndromic hearing loss (Rost et al. 2014).

5.8 Disease Mechanisms

5.8.1 Developmental Origin to Adult Disease

A feature of both AS and COL4A1 syndrome is the early onset of the disease and evidence has been gathered pointing to a significant role for developmental defects or origin for these diseases.

For AS the maintenance of the embryonic α1α1α2(IV) network in the GBM renders the GBM susceptible to biomechanical strain (Kalluri et al. 1997). In the case of COL4A1 syndrome, mice and patients develop ICH at birth or during embryonic development, and congenital or prenatal porencephaly (Gould et al. 2005; Favor et al. 2007; Colin et al. 2014; Meuwissen et al. 2011; Bilguvar et al. 2009; Van Agtmael et al. 2010; Vermeulen et al. 2011; Khalid et al. 2018). The cortical malformation (e.g. lissencephaly) (Zagaglia et al. 2018) is also associated with developmental neuronal migration defects (Labelle-Dumais et al. 2011; Moon and Wynshaw-Boris 2013). Furthermore, induced expression of mutant *Col4a1* in a conditional mouse model revealed that expression of the mutant allele pre-weaning was necessary to induce adult ICH (Jeanne et al. 2015) and for the anterior segment dysgenesis in the eye, around or before 10.5–12.5 days postcoitum (Mao et al. 2017).

Angiogenesis is the biological process of forming new blood vessels from pre-existing capillary networks. During sprouting angiogenesis, endothelial cells (EC) degrade the vascular BM, invade into the surrounding tissue and proliferate in response to angiogenic factors to form tubes (Senger and Davis 2011; Hogan and Schulte-Merker 2017). This process of EC activation, proliferation, and survival is dependent on the ECM (Mundel and Kalluri 2007). Altered angiogenesis is known to lead to defects in vascular patterning (Hogan and Schulte-Merker 2017), which can cause arterial tortuosity syndrome (Coucke et al. 2006), and VEGF (vascular endothelial growth factor) signalling, a key player in angiogenesis is implicated in retinal vascular tortuosity (Hartnett et al. 2008), a feature of COL4A1 syndrome

(Van Agtmael et al. 2005; Gould et al. 2007; Coupry et al. 2010; Alavi et al. 2016). *Col4a1* mutant mice exhibit increased levels of pro-angiogenic growth factors including VEGF, and increased angiogenesis in the brain during embryonic development (Jeanne et al. 2015; Alavi et al. 2016). This strongly argues for a role of altered embryonic—post-natal angiogenesis in COL4A1 syndrome.

The mechanisms by which *Col4a1* mutations affect angiogenesis remain unclear, but collagen IV is implicated in the regulation of angiogenesis. For example, endothelial cell migration and proliferation are modulated by collagen IV (Pedchenko et al. 2004) with collagen IV levels correlating with the amount of tube formation, e.g. the addition of exogenous collagen IV promoted tube formation (Grant et al. 1991). Collagen IV can also exert anti-angiogenic effects through the release of cryptic peptide (Mundel and Kalluri 2007; Silva et al. 2017). This includes the anti-angiogenic peptides, arresten and canstatin, generated by proteolytic cleavage of NC1 domains of α1(IV) and α2(IV), respectively (Mundel and Kalluri 2007). These molecules can bind integrin $\alpha_1\beta_1$, $\alpha_v\beta_3$, and $\alpha_v\beta_5$, and transmembrane integrin like receptor CMG2 (capillary morphogenesis gene 2) and may alter angiogenesis by affecting MAPK signalling (Mundel and Kalluri 2007; Finnell et al. 2020). Therefore, the reduced levels of these peptides due to lower levels of collagen IV in the ECM of *Col4a1* mutant mice (Gould et al. 2005, 2007; Jeanne et al. 2015; Jones et al. 2016; Van Agtmael et al. 2010), represents one possible mechanism by which these mutations affect angiogenesis. In addition, collagen IV binds BMP (bone morphogenic protein) (Wang et al. 2008; Bunt et al. 2010), which is a member of the TGFβ family of proteins, known to affect angiogenesis (Goumans et al. 2018). Thus, the effects of the mutation on growth factor signalling mediated by collagen IV also represent an attractive hypothesis.

5.8.2 Cellular Origins of Collagen IV Diseases

Both COL4A1 and Alport syndromes are multi-systemic disorders and the collagen IV proteins are expressed by multiple different cell types. A key aspect in terms of elucidating disease mechanism is therefore determining the cellular origin of the disease. This is important as it will directly establish which cell types will need to be targeted by any treatments.

Using an elegant conditional *Col4a1* mutant mouse model, Gould and colleagues identified that expression of mutant *Col4a1* in EC or smooth muscle alone is sufficient to cause ICH, with a more minor effect from expression in astrocytes (Jeanne et al. 2015). Interestingly, altered vascular function is implicated in CSVD (Berry et al. 2019) and mutant *Col4a1* expression affects endothelial and vascular function with altered vasodilation of the peripheral vasculature (Van Agtmael et al. 2010). The smooth muscle cell defects in *Col4a1* mutant mice include hypermuscularisation of the transitional segment of the vasculature (a segment between arterioles and capillaries), predicted to increase vascular resistance. Combined with smooth muscle cell apoptosis upstream of this segment, which weakens the vascular wall, this then, could lead to haemorrhage at this weakened site (Ratelade et al. 2020). Importantly the same mechanism was observed in sporadic patients with intracerebral haemorrhage (Ratelade et al. 2020). Combined, these data highlight that both structural and functional vascular defects may contribute to vascular disease.

The vascular defects also contribute to the myopathy (Guiraud et al. 2017) and retinal defects in the eye, while the lens defects play a key role in the development of the ASD defect in the eye including increased intra-ocular pressure (Alavi et al. 2016; Mao et al. 2017). This confirms that COL4A1 syndrome is more than a vascular disorder and thus that the mechanisms of disease will need to be established for the different cell types and tissues.

For AS recent genetic evidence indicated that induced expression of *Col4a3* in EC of *Col4a3*$^{-/-}$ mice, did not ameliorate disease (Funk et al. 2019). This not only indicates that ECs do not contribute to the deposition of the $\alpha3\alpha4\alpha5$(IV) in the GBM but also that treatments will need to be aimed at the podocyte.

5.8.3 Molecular Disease Mechanisms

Collagen IV is folded in the ER and secreted into the ECM. It is therefore logical to assume that ECM defects are a key feature of collagen pathologies, either due to reduced levels of collagen or secretion of mutant protein. It has been shown that delay in protein folding due to collagen missense mutations can lead to excessive post-translational modification of the protomer and disease (Raghunath et al. 1994; Cabral et al. 2007) (Fig. 5.3). However, missense mutations in secreted protein can lead to ER retention and ER stress, which can become pathogenic and has been implicated in several ECM disorders (Bateman et al. 2009; Mullan et al. 2017; Gatseva et al. 2019) (Fig. 5.4).

Thus, conceptually collagen IV mutations could act via ER stress and/or ECM defects. In the following section, we will describe in more detail the different molecular disease mechanisms of the collagen IV pathologies.

5.8.3.1 ECM Defects and Quantity and Quality of the Collagen Network

COL4A1 Syndrome

A central feature of mouse models and patients with *COL4A1/COL4A2* mutations are BM structural defects that affect every tissue analysed to date including eye, kidney, muscle, and vasculature (Guiraud et al. 2017; Gould et al. 2005, 2006, 2007; Van Agtmael et al. 2005, 2010; Ratelade et al. 2018; Chen et al. 2016; Jones et al. 2016; Taylor et al. 2011). These defects are associated with and can be due to severely reduced collagen IV levels in the BM (Poschl et al. 2004; Reissig et al. 2019; Gould et al. 2005; Jeanne et al. 2015; Murray et al. 2014; Verbeek et al. 2012; Labelle-Dumais et al. 2011; Van Agtmael et al. 2010). The reduced level of collagen

IV is a clear mechanism for the nonsense mutations that have been identified in patients (Meuwissen et al. 2015; Lemmens et al. 2013) (Fig. 5.4). However, it also applies to missense mutations. Analysis of mice and patients cells with *Col4a1/COL4A1* or *Col4a2/COL4A2* missense mutations respectively revealed that incorporation of a single mutant α-chain in α1α1α2(IV) is sufficient for the mutant protein to be retained in the ER (Guiraud et al. 2017; Gould et al. 2005, 2007; Murray et al. 2014; Chen et al. 2016) (Fig. 5.4). The severity of this retention is influenced by the position of the mutation, whereby in general more C-terminal mutations cause a higher degree of intracellular retention, and more severe ICH (Jeanne et al. 2015; Kuo et al. 2014). This indicates that at least in some genetic backgrounds reduced levels of collagen IV in the BM can cause cerebrovascular disease. The quantitative differences are not limited to reduced levels, however, increased collagen IV levels can also be pathogenic, for example in PADMAL (Verdura et al. 2016).

Besides these quantitative differences, missense mutations will also affect the quality of the protomer, providing an additional disease mechanism through the possible secretion of mutant α1α1α2(IV) (Fig. 5.4). For example, the network can be less stable as mutant α1α1α2(IV) displays reduced thermal stability reflected by the temperature-sensitive nature of some phenotypes in *C. elegans*, *Drosophila*, and cell culture models (Kuo et al. 2014; Gupta et al. 1997; Kiss et al. 2016, 2019). The mutation could also affect a functional domain within the protomer that impacts on a specific function such as ECM-cell signalling. This could explain how some missense mutations may cause a severe phenotype despite being located in the N-terminal end of the protein, and provide a mechanism for some of the clinical variability with COL4A1 syndrome. For example, a N-terminal *Col4a1* glycine mutation in mice caused a mild cerebrovascular phenotype but a severe myopathy, which was associated with altered GPCR signalling leading to the myelination defects and neuromuscular component of COL4A1 syndrome (Kuo et al. 2014; Labelle-Dumais et al. 2019). Similarly altered integrin signalling occurs in Drosophila (Kiss et al. 2019) and mice (Chen et al. 2016), and has been postulated to play a key role in HANAC (Plaisier et al. 2010). This indicates that aberrant signalling is likely a key feature of COL4A1 syndrome.

ECM composition and turnover are key features for the BM to carry out its functional and biomechanical roles. Proteases such as MMPs and their inhibitors including TIMPs influence ECM composition and turnover (Arpino et al. 2015). In kidneys of the HANAC mouse model, the ECM and Bowman's capsule defects are associated with increased MMP2/9 activity, which are major collagen IV degrading MMPs, and activation of the epithelial cells, illustrating effects on altered cell state (Chen et al. 2016).

These data illustrate the large variety of qualitative and quantitative effects by which *COL4A1/COL4A2* mutations cause compositional and structural matrix defects in COL4A1 syndrome.

Alport Syndrome

ECM defects are also a constant feature of AS and the increased severity of nonsense mutations provides compelling evidence for reduced levels or absence of α3α4α5

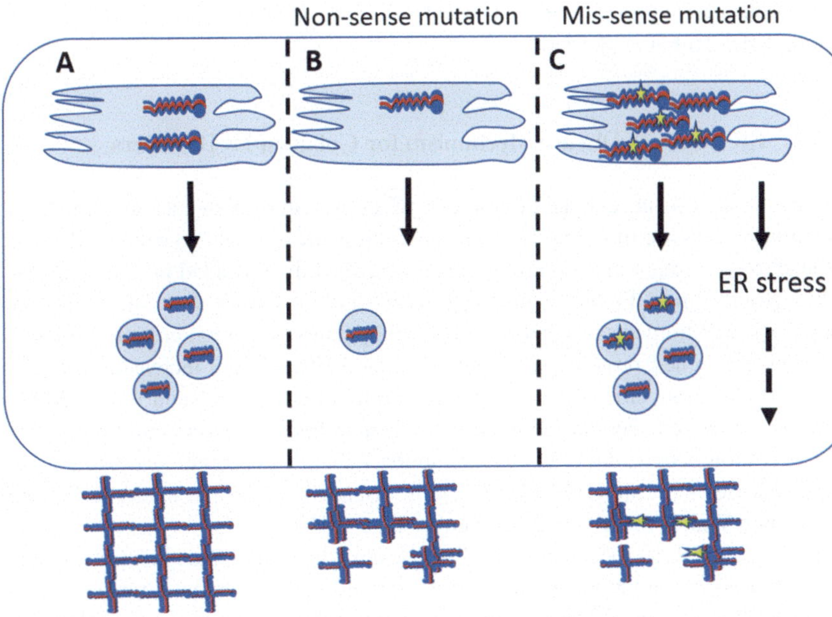

Fig. 5.4 Collagen IV disease mechanisms. (**a**) Under 'correct/normal' physiological conditions, collagen IV protomer is made in the ER, transported out of the cell to be incorporated into the ECM BM. (**b**) Non-sense mutations (star) reduce protein levels leading to reduced levels of collagen IV in the BM, causing BM defects. (**c**) Mis-sense mutations can cause mutant protein (yellow star) to be secreted from the cell but mutant protein may also accumulate in the ER. This can cause ER stress and reduced secretion, meaning ECM defects can be due to reduced protein levels and/or mutant protein incorporation

(IV) and associated structural GBM defects being a major causative mechanism (Fig. 5.4). The persistence of α1α1α2(IV), which is less resistant to proteolytic cleavage (Kalluri et al. 1997), also highlights the altered and reduced biomechanical stability, leading to BM structural defects as the GBM is unable to deal with the biomechanical strain. In addition to structural defects, the ECM composition is also altered with aberrant laminin α2 expression (Kashtan et al. 2001).

These structural and compositional defects affect cell phenotype within the affected kidney. They cause endothelin A receptor expression in glomerular cells and an altered more invasive behaviour of mesangial cells (Dufek et al. 2016). Podocytes also adopt a more invasive phenotype with loss of foot processes (Randles et al. 2016), associated with focal adhesion kinase activation, cytokine (e.g. TGFβ) production indicating an inflammatory state and MMP activation (Delimont et al. 2014), highlighting ECM remodelling and onset of fibrosis. The observed fibrosis and MMP activation are influenced by LOXL2 (lysyl oxidase-like 2) collagen cross-linking activity (Cosgrove et al. 2018). The altered cell phenotypes and their responses may also be determined in part by altered ECM–cell signalling, for example *via* integrins α1β1 and α2β1, as well as DDR receptors (Chew and Lennon 2018; Rubel et al. 2014).

5.8.3.2 ER Stress as Disease Mechanism for Collagen IV Disorders

Missense mutations in secreted protein can affect protein folding such as triple helix formation in the case of collagen. This can elicit a stress response called ER stress that leads to activation of a signalling response called the unfolded protein response (UPR). Briefly, the UPR is a conserved homeostatic response that aims to restore proteostasis, ER homeostasis and cell function by pausing general protein synthesis, degrading misfolded proteins *via* a process called ERAD, and increasing the production of molecular chaperones involved in protein folding (Ron and Walter 2007). However, when cells and the UPR fails to reinstate ER homeostasis and prolonged UPR activation occurs, the UPR triggers apoptosis and can become pathogenic. For excellent reviews on the UPR and its role in disease we refer the reader to (Costa-Mattioli and Walter 2020; Gonzalez-Teuber et al. 2019).

In terms of COL4A1 syndrome, analysis by ourselves and others identified that in mice with *Col4a1* mutations, reduced secretion and collagen retention were associated with ER stress and UPR activation in the vasculature (Jeanne et al. 2015; Van Agtmael et al. 2010) (Fig. 5.4). Importantly, an allelic series of *Col4a1* mouse models revealed that ICH severity correlated with levels of ER retention, which broadly corresponded to the position of the mutation in the collagen domain, rather than extracellular collagen IV levels (Jeanne et al. 2015; Kuo et al. 2014). We described that in a family with a *COL4A2* mutation, ER stress and not ECM defects was associated with ICH (Murray et al. 2014). In this case, ECM defects were detected in an unaffected carrier of the disease (Murray et al. 2014), providing evidence that ECM defects *per se* may not always be sufficient. It should, however, be noted that not every glycine mutation induces ER retention and stress, in particular those that are more located towards the N-terminal end of the collagen domain (Jeanne et al. 2015; Kuo et al. 2014), and that this genotype–phenotype relation did not apply to the myopathy (Kuo et al. 2014). There is less information for non-glycine mutations although a lysine mutation affecting the Yaa position of the collagen triplet can also induce ER stress in the vasculature (Van Agtmael et al. 2010).

Overall these data highlight a trend with more severe protein folding defects associated with mutation located towards the C-terminal end of the collagen domain, and indicate that ER retention and stress can have a strong modifier effect for ICH.

Comparative analysis between ECM defects and ER stress revealed tissue-specific mechanisms occur as in mice the myopathy and Bowman's Capsule defects were associated with ECM defects, and ICH with ER stress (Jeanne et al. 2015; Labelle-Dumais et al. 2019; Jones et al. 2016). Analysis of the kidney defects also provided evidence for cell-specific mechanism within the same tissue whereby the tubular disease was associated with ER stress and the defects in Bowman's capsule with ECM defects (Jones et al. 2016). This may reflect the sensitivity of some cell types or tissues to ER retention and/or BM defects, but the nature and position of the mutation are also likely to affect this (see also Sect. 5.8.3.1). Mutations affecting the Yaa residue of the collagen repeat and a mutation in the NC1 domain caused a milder eye disease and ICH, respectively, than glycine mutations (Van Agtmael et al. 2005; Kuo et al. 2014).

Overall these data highlight a complexity whereby the nature of the mutation and cell-specific effects combine to cause disease. However, it remains unclear what the relative contribution of ER stress (and/or ECM) defects to the different COL4A1 disease mechanisms is. Is it pathogenic or an epiphenomenon? This represents an important knowledge gap to guide treatment development.

For AS, similar questions are being raised given the presence of missense mutations in X-linked and autosomal dominant forms of AS. A key advance was made by transfection of a *COL4A3* cDNA containing a glycine mutation in cultured podocyte cell line, which revealed ER stress induction (Pieri et al. 2014). Importantly, this was confirmed in a novel mouse model harbouring the same *Col4a3* mutation as well as patient podocytes. Intriguingly, the UPR was also activated when *Col4a3* was either knocked-down or overexpressed, suggesting an imbalance between the different collagen IV alpha chains may be sufficient in some cases to induce ER stress (Pieri et al. 2014). Investigation of *COL4A5* glycine mutations in primary patient dermal fibroblasts confirmed induction of ER stress and autophagy (Wang et al. 2017b). However, how these intracellular effects contribute to AS remains unclear.

5.8.4 Environmental and Genetic Modifiers

Investigating genetic and environmental modifiers of disease provides important insight into disease mechanism and potential therapeutic targets and approaches, but can also inform on clinical management of patients to reduce the risk of disease progression. Genetic analysis of families and mouse models with *COL4A1/COL4A2* mutations supports that eye, muscular, and cerebrovascular diseases are dependent on genetic background. In mice the CAST genetic background was able to effectively suppress the severe defects caused by *Col4a1* mutations (Jeanne et al. 2015; Labelle-Dumais et al. 2011; Gould et al. 2007), while our analysis in patients revealed that a *COL4A2* glycine mutation that causes ICH and porencephaly is also subject to genetic modifiers (Murray et al. 2014). The identity of these modifiers in vertebrates remains unclear although for the ocular disease a locus was identified on mouse chromosome 1 (Gould et al. 2007) while it can also involve ER retention and ER stress in patients (Murray et al. 2014). In *C. elegans* other BM components modify *Col4a1/emb9* phenotypes (Gotenstein et al. 2018).

Data from animal studies also uncovered environmental modifiers on cerebrovascular disease. This includes vaginal birth as a risk factor with Caesarian delivery reducing neonatal ICH (Gould et al. 2006), while exercise and use of blood thinners increased disease severity (Jeanne et al. 2015). In addition, age can also be considered as a modifier given the age-dependent nature of the kidney disease, vascular dysfunction, and cerebrovascular disease (Jeanne et al. 2015; Ratelade et al. 2018; Jones et al. 2016; Van Agtmael et al. 2010).

These modifiers underlie the reduced penetrance and clinical variability of COL4A1 syndrome. This implies that absence of disease in one individual with a particular *COL4A1* or *COL4A2* variant does not exclude this variant from being pathogenic per se. Consequently, the contribution of rare *COL4A1/COL4A2* variants to disease in the general population may be higher than anticipated and at least some variants of unknown significance may be pathogenic.

AS development is also influenced by modifier effects of other BM components. This was nicely illustrated by inducing α5α5α6(IV) expression in the GBM of *Col4a3*$^{-/-}$ mice to reduce phenotype severity with a delayed age of onset of renal failure (Kang et al. 2006). In contrast, increased levels of laminin α5 and α1 exacerbated disease (Abrahamson et al. 2007), while a laminin β2 missense variant, which was not pathogenic itself, increased progression to kidney failure in *Col4a3*$^{-/-}$ mice and proteinuria in female *Col4a5*$^{+/-}$ mice (Funk et al. 2018b).

In conclusion, the data of AS and COL4A1 syndrome provide compelling evidence that BM, and potentially other ECM components as well as intracellular pathways can act as genetic modifiers of collagen IV disorders. This also argues for a more complete sequencing analysis of BM/ECM components in patients to provide a more accurate molecular diagnosis.

5.9 Therapeutic Strategies

Treatment for hereditary disorders can be divided into approaches aimed at managing disease progression and providing symptom relief, strategies to modulate disease mechanisms, and gene therapy approaches (Gatseva et al. 2019) (Table 5.3). Gene therapy approaches are appealing conceptually as they are independent of the disease mechanism and provide an actual cure. However, the multi-systemic nature of collagen IV pathologies and the fact that different cell types underlie the different clinical features, e.g. EC for ICH and lens capsule for ASD defects in COL4A1 syndrome, complicates the use of these approaches as multiple cell types require to be targeted.

5.9.1 Gene Therapy and Therapies Targeting Downstream Mechanisms

To date, the multi-systemic nature of COL4A1 syndrome combined with a very young age of onset complicates disease management which focuses on symptom relief including a shunt to drain excess fluid for hydrocephalus, anti-convulsing medicine for seizures etc. Given that hypertension is the major risk factor for haemorrhage (An et al. 2017), blood pressure-lowering drugs can reduce the high risk of stroke. Finally, physical and speech impediments are managed by physiotherapy and speech therapy, respectively.

For AS patients, there has been more success in managing disease progression and treatment (Gross et al. 2014; Omachi and Miner 2019). Following initial small clinical trials targeting the renin–angiotensin system to reduce blood pressure in AS (Proesmans et al. 2000; Webb et al. 2011, 2013), a recent large multicentre phase 3 clinical trial using ramipril with a follow-up period of 6 years indicated reduced progression of proteinuria, lower decline in filtration rate, and slower progression to renal failure in AS children (Gross et al. 2020). This provides compelling data as to the clinical benefit of angiotensin-converting enzyme (ACE) inhibitors for AS.

Fibrosis is a feature of AS, and many other ECM disorders and several approaches have been adopted to target fibrosis in AS mouse models. This includes cerivastatin—a HMG-CoA reductase inhibitor that targets the TGFβ1 pathway, and vasopeptidase inhibitor AVE7688, which had anti-fibrotic and anti-inflammatory effects (Koepke et al. 2007; Gross et al. 2005), while targeting STAT3 signalling also attenuates disease progression in mice (Yokota et al. 2018). Finally, phase 3 clinical trials are progressing with bardoxolone (BARD) (Gross et al. 2018), which has anti-fibrotic and anti-inflammatory effects. BARD does increase glomerular filtration rates in AS patients, which has raised debate as to whether this represents a potential risk (Baigent and Lennon 2018).

Finally, gene therapy-based approaches are also being explored for AS (Table 5.3). This has included reducing fibrosis by targeting miR-21, which promotes fibrosis (Chau et al. 2012), as its levels are increased in kidneys of AS patients and correlate with disease severity (Guo et al. 2019). Initial support came from the administration of an anti-miR21 in *Col4a3*$^{-/-}$ mice that attenuate kidney disease progression (Guo et al. 2019; Gomez et al. 2015). This has formed the basis for the currently ongoing clinical trial using RG-012, also known as lademirsen, designed to block the activity of miR-21 (clinicaltrials.gov Identifier NCT03373786).

In mice proof-of-concept gene therapy approaches targeting the actual GBM included a transgene system to induce α3α4α5(IV) expression by podocytes and promote its incorporation into a defective GBM. This not only restored the missing α3α4α5(IV) network but also slowed kidney disease progression and extended lifespan (Lin et al. 2014). Induction of α3α4α5(IV) expression has also been explored through transplanting amniotic fluid stem cells into *Col4a5*$^{-/-}$ mice which delayed interstitial fibrosis, kidney decline, and prolonged animal survival (Sedrakyan et al. 2012). Transfer of bone marrow-derived stem cells post-irradiation also appeared to repair BM defects in *Col4a3*$^{-/-}$ mice (Sugimoto et al. 2006), but this has been debated as irradiation itself also prolonged lifespan (Katayama et al. 2008; Gross et al. 2009). Further research will be required to confirm the potential of stem cell transplantation and reconcile these differences.

Table 5.3 Mechanism-based therapeutic strategies for collagen-related diseases

Disease	Mechanism target	Treatment	References
COL4A1 Syndrome	ER retained protein, ER stress pathway	Chemical chaperones, e.g. 4-phenylbutyrate (4PBA)	Jeanne et al. (2015), Murray et al. (2014), Jones et al. (2019), Labelle-Dumais et al. (2019)
Alport Syndrome	ER retained protein, ER stress	4PBA	Wang et al. (2017b, #58)
	Blood pressure	Angiotensin-converting enzyme inhibitors, e.g. **Ramipri**l Angiotensin II type 1 receptor blockers, e.g. **Losartan**	Gross et al. (2014), Webb et al. (2011), Webb et al. (2013), Gross et al. (2020)
	Fibrosis	HMG-CoA-reductase inhibitor (cerivastatin) Vasopeptidase inhibitor AVE7688 anti–miR-21 oligonucleotides STAT3 inhibitor, e.g. stattic	Koepke et al. (2007), Gross et al. (2005), Gomez et al. (2015), Yokota et al. (2018)
	Oxidative stress, inflammation, and fibrosis	Nrf2 activator, e.g. bardoxolone methyl (BARD)	Gross et al. (2018)
	Functional replacement	Gene therapy	Lin et al. (2014)
		Stem cell Bone marrow-derived stem cells Amniotic fluid stem cells	Sugimoto et al. (2006), Sedrakyan et al. (2012)

Clinical trials are indicated in bold

5.9.2 *Targeting ER Stress*

Dominant mutations in AS and COL4A1 syndromes can cause intracellular retention with ER stress and UPR activation (Gould et al. 2007; Kuo et al. 2014; Jones et al. 2016; Van Agtmael et al. 2010; Gatseva et al. 2019; Pieri et al. 2014; Wang et al. 2017b). This suggests that enabling protein folding could alleviate intracellular pathology and be beneficial to the ECM by promoting collagen IV secretion. This could be achieved through chemical chaperones, which enhance protein folding or stability by mimicking the molecular chaperones in the ER (Engin and Hotamisligil 2010) or via promoting misfolded protein degradation, as has been applied for *Col10a1* via carbamazepine in mouse models of metaphyseal chrondrodysplasia type Schmid (Mullan et al. 2017). As several chemical chaperones and carbamazepine are FDA approved, these would represent attractive therapeutic approach for transition into clinical trials.

For collagen IV disease, the emphasis has been on the compound sodium 4-phenylbutyrate (4PBA), which is FDA approved for treatment of urea cycle disorders, but also possesses chemical chaperone activity (Engin and Hotamisligil 2010). 4PBA increases ER folding capacity, stabilising folded protein, and increasing ER-associated degradation (ERAD) (Engin and Hotamisligil 2010). In vitro treatment of primary patient cells harbouring a *COL4A2* mutation established that 4PBA reduced ER retention and ER stress (Murray et al. 2014), which was confirmed for an allelic series of mutations in mouse embryonic fibroblasts (Kuo et al. 2014). Importantly in mice chronic preventative 4PBA treatment reduced adult and paediatric ICH for some missense mutations (Jeanne et al. 2015; Jones et al. 2019; Hayashi et al. 2018). 4PBA was also effective in reducing ICH severity once the disease was established (Jones et al. 2019), suggesting 4PBA could be beneficial for patients post-diagnosis with COL4A1 syndrome. Furthermore, a short-term transient treatment also reduced severity with lasting effect, but it should be noted that increased efficacy was obtained when treatment commences early in life (Hayashi et al. 2018). However, the efficacy of PBA was dose-dependent (Hayashi et al. 2018) and did not prevent or treat the ASD and Bowman's capsule defects (Jones et al. 2019). This suggests that for patients with multi-systemic features a combinatorial treatment targeting both ECM defects and ER stress will be required (Jones et al. 2019).

4PBA not only reduces ER stress but also promotes secretion of collagen IV (Kuo et al. 2014; Labelle-Dumais et al. 2019; Jones et al. 2019). Worryingly, 4PBA treatment of mice with a glycine mutation reduced the ability of the dermal BM to withstand mechanical stress (Jones et al. 2019). Moreover, 4PBA treatment increased severity of myopathy in a mouse model with a more N-terminal glycine mutation that did not cause ER stress (Labelle-Dumais et al. 2019). One explanation may be that PBA caused secretion of mutant protein. PBA treatment may thus be contra-indicative for clinical features of COL4A1 syndrome due to ECM defects and missense mutations that do not induce ER stress.

This is also a consideration for AS due to missense mutations where treatment of patient cells with 4PBA was associated with reduced ER stress (Wang et al. 2017b). Given the mechanical stress to which the GBM is exposed, it is now key to determine if PBA has efficacy in vivo or whether it also negatively affects the GBM due to increased deposition of mutant protein.

In conclusion, these intervention studies highlighted that a detailed understanding of the disease mechanisms will be required. A one size fits all therapeutic approach will be unlikely and patient stratification based on the molecular mechanisms of the mutations will be necessary.

5.10 Concluding Remarks

It has been 30 years since the identification of the first collagen IV mutation in disease. Since then it has become apparent that these "simple" Mendelian diseases are complex multi-systemic disorders characterised by tissue and cell-specific disease mechanisms as well as mutation-specific quantitative and qualitative effects on the collagen network. While it may seem that we have come a long way in increasing our understanding of the mechanisms of collagen IV biosynthesis and collagen IV disease, the hard truth is that major gaps in our knowledge remain. These range from gaps in our knowledge relating to fundamental aspects of collagen secretion and its regulation (e.g. is collagen IV biosynthesis, like collagen I (Chang et al. 2020), under the control of circadian rhythm? What is the half-life of collagen IV in the ECM and how is BM degradation/turnover regulated?) to the disease mechanisms (e.g. what is the relative contribution of ER stress and ECM defects? What are the genetic modifiers?).

We are still in desperate need of treatment for collagen IV disorders and the role of collagen IV in human physiology and pathophysiology remains incomplete. In particular, besides Mendelian disorders many questions remain about the role of these molecules in common forms of disease. Large population cohorts will undoubtedly be a powerful resource in addressing these. But even when it comes to Alport and COL4A1 syndromes, increased in-depth molecular mechanistic insight will be needed to develop the effective treatments that patients deserve and need. Key questions remain regarding the role of ER stress and ECM defects in disease and the mechanisms by which these defects exert their effects. Comparison across different diseases will be powerful and may allow us to transition from treatments based on clinical features to phenotype-agnostic therapeutics that target key mechanisms. In many ways, we have only just begun to lay the foundations for an exciting time in ECM research.

This chapter was written during the lockdown in the United Kingdom caused by the COVID-19 pandemic and we hope that you are all safe and healthy.

Acknowledgements We would like to apologise to our colleagues whose work we could not cite due to space constraints. This work was made possible through funding by the MRC (MR/R005567-1), BHF (PG/15/92/31813), Stroke Association (PPA 2016/02), and Heart Research UK (RG 2664/17/20) to TVA. The authors declare no conflict of interest.

References

Abe H, Matsubara T, Iehara N, Nagai K, Takahashi T, Arai H et al (2004) Type IV collagen is transcriptionally regulated by Smad1 under advanced glycation end product (AGE) stimulation. J Biol Chem 279(14):14201–14206

Abrahamson DR, Isom K, Roach E, Stroganova L, Zelenchuk A, Miner JH et al (2007) Laminin compensation in collagen alpha3(IV) knockout (Alport) glomeruli contributes to permeability defects. J Am Soc Nephrol 18(9):2465–2472

Alamowitch S, Plaisier E, Favrole P, Prost C, Chen Z, Van Agtmael T et al (2009) Cerebrovascular disease related to COL4A1 mutations in HANAC syndrome. Neurology 73(22):1873–1882

Alavi MV, Mao M, Pawlikowski BT, Kvezereli M, Duncan JL, Libby RT et al (2016) Col4a1 mutations cause progressive retinal neovascular defects and retinopathy. Sci Rep 6:18602

An SJ, Kim TJ, Yoon B-W (2017) Epidemiology, risk factors, and clinical features of intracerebral hemorrhage: an update. J Stroke 19(1):3–10

Anazco C, Lopez-Jimenez AJ, Rafi M, Vega-Montoto L, Zhang MZ, Hudson BG et al (2016) Lysyl oxidase-like-2 cross-links collagen IV of glomerular basement membrane. J Biol Chem 291 (50):25999–26012

Aouacheria A, Geourjon C, Aghajari N, Navratil V, Deleage G, Lethias C et al (2006) Insights into early extracellular matrix evolution: spongin short chain collagen-related proteins are homologous to basement membrane type IV collagens and form a novel family widely distributed in invertebrates. Mol Biol Evol 23(12):2288–2302

Arpino V, Brock M, Gill SE (2015) The role of TIMPs in regulation of extracellular matrix proteolysis. Matrix Biol 44–46:247–254

Baigent C, Lennon R (2018) Should we increase GFR with bardoxolone in Alport syndrome? J Am Soc Nephrol 29(2):357–359

Bateman JF, Boot-Handford RP, Lamande SR (2009) Genetic diseases of connective tissues: cellular and extracellular effects of ECM mutations. Nat Rev Genet 10(3):173–183

Beirowski B, Weber M, Gross O (2006) Chronic renal failure and shortened lifespan in COL4A3+/- mice: an animal model for thin basement membrane nephropathy. J Am Soc Nephrol 17 (7):1986–1994

Berry C, Sidik N, Pereira AC, Ford TJ, Touyz RM, Kaski JC et al (2019) Small-vessel disease in the heart and brain: current knowledge, unmet therapeutic need, and future directions. J Am Heart Assoc 8(3):e011104

Bhave G, Cummings CF, Vanacore RM, Kumagai-Cresse C, Ero-Tolliver IA, Rafi M et al (2012) Peroxidasin forms sulfilimine chemical bonds using hypohalous acids in tissue genesis. Nat Chem Biol 8(9):784–790

Bignon M, Pichol-Thievend C, Hardouin J, Malbouyres M, Brechot N, Nasciutti L et al (2011) Lysyl oxidase-like protein-2 regulates sprouting angiogenesis and type IV collagen assembly in the endothelial basement membrane. Blood 118(14):3979–3989

Bilguvar K, DiLuna ML, Bizzarro MJ, Bayri Y, Schneider KC, Lifton RP et al (2009) COL4A1 mutation in preterm intraventricular hemorrhage. J Pediatr 155(5):743–745

Bolton WK (1996) Goodpasture's syndrome. Kidney Int 50(5):1753–1766

Boosani CS, Sudhakar A (2006) Cloning, purification, and characterization of a non-collagenous anti-angiogenic protein domain from human alpha1 type IV collagen expressed in Sf9 cells. Protein Expr Purif 49(2):211–218

Borza DB, Bondar O, Colon S, Todd P, Sado Y, Neilson EG et al (2005) Goodpasture autoantibodies unmask cryptic epitopes by selectively dissociating autoantigen complexes lacking structural reinforcement: novel mechanisms for immune privilege and autoimmune pathogenesis. J Biol Chem 280(29):27147–27154

Bouchireb K, Boyer O, Gribouval O, Nevo F, Huynh-Cong E, Moriniere V et al (2014) NPHS2 mutations in steroid-resistant nephrotic syndrome: a mutation update and the associated phenotypic spectrum. Hum Mutat 35(2):178–186

Boute N, Exposito JY, Boury-Esnault N, Vacelet J, Noro N, Miyazaki K et al (1996) Type IV collagen in sponges, the missing link in basement membrane ubiquity. Biol Cell 88(1–2):37–44

Bradshaw AD (2009) The role of SPARC in extracellular matrix assembly. J Cell Commun Signal 3 (3-4):239–246

Brainwood D, Kashtan C, Gubler MC, Turner AN (1998) Targets of alloantibodies in Alport anti-glomerular basement membrane disease after renal transplantation. Kidney Int 53(3):762–766

Brassart-Pasco S, Sénéchal K, Thevenard J, Ramont L, Devy J, Di Stefano L et al (2012) Tetrastatin, the NC1 domain of the α4(IV) collagen chain: a novel potent anti-tumor matrikine. PLoS One 7(4):e29587

Brazel D, Oberbaumer I, Dieringer H, Babel W, Glanville RW, Deutzmann R et al (1987) Completion of the amino acid sequence of the alpha 1 chain of human basement membrane collagen (type IV) reveals 21 non-triplet interruptions located within the collagenous domain. Eur J Biochem 168(3):529–536

Brazel D, Pollner R, Oberbaumer I, Kuhn K (1988) Human basement membrane collagen (type IV). The amino acid sequence of the alpha 2(IV) chain and its comparison with the alpha 1(IV) chain reveals deletions in the alpha 1(IV) chain. Eur J Biochem 172(1):35–42

Brown KL, Cummings CF, Vanacore RM, Hudson BG (2017) Building collagen IV smart scaffolds on the outside of cells. Protein Sci 26(11):2151–2161

Bullich G, Trujillano D, Santin S, Ossowski S, Mendizabal S, Fraga G et al (2015) Targeted next-generation sequencing in steroid-resistant nephrotic syndrome: mutations in multiple glomerular genes may influence disease severity. Eur J Hum Genet 23(9):1192–1199
Bunt S, Hooley C, Hu N, Scahill C, Weavers H, Skaer H (2010) Hemocyte-secreted type IV collagen enhances BMP signaling to guide renal tubule morphogenesis in Drosophila. Dev Cell 19(2):296–306
Buttice G, Kaytes P, D'Armiento J, Vogeli G, Kurkinen M (1990) Evolution of collagen IV genes from a 54-base pair exon: a role for introns in gene evolution. J Mol Evol 30(6):479–488
Cabral WA, Chang W, Barnes AM, Weis M, Scott MA, Leikin S et al (2007) Prolyl 3-hydroxylase 1 deficiency causes a recessive metabolic bone disorder resembling lethal/severe osteogenesis imperfecta. Nat Genet 39(3):359–365
Campbell ID, Humphries MJ (2011) Integrin structure, activation, and interactions. Cold Spring Harb Perspect Biol 3(3)
Chang J, Garva R, Pickard A, Yeung CC, Mallikarjun V, Swift J et al (2020) Circadian control of the secretory pathway maintains collagen homeostasis. Nat Cell Biol 22(1):74–86
Chau BN, Xin CY, Hartner J, Ren SY, Castano AP, Linn G et al (2012) MicroRNA-21 promotes fibrosis of the kidney by silencing metabolic pathways. Sci Transl Med 4(121):ARTN 121ra18
Chen Z, Migeon T, Verpont MC, Zaidan M, Sado Y, Kerjaschki D et al (2016) HANAC syndrome Col4a1 mutation causes neonate glomerular hyperpermeability and adult glomerulocystic kidney disease. J Am Soc Nephrol 27(4):1042–1054
Chew C, Lennon R (2018) Basement membrane defects in genetic kidney diseases. Front Pediatr. 6:11
Chioran A, Duncan S, Catalano A, Brown TJ, Ringuette MJ (2017) Collagen IV trafficking: the inside-out and beyond story. Dev Biol 431(2):124–133
Chiusa M, Hu W, Liao HJ, Su Y, Borza CM, de Caestecker MP et al (2019) The extracellular matrix receptor discoidin domain receptor 1 regulates collagen transcription by translocating to the nucleus. J Am Soc Nephrol 30(9):1605–1624
Chung J, Marini S, Pera J, Norrving B, Jimenez-Conde J, Roquer J et al (2019) Genome-wide association study of cerebral small vessel disease reveals established and novel loci. Brain 142 (10):3176–3189
Colin E, Sentilhes L, Sarfati A, Mine M, Guichet A, Ploton C et al (2014) Fetal intracerebral hemorrhage and cataract: think COL4A1. J Perinatol 34(1):75–77
Colorado PC, Torre A, Kamphaus G, Maeshima Y, Hopfer H, Takahashi K et al (2000) Anti-angiogenic cues from vascular basement membrane collagen. Cancer Res 60(9):2520–2526
Cosgrove D, Liu S (2017) Collagen IV diseases: a focus on the glomerular basement membrane in Alport syndrome. Matrix Biol 57–58:45–54
Cosgrove D, Meehan DT, Grunkemeyer JA, Kornak JM, Sayers R, Hunter WJ et al (1996) Collagen COL4A3 knockout: a mouse model for autosomal Alport syndrome. Genes Dev 10 (23):2981–2992
Cosgrove D, Dufek B, Meehan DT, Delimont D, Hartnett M, Samuelson G et al (2018) Lysyl oxidase like-2 contributes to renal fibrosis in Col4alpha3/Alport mice. Kidney Int 94 (2):303–314
Costa-Mattioli M, Walter P (2020) The integrated stress response: from mechanism to disease. Science 368(6489)
Coucke PJ, Willaert A, Wessels MW, Callewaert B, Zoppi N, De Backer J et al (2006) Mutations in the facilitative glucose transporter GLUT10 alter angiogenesis and cause arterial tortuosity syndrome. Nat Genet 38(4):452–457
Coupry I, Sibon I, Mortemousque B, Rouanet F, Mine M, Goizet C (2010) Ophthalmological features associated with COL4A1 mutations. Arch Ophthalmol 128(4):483–489
Cummings CF, Pedchenko V, Brown KL, Colon S, Rafi M, Jones-Paris C et al (2016) Extracellular chloride signals collagen IV network assembly during basement membrane formation. J Cell Biol 213(4):479–494

Daga S, Fallerini C, Furini S, Pecoraro C, Scolari F, Ariani F et al (2019) Non-collagen genes role in digenic Alport syndrome. BMC Nephrol 20(1):70

Dai J, Ma M, Feng Z, Pastor-Pareja JC (2017) Inter-adipocyte Adhesion and Signaling by Collagen IV Intercellular Concentrations in Drosophila. Curr Biol 27(18):2729–2740. e4

Dai J, Estrada B, Jacobs S, Sánchez-Sánchez BJ, Tang J, Ma M et al (2018) Dissection of Nidogen function in Drosophila reveals tissue-specific mechanisms of basement membrane assembly. PLoS Genet 14(9):e1007483

Dedhar S, Saulnier R, Nagle R, Overall CM (1993) Specific alterations in the expression of alpha 3 beta 1 and alpha 6 beta 4 integrins in highly invasive and metastatic variants of human prostate carcinoma cells selected by in vitro invasion through reconstituted basement membrane. Clin Exp Metastasis 11(5):391–400

Delimont D, Dufek BM, Meehan DT, Zallocchi M, Gratton MA, Phillips G et al (2014) Laminin alpha2-mediated focal adhesion kinase activation triggers Alport glomerular pathogenesis. PLoS One 9(6):e99083

Deml B, Reis LM, Maheshwari M, Griffis C, Bick D, Semina EV (2014) Whole exome analysis identifies dominant COL4A1 mutations in patients with complex ocular phenotypes involving microphthalmia. Clin Genet 86(5):475–481

Donald JE, Kulp DW, DeGrado WF (2011) Salt bridges: geometrically specific, designable interactions. Proteins 79(3):898–915

Draper GW, Shoemark DK, Adams JC (2019) Modelling the early evolution of extracellular matrix from modern Ctenophores and Sponges. Essays Biochem 63(3):389–405

Dufek B, Meehan DT, Delimont D, Cheung L, Gratton MA, Phillips G et al (2016) Endothelin A receptor activation on mesangial cells initiates Alport glomerular disease. Kidney Int 90 (2):300–310.

Elices MJ, Urry LA, Hemler ME (1991) Receptor functions for the integrin VLA-3: fibronectin, collagen, and laminin binding are differentially influenced by Arg-Gly-Asp peptide and by divalent cations. J Cell Biol 112(1):169–181

Engin F, Hotamisligil GS (2010) Restoring endoplasmic reticulum function by chemical chaperones: an emerging therapeutic approach for metabolic diseases. Diabetes Obes Metab 12(Suppl 2):108–115

Exposito JY, Le Guellec D, Lu Q, Garrone R (1991) Short chain collagens in sponges are encoded by a family of closely related genes. J Biol Chem 266(32):21923–21928

Fallerini C, Baldassarri M, Trevisson E, Morbidoni V, La Manna A, Lazzarin R et al (2017) Alport syndrome: impact of digenic inheritance in patients management. Clin Genet 92(1):34–44

Favor J, Gloeckner CJ, Janik D, Klempt M, Neuhauser-Klaus A, Pretsch W et al (2007) Type IV procollagen missense mutations associated with defects of the eye, vascular stability, the brain, kidney function and embryonic or postnatal viability in the mouse, Mus musculus: an extension of the Col4a1 allelic series and the identification of the first two Col4a2 mutant alleles. Genetics 175(2):725–736

Fidler AL, Darris CE, Chetyrkin SV, Pedchenko VK, Boudko SP, Brown KL et al (2017) Collagen IV and basement membrane at the evolutionary dawn of metazoan tissues. Elife 6

Fidler AL, Boudko SP, Rokas A, Hudson BG (2018) The triple helix of collagens – an ancient protein structure that enabled animal multicellularity and tissue evolution. J Cell Sci 131(7)

Finnell JG, Tsang TM, Cryan L, Garrard S, Lee SL, Ackroyd PC et al (2020) A canstatin-derived peptide provides insight into the role of capillary morphogenesis gene 2 in angiogenic regulation and matrix uptake. ACS Chem Biol 15(2):587–596

Fischer G, Schmidt C, Opitz J, Cully Z, Kuhn K, Poschl E (1993) Identification of a novel sequence element in the common promoter region of human collagen type IV genes, involved in the regulation of divergent transcription. Biochem J 292(Pt 3):687–695

Forlino A, Marini JC (2016) Osteogenesis imperfecta. Lancet 387(10028):1657–1671

Fox MA, Sanes JR, Borza DB, Eswarakumar VP, Fassler R, Hudson BG et al (2007) Distinct target-derived signals organize formation, maturation, and maintenance of motor nerve terminals. Cell 129(1):179–193

Friedman L, Higgin JJ, Moulder G, Barstead R, Raines RT, Kimble J (2000) Prolyl 4-hydroxylase is required for viability and morphogenesis in *Caenorhabditis elegans*. Proc Natl Acad Sci USA 97(9):4736–4741

Fujii KK, Taga Y, Sakai T, Ito S, Hattori S, Nagata K et al (2019) Lowering the culture temperature corrects collagen abnormalities caused by HSP47 gene knockout. Sci Rep 9(1):17433

Funk SD, Lin M-H, Miner JH (2018a) Alport syndrome and Pierson syndrome: diseases of the glomerular basement membrane. Matrix Biol 71–72:250–261

Funk SD, Bayer RH, Malone AF, McKee KK, Yurchenco PD, Miner JH (2018b) Pathogenicity of a human laminin beta2 mutation revealed in models of Alport syndrome. J Am Soc Nephrol 29 (3):949–960

Funk SD, Bayer RH, Miner JH (2019) Endothelial cell-specific collagen type IV-alpha3 expression does not rescue Alport syndrome in Col4a3(-)(/-) mice. Am J Physiol Renal Physiol 316(5): F830–F8F7

Gardner H, Kreidberg J, Koteliansky V, Jaenisch R (1996) Deletion of integrin alpha 1 by homologous recombination permits normal murine development but gives rise to a specific deficit in cell adhesion. Dev Biol 175(2):301–313

Gatseva A, Sin YY, Brezzo G, Van Agtmael T (2019) Basement membrane collagens and disease mechanisms. Essays Biochem 63(3):297–312

Gomez IG, MacKenna DA, Johnson BG, Kaimal V, Roach AM, Ren SY et al (2015) Anti-microRNA-21 oligonucleotides prevent Alport nephropathy progression by stimulating metabolic pathways. J Clin Invest 125(1):141–156

Gonzalez-Teuber V, Albert-Gasco H, Auyeung VC, Papa FR, Mallucci GR, Hetz C (2019) Small molecules to improve ER proteostasis in disease. Trends Pharmacol Sci 40(9):684–695

Gotenstein JR, Koo CC, Ho TW, Chisholm AD (2018) Genetic suppression of basement membrane defects in *Caenorhabditis elegans* by gain of function in extracellular matrix and cell-matrix attachment genes. Genetics 208(4):1499–1512

Gould DB, John SW (2002) Anterior segment dysgenesis and the developmental glaucomas are complex traits. Hum Mol Genet 11(10):1185–1193

Gould DB, Phalan FC, Breedveld GJ, van Mil SE, Smith RS, Schimenti JC et al (2005) Mutations in Col4a1 cause perinatal cerebral hemorrhage and porencephaly. Science. 308(5725):1167–1171

Gould DB, Phalan FC, van Mil SE, Sundberg JP, Vahedi K, Massin P et al (2006) Role of COL4A1 in small-vessel disease and hemorrhagic stroke. N Engl J Med 354(14):1489–1496

Gould DB, Marchant JK, Savinova OV, Smith RS, John SW (2007) Col4a1 mutation causes endoplasmic reticulum stress and genetically modifiable ocular dysgenesis. Hum Mol Genet 16(7):798–807

Goumans MJ, Zwijsen A, Ten Dijke P, Bailly S (2018) Bone morphogenetic proteins in vascular homeostasis and disease. Cold Spring Harb Perspect Biol 10(2)

Grami N, Chong M, Lali R, Mohammadi-Shemirani P, Henshall DE, Rannikmae K et al (2020) Global assessment of mendelian stroke genetic prevalence in 101 635 individuals from 7 ethnic groups. Stroke 51(4):1290–1293

Grant DS, Lelkes PI, Fukuda K, Kleinman HK (1991) Intracellular mechanisms involved in basement membrane induced blood vessel differentiation in vitro. In Vitro Cell Dev Biol 27A (4):327–336

Griffiths M, Van Sinderen M, Rainczuk K, Dimitriadis E (2019) miR-29c overexpression and COL4A1 downregulation in infertile human endometrium reduces endometrial epithelial cell adhesive capacity in vitro implying roles in receptivity. Sci Rep 9(1):8644

Gross O, Koepke ML, Beirowski B, Schulze-Lohoff E, Segerer S, Weber M (2005) Nephroprotection by antifibrotic and anti-inflammatory effects of the vasopeptidase inhibitor AVE7688. Kidney Int 68(2):456–463

Gross O, Borza DB, Anders HJ, Licht C, Weber M, Segerer S et al (2009) Stem cell therapy for Alport syndrome: the hope beyond the hype. Nephrol Dial Transplant 24(3):731–734

Gross O, Perin L, Deltas C (2014) Alport syndrome from bench to bedside: the potential of current treatment beyond RAAS blockade and the horizon of future therapies. Nephrol Dial Transplant 29(Suppl 4):iv124-30

Gross O, Appel G, Block G, Chin M, Goldsberry A, Inker L et al (2018) A phase 2/3 study of the efficacy and safety of bardoxolone methyl in patients with Alport syndrome. Nephrol Dial Transplant 33

Gross O, Tonshoff B, Weber LT, Pape L, Latta K, Fehrenbach H et al (2020) A multicenter, randomized, placebo-controlled, double-blind phase 3 trial with open-arm comparison indicates safety and efficacy of nephroprotective therapy with ramipril in children with Alport's syndrome. Kidney Int 97(6):1275–1286

Guan M, Ma J, Keaton JM, Dimitrov L, Mudgal P, Stromberg M et al (2016) Association of kidney structure-related gene variants with type 2 diabetes-attributed end-stage kidney disease in African Americans. Hum Genet 135(11):1251–1262

Gubler MC (2008) Inherited diseases of the glomerular basement membrane. Nat Clin Pract Nephrol 4(1):24–37

Guiraud S, Migeon T, Ferry A, Chen Z, Ouchelouche S, Verpont MC et al (2017) HANAC Col4a1 mutation in mice leads to skeletal muscle alterations due to a primary vascular defect. Am J Pathol 187(3):505–516

Guo J, Song W, Boulanger J, Xu EY, Wang F, Zhang Y et al (2019) Dysregulated expression of microRNA-21 and disease-related genes in human patients and in a mouse model of alport syndrome. Hum Gene Ther 30(7):865–881

Gupta MC, Graham PL, Kramer JM (1997) Characterization of alpha1(IV) collagen mutations in Caenorhabditis elegans and the effects of alpha1 and alpha2(IV) mutations on type IV collagen distribution. J Cell Biol 137(5):1185–1196

Haniel A, Welge-Lussen U, Kuhn K, Poschl E (1995) Identification and characterization of a novel transcriptional silencer in the human collagen type IV gene COL4A2. J Biol Chem 270 (19):11209–11215

Haraida S, Nerlich AG, Wiest I, Schleicher E, Lohrs U (1996) Distribution of basement membrane components in normal adipose tissue and in benign and malignant tumors of lipomatous origin. Mod Pathol 9(2):137–144

Hartnett ME, Martiniuk D, Byfield G, Geisen P, Zeng G, Bautch VL (2008) Neutralizing VEGF decreases tortuosity and alters endothelial cell division orientation in arterioles and veins in a rat model of ROP: relevance to plus disease. Invest Ophthalmol Vis Sci 49(7):3107–3114

Hashikami K, Asahina M, Nozu K, Iijima K, Nagata M, Takeyama M (2019) Establishment of X-linked Alport syndrome model mice with a Col4a5 R471X mutation. Biochem Biophys Rep 17:81–86

Hayashi G, Labelle-Dumais C, Gould DB (2018) Use of sodium 4-phenylbutyrate to define therapeutic parameters for reducing intracerebral hemorrhage and myopathy in *Col4a1* mutant mice. Dis Models Mech 11(7):dmm034157

He G-A, Luo J-X, Zhang T-Y, Wang F-Y, Li R-F (2003) Canstatin-N fragment inhibits in vitro endothelial cell proliferation and suppresses in vivo tumor growth. Biochem Biophys Res Commun 312(3):801–805

Hedin U, Roy J, Tran PK, Lundmark K, Rahman A (1999) Control of smooth muscle cell proliferation—the role of the basement membrane. Thromb Haemost 82(Suppl 1):23–26

Heidet L, Dahan K, Zhou J, Xu Z, Cochat P, Gould JD et al (1995) Deletions of both alpha 5 (IV) and alpha 6(IV) collagen genes in Alport syndrome and in Alport syndrome associated with smooth muscle tumours. Hum Mol Genet 4(1):99–108

Heidet L, Borza DB, Jouin M, Sich M, Mattei MG, Sado Y et al (2003) A human-mouse chimera of the alpha3alpha4alpha5(IV) collagen protomer rescues the renal phenotype in Col4a3-/- Alport mice. Am J Pathol 163(4):1633–1644

Heikkila P, Soininen R, Tryggvason K (1993) Directional regulatory activity of cis-acting elements in the bidirectional alpha 1(IV) and alpha 2(IV) collagen gene promoter. J Biol Chem 268 (33):24677–24682

Henderson SR, Salama AD (2018) Diagnostic and management challenges in Goodpasture's (anti-glomerular basement membrane) disease. Nephrol Dial Transplant 33(2):196–202
Herbst TJ, McCarthy JB, Tsilibary EC, Furcht LT (1988) Differential effects of laminin, intact type IV collagen, and specific domains of type IV collagen on endothelial cell adhesion and migration. J Cell Biol 106(4):1365–1373
Hogan BM, Schulte-Merker S (2017) How to plumb a pisces: understanding vascular development and disease using zebrafish embryos. Dev Cell 42(6):567–583
Hohenester E, Maurer P, Timpl R (1997) Crystal structure of a pair of follistatin-like and EF-hand calcium-binding domains in BM-40. EMBO J 16(13):3778–3786
Holster T, Pakkanen O, Soininen R, Sormunen R, Nokelainen M, Kivirikko KI et al (2007) Loss of assembly of the main basement membrane collagen, type IV, but not fibril-forming collagens and embryonic death in collagen prolyl 4-hydroxylase I null mice. J Biol Chem 282 (4):2512–2519
Hostikka SL, Tryggvason K (1988) The complete primary structure of the alpha 2 chain of human type IV collagen and comparison with the alpha 1(IV) chain. J Biol Chem 263 (36):19488–19493
Hostikka SL, Eddy RL, Byers MG, Hoyhtya M, Shows TB, Tryggvason K (1990) Identification of a distinct type IV collagen alpha chain with restricted kidney distribution and assignment of its gene to the locus of X chromosome-linked Alport syndrome. Proc Natl Acad Sci USA 87 (4):1606–1610
Hudson BG, Tryggvason K, Sundaramoorthy M, Neilson EG (2003) Alport's syndrome, Goodpasture's syndrome, and type IV collagen. N Engl J Med 348(25):2543–2556
Hudson DM, Joeng KS, Werther R, Rajagopal A, Weis M, Lee BH et al (2015) Post-translationally abnormal collagens of prolyl 3-hydroxylase-2 null mice offer a pathobiological mechanism for the high myopia linked to human LEPREL1 mutations. J Biol Chem 290(13):8613–8622
Humphries JD, Chastney MR, Askari JA, Humphries MJ (2019) Signal transduction via integrin adhesion complexes. Curr Opin Cell Biol 56:14–21
Hynes RO (2012) The evolution of metazoan extracellular matrix. J Cell Biol. 196(6):671–679
Isabella AJ, Horne-Badovinac S (2015) Dynamic regulation of basement membrane protein levels promotes egg chamber elongation in Drosophila. Dev Biol 406(2):212–221
Ishikawa Y, Wirz J, Vranka JA, Nagata K, Bachinger HP (2009) Biochemical characterization of the prolyl 3-hydroxylase 1.cartilage-associated protein.cyclophilin B complex. J Biol Chem 284 (26):17641–17647
Ishikawa Y, Ito S, Nagata K, Sakai LY, Bachinger HP (2016) Intracellular mechanisms of molecular recognition and sorting for transport of large extracellular matrix molecules. Proc Natl Acad Sci USA 113(41):E6036–E6E44
Ito S, Nagata K (2019) Roles of the endoplasmic reticulum–resident, collagen-specific molecular chaperone Hsp47 in vertebrate cells and human disease. J Biol Chem 294(6):2133–2141
Itoh Y (2015) Membrane-type matrix metalloproteinases: their functions and regulations. Matrix Biol 44–46:207–223
Jayadev R, Sherwood DR (2017) Basement membranes. Curr Biol 27(6):R207–RR11
Jayadev R, Chi Q, Keeley DP, Hastie EL, Kelley LC, Sherwood DR (2019) alpha-Integrins dictate distinct modes of type IV collagen recruitment to basement membranes. J Cell Biol 218 (9):3098–3116
Jeanne M, Gould DB (2017) Genotype-phenotype correlations in pathology caused by collagen type IV alpha 1 and 2 mutations. Matrix Biol 57–58:29–44
Jeanne M, Labelle-Dumais C, Jorgensen J, Kauffman WB, Mancini GM, Favor J et al (2012) COL4A2 mutations impair COL4A1 and COL4A2 secretion and cause hemorrhagic stroke. Am J Hum Genet 90(1):91–101
Jeanne M, Jorgensen J, Gould DB (2015) Molecular and genetic analyses of collagen type IV mutant mouse models of spontaneous intracerebral hemorrhage identify mechanisms for stroke prevention. Circulation 131(18):1555–1565

Jiang W, Zhang Y, Wu H, Zhang X, Gan H, Sun J et al (2010) Role of cross-talk between the Smad2 and MAPK pathways in TGF-beta1-induced collagen IV expression in mesangial cells. Int J Mol Med 26(4):571–576

Jones FE, Bailey MA, Murray LS, Lu Y, McNeilly S, Schlotzer-Schrehardt U et al (2016) ER stress and basement membrane defects combine to cause glomerular and tubular renal disease resulting from Col4a1 mutations in mice. Dis Model Mech 9(2):165–176

Jones FE, Murray LS, McNeilly S, Dean A, Aman A, Lu Y et al (2019) 4-Sodium phenyl butyric acid has both efficacy and counter-indicative effects in the treatment of Col4a1 disease. Hum Mol Genet 28(4):628–638

Kaido T, Yebra M, Cirulli V, Montgomery AM (2004) Regulation of human beta-cell adhesion, motility, and insulin secretion by collagen IV and its receptor alpha1beta1. J Biol Chem 279 (51):53762–53769

Kalluri R, Danoff TM, Okada H, Neilson EG (1997) Susceptibility to anti-glomerular basement membrane disease and Goodpasture syndrome is linked to MHC class II genes and the emergence of T cell-mediated immunity in mice. J Clin Invest 100(9):2263–2275

Kalluri R, Cantley LG, Kerjaschki D, Neilson EG (2000) Reactive oxygen species expose cryptic epitopes associated with autoimmune goodpasture syndrome. J Biol Chem 275 (26):20027–20032

Kamagata Y, Mattei MG, Ninomiya Y (1992) Isolation and sequencing of cDNAs and genomic DNAs encoding the alpha 4 chain of basement membrane collagen type IV and assignment of the gene to the distal long arm of human chromosome 2. J Biol Chem 267(33):23753–23758

Kang JS, Wang XP, Miner JH, Morello R, Sado Y, Abrahamson DR et al (2006) Loss of alpha3/alpha4(IV) collagen from the glomerular basement membrane induces a strain-dependent isoform switch to alpha5alpha6(IV) collagen associated with longer renal survival in Col4a3-/- Alport mice. J Am Soc Nephrol 17(7):1962–1969

Kang JS, Kashtan CE, Turner AN, Heidet L, Hudson BG, Borza DB (2007) The alloantigenic sites of alpha3alpha4alpha5(IV) collagen: pathogenic X-linked alport alloantibodies target two accessible conformational epitopes in the alpha5NC1 domain. J Biol Chem 282 (14):10670–10677

Kashtan CE, Kim Y, Lees GE, Thorner PS, Virtanen I, Miner JH (2001) Abnormal glomerular basement membrane laminins in murine, canine, and human Alport syndrome: aberrant laminin alpha2 deposition is species independent. J Am Soc Nephrol 12(2):252–260

Kashtan CE, Ding J, Garosi G, Heidet L, Massella L, Nakanishi K et al (2018) Alport syndrome: a unified classification of genetic disorders of collagen IV alpha345: a position paper of the Alport Syndrome Classification Working Group. Kidney Int 93(5):1045–1051

Katayama K, Kawano M, Naito I, Ishikawa H, Sado Y, Asakawa N et al (2008) Irradiation prolongs survival of Alport mice. J Am Soc Nephrol 19(9):1692–1700

Keely PJ, Fong AM, Zutter MM, Santoro SA (1995) Alteration of collagen-dependent adhesion, motility, and morphogenesis by the expression of antisense alpha(2) integrin messenger-Rna in mammary cells. J Cell Sci 108:595–607

Kefalides NA (1966) A collagen of unusual composition and a glycoprotein isolated from canine glomerular basement membrane. Biochem Biophys Res Commun 22(1):26–32

Kefalides NA (1973) Structure and biosynthesis of basement membranes. Int Rev Connect Tissue Res 6:63–104

Kelemen-Valkony I, Kiss M, Csiha J, Kiss A, Bircher U, Szidonya J et al (2012) Drosophila basement membrane collagen col4a1 mutations cause severe myopathy. Matrix Biol 31 (1):29–37

Kelley PB, Sado Y, Duncan MK (2002) Collagen IV in the developing lens capsule. Matrix Biol 21 (5):415–423

Kern A, Eble J, Golbik R, Kuhn K (1993) Interaction of type-Iv collagen with the isolated integrin-alpha-1-beta-1 and integrin-alpha-2-beta-1. Eur J Biochem 215(1):151–159

Khalid R, Krishnan P, Andres K, Blaser S, Miller S, Moharir M et al (2018) COL4A1 and fetal vascular origins of schizencephaly. Neurology 90(5):232–234

Khoshnoodi J, Sigmundsson K, Cartailler JP, Bondar O, Sundaramoorthy M, Hudson BG (2006a) Mechanism of chain selection in the assembly of collagen IV: a prominent role for the alpha2 chain. J Biol Chem 281(9):6058–6069

Khoshnoodi J, Cartailler JP, Alvares K, Veis A, Hudson BG (2006b) Molecular recognition in the assembly of collagens: terminal noncollagenous domains are key recognition modules in the formation of triple helical protomers. J Biol Chem. 281(50):38117–38121

Kinoshita K, Ishizaki Y, Yamamoto H, Sonoda M, Yonemoto K, Kira R et al (2020) De novo p. G696S mutation in COL4A1 causes intracranial calcification and late-onset cerebral hemorrhage: A case report and review of the literature. Eur J Med Genet 63(4):103825

Kiss M, Kiss AA, Radics M, Popovics N, Hermesz E, Csiszar K et al (2016) Drosophila type IV collagen mutation associates with immune system activation and intestinal dysfunction. Matrix Biol 49:120–131

Kiss AA, Somlyai-Popovics N, Kiss M, Boldogkoi Z, Csiszar K, Mink M (2019) Type IV collagen is essential for proper function of integrin-mediated adhesion in drosophila muscle fibers. Int J Mol Sci 20(20)

Kitzler TM, Schneider R, Kohl S, Kolvenbach CM, Connaughton DM, Dai R et al (2019) COL4A1 mutations as a potential novel cause of autosomal dominant CAKUT in humans. Hum Genet 138(10):1105–1115

Kivirikko KI, Pihlajaniemi T (1998) Collagen hydroxylases and the protein disulfide isomerase subunit of prolyl 4-hydroxylases. Adv Enzymol Relat Areas Mol Biol 72:325–398

Knight CG, Morton LF, Peachey AR, Tuckwell DS, Farndale RW, Barnes MJ (2000) The collagen-binding A-domains of integrins alpha(1)beta(1) and alpha(2)beta(1) recognize the same specific amino acid sequence, GFOGER, in native (triple-helical) collagens. J Biol Chem 275(1):35–40

Koepke ML, Weber M, Schulze-Lohoff E, Beirowski B, Segerer S, Gross O (2007) Nephroprotective effect of the HMG-CoA-reductase inhibitor cerivastatin in a mouse model of progressive renal fibrosis in Alport syndrome. Nephrol Dial Transplant 22(4):1062–1069

Korstanje R, Caputo CR, Doty RA, Cook SA, Bronson RT, Davisson MT et al (2014) A mouse *Col4a4* mutation causing Alport glomerulosclerosis with abnormal collagen α3α4α5 (IV) trimers. Kidney Int 85(6):1461–1468

Koskimaki JE, Karagiannis ED, Tang BC, Hammers H, Watkins DN, Pili R et al (2010) Pentastatin-1, a collagen IV derived 20-mer peptide, suppresses tumor growth in a small cell lung cancer xenograft model. BMC Cancer 10(1):29

Kuo DS, Labelle-Dumais C, Mao M, Jeanne M, Kauffman WB, Allen J et al (2014) Allelic heterogeneity contributes to variability in ocular dysgenesis, myopathy and brain malformations caused by Col4a1 and Col4a2 mutations. Hum Mol Genet 23(7):1709–1722

Labelle-Dumais C, Dilworth DJ, Harrington EP, de Leau M, Lyons D, Kabaeva Z et al (2011) COL4A1 mutations cause ocular dysgenesis, neuronal localization defects, and myopathy in mice and Walker-Warburg syndrome in humans. PLoS Genet 7(5):e1002062

Labelle-Dumais C, Schuitema V, Hayashi G, Hoff K, Gong W, Dao DQ et al (2019) COL4A1 mutations cause neuromuscular disease with tissue-specific mechanistic heterogeneity. Am J Hum Genet 104(5):847–860

Leinonen A, Mariyama M, Mochizuki T, Tryggvason K, Reeders ST (1994) Complete primary structure of the human type IV collagen alpha 4(IV) chain. Comparison with structure and expression of the other alpha (IV) chains. J Biol Chem 269(42):26172–26177

Leitinger B, Hohenester E (2007) Mammalian collagen receptors. Matrix Biol 26(3):146–155

Lemmens R, Maugeri A, Niessen HW, Goris A, Tousseyn T, Demaerel P et al (2013) Novel COL4A1 mutations cause cerebral small vessel disease by haploinsufficiency. Hum Mol Genet 22(2):391–397

Licholai S, Szczeklik W, Sanak M (2016) miR-29c-3p is an effective biomarker of abdominal aortic aneurysm in patients undergoing elective surgery. Microrna 5(2):124–131

Lin X, Suh JH, Go G, Miner JH (2014) Feasibility of repairing glomerular basement membrane defects in Alport syndrome. J Am Soc Nephrol 25(4):687–692

Lu W, Phillips CL, Killen PD, Hlaing T, Harrison WR, Elder FFB et al (1999) Insertional mutation of the collagen genes Col4a3 and Col4a4 in a mouse model of Alport syndrome. Genomics 61 (2):113–124

Maeshima Y, Colorado PC, Torre A, Holthaus KA, Grunkemeyer JA, Ericksen MB et al (2000) Distinct antitumor properties of a type IV collagen domain derived from basement membrane. J Biol Chem 275(28):21340–21348

Makareeva E, Aviles NA, Leikin S (2011) Chaperoning osteogenesis: new protein-folding disease paradigms. Trends Cell Biol 21(3):168–176

Malhotra V, Erlmann P (2015) The pathway of collagen secretion. Annu Rev Cell Dev Biol 31:109–124

Malik R, Chauhan G, Traylor M, Sargurupremraj M, Okada Y, Mishra A et al (2018) Multiancestry genome-wide association study of 520,000 subjects identifies 32 loci associated with stroke and stroke subtypes. Nat Genet 50(4):524–537

Mao M, Kiss M, Ou Y, Gould DB (2017) Genetic dissection of anterior segment dysgenesis caused by a Col4a1 mutation in mouse. Dis Model Mech 10(4):475–485

Marcocci E, Uliana V, Bruttini M, Artuso R, Silengo MC, Zerial M et al (2009) Autosomal dominant Alport syndrome: molecular analysis of the COL4A4 gene and clinical outcome. Nephrol Dial Transplant. 24(5):1464–1471

Mariyama M, Leinonen A, Mochizuki T, Tryggvason K, Reeders ST (1994) Complete primary structure of the human alpha 3(IV) collagen chain. Coexpression of the alpha 3(IV) and alpha 4 (IV) collagen chains in human tissues. J Biol Chem 269(37):23013–23017

Matsubayashi Y, Louani A, Dragu A, Sanchez-Sanchez BJ, Serna-Morales E, Yolland L et al (2017) A moving source of matrix components is essential for de novo basement membrane formation. Curr Biol 27(22):3526–3534. e4

Matsuoka Y, Kubota H, Adachi E, Nagai N, Marutani T, Hosokawa N et al (2004) Insufficient folding of type IV collagen and formation of abnormal basement membrane-like structure in embryoid bodies derived from Hsp47-null embryonic stem cells. Mol Biol Cell 15 (10):4467–4475

McCall AS, Cummings CF, Bhave G, Vanacore R, Page-McCaw A, Hudson BG (2014) Bromine is an essential trace element for assembly of collagen IV scaffolds in tissue development and architecture. Cell 157(6):1380–1392

McCaughey J, Stephens DJ (2019) ER-to-golgi transport: a sizeable problem. Trends Cell Biol 29 (12):940–953

McCaughey J, Stevenson NL, Cross S, Stephens DJ (2019) ER-to-golgi trafficking of procollagen in the absence of large carriers. J Cell Biol 218(3):929–948

McMahon CJ, Breathnach C, Betts DR, Sharkey FH, Greally MT (2015) De Novo interstitial deletion 13q33.3q34 in a male patient with double outlet right ventricle, microcephaly, dysmorphic craniofacial findings, and motor and developmental delay. Am J Med Genet A 167A (5):1134–1141

Mencarelli MA, Heidet L, Storey H, van Geel M, Knebelmann B, Fallerini C et al (2015) Evidence of digenic inheritance in Alport syndrome. J Med Genet 52(3):163–174

Meuwissen ME, de Vries LS, Verbeek HA, Lequin MH, Govaert PP, Schot R et al (2011) Sporadic COL4A1 mutations with extensive prenatal porencephaly resembling hydranencephaly. Neurology 76(9):844–846

Meuwissen ME, Halley DJ, Smit LS, Lequin MH, Cobben JM, de Coo R et al (2015) The expanding phenotype of COL4A1 and COL4A2 mutations: clinical data on 13 newly identified families and a review of the literature. Genet Med 17(11):843–853

Miner JH, Sanes JR (1996) Molecular and functional defects in kidneys of mice lacking collagen alpha 3(IV): implications for Alport syndrome. J Cell Biol 135(5):1403–1413

Minor RR, Clark CC, Strause EL, Koszalka TR, Brent RL, Kefalides NA (1976) Basement membrane procollagen is not converted to collagen in organ cultures of parietal yolk sac endoderm. J Biol Chem 251(6):1789–1794

Miyatake S, Schneeberger S, Koyama N, Yokochi K, Ohmura K, Shiina M et al (2018) Biallelic COLGALT1 variants are associated with cerebral small vessel disease. Ann Neurol 84 (6):843–853

Mizuno K, Hayashi T, Peyton DH, Bachinger HP (2004) The peptides acetyl-(Gly-3(S)Hyp-4 (R)Hyp)10-NH2 and acetyl-(Gly-Pro-3(S)Hyp)10-NH2 do not form a collagen triple helix. J Biol Chem 279(1):282–287

Momota R, Sugimoto M, Oohashi T, Kigasawa K, Yoshioka H, Ninomiya Y (1998) Two genes, COL4A3 and COL4A4 coding for the human alpha3(IV) and alpha4(IV) collagen chains are arranged head-to-head on chromosome 2q36. FEBS Lett 424(1–2):11–16

Montgomery NT, Zientek KD, Pokidysheva EN, Bachinger HP (2018) Post-translational modification of type IV collagen with 3-hydroxyproline affects its interactions with glycoprotein VI and nidogens 1 and 2. J Biol Chem 293(16):5987–5999

Moon HM, Wynshaw-Boris A (2013) Cytoskeleton in action: lissencephaly, a neuronal migration disorder. Wiley Interdiscip Rev Dev Biol 2(2):229–245

Moroz LL, Kocot KM, Citarella MR, Dosung S, Norekian TP, Povolotskaya IS et al (2014) The ctenophore genome and the evolutionary origins of neural systems. Nature 510(7503):109–114

Morrissey MA, Jayadev R, Miley GR, Blebea CA, Chi Q, Ihara S et al (2016) SPARC promotes cell invasion in vivo by decreasing type IV collagen levels in the basement membrane. PLoS Genet 12(2):e1005905

Mullan LA, Mularczyk EJ, Kung LH, Forouhan M, Wragg JM, Goodacre R et al (2017) Increased intracellular proteolysis reduces disease severity in an ER stress-associated dwarfism. J Clin Invest 127(10):3861–3865

Mundel TM, Kalluri R (2007) Type IV collagen-derived angiogenesis inhibitors. Microvasc Res 74 (2-3):85–89

Murata T, Katayama K, Oohashi T, Jahnukainen T, Yonezawa T, Sado Y et al (2016) COL4A6 is dispensable for autosomal recessive Alport syndrome. Sci Rep 6:29450

Murray JC, Stingl G, Kleinman HK, Martin GR, Katz SI (1979) Epidermal cells adhere preferentially to type IV (basement membrane) collagen. J Cell Biol 80(1):197–202

Murray LS, Lu Y, Taggart A, Van Regemorter N, Vilain C, Abramowicz M et al (2014) Chemical chaperone treatment reduces intracellular accumulation of mutant collagen IV and ameliorates the cellular phenotype of a COL4A2 mutation that causes haemorrhagic stroke. Hum Mol Genet 23(2):283–292

Myllyharju J (2008) Prolyl 4-hydroxylases, key enzymes in the synthesis of collagens and regulation of the response to hypoxia, and their roles as treatment targets. Ann Med 40(6):402–417

Myllyharju J, Kivirikko KI (2004) Collagens, modifying enzymes and their mutations in humans, flies and worms. Trends Genet 20(1):33–43

Nagai N, Hosokawa M, Itohara S, Adachi E, Matsushita T, Hosokawa N et al (2000) Embryonic lethality of molecular chaperone hsp47 knockout mice is associated with defects in collagen biosynthesis. J Cell Biol 150(6):1499–1506

Nauroy P, Hughes S, Naba A, Ruggiero F (2018) The in-silico zebrafish matrisome: a new tool to study extracellular matrix gene and protein functions. Matrix Biol 65:5–13

Netzer KO, Suzuki K, Itoh Y, Hudson BG, Khalifah RG (1998) Comparative analysis of the noncollagenous NC1 domain of type IV collagen: identification of structural features important for assembly, function, and pathogenesis. Protein Sci 7(6):1340–1351

Nguyen H, Ostendorf AP, Satz JS, Westra S, Ross-Barta SE, Campbell KP et al (2013) Glial scaffold required for cerebellar granule cell migration is dependent on dystroglycan function as a receptor for basement membrane proteins. Acta Neuropathol Commun 1:58

Ninomiya Y, Kagawa M, Iyama K, Naito I, Kishiro Y, Seyer JM et al (1995) Differential expression of two basement membrane collagen genes, COL4A6 and COL4A5, demonstrated by immunofluorescence staining using peptide-specific monoclonal antibodies. J Cell Biol 130 (5):1219–1229

Norman KR, Moerman DG (2000) The let-268 locus of *Caenorhabditis elegans* encodes a procollagen lysyl hydroxylase that is essential for type IV collagen secretion. Dev Biol 227 (2):690–705

Noro A, Sillat T, Virtanen I, Ingerpuu S, Back N, Konttinen YT et al (2013) Laminin production and basement membrane deposition by mesenchymal stem cells upon adipogenic differentiation. J Histochem Cytochem 61(10):719–730

Nozu K, Minamikawa S, Yamada S, Oka M, Yanagita M, Morisada N et al (2017) Characterization of contiguous gene deletions in COL4A6 and COL4A5 in Alport syndrome-diffuse leiomyomatosis. J Hum Genet 62(7):733–735

Nozu K, Nakanishi K, Abe Y, Udagawa T, Okada S, Okamoto T et al (2019) A review of clinical characteristics and genetic backgrounds in Alport syndrome. Clin Exp Nephrol 23(2):158–168

Oberbaumer I, Laurent M, Schwarz U, Sakurai Y, Yamada Y, Vogeli G et al (1985) Amino acid sequence of the non-collagenous globular domain (NC1) of the alpha 1(IV) chain of basement membrane collagen as derived from complementary DNA. Eur J Biochem 147(2):217–224

O'Donnell CJ, Kavousi M, Smith AV, Kardia SL, Feitosa MF, Hwang SJ et al (2011) Genome-wide association study for coronary artery calcification with follow-up in myocardial infarction. Circulation 124(25):2855–2864

Omachi K, Miner JH (2019) Alport syndrome therapeutics: ready for prime-time players. Trends Pharmacol Sci 40(11):803–806

Omari S, Makareeva E, Roberts-Pilgrim A, Mirigian L, Jarnik M, Ott C et al (2018) Noncanonical autophagy at ER exit sites regulates procollagen turnover. Proc Natl Acad Sci USA.115(43): E10099–E1E108

Oohashi T, Sugimoto M, Mattei MG, Ninomiya Y (1994) Identification of a new collagen IV chain, alpha 6(IV), by cDNA isolation and assignment of the gene to chromosome Xq22, which is the same locus for COL4A5. J Biol Chem 269(10):7520–7526

Paavola KJ, Sidik H, Zuchero JB, Eckart M, Talbot WS (2014) Type IV collagen is an activating ligand for the adhesion G protein-coupled receptor GPR126. Sci Signal 7(338):ra76

Parkin JD, San Antonio JD, Pedchenko V, Hudson B, Jensen ST, Savige J (2011) Mapping structural landmarks, ligand binding sites, and missense mutations to the collagen IV heterotrimers predicts major functional domains, novel interactions, and variation in phenotypes in inherited diseases affecting basement membranes. Hum Mutat 32(2):127–143

Pastor-Pareja JC, Xu T (2011) Shaping cells and organs in Drosophila by opposing roles of fat body-secreted Collagen IV and perlecan. Dev Cell 21(2):245–256

Pedchenko V, Zent R, Hudson BG (2004) Alpha(v)beta3 and alpha(v)beta5 integrins bind both the proximal RGD site and non-RGD motifs within noncollagenous (NC1) domain of the alpha3 chain of type IV collagen: implication for the mechanism of endothelia cell adhesion. J Biol Chem 279(4):2772–2780

Pedchenko V, Kitching AR, Hudson BG (2018) Goodpasture's autoimmune disease—a collagen IV disorder. Matrix Biol 71–72:240–249

Perrucci GL, Gowran A, Zanobini M, Capogrossi MC, Pompilio G, Nigro P (2015) Peptidyl-prolyl isomerases: a full cast of critical actors in cardiovascular diseases. Cardiovasc Res 106 (3):353–364

Petitclerc E, Boutaud A, Prestayko A, Xu J, Sado Y, Ninomiya Y et al (2000) New functions for non-collagenous domains of human collagen type IV: novel integrin ligands inhibiting angiogenesis and tumor growthin vivo. J Biol Chem 275(11):8051–8061

Pieri M, Stefanou C, Zaravinos A, Erguler K, Stylianou K, Lapathitis G et al (2014) Evidence for activation of the unfolded protein response in collagen IV nephropathies. J Am Soc Nephrol 25 (2):260–275

Pihlajaniemi T, Tryggvason K, Myers JC, Kurkinen M, Lebo R, Cheung MC et al (1985) cDNA clones coding for the pro-alpha1(IV) chain of human type IV procollagen reveal an unusual homology of amino acid sequences in two halves of the carboxyl-terminal domain. J Biol Chem 260(12):7681–7687

Plaisier E, Gribouval O, Alamowitch S, Mougenot B, Prost C, Verpont MC et al (2007) COL4A1 mutations and hereditary angiopathy, nephropathy, aneurysms, and muscle cramps. N Engl J Med 357(26):2687–2695

Plaisier E, Chen Z, Gekeler F, Benhassine S, Dahan K, Marro B et al (2010) Novel COL4A1 mutations associated with HANAC syndrome: a role for the triple helical CB3[IV] domain. Am J Med Genet A 152A(10):2550–2555

Poduval P, Sillat T, Beklen A, Kouri VP, Virtanen I, Konttinen YT (2007) Type IV collagen alpha-chain composition in synovial lining from trauma patients and patients with rheumatoid arthritis. Arthritis Rheum 56(12):3959–3967

Pokidysheva E, Boudko S, Vranka J, Zientek K, Maddox K, Moser M et al (2014) Biological role of prolyl 3-hydroxylation in type IV collagen. Proc Natl Acad Sci USA 111(1):161–166

Pollner R, Schmidt C, Fischer G, Kuhn K, Poschl E (1997) Cooperative and competitive interactions of regulatory elements are involved in the control of divergent transcription of human Col4A1 and Col4A2 genes. FEBS Lett 405(1):31–36

Poschl E, Pollner R, Kuhn K (1988) The genes for the alpha 1(IV) and alpha 2(IV) chains of human basement membrane collagen type IV are arranged head-to-head and separated by a bidirectional promoter of unique structure. EMBO J 7(9):2687–2695

Poschl E, Schlotzer-Schrehardt U, Brachvogel B, Saito K, Ninomiya Y, Mayer U (2004) Collagen IV is essential for basement membrane stability but dispensable for initiation of its assembly during early development. Development 131(7):1619–1628

Pozzi A, Yurchenco PD, Iozzo RV (2017) The nature and biology of basement membranes. Matrix Biol 57–58:1–11

Proesmans W, Knockaert H, Trouet D (2000) Enalapril in paediatric patients with Alport syndrome: 2 years' experience. Eur J Pediatr 159(6):430–433

Raghunath M, Bruckner P, Steinmann B (1994) Delayed triple helix formation of mutant collagen from patients with osteogenesis imperfecta. J Mol Biol 236(3):940–949

Raija S, Haka-Risku T, Prockop DJ, Tryggvason K (1987) Complete primary structure of the α1-chain of human basement membrane (type IV) collagen. FEBS Lett 225(1-2):188–194

Randles MJ, Collinson S, Starborg T, Mironov A, Krendel M, Konigshausen E et al (2016) Three-dimensional electron microscopy reveals the evolution of glomerular barrier injury. Sci Rep 6:35068

Rannikmae K, Davies G, Thomson PA, Bevan S, Devan WJ, Falcone GJ et al (2015) Common variation in COL4A1/COL4A2 is associated with sporadic cerebral small vessel disease. Neurology 84(9):918–926

Rannikmae K, Sivakumaran V, Millar H, Malik R, Anderson CD, Chong M et al (2017) COL4A2 is associated with lacunar ischemic stroke and deep ICH: Meta-analyses among 21,500 cases and 40,600 controls. Neurology 89(17):1829–1839

Raote I, Malhotra V (2019) Protein transport by vesicles and tunnels. J Cell Biol. 218(3):737–739

Rappu P, Salo AM, Myllyharju J, Heino J (2019) Role of prolyl hydroxylation in the molecular interactions of collagens. Essays Biochem 63(3):325–335

Ratelade J, Mezouar N, Domenga-Denier V, Rochey A, Plaisier E, Joutel A (2018) Severity of arterial defects in the retina correlates with the burden of intracerebral haemorrhage in COL4A1-related stroke. J Pathol 244(4):408–420

Ratelade J, Klug NR, Lombardi D, Angelim M, Dabertrand F, Domenga-Denier V et al (2020) Reducing hypermuscularization of the transitional segment between arterioles and capillaries protects against spontaneous intracerebral hemorrhage. Circulation 141(25):2078–2094

Rautavuoma K, Takaluoma K, Sormunen R, Myllyharju J, Kivirikko KI, Soininen R (2004) Premature aggregation of type IV collagen and early lethality in lysyl hydroxylase 3 null mice. Proc Natl Acad Sci USA 101(39):14120–14125

Raya A, Revert-Ros F, Martinez-Martinez P, Navarro S, Roselló E, Vieites B et al (2000) Goodpasture antigen-binding protein, the kinase that phosphorylates the goodpasture antigen, is an alternatively spliced variant implicated in autoimmune pathogenesis. J Biol Chem 275 (51):40392–40399

Reissig LF, Herdina AN, Rose J, Maurer-Gesek B, Lane JL, Prin F et al (2019) The Col4a2(em1 (IMPC)Wtsi) mouse line: lessons from the deciphering the mechanisms of developmental disorders program. Biol Open 8(8)

Renard D, Miné M, Pipiras E, Labauge P, Delahaye A, Benzacken B et al (2014) Cerebral small-vessel disease associated with *COL4A1* and *COL4A2* gene duplications. Neurology 83 (11):1029–1031

Rheault MN (2012) Women and Alport syndrome. Pediatr Nephrol 27(1):41–46

Rheault MN, Kren SM, Thielen BK, Mesa HA, Crosson JT, Thomas W et al (2004) Mouse model of X-linked Alport syndrome. J Am Soc Nephrol 15(6):1466–1474

Ricard-Blum S (2011) The collagen family. Cold Spring Harb Perspect Biol 3(1):a004978

Risteli J, Bachinger HP, Engel J, Furthmayr H, Timpl R (1980) 7-S collagen: characterization of an unusual basement membrane structure. Eur J Biochem 108(1):239–250

Rodahl E, Knappskog PM, Majewski J, Johansson S, Telstad W, Krakenes J et al (2013) Variants of anterior segment dysgenesis and cerebral involvement in a large family with a novel COL4A1 mutation. Am J Ophthalmol 155(5):946–953

Ron D, Walter P (2007) Signal integration in the endoplasmic reticulum unfolded protein response. Nat Rev Mol Cell Biol 8(7):519–529

Rost S, Bach E, Neuner C, Nanda I, Dysek S, Bittner RE et al (2014) Novel form of X-linked nonsyndromic hearing loss with cochlear malformation caused by a mutation in the type IV collagen gene COL4A6. Eur J Hum Genet 22(2):208–215

Rouaud T, Labauge P, Lasserve ET, Mine M, Coustans M, Deburghgraeve V et al (2010) Acute urinary retention due to a novel collagen Col4a1 mutation. Neurology 75(8):747–749

Rubel D, Frese J, Martin M, Leibnitz A, Girgert R, Miosge N et al (2014) Collagen receptors integrin alpha2beta1 and discoidin domain receptor 1 regulate maturation of the glomerular basement membrane and loss of integrin alpha2beta1 delays kidney fibrosis in COL4A3 knockout mice. Matrix Biol 34:13–21

Rubin K, Hook M, Obrink B, Timpl R (1981) Substrate adhesion of rat hepatocytes: mechanism of attachment to collagen substrates. Cell 24(2):463–470

Ruotsalainen H, Sipila L, Vapola M, Sormunen R, Salo AM, Uitto L et al (2006) Glycosylation catalyzed by lysyl hydroxylase 3 is essential for basement membranes. J Cell Sci 119 (Pt 4):625–635

Ryan JF, Pang K, Schnitzler CE, Nguyen AD, Moreland RT, Simmons DK et al (2013) The genome of the ctenophore *Mnemiopsis leidyi* and its implications for cell type evolution. Science 342 (6164):1242592

Sá MJN, Fieremans N, de Brouwer APM, Sousa R, Costa FT, Brito MJ et al (2013) Deletion of the 5′exons of *COL4A6* is not needed for the development of diffuse leiomyomatosis in patients with Alport syndrome. J Med Genet 50(11):745–753

Saito K, Chen M, Bard F, Chen S, Zhou H, Woodley D et al (2009) TANGO1 facilitates cargo loading at endoplasmic reticulum exit sites. Cell 136(5):891–902

Sanes JR (2003) The basement membrane/basal lamina of skeletal muscle. J Biol Chem 278 (15):12601–12604

Santoro SA (1986) Identification of a 160,000 dalton platelet membrane protein that mediates the initial divalent cation-dependent adhesion of platelets to collagen. Cell 46(6):913–920

Saskin A, Sillon G, Palfreeman N, Buhas D (2018) COL4A1/2 CNVs and cerebral small vessel disease: Narrowing in on the critical chromosomal region. Neurology 90(22):1026–1028

Savige J, Sheth S, Leys A, Nicholson A, Mack HG, Colville D (2015) Ocular features in Alport syndrome: pathogenesis and clinical significance. Clin J Am Soc Nephrol 10(4):703–709

Schmidt C, Pollner R, Poschl E, Kuhn K (1992) Expression of human collagen type IV genes is regulated by transcriptional and post-transcriptional mechanisms. FEBS Lett 312(2-3):174–178

Schmidt C, Fischer G, Kadner H, Genersch E, Kuhn K, Poschl E (1993) Differential effects of DNA-binding proteins on bidirectional transcription from the common promoter region of human collagen type IV genes COL4A1 and COL4A2. Biochim Biophys Acta 1174(1):1–10

Schunkert H, Konig IR, Kathiresan S, Reilly MP, Assimes TL, Holm H et al (2011) Large-scale association analysis identifies 13 new susceptibility loci for coronary artery disease. Nat Genet 43(4):333–338

Schwarz-Magdolen U, Oberbaumer I, Kuhn K (1986) cDNA and protein sequence of the NC1 domain of the alpha 2-chain of collagen IV and its comparison with alpha 1(IV). FEBS Lett 208 (2):203–207

Sedrakyan S, Da Sacco S, Milanesi A, Shiri L, Petrosyan A, Varimezova R et al (2012) Injection of amniotic fluid stem cells delays progression of renal fibrosis. J Am Soc Nephrol 23(4):661–673

Segal Y, Zhuang L, Rondeau E, Sraer JD, Zhou J (2001) Regulation of the paired type IV collagen genes COL4A5 and COL4A6. Role of the proximal promoter region. J Biol Chem 276 (15):11791–11797

Senger DR, Davis GE (2011) Angiogenesis. Cold Spring Harb Perspect Biol 3(8):a005090

Setty S, Kim Y, Fields GB, Clegg DO, Wayner EA, Tsilibary EC (1998) Interactions of type IV collagen and its domains with human mesangial cells. J Biol Chem 273(20):12244–12249.

Shah S, Ellard S, Kneen R, Lim M, Osborne N, Rankin J et al (2012) Childhood presentation of COL4A1 mutations. Dev Med Child Neurol 54(6):569–574

Shahab J, Baratta C, Scuric B, Godt D, Venken KJT, Ringuette MJ (2015) Loss of SPARC dysregulates basal lamina assembly to disrupt larval fat body homeostasis in *Drosophila melanogaster*. Dev Dyn 244(4):540–552

Shoulders MD, Raines RT (2009) Collagen structure and stability. Annu Rev Biochem 78:929–958

Sibon I, Coupry I, Menegon P, Bouchet JP, Gorry P, Burgelin I et al (2007) COL4A1 mutation in Axenfeld-Rieger anomaly with leukoencephalopathy and stroke. Ann Neurol 62(2):177–184

Siebold B, Deutzmann R, Kuhn K (1988) The arrangement of intra- and intermolecular disulfide bonds in the carboxyterminal, non-collagenous aggregation and cross-linking domain of basement-membrane type IV collagen. Eur J Biochem 176(3):617–624

Siitonen M, Börjesson-Hanson A, Pöyhönen M, Ora A, Pasanen P, Bras J et al (2017) Multi-infarct dementia of Swedish type is caused by a 3'UTR mutation of COL4A1. Brain 140(5):e29-e

Silva RLE, Kanan Y, Mirando AC, Kim J, Shmueli RB, Lorenc VE et al (2017) Tyrosine kinase blocking collagen IV-derived peptide suppresses ocular neovascularization and vascular leakage. Sci Transl Med 9(373)

Smyth N, Vatansever HS, Murray P, Meyer M, Frie C, Paulsson M et al (1999) Absence of basement membranes after targeting the LAMC1 gene results in embryonic lethality due to failure of endoderm differentiation. J Cell Biol 144(1):151–160

Soder S, Poschl E (2004) The NC1 domain of human collagen IV is necessary to initiate triple helix formation. Biochem Biophys Res Commun 325(1):276–280

Soininen R, Huotari M, Hostikka SL, Prockop DJ, Tryggvason K (1988) The structural genes for alpha 1 and alpha 2 chains of human type IV collagen are divergently encoded on opposite DNA strands and have an overlapping promoter region. J Biol Chem 263(33):17217–17220

Sugimoto M, Oohashi T, Ninomiya Y (1994) The genes COL4A5 and COL4A6, coding for basement membrane collagen chains alpha 5(IV) and alpha 6(IV), are located head-to-head in close proximity on human chromosome Xq22 and COL4A6 is transcribed from two alternative promoters. Proc Natl Acad Sci USA 91(24):11679–11683

Sugimoto H, Mundel TM, Sund M, Xie L, Cosgrove D, Kalluri R (2006) Bone-marrow-derived stem cells repair basement membrane collagen defects and reverse genetic kidney disease. Proc Natl Acad Sci USA 103(19):7321–7326

Sund M, Maeshima Y, Kalluri R (2005) Bifunctional promoter of type IV collagen COL4A5 and COL4A6 genes regulates the expression of alpha5 and alpha6 chains in a distinct cell-specific fashion. Biochem J 387(Pt 3):755–761

Sundaramoorthy M, Meiyappan M, Todd P, Hudson BG (2002) Crystal structure of NC1 domains. Structural basis for type IV collagen assembly in basement membranes. J Biol Chem 277 (34):31142–31153

Takahashi M, Eda A, Fukushima T, Hohjoh H (2012) Reduction of type IV collagen by upregulated miR-29 in normal elderly mouse and klotho-deficient, senescence-model mouse. PLoS One 7 (11):e48974

Takeuchi M, Yamaguchi S, Yonemura S, Kakiguchi K, Sato Y, Higashiyama T et al (2015) Type IV collagen controls the axogenesis of cerebellar granule cells by regulating basement membrane integrity in zebrafish. PLoS Genet 11(10):e1005587

Tarasov KV, Sanna S, Scuteri A, Strait JB, Orrù M, Parsa A et al (2009) COL4A1 is associated with arterial stiffness by genome-wide association scan. Circ Cardiovasc Genet 2(2):151–158

Taylor SH, Al-Youha S, Van Agtmael T, Lu Y, Wong J, McGrouther DA et al (2011) Tendon is covered by a basement membrane epithelium that is required for cell retention and the prevention of adhesion formation. PLoS One 6(1):e16337

Than ME, Henrich S, Huber R, Ries A, Mann K, Kuhn K et al (2002) The 1.9-A crystal structure of the noncollagenous (NC1) domain of human placenta collagen IV shows stabilization via a novel type of covalent Met-Lys cross-link. Proc Natl Acad Sci U S A 99(10):6607–6612

Tiainen P, Pasanen A, Sormunen R, Myllyharju J (2008) Characterization of recombinant human prolyl 3-hydroxylase isoenzyme 2, an enzyme modifying the basement membrane collagen IV. J Biol Chem 283(28):19432–19439

Tiger C-F, Fougerousse F, Grundström G, Velling T, Gullberg D (2001) $\alpha 11\beta 1$ integrin is a receptor for interstitial collagens involved in cell migration and collagen reorganization on mesenchymal nonmuscle cells. Dev Biol 237(1):116–129

Timpl R, Wiedemann H, van Delden V, Furthmayr H, Kuhn K (1981) A network model for the organization of type IV collagen molecules in basement membranes. Eur J Biochem 120 (2):203–211

Tonduti D, Pichiecchio A, La Piana R, Livingston JH, Doherty DA, Majumdar A et al (2012) COL4A1-related disease: raised creatine kinase and cerebral calcification as useful pointers. Neuropediatrics 43(5):283–288

Traylor M, Zhang CR, Adib-Samii P, Devan WJ, Parsons OE, Lanfranconi S et al (2016) Genome-wide meta-analysis of cerebral white matter hyperintensities in patients with stroke. Neurology 86(2):146–153

Traylor M, Malik R, Nalls MA, Cotlarciuc I, Radmanesh F, Thorleifsson G et al (2017) Genetic variation at 16q24.2 is associated with small vessel stroke. Ann Neurol 81(3):383–394

Tulla M, Pentikäinen OT, Viitasalo T, Käpylä J, Impola U, Nykvist P et al (2001) Selective binding of collagen subtypes by integrin α1I, α2I, and α10I domains. J Biol Chem 276 (51):48206–48212

Turner N, Mason PJ, Brown R, Fox M, Povey S, Rees A et al (1992) Molecular cloning of the human Goodpasture antigen demonstrates it to be the alpha 3 chain of type IV collagen. J Clin Invest 89(2):592–601

Van Agtmael T, Bruckner-Tuderman L (2010) Basement membranes and human disease. Cell Tissue Res 339(1):167–188

Van Agtmael T, Schlotzer-Schrehardt U, McKie L, Brownstein DG, Lee AW, Cross SH et al (2005) Dominant mutations of Col4a1 result in basement membrane defects which lead to anterior segment dysgenesis and glomerulopathy. Hum Mol Genet 14(21):3161–3168

Van Agtmael T, Bailey MA, Schlötzer-Schrehardt U, Craigie E, Jackson IJ, Brownstein DG et al (2010) Col4a1 mutation in mice causes defects in vascular function and low blood pressure associated with reduced red blood cell volume. Hum Mol Genet 19(6):1119–1128

Vandenberg P, Kern A, Ries A, Luckenbill-Edds L, Mann K, Kuhn K (1991) Characterization of a type IV collagen major cell binding site with affinity to the alpha 1 beta 1 and the alpha 2 beta 1 integrins. J Cell Biol 113(6):1475–1483

Verbeek E, Meuwissen ME, Verheijen FW, Govaert PP, Licht DJ, Kuo DS et al (2012) COL4A2 mutation associated with familial porencephaly and small-vessel disease. Eur J Hum Genet 20 (8):844–851

Verdura E, Herve D, Bergametti F, Jacquet C, Morvan T, Prieto-Morin C et al (2016) Disruption of a miR-29 binding site leading to COL4A1 upregulation causes pontine autosomal dominant microangiopathy with leukoencephalopathy. Ann Neurol 80(5):741–753

Vermeulen RJ, Peeters-Scholte C, Van Vugt JJ, Barkhof F, Rizzu P, van der Schoor SR et al (2011) Fetal origin of brain damage in 2 infants with a COL4A1 mutation: fetal and neonatal MRI. Neuropediatrics 42(1):1–3

Vitale G, Pichiecchio A, Ormitti F, Tonduti D, Asaro A, Farina L et al (2019) Cortical malformations and COL4A1 mutation: three new cases. Eur J Paediatr Neurol 23(3):410–417

Voskarides K, Damianou L, Neocleous V, Zouvani I, Christodoulidou S, Hadjiconstantinou V et al (2007) COL4A3/COL4A4 mutations producing focal segmental glomerulosclerosis and renal failure in thin basement membrane nephropathy. J Am Soc Nephrol 18(11):3004–3016

Vracko R (1974) Basal lamina scaffold-anatomy and significance for maintenance of orderly tissue structure. Am J Pathol 77(2):314–346

Vranka J, Stadler HS (2009) auml, chinger HP. expression of prolyl 3-hydroxylase genes in embryonic and adult mouse tissues. Cell Struct Funct 34(2):97–104

Walker CA, Spinale FG (1999) The structure and function of the cardiac myocyte: a review of fundamental concepts. J Thorac Cardiovasc Surg 118(2):375–382

Wang XP, Fogo AB, Colon S, Giannico G, Abul-Ezz SR, Miner JH et al (2005) Distinct epitopes for anti-glomerular basement membrane alport alloantibodies and goodpasture autoantibodies within the noncollagenous domain of alpha3(IV) collagen: a janus-faced antigen. J Am Soc Nephrol 16(12):3563–3571

Wang X, Harris RE, Bayston LJ, Ashe HL (2008) Type IV collagens regulate BMP signalling in Drosophila. Nature 455(7209):72–77

Wang YP, Wang DJ, Niu ZB, Cui WT (2017a) Chromosome 13q deletion syndrome involving 13q31qter: a case report. Mol Med Rep 15(6):3658–3664

Wang D, Mohammad M, Wang Y, Tan R, Murray LS, Ricardo S et al (2017b) The chemical chaperone, PBA, reduces ER stress and autophagy and increases collagen IV α5 expression in cultured fibroblasts from men with X-linked Alport syndrome and Missense mutations. Kidney Int Rep 2(4):739–748

Wang Y, Zhang J, Zhao Y, Wang S, Zhang J, Han Q et al (2018) COL4A3 gene variants and diabetic kidney disease in MODY. Clin J Am Soc Nephrol 13(8):1162–1171

Wardlaw JM, Smith C, Dichgans M (2013) Mechanisms of sporadic cerebral small vessel disease: insights from neuroimaging. Lancet Neurol 12(5):483–497

Wardlaw JM, Smith C, Dichgans M (2019) Small vessel disease: mechanisms and clinical implications. Lancet Neurol 18(7):684–696

Webb NJ, Lam C, Shahinfar S, Strehlau J, Wells TG, Gleim GW et al (2011) Efficacy and safety of losartan in children with Alport syndrome—results from a subgroup analysis of a prospective, randomized, placebo- or amlodipine-controlled trial. Nephrol Dial Transplant 26(8):2521–2526

Webb NJ, Shahinfar S, Wells TG, Massaad R, Gleim GW, McCrary Sisk C et al (2013) Losartan and enalapril are comparable in reducing proteinuria in children with Alport syndrome. Pediatr Nephrol 28(5):737–743

Weng YC, Sonni A, Labelle-Dumais C, de Leau M, Kauffman WB, Jeanne M et al (2012) COL4A1 mutations in patients with sporadic late-onset intracerebral hemorrhage. Ann Neurol 71 (4):470–477

Xiao T, Baier H (2007) Lamina-specific axonal projections in the zebrafish tectum require the type IV collagen Dragnet. Nat Neurosci 10:1529

Yamada Y, Avvedimento VE, Mudryj M, Ohkubo H, Vogeli G, Irani M et al (1980) The collagen gene: evidence for its evolutinary assembly by amplification of a DNA segment containing an exon of 54 bp. Cell 22(3):887–892

Yamada Y, Kato K, Oguri M, Fujimaki T, Yokoi K, Matsuo H et al (2008) Genetic risk for myocardial infarction determined by polymorphisms of candidate genes in a Japanese population. J Med Genet 45(4):216–221

Yang W, Ng FL, Chan K, Pu X, Poston RN, Ren M et al (2016) Coronary-heart-disease-associated genetic variant at the COL4A1/COL4A2 locus affects COL4A1/COL4A2 expression, vascular cell survival, atherosclerotic plaque stability and risk of myocardial infarction. PLoS Genet 12 (7):e1006127

Yaramis A, Lochmuller H, Topf A, Sonmezler E, Yilmaz E, Hiz S et al (2020) COL4A1-related autosomal recessive encephalopathy in 2 Turkish children. Neurol Genet 6(1):e392

Yokota T, Omachi K, Suico MA, Kamura M, Kojima H, Fukuda R et al (2018) STAT3 inhibition attenuates the progressive phenotypes of Alport syndrome mouse model. Nephrol Dial Transplant 33(2):214–223

Yoneda Y, Haginoya K, Kato M, Osaka H, Yokochi K, Arai H et al (2013) Phenotypic spectrum of COL4A1 mutations: porencephaly to schizencephaly. Ann Neurol 73(1):48–57

Yurchenco PD (2011) Basement membranes: cell scaffoldings and signaling platforms. Cold Spring Harb Perspect Biol 3(2)

Yurchenco PD, Furthmayr H (1984) Self-assembly of basement membrane collagen. Biochemistry 23(8):1839–1850

Yurchenco PD, Ruben GC (1987) Basement membrane structure in situ: evidence for lateral associations in the type IV collagen network. J Cell Biol 105(6 Pt 1):2559–2568

Yurchenco PD, Ruben GC (1988) Type IV collagen lateral associations in the EHS tumor matrix. Comparison with amniotic and in vitro networks. Am J Pathol 132(2):278–291

Zagaglia S, Selch C, Nisevic JR, Mei D, Michalak Z, Hernandez-Hernandez L et al (2018) Neurologic phenotypes associated with COL4A1/2 mutations: expanding the spectrum of disease. Neurology 91(22):e2078–e2e88.

Zeltz C, Gullberg D (2016) The integrin-collagen connection – a glue for tissue repair? J Cell Sci 129(6):1284

Zhang X, Zhou J, Reeders ST, Tryggvason K (1996) Structure of the human type IV collagen COL4A6 gene, which is mutated in Alport syndrome-associated leiomyomatosis. Genomics 33 (3):473–479

Zhao YY, Duan RN, Ji L, Liu QJ, Yan CZ (2019a) Cervical spinal involvement in a Chinese pedigree with pontine autosomal dominant microangiopathy and leukoencephalopathy caused by a 3′ untranslated region mutation of COL4A1 gene. Stroke 50(9):2307–2313

Zhao X, Chen C, Wei Y, Zhao G, Liu L, Wang C et al (2019b) Novel mutations of COL4A3, COL4A4, and COL4A5 genes in Chinese patients with Alport Syndrome using next generation sequence technique. Mol Genet Genomic Med 7(6):e653

Zhou J, Hertz JM, Tryggvason K (1992) Mutation in the alpha 5(IV) collagen chain in juvenile-onset Alport syndrome without hearing loss or ocular lesions: detection by denaturing gradient gel electrophoresis of a PCR product. Am J Hum Genet 50(6):1291–1300

Zhou J, Mochizuki T, Smeets H, Antignac C, Laurila P, de Paepe A et al (1993) Deletion of the paired alpha 5(IV) and alpha 6(IV) collagen genes in inherited smooth muscle tumors. Science 261(5125):1167–1169

Zhou J, Leinonen A, Tryggvason K (1994a) Structure of the human type IV collagen COL4A5 gene. J Biol Chem 269(9):6608–6614

Zhou J, Ding M, Zhao Z, Reeders ST (1994b) Complete primary structure of the sixth chain of human basement membrane collagen, alpha 6(IV). Isolation of the cDNAs for alpha 6(IV) and comparison with five other type IV collagen chains. J Biol Chem 269(18):13193–13199

Ziomber A, Machnik A, Dahlmann A, Dietsch P, Beck FX, Wagner H et al (2008) Sodium-, potassium-, chloride-, and bicarbonate-related effects on blood pressure and electrolyte homeostasis in deoxycorticosterone acetate-treated rats. Am J Physiol Renal Physiol 295(6):F1752–F1763

Zou Y, Donkervoort S, Salo AM, Foley AR, Barnes AM, Hu Y et al (2017) P4HA1 mutations cause a unique congenital disorder of connective tissue involving tendon, bone, muscle and the eye. Hum Mol Genet 26(12):2207–2217

Chapter 6
Collagens and Muscle Diseases: A Focus on Collagen VI

Valentina Tonelotto, Silvia Castagnaro, Matilde Cescon, and Paolo Bonaldo

Abstract The extracellular matrix is a three-dimensional network providing the proper microenvironment for muscle development and function. Different types of collagens play critical roles in skeletal muscle homeostasis, as revealed by the fact that mutations in distinct collagen genes are linked to inherited muscle disorders, including some forms of myopathies and congenital muscular dystrophies. The following chapter discusses the current knowledge of collagens and their roles in muscle disorders, mostly focusing on collagen VI since it is the collagen type with the major known impact in muscle physiology and whose deficiency is causative for a distinctive group of muscle diseases. Collagen VI is a unique member of the collagen superfamily, displaying distinctive protein domains, supramolecular assembly, and tissue distribution. Moreover, it exerts a broad range of essential functions in the skeletal muscle, and mutations of its genes are linked to congenital muscular dystrophies and congenital myopathies. Mutant and knockout animal models for collagen VI and other collagen types provided a tremendous tool for the mechanistic understanding of human muscle diseases linked to collagens' defects and, thus, for translational studies. Indeed, the increasing knowledge on the underlying pathomolecular mechanisms and the identification of suitable targets represent invaluable benefits for proper clinical diagnosis and for developing treatments and therapeutic opportunities to counteract the onset and progression of clinical symptoms of such severe and often life-threatening muscle pathologies.

Abbreviations

$\Delta\psi_m$	Mitochondrial membrane potential
AFM	Atomic force microscopy
AON	Antisense oligoribonucleotides
BM	Bethlem myopathy

V. Tonelotto · S. Castagnaro · M. Cescon · P. Bonaldo (✉)
Department of Molecular Medicine, University of Padova, Padova, Italy
e-mail: bonaldo@bio.unipd.it

F. Ruggiero (ed.), *The Collagen Superfamily and Collagenopathies*, Biology of Extracellular Matrix 8, https://doi.org/10.1007/978-3-030-67592-9_6

CsA	Cyclosporin A
ECM	Extracellular matrix
EDS	Ehlers-Danlos syndrome
LPD	Low-protein diet
MAO	Monoamine oxidase
MDC1A	Congenital muscular dystrophy type 1A
MTJ	*Myotendinous junction*
NMJ	Neuromuscular junction
PTC	Premature termination codon
PTP	Permeability transition pore
ROS	*Reactive oxygen species*
SCs	Satellite cells
SR	Sarcoplasmic reticulum
TALEN	Transcription activator-like effector nuclease
TCA	Tricarboxylic acid
UCMD	Ullrich congenital muscular dystrophy
vWF-A	Willebrand factor A

6.1 Introduction

Various types of collagens are abundantly present in the extracellular matrix (ECM) of skeletal muscle, most often with a distinctive and/or restricted pattern of deposition in specialized region of muscle ECM. These include collagen types I, III, IV, V, VI, XII, XIV, and XV, as well as some distinct types and isoforms expressed in the specialized ECM of the neuromuscular junction (NMJ) and/or of the myotendinous junction (MTJ), such as types IV, VI, XIII, and XXII (for detailed overviews, see Gillies and Lieber 2011; Csapo et al. 2020). The diversity of the ECM composition of skeletal muscle reflects its structural organization into different sheaths of connective tissue layers, from the epimysium enveloping the entire muscle, to the perimysium which bundles myofibers into fascicles, and the endomysium that includes the basal lamina and ensheaths individual myofibers. The organized distribution of different collagen types in the endomysium, perimysium, and epimysium is thought to play critical roles not only for binding single myofibers together and ensuring their proper alignment, but also for the coordinated transmission of forces from single myofibers to the entire muscle and for the integrated function of the muscle-tendon-bone unit (Gillies and Lieber 2011; Csapo et al. 2020). Moreover, the regulated synthesis and deposition of different collagens types in muscle ECM compartments contribute to the fine regulation of the mechanical properties of skeletal muscles, which in turn influences various cellular aspects, including the satellite cell niche (Gattazzo et al. 2014b; Thomas et al. 2015).

Fibrillar collagens I and III, together with associated collagen V, are found in the epimysium, perimysium, and endomysium (Bailey et al. 1979; Light and Champion

1984). The perimysium is also enriched with collagen types XII and XIV, which belong to the family of fibril-associated collagens with interrupted triple helices (FACIT) and have a role in connecting fibrillar collagens to other ECM components (Listrat et al. 2000). Besides collagen types IV and VI, which represent the most abundant collagens of muscle basal lamina and endomysium, two multiplexins (multiple triple-helix domains with interruptions) are present in the muscle basement membrane, namely collagen types XV and XVIII (Sanes 1982; Marvulli et al. 1996; Myers et al. 1996; Halfter et al. 1998; Gatseva et al. 2019). The NMJ found in the neuromuscular synapse areas has a specialized basement membrane containing distinct isoforms of collagen IV, as well as collagen types VI and XIII (Singhal and Martin 2011; Cescon et al. 2018). Finally, collagen XXII is a major component of the MTJ, a specialized region located at the muscle-tendon interface which represents the primary site of force transmission to bones (Charvet et al. 2012; Subramanian and Schilling 2015).

Overall, collagens fulfill several critical functions in skeletal muscle, which span from conferring tensile strength and elasticity to transmitting forces to bones and regulating cell attachment and differentiation. The critical role played by different collagen types in muscles is highlighted by the fact that mutations of certain collagen genes are causative for specific forms of inherited muscle disorders, including myopathies and muscular dystrophies. Furthermore, studies in a range of mutant and knockout animal models for these key ECM components provided a wealth of peerless information on their specific roles in skeletal muscles, as well as in the underlying cellular and molecular mechanisms linking collagen deficiency with muscle pathology. Among them, collagen VI represents the most prominent and best-studied example, as mutations of its genes are characteristic of a distinct class of congenital muscular dystrophies and congenital myopathies primarily affecting skeletal muscles, but some other collagens are also involved in a number of diseases affecting the neuromuscular system.

6.2 Collagen VI: Structure, Expression, and Binding Partners

6.2.1 Structure

Collagen VI (ColVI) is a quite distinctive member of the collagen superfamily, as it exerts a remarkably large number of critical functions for cells and displays unique properties in terms of protein domains, supramolecular assembly, tissue distribution, and physiological roles (Cescon et al. 2015; Lamandé and Bateman 2018). Six genes are known to encode for the different polypeptide subunits forming ColVI assemblies, respectively indicated as α1(VI)–α6(VI) chains (Fig. 6.1). These genes include: *COL6A1* and *COL6A2*, both mapping on human chromosome 21q22.3; *COL6A3*, located on human chromosome 2q37.3; *COL6A4*, *COL6A5*, and *COL6A6*,

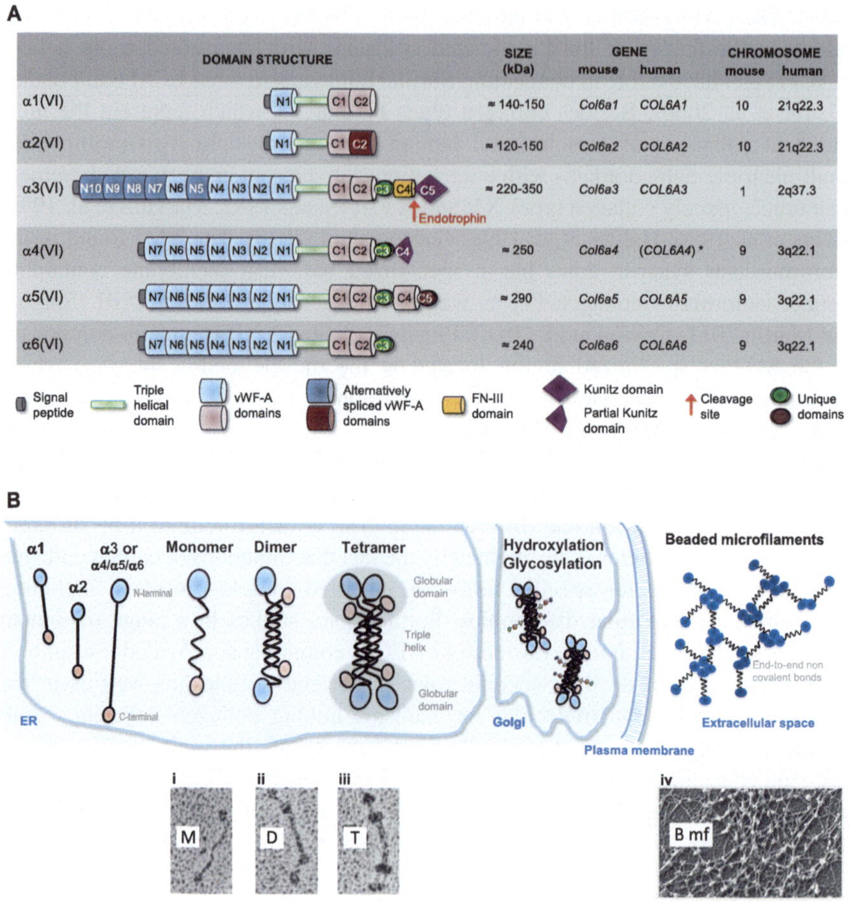

Fig. 6.1 ColVI structure and assembly. (**a**) Schematic representation of the domain structure of ColVI subunits. Six genes coding for ColVI chains (α1–α6) were identified in mammals. Each chain is made of a short collagenous region flanked by a variable number of different modules sharing similarities with the von Willebrand factor type A (vWF-A) domain (in the figure the modules are shown in blue or pink depending on their N- or C-terminal position, respectively), including some undergoing alternative splicing. The longer chains contain additional domains at their C-terminal end (see text for details). The *COL6A1* and *COL6A2* genes are organized in a head-to-tail fashion on chromosome 21q22.3 in humans and chromosome 10 in mouse, whereas the *COL6A3* gene maps to chromosome 2q37 in humans and chromosome 1 mouse. The *COL6A4*, *COL6A5* and *COL6A6* genes are organized in tandem, mapping on chromosome 3q21 in humans and on chromosome 9 in mice. In humans, *COL4A4* is a pseudogene (asterisk) split into two separate parts and therefore unable to code for a functional COL6 subunit. (**b**) Diagram displaying ColVI intracellular assembly and secretion. Three different chains—most frequently α1(VI), α2(VI) and α3(VI)—associate in the ER to form triple helical monomers. Monomers then form disulfide-bonded dimers, by lateral association in an antiparallel fashion, and dimers associate into tetramers, also stabilized by disulfide bonds. Tetramers are finally secreted in the extracellular space where they associate through non-covalent bonds to form the characteristic network of beaded microfilaments in the ECM. The images in the lower part of the figure show isolated ColVI monomers (M), dimers (D) and tetramers (T), and beaded microfilaments (B mf), visualized by rotary shadowing electron microscopy *(i–iii: adopted from Engel et al.* 1985, *with permission from John Wiley & Sons; iv: adopted from Bovolenta et al. 2010, with permission from Springer Nature)*

arranged in tandem on human chromosome 3q22.1 (Cescon et al. 2015; Lamandé and Bateman 2018). As a result of a pericentric inversion of chromosome 3, human *COL6A4* is split into two pseudogenes, *COL6A4P1* and *COL6A4P2*, that are unable to code for a functional ColVI α-chain, whereas in mice and other mammals *Col6a4* is coding for a functional polypeptide expressed in a developmentally regulated fashion and that can be incorporated into ColVI molecules (Fitzgerald et al. 2008; Gara et al. 2008). Human *COL6A5* was first annotated as the supposed new gene *COL29A1* (Söderhäll et al. 2007), but later work showed its identity with *COL6A5*, allowing its correct annotation as part of the ColVI genes (Gara et al. 2008).

A range of studies demonstrated that the triple-helical monomer of ColVI is formed by three distinct polypeptide chains, each assembled in equimolar ratio and encoded by a different ColVI gene (*COL6*) (Fig. 6.1). The α1(VI) and α2(VI) chains have a similar size of about 130–150 kDa and are always incorporated into ColVI triple-helical monomers, but they are not competent to form the triple helix alone (Colombatti et al. 1987, 1995; Colombatti and Bonaldo 1987; Lamandé et al. 1998). The α1(VI) and α2(VI) chains associate with a third, longer ColVI subunit more often represented by α3(VI) and having a size of about 250–300 kDa (Engel et al. 1985; Colombatti et al. 1987, 1995; Colombatti and Bonaldo 1987; Lamandé et al. 1998), or alternatively by one among α4(VI), α5(VI) or α6(VI) (Gara et al. 2008, 2011). Originally, it was described that ColVI consists of α1(VI):α2(VI):α3(VI), and indeed in most cells and tissues α3(VI) displays an overlapping pattern with the other two ColVI chains (Colombatti and Bonaldo 1987; Kuo et al. 1997; Cescon et al. 2015). The subsequent discovery of genes coding for three additional ColVI subunits showed that in some districts they can substitute for α3(VI), as indeed these alternative chains are detectable in a more restricted set of tissues, with highly specialized and sometimes complementary expression patterns (Gara et al. 2011; Sabatelli et al. 2011, 2012).

ColVI is quite distinctive among other collagens for different reasons. First, the unique ultrastructural organization of its extracellular mesh, showing a branched network of beaded microfilaments that interact with various other ECM components and with the surface of different cell types. Second, a characteristic multimodular primary structure, in which Gly-Xaa-Yaa motifs, forming an uninterrupted triple-helical region, represent less than one-fourth of the molecule and are flanked by large globular regions that contain a series of different domains sharing homology with various protein modules. Third, its complex biosynthesis that accounts for a unique intracellular mode of assembly and that generates very large multimers before secretion. When *COL6* genes are transcribed and translated, the nascent polypeptide α-chains enter the endoplasmic reticulum where they associate in 1:1:1 ratio to form triple-helical monomer of about 500 kDa, stabilized by disulfide bonds. These monomers then assemble into disulfide-bonded antiparallel dimers (about 1000 kDa), which subsequently align by the lateral association to form tetramers (about 2000 kDa), linked by other disulfide bonds, and that are finally secreted in the extracellular space (Engvall et al. 1986; Colombatti et al. 1987, 1995; Colombatti and Bonaldo 1987). Outside the cell, ColVI tetramers associate end-to-end with an axial periodicity of about 105 nm, giving rise to larger non-fibrillar aggregates by non-covalent bonds and finally forming a network of beaded microfilaments that are deposited into the ECM (Fig. 6.1). In these characteristic "bead-on-a-string" microfilaments with 105-nm periodicity, the beads correspond to the N- and C-terminal globular regions, whereas the linear rod coincides with the triple-helical region (Bruns et al. 1986; Baldock et al. 2003; Beecher et al. 2011). Post-translational modifications of ColVI chains in the endoplasmic reticulum and Golgi also play a role in the biosynthesis and in the proper activity of this ECM protein. Besides the extensive formation of disulfide bonds, which are required for the proper chain assembly into monomers, dimers, and tetramers, during its synthesis ColVI is subjected to extensive N-glycosylation and to hydroxylation of several proline and lysine residues, with further glycosylation of hydroxylated lysyl residues to galactosyl-hydroxylysyl and glucosyl-galactosyl-hydroxylysyl residues (Colombatti et al. 1987; Colombatti and Bonaldo 1987; Sipilä et al. 2007).

ColVI primary structure exhibits a complex multimodular organization, sharing homology with various non-collagenous proteins. All the ColVI α-chains have a central collagenous region made of 335–336 Gly–Xaa–Yaa repeats, flanked by globular regions containing repeated modules of about 200 amino acid residues in length and sharing similarity with the von Willebrand factor type A (vWF-A) domains (Colombatti and Bonaldo 1991; Colombatti et al. 1993) (Fig. 6.1). In particular, α1(VI) and α2(VI) polypeptides contain one N-terminal vWF-A domain (named N1), followed by the triple-helical domain and two C-terminal vWF-A domains (C1 and C2) (Bonaldo et al. 1989; Chu et al. 1989). The α3(VI), α4(VI), α5(VI) and α6(VI) chains are larger but, in a similar way, always display a triple-helical collagenous domain made of 336 Gly–Xaa–Yaa repeats, flanked by two vWF-A domains (again named C1-C2) and a variable number of vWF-A domains at the N-terminal end—up to ten for α3, and seven for the other chains—(named respectively N1–N10) (Fig. 6.1). In addition, the α3-α6(VI) chains differ from α1 (VI) and α2(VI) by the presence of additional domains at the C-terminal end, in particular:

- The α3-α6(VI) chains contain a unique proline-rich sequence without any homology with other known protein module (C3).
- α3(VI) has a fibronectin type III domain (C4) and a Kunitz-like domain (C5).
- α4(VI) ends with a short sequence that corresponds to a partial Kunitz-like domain (C4).
- α5(VI) has an additional vWF-A C-terminal module (C4) and another unique domain (C5).

(Bonaldo and Colombatti 1989; Bonaldo et al. 1990; Chu et al. 1990; Fitzgerald et al. 2008; Gara et al. 2008).

The molecular heterogeneity and structural complexity of ColVI are increased by the presence of isoforms generated by alternative splicing at multiple sites. Indeed, several splicing variants involving certain vWF-A domains were reported for the α2 (VI) and α3(VI) chains in different species (Fig. 6.1). For example, alternative mRNA variants of *COL6A2* were found in human cells, coding for three α2 (VI) isoforms that differ in their C-terminal end (Saitta et al. 1990). Interestingly, these α2(VI) splicing isoforms were suggested to influence the higher order assembly of ColVI monomers (Ball et al. 2001). In addition, the α3(VI) transcript undergoes multiple alternative splicing, generating different polypeptide variants characterized by a variable number of vWF-A modules at the N-terminal side. Indeed, high levels of α3(VI) isoforms lacking one or more domains in the N7–N10 region were found in both human cells and mouse tissues, and the generation of these isoforms appears to be regulated in a cell- and tissue-specific manner (Doliana et al. 1990; Stokes et al. 1991; Zanussi et al. 1992; Dziadek et al. 2002).

Moreover, a number of evidence indicates that α3(VI) undergoes proteolytic maturation in the extracellular space. In particular, the C5 domain at the C-terminus of α3(VI) is cut off from ColVI microfibrils by proteolytic cleavage immediately after secretion (Aigner et al. 2002). Further studies indicated that the short protein product derived from this cleavage—also named endotrophin—plays an important role in adipose tissue and in the tumor microenvironment (Iyengar et al. 2005). In particular, ColVI-derived endotrophin was found to promote tumor progression, being also involved in adipose tissue fibrosis and metabolic dysfunction and in chronic liver diseases (Park and Scherer 2012; Sun et al. 2014; Lee et al. 2019). A recent work demonstrated that the C-terminal end of α3(VI), containing the C2–C5 domains, undergoes multiple proteolytic cleavages through the action of BMP-1 metalloproteinase and furin-like proprotein convertases, releasing fragments of different sizes. These maturations occur during microfibril assembly and the cleavage products were found to be deposited in the ECM (Heumüller et al. 2019).

6.2.2 *Expression*

The expression and ECM deposition of ColVI has been reported in a large variety of tissues. ColVI is present in the interstitial connective tissues of most organs, including skin, muscles, tendons, cartilage, bone, nerves, blood vessels, lung, liver, kidney, and cornea, both during development and adult life (von der Mark et al. 1984; Marvulli et al. 1996; for an overview, see Cescon et al. 2015). ColVI microfibrils are particularly abundant in the pericellular matrix surrounding the surface of chondrocytes in cartilage (Zelenski et al. 2015). ColVI is also abundant in the basement membrane of skeletal muscle and in the epidermal-dermal region (Keene et al. 1988; Kuo et al. 1997). Despite the broad distribution of ColVI in the ECM of most tissue, its spatio-temporal expression in the different tissues is finely regulated both during embryogenesis and adult life. In particular, a number of studies that made use of transgenic mice expressing different regions of the *Col6a1* gene revealed that the regulation of *Col6a1* transcription is surprisingly very elaborate and under the control of a number of *cis*-acting regulatory elements spread over a very large portion of the 5′-flanking genomic region, including the core promoter and several enhancers and silencers (Braghetta et al. 1996, 1997; Fabbro et al. 1999; Girotto et al. 2000). These *cis*-acting elements confer high tissue-specific activation of the *Col6a1* gene, with a very dynamic ability to adapt temporally and spatially during development. In particular, these studies demonstrated that the tissue-regulated expression of *Col6a1* relies on different enhancer elements spread over a region extending for >12 kb upstream the transcription start site and controlling expression in skin, tendons, joints, peripheral nerves, and skeletal muscles (Braghetta et al. 1996; Girotto et al. 2000).

The major producers of ColVI in connective tissues are fibroblasts, which express ColVI in a regulated manner in response to various conditions and extracellular signals (Hatamochi et al. 1989; Olsen et al. 1989; Sardone et al. 2014, 2016; Sabatelli et al. 2016). The expression and synthesis of ColVI are also finely regulated in other cell types, including chondrocytes, Schwann cells, and adipocytes, where it is part of their differentiation program (Dani et al. 1989; Quarto et al. 1993; Vitale et al. 2001). Schwann cells produce and secrete ColVI in peripheral nerves, and the transcription of *Col6a1* gene is precisely activated during their commitment toward differentiation from the neural crest, upon neuregulin stimulation (Vitale et al. 2001). In peripheral nerves, *Col6a1* activation is linked with the differentiation of immature Schwann cells into myelinating ones (Braghetta et al. 1996; Vitale et al. 2001). In addition, ColVI also affects the myelinating functions of Schwann cells by acting on multiple regulatory signals for myelination (Chen et al. 2014).

In skeletal muscle, ColVI is a major component of the ECM of epimysium, perimysium, and endomysium, being particularly abundant in the basement membrane surrounding muscle fibers (von der Mark et al. 1984; Kuo et al. 1997). The alternative long chains α5(VI) and α6(VI) display restricted and partially complementary distributions in different regions of skeletal muscle, in which α5(VI) is found in the perimysium and at the level of MTJ and NMJ, while α6(VI) is more abundant in the endomysium and perimysium (Sabatelli et al. 2012; Gara et al. 2011). The main cell type producing ColVI in skeletal muscle are interstitial fibroblasts (Braghetta et al. 2008; Zou et al. 2008). Much interestingly, a muscle-specific enhancer, strictly required for ColVI expression by interstitial fibroblasts, was identified in the *Col6a1* gene (Braghetta et al. 1996, 2008). This enhancer is essential for activating *Col6a1* transcription during embryonic development and myogenesis, and the presence of cells of the myogenic lineage is needed for the enhancer activation in differentiating fibroblasts cells. In agreement with this, the lack of myogenic cells in limb buds of $met^{D/D}$ mutant mouse embryos markedly reduces ColVI deposition in developing muscles (Braghetta et al. 1997). This highlights an intriguing non-cell autonomous mechanism for ColVI expression and deposition in muscles, where signals relayed by myogenic cells are required for the activation of ColVI expression by fibroblasts, which in turn secrete the protein allowing its organization in the basal lamina of differentiating myofibers. Although mature myofibers do not express ColVI transcription of the *Col6a1* gene by cells of the myogenic lineage was found during early development and during in vitro myoblast differentiation (Piccolo et al. 1995; Braghetta et al. 2008). Additionally, as discussed in the next section, satellite cells express *Col6a1* in a distinctive and highly regulated manner (Urciuolo et al. 2013), further highlighting the sophisticated mechanisms underlying the tight regulation of ColVI expression in response to different conditions and stimuli in different tissues, including skeletal muscle.

6.2.3 Binding Partners

One of the most complex and still partially elucidated aspects of ColVI biology and function concerns the surprisingly large number of interactions it has with other

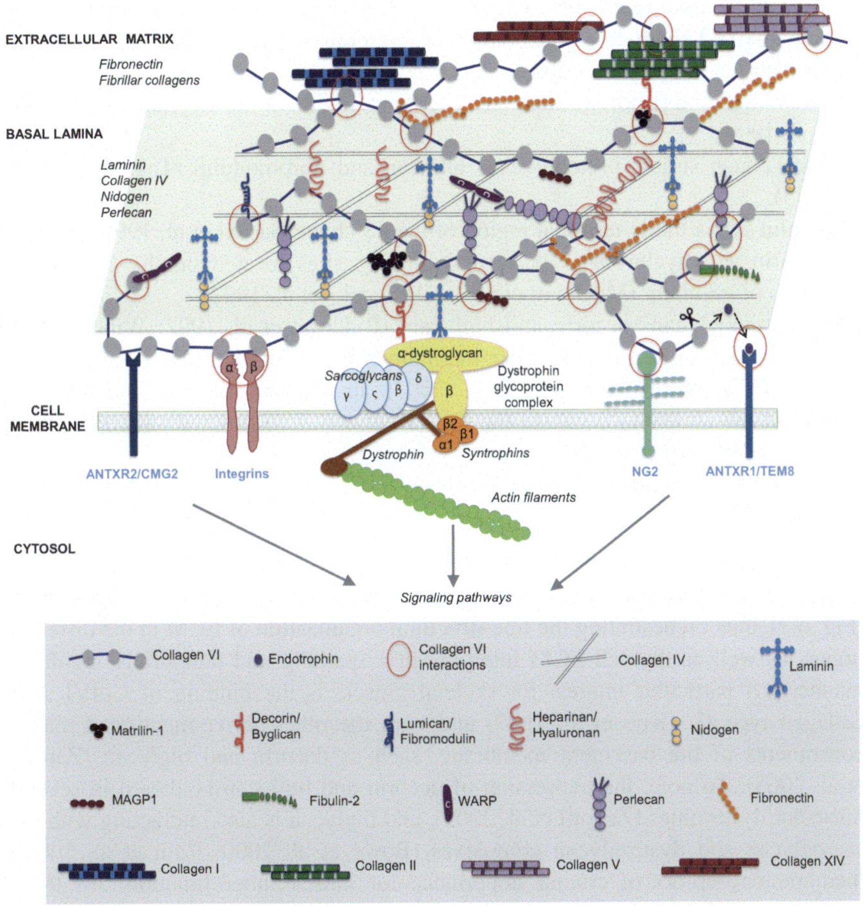

Fig. 6.2 Collagen VI binding partners and interactions. Diagram showing ColVI interactions with other collagens and with non-collagenous proteins of the basal lamina and the ECM. ColVI also interacts with different cell surface receptors, including NG2/CSPG4, TEM8/ANTXR1, CMG2/ANTXR2, and several integrin α/β isoforms (in light blue font). The C-terminal domain of the α3 (VI) chain is cleaved outside the cell (the cleavage is indicated by scissors) and the short protein product derived from this cleavage—named endotrophin—can bind to TEM8/ANTXR1

extracellular components and cell surface receptors, not only found in muscle, but broadly distributed in other tissues. Indeed, ColVI was found to directly bind various ECM components (Fig. 6.2), including:

- Fibrillar collagens types I, II, and V (Bonaldo et al. 1990; Bidanset et al. 1992; Symoens et al. 2011), basement membrane collagen IV (Kuo et al. 1997), and collagen XIV (Brown et al. 1994).
- Fibronectin, by interacting at discrete sites in the ECM, and ColVI deficiency impinges on the organization of fibronectin network (Tillet et al. 1994; Kuo et al. 1997; Sabatelli et al. 2001).
- Perlecan, via the triple-helical domain of α2 chain (Tillet et al. 1994).
- Microfibril associated glycoprotein 1 (MAGP1), through the triple-helical domain of α3 chain, likely mediating the interconnection with fibrillin-containing microfibrils (Finnis and Gibson 1997).
- von Willebrand factor, through the N-terminal region of the α3 chain (Mazzucato et al. 1999).
- The keratan sulfate proteoglycans lumican and fibromodulin (Takahashi et al. 1993).
- Fibulin-2, via the N-terminal region of the α3 chain (Sasaki et al. 1995).
- Heparin and hyaluronan, by several binding sites located on the N-terminal region of α3 chain (McDevitt et al. 1991; Specks et al. 1992).
- The proteoglycans decorin and biglycan, (Bidanset et al. 1992; Wiberg et al. 2001) which in turn may mediate the indirect link with matrilins, aggrecan, and other ECM proteins (Wiberg et al. 2003). In addition, decorin and biglycan exhibit the ability to organize the ColVI network into ordered structures upon binding (Wiberg et al. 2002).
- von Willebrand factor A-domain related protein (WARP), through the triple-helical domain (Hansen et al. 2012).

In turn, such a large number of binding partners could mediate themselves several indirect connections of ColVI with other extracellular and membrane molecules (Fig. 6.2), thus orchestrating the fine structural organization of ECM in the different organs as well as the cell-ECM interplay in a spatially and temporally regulated manner. Of particular interest for skeletal muscle is the binding of ColVI with collagen type IV (Kuo et al. 1997) and with the other above-mentioned matrix constituents of the basement membrane, such as decorin and biglycan (Zanotti et al. 2005). To note, the expression of decorin and biglycan is altered in several muscular dystrophies (Zanotti et al. 2005), and biglycan is also interacting with the sarcoglycan and dystroglycan complexes (Bowe et al. 2000; Rafii et al. 2006), membrane receptors of crucial importance for muscle fiber function. All these interactors are thus underlining a possible molecular link for the pathological defects found in ColVI-related muscle diseases, as discussed below.

ColVI interacts with several cell surface receptors. In vitro studies identified several integrin subunits that are able to bind ColVI through direct interaction with its triple-helical region, including $\alpha_1\beta_1$, $\alpha_2\beta_1$, $\alpha_3\beta_1$, $\alpha_{10}\beta_1$ and $\alpha_V\beta_3$ integrins (Aumailley et al. 1989; Pfaff et al. 1993; Tulla et al. 2001). ColVI was also shown to bind to the chondroitin sulfate proteoglycan-4 (CSPG4 or NG2), and this interaction triggers signaling events leading to rearrangement of the actin cytoskeleton (Burg et al. 1996; Tillet et al. 1997, 2002). Interestingly, further studies showed that NG2 is affected in ColVI-deficient muscles (Petrini et al. 2005) and that the ColVI-NG2 axis plays a role in tendon repair (Sardone et al. 2016). Another class of receptors found to interact with ColVI is represented by the anthrax toxin receptors 1 (ANTXR1/TEM8) and 2 (ANTXR2/CMG2) (Nanda et al. 2004; Bürgi et al. 2017). The biological meanings of this broad range of receptor interactions and the downstream pathways transducing the signals elicited by ColVI-receptor interactions remain largely unknown, both in the context of their relevance for skeletal muscle as well as of their significance for the various functions exerted by ColVI in different tissues and cell types.

6.3 Collagen VI-Related Myopathies

The importance of ColVI for the proper homeostasis of skeletal muscle is underlined by the evidence that altered expression or mutations in the genes encoding its chains are causative for a distinct group of congenital muscular dystrophies, collectively known as "ColVI-related myopathies" (Fig. 6.3). These are muscular diseases with varying severity, ranging from mild Bethlem myopathy to severe Ullrich congenital muscular dystrophy, with several intermediate phenotypes in between. Clinically, these are hybrid diseases displaying features of both muscular dystrophies and connective tissue disorders (Bönnemann 2011a; Lamandé and Bateman 2018).

6.3.1 UCMD Typical Clinical Features

Ullrich congenital muscular dystrophy (UCMD, OMIM #254090) was initially described by Otto Ullrich as a condition characterized by early-onset weakness and joint laxity, together with progressive contractures in the proximal joints, which he termed scleroatonic muscular dystrophy (Skleratonische Muskeldystrophie; Ullrich 1930). Subsequently, the disease was discussed in a number of clinical studies in Japanese and German patients (Furukawa and Toyokura 1977; Nonaka et al. 1981; Voit 1998). Nowadays, UCMD is recognized as one of the most common types of congenital muscular dystrophies of the North American and Japanese populations (Okada et al. 2007; Bönnemann 2011b). The disease presents with symptoms that are readily evident at birth or during the first year of life. In some cases, reduced prenatal movements are perceived during pregnancy (Nonaka et al. 1981; Voit 1998; Lampe and Bushby 2005). Clinical signs at birth include hypotonia and weakness, in association with hyperlaxity of the distal joints (Fig. 6.3). The hands, feet, and fingers are extremely flexible, allowing the fingers to bend backwards against the forearm and the feet to dorsiflect

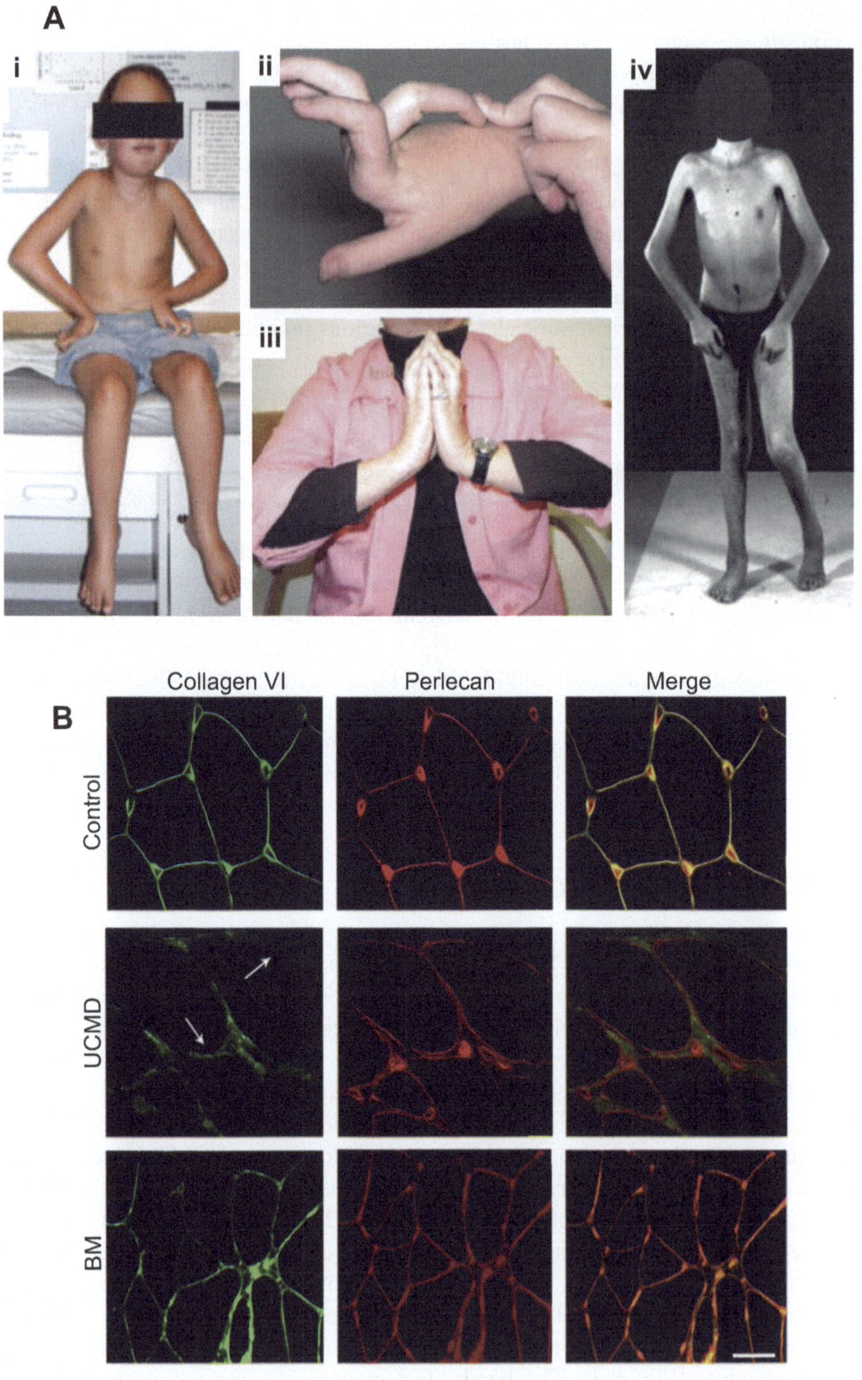

Fig. 6.3 ColVI-related myopathies. (**a**) Typical clinical features of UCMD (panels i and ii), BM (panel iii) and myosclerosis (iv). Proximal joint contractures and distal hyperlaxity, particularly in fingers, is usually seen in children with UCMD (i, ii). An adult BM patient displaying incomplete

against the shin. At the same time, the hand can drop down at the wrist, thus recalling some clinical manifestations of Ehlers-Danlos syndrome (EDS) (for a detailed overview see Chap. 4). There may be congenital pes adductus, while a posterior protruding calcaneus is often evident, representing a distinct feature in a patient with typical UCMD. Further symptoms at birth may include congenital dislocated hips, torticollis, and kyphoscoliosis, as well as contractures affecting elbows, hips, and knees. The congenital joint contractures may be present at birth and improve over the first months of life, but they frequently recur later, occasionally together with new contractures (Lampe and Bushby 2005; Bönnemann 2011a). Although in the most severe cases patients never achieve the ability to walk, the majority of UCMD patients acquires the ability to ambulate, often with a delay of up to two years and sometimes only by using assistive devices (Voit 1998). However, they then lose ambulation by around 10 years (Nadeau et al. 2009), as a consequence of increased weakness and worsening of contractures, particularly at the level of knees and hips. The muscle weakness is slowly progressive, but the resulting disability is aggravated by progressive contractures of the large joints, in particular affecting the shoulder, elbows, hips, knees, ankles, and spine. Scoliosis is often a serious problem, which can develop from kyphoscoliosis present at birth or start to appear before the loss of ambulation (Nadeau et al. 2009). The presence of contractures is accompanied by hyperlaxity of distal joints, usually present in interphalangeal joints, while long finger flexors display severe contractures. Respiratory insufficiency is another critical aspect of the disease as the condition progresses, and it results from weakness of the diaphragm together with the stiffness of the chest wall (Nadeau et al. 2009; Bönnemann 2011a). The use of nocturnal non-invasive ventilation is usually sufficient to treat this situation for a number of years, but failure in providing adequate support to patients with signs of respiratory failure leads to death in a relatively short amount of time, usually in the second decade of life (Nadeau et al. 2009). Skin is another affected tissue in UCMD. Indeed, excessive scar formation, including large keloids, is observed in several patients, while follicular hyperkeratosis is often found over the extensor surfaces of upper and lower limbs. This latter dermatological finding is usually found in severely affected patients, representing a useful diagnostic parameter in the overall context of the clinical symptoms. Soft velvety skin on the palms and soles was also described (Kirschner et al. 2005; Bönnemann 2011a).

Fig. 6.3 (continued) finger extension due to contractures affecting long finger flexors, a typical hallmark of BM ('Bethlem sign') (iii). A patient affected by myosclerosis, showing diffuse muscle wasting and extensive flexion contractures of elbows and fingers (iv) *(i–iii: adopted from Bönnemann* 2011b, *with permission from Springer Nature; iv: adopted from Merlini et al.* 2008b, *with permission from Wolters Kluver Health, Inc.).* (**b**) Immunofluorescence microscopy of cross sections of muscle biopsies from unaffected control (top panels), one UCMD patient (middle panels) and one BM patient (bottom panels). In normal conditions, ColVI (green) is abundant in the endomysial basement membrane, showing co-localization with the basement membrane marker perlecan (red). In the UCMD patient, ColVI is markedly decreased in the basement membrane, as confirmed by the lack of colocalization with perlecan labeling in several areas (arrows). In the BM patient, ColVI labeling is moderately decreased. Scale bar, 50 μm *(adopted from Tagliavini et al.* 2014, *with permission from Elsevier)*

Some transient feeding and swallowing difficulties might be observed in the neonatal period (Nadeau et al. 2009).

6.3.2 BM Typical Clinical Features

Bethlem myopathy (BM, OMIM #158810) was first described in 1976 by Jaap Bethlem and George K. van Wijngaarden as an autosomal dominantly inherited benign neuromuscular disease affecting 28 individuals of Dutch pedigree (Bethlem and Wijngaarden 1976). The first symptoms occurred around the fifth year of life, and they could include moderate muscle weakness and atrophy, torticollis and flexion contracture of the elbows, the interphalangeal joints, and the ankles. Despite the early manifestation, the disease was characterized by a mild course. Similarly affected families were then described by other studies, allowing to emphasize the characteristic clinical features of early onset and slow progression of the syndrome (Mohire et al. 1988). Data collected from worldwide studies allowed to establish the unique clinical features of this disease. Although BM was originally described as a mild myopathy with its major impact during adulthood, cases of diminished fetal movements, neonatal hypotonia or torticollis, congenital contractures, and delayed motor milestones were also reported (Jöbsis et al. 1999). Usually the onset of symptoms is in the first or second decade of life, however, some adult patients are not even aware of the presence of mild contractures or moderate weakness (Merlini et al. 1994), therefore the age of onset cannot be precisely established in some cases. The development of contractures is a distinct hallmark of this condition (Fig. 6.3). The congenital contractures usually tend to resolve in the first two years of life, and young children show distal joint hyperlaxity rather than contractures. New contractures then tend to present later in the first decade of life and during teenage years, affecting wrists, elbows, long finger flexors, shoulders, and Achilles tendons (Jöbsis et al. 1999; Lampe and Bushby 2005; Bönnemann 2011a). Once developed, these contractures often remain stable, but they can worsen becoming disabling. For instance, contractures of long finger flexors can lead to a restricted hand function. The prevented complete finger extension is a typical 'Bethlem sign' (Fig. 6.3). Contractures can also affect the spine and cause spinal rigidity in some patients, although this condition is moderate when compared to UCMD (Jöbsis et al. 1999; Bönnemann 2011a). Patients typically display moderate predominantly proximal weakness and atrophy, and distal weakness may be present as well (Jöbsis et al. 1999). During childhood, muscle weakness may remain stable, but a slowly progressive increase in weakness starts in the third to fourth decades of life. Therefore, the combination of weakness and contractures can cause walking difficulties, with more than two-thirds of patients over the age of 60 years requiring aids for ambulation (Jöbsis et al. 1999). A potential complication in BM is the development of respiratory insufficiency, resulting from the combination of the stiffness of the rib cage and diaphragmatic muscle weakness (Haq et al. 1999). Thus, some patients with respiratory symptoms necessitate nocturnal respiratory support (Jöbsis et al.

1999). Skin involvement is similar to that described in UCMD, although features such as follicular hyperkeratosis and keloid formation or "cigarette paper" scarring are more frequently observed than soft and velvety skin (Lampe and Bushby 2005; Nadeau and Muntoni 2008).

6.3.3 *Myosclerosis and Other Phenotypes*

Despite some distinctive clinical manifestations, the phenotype of ColVI-related myopathies can be very variable. Some patients may display mostly proximal weakness and very few contractures, thus resembling limb-girdle muscular dystrophy (Scacheri et al. 2002; Jokela et al. 2019). Other patients may show relatively mild weakness but a more severe and diffuse contracture phenotype. In particular, Merlini and coll. described two siblings with a clinical diagnosis of autosomal recessive myosclerosis (OMIM #255600), who were homozygous for a mutation in the *COL6A2* gene. The patients displayed moderate weakness, muscles with "woody" consistence and severe contractures of multiple joints, among which masseter muscles, neck, shoulders, elbows, fingers, knees, and Achilles tendons (Merlini et al. 2008b). Therefore, particular attention to these connective tissue-related aspects of the phenotype is needed to achieve unequivocal clinical diagnosis in this heterogeneous group of conditions.

6.3.4 *Mutations of* COL6 *Genes and Genotype-Phenotype Correlation*

A range of mutations linked with BM and UCMD were described in the *COL6A1*, *COL6A2*, *COL6A3,* and *COL6A6* genes (Bushby et al. 2014; Hunter et al. 2015; Lamandé and Bateman, 2018). A wide genetic spectrum underlies the clinical variability displayed by these disorders. The causative role of *COL6* gene mutations for inherited muscle disorders was first reported in 1996 by Jöbsis and coll. for BM (Jöbsis et al. 1996) and in 2001 by Bertini and coll. for UCMD (Camacho Vanegas et al. 2001). Although initially it was thought that recessive mutations of *COL6* genes cause UCMD and those dominant mutations underlie BM, this distinction is no longer valid. Many of the mutations have been summarized in detail by Lampe and Bushby (Lampe and Bushby 2005) and, more recently, by Lamandé and Bateman (2018).

For UCMD, the vast majority of recessively acting mutations appears to result in premature termination codons (PTC), thus causing nonsense-mediated mRNA decay and loss of the mutated chain. Since heterozygous carriers of *COL6A2* and *COL6A3* PTC mutations have no evident pathology, it seems that α2(VI) or α3 (VI) haploinsufficiency is well tolerated, whereas α1(VI) haploinsufficiency was reported to cause BM in some cases (Peat et al. 2007; Baker et al. 2007) or having no deleterious effects in some other cases (Giusti et al. 2005; Foley et al. 2011). When PTC mutations occur in alternatively spliced exons, clinical manifestations are usually less severe (Demir et al. 2002; Lampe et al. 2005; Giusti et al. 2005; Okada et al. 2007). Additional recessive mutations include changes of splice sites that cause out-of-frame exon skipping (Camacho Vanegas et al. 2001; Ishikawa et al. 2002; Lucarini et al. 2005). A significant proportion of patients with sporadic UCMD present *de novo* dominant mutations in *COL6* genes (Baker et al. 2005; Pan et al. 2003; Okada et al. 2007; Lampe et al. 2005, 2008). These mutations are typically splice site mutations or genomic deletions that lead to the in-frame skipping of exons coding for the N-terminal region of the triple-helical domain, without affecting the cysteine residues responsible for dimer formation (Pan et al. 2003; Pepe et al. 2006; Lampe et al. 2008). As a consequence, the chain with the deleted sequence in the triple-helical region is effectively incorporated into the monomer, and since the cysteine residue is still present, the higher order assembly of dimers and tetramers is allowed. This leads to abnormal tetramers being secreted in the ECM, with a consequent dominant-negative effect on microfibrillar assembly (Baker et al. 2005; Pan et al. 2003; Lampe et al. 2008). In contrast, in-frame exon skipping mutations occurring towards the C-terminus of the triple-helical domain are not incorporated into the heterotrimeric monomer, therefore acting in a recessive way (Baker et al. 2005; Demir et al. 2002; Ishikawa et al. 2004; Lampe et al. 2005, 2008). Interestingly, patients with dominant-negative mutations may acquire the ability to ambulate for certain a period of time, compared with those with a total null situation.

Concerning BM, mutations of *COL6* genes identified thus far are mostly dominant. The typically observed changes are missense mutations of the glycine residue in the Gly-Xaa-Yaa motifs of the N-terminal end of the triple-helical domain of either the α1(VI), α2(VI) or α3(VI) chains (Jöbsis et al. 1996; Pepe et al. 1999a; Scacheri et al. 2002; Lampe et al. 2005; Lucioli et al. 2005). These mutations act in a dominant-negative way, as they introduce a kink into the triple-helical domain of tetramers (Lamandé et al. 2002). The effects of autosomal dominant substitutions in Gly-Xaa-Yaa motifs are variable, depending on the precise sequence position in which they occur. Therefore, they can generate variable clinical phenotypes, and if the effect on ColVI assembly is more pronounced, the phenotype can even fall into the UCMD range of severity (Okada et al. 2007; Pace et al. 2008). Another frequent type of mutation in BM is the heterozygous in-frame skipping of exon 14 of the *COL6A1* gene, which leads to the production of an aberrant α1(VI) chain lacking a cysteine residue necessary for ColVI assembly, thus resulting in the secretion of half amounts of COL6 (Lamandé et al. 1999; Pepe et al. 1999b; Pan et al. 2003; Lucioli et al. 2005). Further mutations described in BM include missense mutations other than glycine substitutions within the triple-helical domain (Scacheri et al. 2002; Lucioli et al. 2005). Of note, not all the mutations reported so far to be associated with BM have been completely analyzed for their pathogenic effect on ColVI. Therefore, caution is needed since many polymorphisms in *COL6* genes have not been fully cataloged, and it can be difficult to establish whether a newly reported sequence change in a patient is a polymorphism or if it is pathogenic (Bönnemann 2011a; Lamandé and Bateman 2018).

6.3.5 Diagnostic Tools

The clinical diagnosis of ColVI-related myopathies is based on the recognition of the typical clinical features that can suggest a picture compatible with either UCMD, BM, or a pathology of intermediate severity. Muscle biopsies can be very helpful in the clinical differentiation of these disorders, since there are distinctive muscular alterations associated with each of the conditions (Mercuri et al. 2010; Deconinck et al. 2010). For instance, immunohistochemical analysis for the amount and localization of ColVI in muscle tissue represents a useful diagnostic tool (Fig. 6.3). In recessive cases of UCMD, ColVI staining is usually absent or greatly reduced (Ishikawa et al. 2002), while dominant UCMD mutations lead to strong labeling for ColVI immunoreactivity in the ECM, but the protein is not properly localized in the basement membrane (Pan et al. 2003; Ishikawa et al. 2004). This mislocalization, which suggests that ColVI is secreted but does not have a proper deposition, can be highlighted by double labeling with basement membrane markers, such as collagen IV, laminin, or perlecan. However, this is not the case for all dominant mutations, as in the milder BM, ColVI is not depleted from the basement membrane, even though it appears disorganized (Pan et al. 2003). Of note, some patients with ColVI disorders in which apparently normal staining for α1(VI)/α2(VI)α/3(VI) is observed, α6(VI) labeling is instead reduced, suggesting that immunostaining for α6(VI) could have diagnostic relevance (Tagliavini et al. 2014). More importantly, analysis of ColVI synthesis, assembly and ECM deposition in cultures of dermal fibroblasts from patients with ColVI mutations represents a useful diagnostic tool, spanning from completely absent or markedly reduced with intracellular retention in UCMD, to milder abnormalities in BM (Lamandé et al. 1999, 2002; Jimenez-Mallebrera et al. 2006; Baker et al. 2007; Hicks et al. 2008; Merlini et al. 2008b; Gualandi et al. 2009; Tooley et al. 2010).

Concerning the histological analysis of muscle biopsy from BM and UCMD patients, the findings can be very variable, depending on the age and severity of the disease. In the first phases, myofiber atrophy is the most evident feature (Schessl et al. 2008), and as the disease progresses dystrophic features, such as fiber size variability, myofiber degeneration, and increased endomysial fibrosis, become prominent (Higuchi et al. 2003). Furthermore, some patients can display type I fiber atrophy and predominance (Schessl et al. 2008). A further useful tool in the diagnosis of ColVI-related myopathies and their differentiation from *LMNA*-related Emery-Dreifuss muscular dystrophy and other myopathies presenting contractures and spinal rigidity is muscle magnetic resonance imaging (Mercuri et al. 2010; Deconinck et al. 2010). Indeed, magnetic resonance imaging reveals a distinct pattern of muscle involvement in ColVI-related diseases, with a characteristic concentric pattern of alterations where degenerative changes and fatty and connective tissue replacement are found mainly in the outer region of muscle, rather than in the central part (Mercuri et al. 2005; Bönnemann 2011a). This pattern is most consistently seen in the rectus femoris and vastus lateralis muscles. The peripheral predominance of pathology can also be appreciated using muscle ultrasound, where degeneration around the central fascia of the rectus femoris originates a characteristic appearance (Bönnemann et al. 2003).

Therefore, a combination of clinical and genetic data, together with immunohistochemical and protein studies, is fundamental to reach a clear diagnosis in the heterogeneous group of ColVI-related myopathies.

6.4 Animal Models for the Study of Collagen VI Pathophysiology

The detailed dissection of the different biological functions and properties of ColVI is fundamental for developing targeted therapeutic approaches for ColVI-related myopathies. In this regard, mutant and knockout animal models provide a useful tool for translational studies and for the mechanistic understanding of human diseases linked to ColVI defects.

6.4.1 Col6a1 *Knockout Mouse*

The best-characterized animal model for ColVI-related myopathies is the *Col6a1* knockout (*Col6a1*$^{-/-}$) mouse. In this mouse, the second exon of the *Col6a1* gene, coding for the signal peptide of the α1(VI) chain, was interrupted by targeted gene disruption, thus preventing the translation of the α1(VI) chain. As a consequence, even if α2(VI) and α3(VI) chains can be translated, the formation of ColVI monomers is prevented, and therefore no triple-helical ColVI molecules can be synthesized and secreted in the ECM (Bonaldo et al. 1998). This mouse model largely contributed to unveil a number of in vivo roles of ColVI, throwing new light on the relevance of this ECM component in regulating several cellular processes.

6.4.1.1 Muscle Structural Defects

Although ColVI is broadly distributed in the ECM of several tissues both during development and adult life, *Col6a1* null mice are born at Mendelian ratios and do not display any overt malformation, suggesting that ablation of ColVI is not hampering embryonic development and does not affect organogenesis. However, *Col6a1*$^{-/-}$ mice display several signs of an early-onset myopathic pathology, such as muscle necrosis and phagocytosis, increased incidence of centrally nucleated myofibers, and a marked variation in myofiber diameter. These phenotypic features are present in the diaphragm and, at lower frequencies, in intercostal, abdominal, and leg muscles. Similar, albeit milder, defects are also displayed by heterozygous *Col6a1*$^{+/-}$ mice, indicating haploinsufficiency for ColVI (Bonaldo et al. 1998). These histological hallmarks are paralleled by a significant decrease of muscle contractile strength in *Col6a1*$^{-/-}$ mice (Irwin et al. 2003). Ultrastructural studies by transmission electron microscopy revealed unexpected and distinctive defects in myofibers of *Col6a1*$^{-/-}$ muscles. Indeed, while the sarcomeric pattern appears normal, ColVI-deficient myofibers display remarkable ultrastructural alterations of sarcoplasmic reticulum (SR) and mitochondria. In particular, SR show pronounced dilations especially at the level of the triadic system, while mitochondrial abnormalities range from altered cristae with tubular shape, to electron-dense matrix inclusions, to swelling (Irwin et al. 2003). In addition, *Col6a1*$^{-/-}$ myofibers have a significantly higher percentage of apoptotic nuclei, suggesting a link between defective organelle and increased incidence of cell death. Atomic force microscopy (AFM) of myofibers from *Col6a1*$^{-/-}$ mice revealed a reduced sarcolemma stiffness and a decreased electrical capacitance, which is paralleled by a significantly lower depolarization rate (Canato et al. 2010). Altogether, these studies revealed that ablation of ColVI in *Col6a1*$^{-/-}$ mice leads to remarkable structural and functional alterations in skeletal muscles.

6.4.1.2 Alterations in Mitochondrial Function and Oxidative Stress Pathways

To assess whether the ultrastructural alterations reflect alterations in mitochondrial function, mitochondrial membrane potential ($\Delta\psi_m$) was measured in isolated myofibers from *Col6a1*$^{-/-}$ mice and in muscle cell cultures from BM/UCMD patients. Although in resting conditions $\Delta\psi_m$ is not affected by lack of ColVI, mitochondria were found to quickly depolarize after addition of oligomycin (an inhibitor of the F_1F_O-ATP synthase) or of rotenone (an inhibitor of the respiratory chain complex I), only in cells from ColVI-deficient muscles (Irwin et al. 2003; Angelin et al. 2007; Tiepolo et al. 2009; Palma et al. 2009). Further experiments revealed that the oligomycin-dependent mitochondrial depolarization involves an increased opening of the permeability transition pore (PTP) in ColVI-deficient myofibers (Fig. 6.4). PTP is a high-conductance channel present in the mitochondrial inner membrane and whose opening is regulated by the proton electrochemical gradient, both through pH (matrix acidification promotes the closure) and voltage (depolarization favors opening). Moreover, the redox state finely tunes PTP, as high levels of reactive oxygen species (ROS) favor pore opening, and mitochondrial Ca^{2+} is a crucial factor sensitizing the PTP to a plethora of inducing factors (Bernardi et al. 2006). The mechanistic interpretation of the mitochondrial defects displayed by ColVI-deficient myofibers is that, at least initially, the threshold voltage for PTP opening is altered, being dangerously close to the resting $\Delta\psi_m$. Oligomycin addition inhibits the ATP synthase, causing an initial hyperpolarization and a further shift of the threshold voltage toward the value of the resting $\Delta\psi_m$. The shift is likely due to

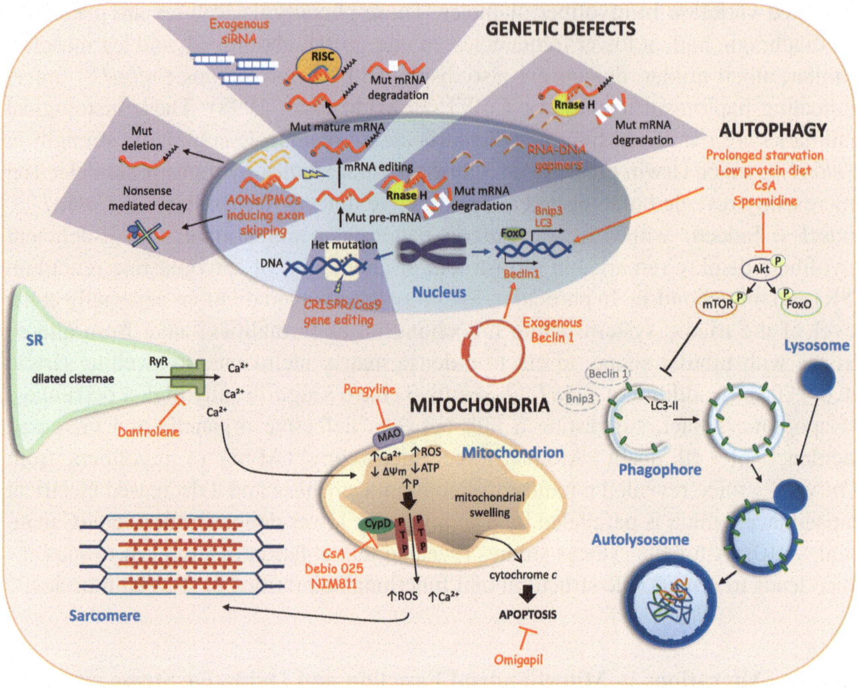

Fig. 6.4 Pathogenic mechanisms underlying ColVI-related myopathies and approaches to restore genetic defects and intracellular alterations. The diagram summarizes the cellular alterations of muscle fibers elicited by ColVI deficiency and involving mitochondria, SR, and autophagy. **MITOCHONDRIA** . ColVI deficiency affects some of the regulators of the PTP, such as mitochondrial transmembrane potential (Δψm) and the concentration of P, ATP, Ca^{2+}, and ROS. In addition, a significant increased activity of MAO leads to an accumulation of mitochondrial ROS. The deregulation of PTP inducers and inhibitors leads to PTP opening. Oxidative (e.g., ROS accumulation) and other cellular stresses promote the recruitment of CyP-D to the mitochondrial inner membrane, where it further favors the opening of the PTP, leading to the release of Ca^{2+} and ROS in the cytoplasm. ROS accumulation in the cytosol can cause oxidative modifications of myofibrillar proteins in the sarcomere, thus hampering contractility. Long-lasting PTP opening, together with cristae remodeling and matrix swelling, promotes the release of cytochrome *c* in the cytoplasm, eliciting apoptosis. Treatment with CsA, Debio 025, and NIM811 blocks CyP-D, thus inhibiting PTP opening, whereas treatment with pargyline reduces ROS accumulation. **SR**. Treatment with oligomycin leads to an increase of intracellular Ca^{2+} in ColVI-deficient myofibers, which is prevented when the cells are incubated with dantrolene, an inhibitor of ryanodine receptor channels. An abnormal regulation of Ca^{2+} homeostasis may cause Ca^{2+} overload, eventually eliciting PTP opening. **AUTOPHAGY** . Abnormal regulation of the Akt/mTOR axis inhibits autophagy in ColVI-deficient myofibers. A persistent Akt phosphorylation maintains mTOR in an active state, retaining FoxO3 transcription factor in the cytoplasm. This in turn prevents the transcription of Bnip3 and LC3B genes. The decreased autophagosome formation leads to the accumulation of defective organelles (mitochondria and SR), causing myofiber degeneration. Transfection of myofibers with a *Becn1* expression construct reinstate Beclin 1 levels and activates autophagy. Prolonged starvation, LPD, CsA, and spermidine elicit Akt dephosphorylation and restore Beclin 1 and Bnip3 protein levels, with reactivation of autophagy and rescue from spontaneous apoptosis. Treatment with a compound that inhibits GAPDH/Siah1-mediated apoptosis (omigapil) may also counteract myofiber apoptosis. **GENETIC DEFECT**. The upper part of the figure summarizes the exploited approaches aimed at directly targeting the genetic defect. Heterozygous mutations in ColVI genes can be corrected by CRISPR/Cas9 gene editing. Other approaches

an increased ROS production, as a consequence of the mitochondrial state 3 to state 4 respiration transition, and a decreased ability of the SR to take up Ca^{2+}, leading to an additional increase of mitochondrial Ca^{2+} (Bernardi and Bonaldo 2013). The consequent PTP opening may then cause the observed mitochondrial depolarization (Irwin et al. 2003; Angelin et al. 2007). Rotenone treatment also causes mitochondrial depolarization, as, after pore opening, ATP hydrolysis by the ATP synthase working in reverse cannot restore $\Delta\psi_m$ as long as the PTP is open. The inappropriate opening of the PTP also leads to the release of cytochrome c, which is a major inductor of apoptosis, and to altered dynamics of intracellular Ca^{2+} fluxes in the presence of oligomycin (Irwin et al. 2003). Of note, mitochondria and SR have combined regulatory roles for Ca^{2+} handling in excitation-contraction coupling (Robert et al. 2001). Thus, hampered contractility and increased cell death, which inevitably lead to muscle wasting, appear to be linked to the ultrastructural defects displayed by *Col6a1*$^{-/-}$ muscle fibers. Notably, the mitochondrial dysfunction and spontaneous apoptosis of *Col6a1*$^{-/-}$ myofibers are reversible and strictly dependent on the absence of extracellular ColVI, as they are recovered when muscle cells are plated onto purified native ColVI, but not other ECM proteins (Irwin et al. 2003).

The involvement of PTP in the pathogenesis of myofiber defects was confirmed by treatment with cyclosporin A (CsA), a drug that desensitizes PTP opening (Fig. 6.4). Indeed, CsA treatment greatly improves the abnormal response of $\Delta\psi_m$ in *Col6a1*$^{-/-}$ myofibers, with a pronounced decrease of apoptosis (Irwin et al. 2003). Based on these findings, *Col6a1*$^{-/-}$ mice were subjected to in vivo CsA treatment and their pathological hallmarks were analyzed. Four days of CsA administration at 5 mg/kg was sufficient to lead to a remarkable recovery of myofiber apoptosis and of mitochondrial and SR defects (Irwin et al. 2003). CsA acts by inhibiting cyclophilin D (CyP-D), a peptidyl-prolyl cis-trans isomerase located in the mitochondrial matrix, and oxidative and other cellular stresses promote the recruitment of CyP-D to the mitochondrial inner membrane, where it favors PTP opening (Bernardi et al. 2006). Of note, genetic ablation of *Ppif*, the gene encoding for CyP-D in mice, has the same beneficial effects on the myopathic phenotype of *Col6a1*$^{-/-}$ mice as pharmacological inhibition of CyP-D by CsA. Indeed, *Col6a1*$^{-/-}$::*Ppif1*$^{-/-}$ double knockout mice show rescue from mitochondrial dysfunction and ultrastructural defects, and normalized incidence of apoptosis, similarly to CsA-treated *Col6a1*$^{-/-}$ mice (Palma et al. 2009).

Fig. 6.4 (continued) (AONs, PMOs, siRNAs) can modify the expression of a mutated ColVI allele by acting on mRNA splicing, stability, or translation, in order to amend the mutation obtaining a shorter but correct mature transcript or to abolish the expression of the mutated allele and favor the expression of the alternative correct one, therefore establishing a form of clinically irrelevant haploinsufficiency. *AON* antisense oligoribonucleotides; *ATP* adenosine triphosphate; *CyP-D* cyclophilin D; *CsA* cyclosporin A; *het* heterozygous; *P* inorganic phosphate; *LPD* low-protein diet; *MAO* monoamine oxidases; *mut* mutant; *PTP* permeability transition pore; *ROS* reactive oxygen species; *RyR* ryanodine receptor; *SR* sarcoplasmic reticulum

The discovery of mitochondrial PTP as a relevant target in the myopathic phenotype of ColVI-deficient mice and the beneficial effects displayed by CsA opened the field to novel unexpected perspectives for understanding the pathogenesis of BM and UCMD and for the prospective treatment of ColVI-related myopathies, which led to further studies and pilot clinical trials, as discussed below. Although CsA displays protective effects on ColVI-deficient cells, its pharmacology and toxicology are very complex, since several cell pathways are affected by this drug. For instance, CsA also interacts with the cytosolic Cyp-A, forming a complex that targets and inhibits the cytosolic phosphatase calcineurin, which in turn prevents the dephosphorylation and nuclear translocation of the Nuclear Factor of Activated T-cells (NFAT), resulting in the well-known immunosuppressive activity of CsA (Liu et al. 1991). Because of this, further work was carried out, aimed at identifying non-immunosuppressive CsA derivatives that maintain the beneficial effects of CsA without exposing to the complications that may arise from its inhibition of calcineurin and immunosuppression. For instance, a study demonstrated that in vivo treatment of *Col6a1*$^{-/-}$ mice with [D-MeAla]^3-[EtVal]^4-cyclosporin (Debio 025 or alisporivir), a cyclophilin inhibitor that does not affect calcineurin, is able to desensitize the PTP, leading to the recovery of ultrastructural lesions and spontaneous apoptosis of ColVI-deficient myofibers (Tiepolo et al. 2009). Another study explored the efficacy of N-methyl-4-isoleucine-cyclosporin (also known as NIM811), a further non-immunosuppressive cyclophilin inhibitor, showing that treatment of *Col6a1*$^{-/-}$ mice with NIM811 normalizes the depolarizing response to oligomycin, prevents apoptosis and ameliorates muscle strength (Zulian et al. 2014).

Importantly, the same pathological features of increased apoptosis, ultrastructural organelle defects, and anomalous PTP-dependent mitochondrial depolarization in response to oligomycin, are found in muscle biopsies and muscle-derived cell cultures of UCMD patients with different genetic mutations of ColVI genes (Angelin et al. 2007). As in the mouse model, treatment of patients' cultures with CsA or its non-immunosuppressive derivatives Debio 025 and NIM811 leads to a marked amelioration of these defects (Angelin et al. 2007; Tiepolo et al. 2009; Zulian et al. 2014). Altogether, these studies allowed to discover a cause-effect relationship between the Cyp-D-dependent PTP regulation and the pathogenesis of the ColVI-related myopathies, and pointed at Cyp-D and PTP as druggable targets for the prospective treatment of these pathologies, as discussed below.

As discussed above, the *Col6a1*$^{-/-}$ mouse model was fundamental to elucidate that mitochondria are part of the pathogenic mechanisms underlying COL6-related myopathies and to disclose the protective role of ColVI for muscle fibers. Indeed, besides counteracting apoptosis, ColVI exerts several cytoprotective functions. A study demonstrated that the activity of monoamine oxidases (MAO), flavoproteins located in the outer mitochondrial and generating hydrogen peroxide, is markedly increased in *Col6a1*$^{-/-}$ mice (Menazza et al. 2010). This in turn leads to accumulation of ROS, with consequent oxidative modifications of myofibrillar protein which may hamper contractility (Fig. 6.4). These findings suggest that oxidative stress may be part of the mechanisms linking ColVI deficiency with mitochondrial dysfunctions, such as PTP opening. Indeed, changes in the composition of ECM may be sensed by integrins, which have an important role in conveying intracellular signals that promote ROS formation (Werner and Werb 2002). Since it has been demonstrated that increased ROS formation increases PTP opening probability and that PTP opening dramatically increases ROS production (Zorov et al. 2000; Bernardi et al. 2015), it is possible that a vicious cycle takes place in ColVI-deficient muscle fibers, ultimately leading to both contractile dysfunction and cell death. Of note, in vivo treatment of *Col6a1*$^{-/-}$ mice with pargyline, a MAO inhibitor, is able to reduce ROS accumulation in ColVI-deficient muscles, leading to decreased myofiber apoptosis and improved muscle strength (Menazza et al. 2010). These findings point at a role for MAO and ROS accumulation in the etiology of ColVI-related myopathies, and encourages to exploit MAO inhibition as a potential strategy for the treatment of these pathologies.

6.4.1.3 Dysregulated Autophagy

A major advance in the understanding of the molecular pathogenesis of ColVI-related myopathies was provided by the finding that the persistence of altered organelles in muscle fibers of *Col6a1*$^{-/-}$ mice is caused by an impaired regulation of the autophagic machinery (Grumati et al. 2010, 2011a). Autophagy is an evolutionarily highly conserved process in which bulk cytoplasm, long-lived proteins, and organelles are sequestered by double-membraned vesicles called autophagosomes and delivered to lysosomes, where they are degraded and recycled for further cellular use (Levine and Kroemer 2019; Pohl and Dikic 2019). One of the "gold standard" approaches currently used to study the autophagic flux is based on the quantification of microtubule-associated protein light chain 3 (LC3B) lipidation, since the lipidated form of LC3B (LC3B-II) is bound to the autophagosome membrane until fusion with the lysosome (Mizushima et al. 2008; Klionsky et al. 2016). In the search for the mechanisms causing accumulation of dysfunctional organelles in ColVI-deficient muscles, it was found that muscles of *Col6a1−/−* mice display decreased lipidation of LC3B both in fed condition and after 24-h fasting, a well-known autophagy-promoting stimulus. Myofibers of starved *Col6a1*$^{-/-}$ mice display markedly lower amounts of double-membrane vesicles when compared to those of wild-type animals, confirming an impaired response to a physiological autophagic stimulus (Grumati et al. 2010). Further experiments revealed defective autophagic flux in ColVI-deficient muscles, with markedly decreased levels of Beclin 1, one of the key regulators of autophagosome formation, and of Bnip3, a mitochondrial protein involved in the selective clearance of mitochondria (mitophagy) (Fig. 6.4). These defects are paralleled by the persistent activation of the Akt/mTOR pathway, which promotes protein synthesis and inhibits autophagy, as this pathway remains active in *Col6a1*$^{-/-}$ muscles even after starvation (Grumati et al. 2010). Interestingly, the impaired autophagic flux of ColVI-deficient muscles can be reactivated, since prolonged starvation is able to normalize LC3B lipidation and Beclin 1 levels, with the formation of autophagic vesicles and a marked amelioration of myofiber defects in *Col6a1*$^{-/-}$ mice subjected to 30-h fasting (Grumati et al. 2010). The autophagic defect is also present in muscle biopsies of patients affected by ColVI-related myopathies, as shown by the decreased amounts of both Beclin 1 and Bnip3 when compared to biopsies from unaffected controls. Of note, it was found that patients with UCMD display very low Beclin 1 levels, while subjects with BM show a less marked decrease in the amount of Beclin 1, suggesting that the extent of autophagy impairment may correlate with the severity of the phenotype (Grumati et al. 2010).

The results obtained in *Col6a1*$^{-/-}$ mice represent a strong evidence that an impairment of the autophagic machinery plays a crucial role in the pathogenesis of ColVI-related diseases, and that the alterations of Beclin 1 complex and of the AKT/mTOR pathway are part of the mechanisms causing the failure of the autophagic process (Fig. 6.4). These intriguing results pointed out the importance of elucidating in further detail the mechanisms that affect the autophagy/lysosome machinery in ColVI-deficient muscle and of identifying the link between ColVI deficiency in the ECM and the abnormal regulation of the Akt/mTOR axis. Of note, physical activity, which has been shown to promote autophagy in normal muscles, is detrimental for ColVI-deficient mice. Both acute and prolonged exercise exacerbate autophagic impairment, apoptosis, and muscle wasting in *Col6a1*$^{-/-}$ mice, thus highlighting that a proper balance of the autophagic flux is fundamental for muscle homeostasis during physical activity (Grumati et al. 2011b). A recent work demonstrated that autophagy regulation is compromised not only in *Col6a1*$^{-/-}$ muscle fibers, but also in *Col6a1*$^{-/-}$ fibroblasts. This is interesting, as fibroblasts are the main producers of ColVI in various tissues, including skeletal muscle. Indeed, fibroblast cultures from ColVI-deficient mice display defects in the autophagy/lysosome machinery, with impaired clearance of autophagosomes and lysosomal dysfunctions. In particular, *Col6a1*$^{-/-}$ fibroblasts display a deregulation in the activity of TFEB, a master regulator of lysosomal function, together with increased activation of the Akt/mTOR pathway. Additionally, *Col6a1*$^{-/-}$ fibroblasts display a failure of Parkin-dependent mitophagy (Castagnaro et al. 2018). These findings suggest that fibroblasts likely contribute to the pathophysiological features of ColVI-related diseases.

Overall, the above studies underlined the importance of modulating autophagy to improve cell homeostasis in ColVI-deficient muscles. Towards this aim, a range of genetic, dietary, and pharmacological approaches have been used to reactivate the autophagic flux in *Col6a1*$^{-/-}$ mice (Fig. 6.4). First of all, in vivo muscle transfection with a *Becn1* expression construct showed that restoration of Beclin 1 expression not only successfully activates autophagy, but also leads to a decreased incidence of apoptotic nuclei in *Col6a1*$^{-/-}$ myofibers (Grumati et al. 2010). A further strategy exploited for promoting autophagy induction in ColVI-deficient muscles is prolonged fasting. As mentioned above, starvation for 30 h triggers LC3B lipidation and increases Beclin 1 and Bnip3 protein levels in *Col6a1*$^{-/-}$ mice together with a marked rescue of ultrastructural alterations of SR and mitochondria and a lower amount of myofibers showing mitochondrial depolarization and apoptotic nuclei (Grumati et al. 2010). However, prolonged starvation represents an acute short-term approach, therefore milder and long-term strategies were also evaluated for their capability to activate autophagy and ameliorate muscle structure and function. Towards this aim, based on the knowledge that amino acid depletion is known to induce autophagy, a low-protein diet (LPD) was designed, and mice were left to freely feed with an *ad hoc* prepared chow, in which the total protein amount was decreased to about one-fourth of its normal content, but containing the same caloric value of a standard diet. This nutritional strategy was highly effective, as one month of feeding with LPD successfully reactivated autophagy in *Col6a1*$^{-/-}$ mice, leading to the clearance of altered organelles, normalization of myofiber ultrastructure, decreased apoptosis, and recovery of muscle strength (Grumati et al. 2010). Further work explored the efficacy of some nutraceuticals in promoting the autophagic flux in the muscles of *Col6a1*$^{-/-}$ muscles. Spermidine is a cationic polyamine present in various foods and living tissues and whose positive effects on the body are various and well known (Eisenberg et al. 2009; Minois et al. 2012; Gupta et al. 2013). A number of studies showed that spermidine administration markedly extends the lifespan of different model organisms, an effect which is linked to enhanced autophagy (Eisenberg et al. 2009; Rubinsztein et al. 2011). Systemic administration of spermidine to *Col6a1*$^{-/-}$ mice promotes a significant increase of LC3B lipidation and of autophagosome formation in muscles, together with a lower incidence of apoptotic nuclei, amelioration of myofiber defects, and improved muscle histology. The beneficial effects of other autophagy-inducing nutraceuticals, such as stilbenoids, is currently under study in *Col6a1*$^{-/-}$ mice and the data obtained thus far indicate that they can have a strong beneficial action in ameliorating the myopathic phenotype and promoting muscle remodeling (Metti et al. 2020). Altogether, the data obtained in the *Col6a1*$^{-/-}$ mouse model indicate that dietary approaches represent an intriguing tool for modulating autophagy and counteracting muscle pathology in ColVI-related myopathies, and they pave the way for exploring their usefulness in the clinical setting by means of pilot clinical trials in BM/UCMD patients, as discussed in further detail in the next section.

The effects of autophagy-promoting drugs were also studied in the ColVI null mouse model. For instance, treatment with rapamycin, a well characterized inducer of autophagy already in use in the in vivo setting (Rubinsztein et al. 2011), is able to counteract myofiber degeneration and promote the removal of abnormal organelles in *Col6a1*$^{-/-}$ muscles (Grumati et al. 2010). Moreover, based on the efficacy of CsA in rescuing the myopathic defects of *Col6a1*$^{-/-}$ mice (Irwin et al. 2003), and knowing that this drug can also modulate autophagy (Pallet et al. 2008), it is conceivable that CsA may exert beneficial effects in promoting autophagy in ColVI-deficient muscle. Indeed, CsA was found to reinstate the autophagic flux by rescuing Beclin 1 levels and promoting Akt dephosphorylation *Col6a1*$^{-/-}$ muscles, thus supporting the concept that the robust beneficial effects of CsA in ColVI-deficient muscles are associated with autophagy induction as well as with amelioration of mitochondrial function (Grumati et al. 2010).

6.4.1.4 Further Processes Affected by ColVI Deficiency

Although the above findings highlighted that autophagy deregulation and mitochondrial dysfunction play a key role in the pathomolecular mechanisms underlying the myopathic phenotype caused by ColVI deficiency, several questions remain to be answered for a better understanding of the metabolic and signaling defects linking ColVI deficiency in the ECM with the onset of muscle fiber defects. Studies aimed at identifying protein changes occurring in muscles of young adult (6-month-old) *Col6a1*$^{-/-}$ mice revealed that metabolic profiles of the gastrocnemius, tibialis anterior, and diaphragm muscles display some common changes in proteins involved in glycolysis, tricarboxylic acid (TCA) cycle, mitochondrial biogenesis, and Ca^{2+} homeostasis. Furthermore, alterations of proteins involved in mechanotransduction, as well as in the costameric and sarcomeric compartments, are also detected in *Col6a1*$^{-/-}$ mice (De Palma et al. 2013). The proteomic changes occurring in muscles of *Col6a1*$^{-/-}$ mice during aging were also investigated. Towards this aim, the metabolic profiles were analyzed in gastrocnemius and diaphragm muscles of 6- (adult), 12- (aged), and 24- (geriatric) month-old wild-type and *Col6a1*$^{-/-}$ mice (Capitanio et al. 2017). This study revealed that at 6 months *Col6a1*$^{-/-}$ diaphragm displays hallmarks of aging, which are maintained in aged and geriatric muscle. Such hallmarks include alterations in gluconeogenesis as well as glycolysis and TCA cycle rewiring, which culminate in lipotoxicity and autophagic impairment. Conversely, the effects of aging in *Col6a1*$^{-/-}$ gastrocnemius are similar but delayed, as they appear at 12 months of age. Interestingly, a similar metabolic imbalance and autophagic impairment are observed in the diaphragm of 24-month-old wild-type mice, suggesting that the changes displayed by *Col6a1*$^{-/-}$ muscles represent signatures of the aging process (Capitanio et al. 2017). A recent study suggests that the metabolic alterations found in *Col6a1*$^{-/-}$ muscles may be linked to a decrease in the local and circulating levels of adiponectin (Gamberi et al. 2019). This protein hormone exerts key roles in metabolism, as it regulates glucose uptake, glycolysis activation, glycogen synthesis, and fatty acid oxidation (Yamauchi et al. 2002), but it has also differentiating functions, as it promotes myogenesis (Fiaschi et al. 2009). Intriguingly, it was found that *Col6a1*$^{-/-}$ mice have decreased plasma adiponectin content and that myoblast cultures from *Col6a1*$^{-/-}$ mice have an impaired autocrine secretion of the hormone, which in turn leads to impaired glucose uptake. In addition, *Col6a1*$^{-/-}$ myoblasts display higher glutamine dependence, which appears fundamental for muscle cell metabolism (Gamberi et al. 2019). Therefore, these alterations may explain at least in part the blunted glycolysis and impaired TCA cycle highlighted by the proteomic studies (De Palma et al. 2013). The oxidative stress displayed by *Col6a1*$^{-/-}$ muscles (Menazza et al. 2010) may be a major cause for the compromised autocrine secretion of adiponectin by ColVI-deficient myoblasts, as ROS have been reported to inhibit adiponectin secretion by adipose tissue (Parola and Marra 2011). Of note, exogenous replenishment of adiponectin reverses the metabolic alterations displayed by *Col6a1*$^{-/-}$ myoblasts (Gamberi et al. 2019). These results highlight the possibility of considering adiponectin as a new tool for the improvement of muscles affected by ColVI deficiency.

In addition to proteomic and metabolic studies, RNA profiling was also carried out in *Col6a1*$^{-/-}$ mice and BM/UCMD patients. Transcriptome analysis in muscle biopsies from UCMD patients revealed significant changes in pathways involved in muscle regeneration, ECM remodeling, and inflammation (Paco et al. 2013). Further transcriptome analysis in diaphragm, tibialis anterior, and gastrocnemius muscles of 6-month-old *Col6a1*$^{-/-}$ mice revealed pronounced changes in the expression of genes involved in the circadian rhythm (Scotton et al. 2016). Intriguingly, *Bmal1* knockout mice, an animal model displaying altered circadian rhythms, shows deregulation of *COL6* genes as well as of some genes coding for autophagy mediators (Scotton et al. 2016). Therefore, clock genes deregulation may be involved in the link between ColVI deficiency and autophagic alterations.

6.4.1.5 ColVI in Muscle Stem Cell Homeostasis

An important step forward in unraveling the roles played by ColVI in muscle homeostasis was provided by studying muscle stem cells, also known as satellite cells (SCs). SCs are present between the basal lamina and the plasma membrane of muscle fibers, a district that represents their niche, where SCs contact myofibers through the apical surface, while the ECM is located on their basal surface (Mashinchian et al. 2018). SCs are normally quiescent in adult muscle and act as a reservoir of stem cells able to proliferate during muscle regeneration, giving rise to both SCs and differentiated muscle cells that can fuse and form new myofibers or fuse with existing fibers (Wang et al. 2014). The remodeling of the local extracellular environment directly influences the quiescence, activation, differentiation, and self-renewal of SCs, as ECM provides instructive cues that direct cellular behaviors (Calve et al. 2010; Li et al. 2018). In parallel, ECM composition affects the transmission of mechanical properties and other biophysical factors to SCs, thus influencing their activity (Li et al. 2018). A study focusing on muscle regeneration and SCs demonstrated that ColVI is a key component of the SC niche (Urciuolo et al. 2013). Of note, SCs express ColVI in a distinctively regulated manner: at difference from other ECM components, whose expression increases after SC activation and differentiation, high levels of ColVI expression are present in quiescent SCs, whereas ColVI transcripts are markedly decreased in activated SCs (Urciuolo et al. 2013). Interestingly, in vitro studies show that SCs maintain a significantly higher level of quiescence when cultured on a substrate containing purified ColVI, but not when they are cultured on other ECM substrates. Work in *Col6a1*$^{-/-}$ mice showed that ColVI is required for the proper SC self-renewal both in physiological conditions and during muscle regeneration. Indeed, lack of ColVI causes impaired self-renewal capabilities of SCs and defective muscle regeneration after injury. Further in vivo and in vitro work in *Col6a1*$^{-/-}$ mice revealed that ColVI is involved in the fine regulation of the biomechanical properties of muscle, and that this represents a critical mechanism through which ColVI influences SC stemness and self-renewal capabilities (Urciuolo et al. 2013). Once more, CsA administration was found to elicit beneficial effects, as it simulates myogenesis in *Col6a1*$^{-/-}$ mice by increasing the amount of differentiated myogenic cells and of regenerating myofibers at the early stages of muscle regeneration (Gattazzo et al. 2014a). CsA treatment also leads to an amelioration in *Col6a1*$^{-/-}$ mice subjected to multiple muscle injuries, with significantly decreased muscle loss and higher maintenance of the SC pool (Gattazzo et al. 2014a). Altogether, these findings underline a critical role for ColVI in the regulation of SC homeostasis and muscle regeneration, opening further perspectives on the etiology and treatment of ColVI-related myopathies.

6.4.1.6 ColVI in the Neuromuscular Junction

Intriguingly, further studies in *Col6a1*$^{-/-}$ mice revealed that ColVI is also a crucial ECM component of the neuromuscular junction (NMJ), the highly specialized structure between motor neurons and muscle fibers and involved in synaptic transmission. Indeed, ColVI deficiency causes the fragmentation of acetylcholine receptor (AChR) clusters, with altered expression of NMJ proteins, such as perlecan, utrophin, dystrophin, and biglycan (Cescon et al. 2018). The latter protein, in particular, is known to take contact with the MuSK/LRP4 heterodimer, a receptor for agrin, and this complex plays a key role in synapse stability. In *Col6a1*$^{-/-}$ mice, both MuSK gene expression and LRP4 protein levels are altered, and this is paralleled by re-expression of fetal AChR γ subunit, pointing at an abnormal neuromuscular transmission. In agreement with this, functional studies in *Col6a1*$^{-/-}$ mice revealed electrophysiological defects and decreased safety factors (Cescon et al. 2018). Thus, these findings revealed a novel role for ColVI in maintaining the structural and functional integrity of the NMJ compartment, pointing at the involvement of NMJ alterations in the clinical features of patients affected by ColVI-related myopathies.

Overall, the translational impact of the studies carried out in the *Col6a1* null mouse model for human ColVI-related myopathies has been very high, since several pathophysiological and molecular features were uncovered in *Col6a1*$^{-/-}$ mice were also detected in UCMD and BM patients. Therefore, *Col6a1* knockout mice represent a valuable tool for the dissection of the pathomolecular mechanisms of ColVI-related myopathies and for the development and evaluation of therapeutic approaches.

6.4.2 Col6a3 *Mutant Mice*

As for the *Col6a1*$^{-/-}$ mouse model, *Col6a3* mutant mice were generated by exploiting targeted gene disruption (Pan et al. 2013, 2014). One mouse model (*Col6a3*$^{hm/hm}$) harbors an in-frame deletion of 147 bp that leads to the exclusion of exon 15 and 16 from the *Col6a3* transcript, thus representing a hypomorphic mutation. The shorter α3(VI) chain lacks a cysteine-rich linker segment, causing an unstable assembly of the triple-helical ColVI monomer. Therefore, *Col6a3*$^{hm/hm}$ mice are deficient in ColVI deposition in the ECM and display intracellular retention of ColVI chains (Pan et al. 2013). These mice display mild myopathic features and connective tissue defects. Histological analysis of different muscles revealed variation in myofiber sizes, increased interstitial connective tissue, and regenerative myofibers with centrally located nuclei. However, and at difference from *Col6a1*$^{-/-}$ mice, *Col6a3*$^{hm/hm}$ muscles do not exhibit a significant increase in the expression of apoptotic markers. Functional analysis on ex vivo extensor digitorum longus muscles shows decreased muscle strength (Pan et al. 2013). Altogether, *Col6a3*$^{hm/hm}$ animals display myopathic features similar to *Col6a1*$^{-/-}$ mice, albeit with milder phenotypic manifestation.

To mimic dominant ColVI mutations, a further mouse model was generated, bearing the most common molecular defect found in dominant UCMD patients and corresponding to the skipping of exon 16 of *COL6A3* gene (Pan et al. 2014). The resulting heterozygous mouse (*Col6a3*$^{+/d16}$) produces both normal *Col6a3* mRNA and a mutant transcript with an in-frame deletion of 54 bp in exon 16. This in turn leads to the production of a mutant α3(VI) chain with an in-frame deletion of six Gly-Xaa-Yaa repeats in the N-terminal region of the triple-helical domain. Triple-helical ColVI molecules containing the mutant α3(VI) chain are still able to assemble into tetramers and to be secreted. However, increased amounts of dimers and of a protein product with a molecular weight lower than that of ColVI monomers were detected in both cell lysates and media of homozygous mutant (*Col6a3*$^{d16/d16}$) fibroblasts and, to a lower degree, in heterozygous *Col6a3*$^{+/d16}$ fibroblasts (Pan et al. 2014). The authors concluded that the product with the lower size may represent mutant monomers with alternative disulfide bonding, suggesting that tetramer assembly from the mutant α3(VI) chain is somehow compromised, leading to the accumulation of dimers. Furthermore, ColVI microfibrils are barely detectable in the ECM of homozygous and heterozygous fibroblasts, indicating that abnormal ColVI tetramers formed by either the mutant α3(VI) chain exclusively or by a mixture of mutant and normal chains could not assemble into long microfibrils. Conversely, fibroblasts from *Col6a3*$^{+/hm}$ mice deposit, albeit at low amounts, ColVI microfibrils, indicating that the heterozygous exon 16 deletions in *Col6a3* has a dominant-negative effect on ColVI microfibril assembly (Pan et al. 2014). Since in the homozygous cells all ColVI tetramers are defective and do not form microfibrils, *Col6a3*$^{d16/d16}$ mice have been proposed to be the equivalent of *Col6a1*$^{-/-}$ mice. Of note, lack of normal α3(VI) chain in *Col6a3*$^{d16/\mathrm{d}16}$ mice does not lead to a compensatory upregulation of the three additional α3(VI)-like chains during development, probably because of the low abundance and restricted tissue-specific distribution of these chains. However, immunoreactivities of α5(VI) and α6(VI) chains are both increased in a subset of muscle fibers of adult mutant mice. Interestingly, and similarly to *Col6a1*$^{-/-}$ mice, *Col6a3*$^{+/d16}$ mice develop histopathological signs of myopathy, together with ultrastructural alterations of mitochondria and SR and significantly decreased muscle strength (Pan et al. 2014). Therefore, the *Col6a3*$^{+/}$

d16 mouse model represents a useful tool for the development and evaluation of treatment strategies for dominant ColVI mutations.

6.4.3 Zebrafish Models of ColVI-Related Diseases

Danio rerio (zebrafish) is widely used for studies of vertebrate development and gene function. Indeed, thanks to its transparency and rapid development, this teleost fish represents a useful tool to visualize the expression pattern of a specific gene or the activation of signaling pathways in the whole organism. Moreover, zebrafish allows the dissection of different aspects related to specific gene functions, providing valuable information for a deep understanding of human development and disease mechanisms, including collagenopathies (Lieschke and Currie 2007; Bretaud et al. 2019). Accordingly, different zebrafish models of ColVI myopathies were generated in the past years.

RNA knockdown by injection of antisense morpholino oligonucleotides in embryos has been used to generate zebrafish models with transient ColVI deficiency (Telfer et al. 2010; Zech et al. 2015; Ramanoudjame et al. 2015). A first study designed two morpholinos to target zebrafish *col6a1* transcript: one targeting exon 13 and mimicking a dominant mutation frequently found in BM patients, and a second one targeting exon 9, mimicking the effects of one of the most common mutations observed in UCMD (Telfer et al. 2010). Although the transient nature of morpholino-mediated knockdown only allows to investigate its effects during the embryonic and early larval stages, characterization of *col6a1* morphant embryos showed ultrastructural defects in skeletal muscles, with swollen mitochondria and dilated ER, and apoptosis (Telfer et al. 2010), similarly to the phenotypic features previously described for *Col6a1* null mice (Irwin et al. 2003). Of interest, the two morpholinos display different levels of severity. While the morpholino targeting exon 13 resulted in a mild myopathy, with late-onset motor abnormalities, exon 9 morphants display a severe myopathy characterized by early-onset motor deficits and severe ultrastructural changes. Based on the findings in *Col6a1* mice, in which treatment with CsA was demonstrated to rescue the myopathic phenotype (Irwin et al. 2003), the authors tested the effects of CsA on ColVI morphants, obtaining a decreased rate of apoptosis and improved mitochondrial morphology in skeletal muscles (Telfer et al. 2010). The validation of PTP as a target for CsA in zebrafish was demonstrated by independent studies, which demonstrated that the features of fish PTP are similar to those of mammals, and its opening is largely prevented by CsA (Azzolin et al. 2010). Of note, a subsequent study revealed that NIM811, a cyclophilin inhibitor without immunosuppressive activity, is significantly more effective than CsA in preventing the structural and functional muscle abnormalities of zebrafish exon 9 morphants (Zulian et al. 2014).

Another work described two zebrafish ColVI genes, named *col6a4a* and *col6a4b*, that were proposed to be homologous to the mammalian gene coding for the α4 (VI) chain, and transient morphant embryos were generated for these two genes as well as for *col6a2,* with the aim of modeling partial and complete ColVI deficiency (Ramanoudjame et al. 2015). All these morphants display defects in muscle structure and impaired motility, paralleled by defective autophagy, and knockdown of *col6a2* transcripts leads to the most severe muscle phenotype, as expected. Intriguingly, neuronal outgrowth defects are displayed by *col6a2* and *col6a4a* morphants, and at lesser extent in *col6a4b* morphants. Similar defects in axonal growth were described also in *col6a3* zebrafish morphants mimicking human early-onset isolated dystonia (Zech et al. 2015), suggesting that ColVI may be involved in axonal guidance, a role already reported for other ECM molecules such as collagen XV (Pagnon-Minot et al. 2008; Guillon et al. 2016).

Although zebrafish morphants can provide useful information on the effects of transient ColVI deficiency in tissues such as skeletal muscle and nervous system, the use of morphant models has several constraints and pitfalls, as the activity of antisense morpholino oligonucleotides lasts very few days in zebrafish embryos, and it may elicit several off-target effects (Kok et al. 2015). A transcription activator-like effector nuclease (TALEN) approach was used to generate a mutant *col6a1* zebrafish line, named *col6a1*^{ama605003} (Radev et al. 2015). This line harbors a mutation disrupting an essential splice site, leading to an in-frame skipping of *col6a1* exon 14, which codes for the N-terminal region of the collagenous domain of the α1(VI) chain. The deletion mimics a dominant mutation frequently observed in BM patients, which inhibits the assembly of ColVI dimers and tetramers and prevents the secretion of tetramers into the ECM. Homozygous and heterozygous *col6a1*^{ama605003} mutant larvae and adults display a strong myopathic phenotype, with altered myofibers, abnormal dilation of SR, altered mitochondria, and misaligned adjacent sarcomeres. In addition, locomotion studies revealed a hypoxia-response behavior in adult *col6a1*^{ama605003} mutants. Of note, the symptoms worsened with aging, similarly to what observed in patients with ColVI deficiency (Radev et al. 2015). Thus, *col6a1*^{ama605003} represents the first zebrafish line that allows to study the pathological changes occurring in ColVI myopathies during development but also in adult life. Therefore, this line represents a valuable model for drug studies in order to identify candidates for the treatment of ColVI-related disorders, also in the light of the fact that zebrafish is an ideal in vivo platform for high-throughput drug screenings.

Altogether, these studies on zebrafish exon skipping morphants and mutants provide new interesting findings, but they also highlight the importance of generating a stable ColVI null zebrafish line in which the protein is permanently ablated, from embryonic to adult stages. Moreover, they underlined the need of extending the very limited data about the expression and distribution of ColVI in fish during development and adult life, to provide the basis for further functional studies. A recent work described the detailed spatio-temporal expression pattern and distribution of ColVI in zebrafish embryos, larvae, and adults (Tonelotto et al. 2019). Bioinformatics and phylogenetic analyses showed that the fish genome contains single genes (*col6a1*, *col6a2*, and *col6a3*) encoding for α1(VI), α2(VI) and α3 (VI) chains, and duplicated genes (*col6a4a* and *col6a4b*) encoding for α4 (VI) polypeptides. Interestingly, although these latter correspond to the fish ColVI genes already identified in another work (Ramanoudjame et al., 2015), they are not ohnolog genes deriving from the whole genome duplication event that occurred in teleosts. Indeed, bioinformatics analyses indicate that only one of the two ohnologs was maintained in zebrafish after the whole genome duplication, whereas a duplication of the *col6a4* gene, independent from whole genome duplication of teleosts, effectively occurred but only in the *Cyprinidae* family (Tonelotto et al. 2019). These findings suggest that caution should be taken when using data obtained by studying the zebrafish α4(VI) chains in order to understand functions of α4(VI) in higher vertebrates, as one of the two chains may have acquired novel functions, unique for the *Cyprinidae* family. Transcript analysis for the five zebrafish ColVI genes indicates that their expression patterns are finely regulated during embryonic development and adult life. Interestingly, throughout zebrafish life, both ColVI transcripts and protein delineate distinct domains in the ECM of several organs, including eye, cartilage, spleen, skeletal muscle and skin (Tonelotto et al. 2019). The pattern of ColVI expression and ECM deposition is very similar to that reported for the mammalian ColVI (Marvulli et al. 1996; Gara et al. 2008), confirming that zebrafish is a valuable model to perform functional studies aimed at elucidating the in vivo roles of ColVI. With the aim of dissecting in detail the impact of ColVI ablation in tissue homeostasis and development, a stable zebrafish *col6a1* null line was recently generated by CRISPR/Cas9-mediated gene inactivation (P. Bonaldo and coll., unpublished data). This zebrafish ColVI null line also represents an ideal tool for the elucidation of ColVI roles in regulating signaling pathways, thus allowing to unveil still unknown connections between molecular/signaling defects and ColVI ablation. Moreover, considering that zebrafish is an ideal in vivo platform for high-throughput drug screenings, *col6a1* null zebrafish will provide a valuable platform for drug studies aimed at the urgent quest for efficacious therapeutic opportunities for ColVI-related diseases.

6.5 Prospective Therapies and Pilot Clinical Trials for ColVI-Related Myopathies

Various physical therapy and surgical interventions have been introduced into the clinical practice in order to counteract the most prominent symptoms of ColVI-related myopathies. These encompass the adoption of forced ventilation to overcome respiratory failure and surgical interventions aiming at releasing the recurring tendon contractures at different sites (the most affected being Achilles tendons) and at treating the progressively impairing scoliosis (Bönnemann 2011b). Such interventions need to be constantly supported by scrupulous cautions in anesthesia practices (Grosu et al. 2012; Martin et al. 2013) and in treating respiratory infections (Lampe and Bushby 2005; Merlini and Bernardi 2008), and should be accompanied by the association of physiotherapy to a balanced nutritional intake (Merlini and Bernardi 2008; Toni et al. 2014).

Despite the relevance of such approaches, a specific therapeutic treatment for ColVI myopathies, translatable into a definite cure, has not been identified yet. Nonetheless, several strategies are currently under investigation and some were also adopted in clinical trials with promising results. Taking into account the rationale of the different approaches designed so far, two major lines of action can be identified:

- Targeting the genetic defect, thereby rescuing the secretion of a functional form of ColVI.
- Targeting the downstream effects of the genetic defect, namely the cellular alterations underlying the pathogenic mechanisms of the disease.

6.5.1 RNA Targeting in COL6 Myopathies

A gene therapy intended as a stable or temporary modification of the genomic DNA within patients' cells has not been pursued so far. This is due on one hand to the high heterogeneity of ColVI mutations found in BM and UCMD patients, spanning four different genes (*COL6A1, COL6A2, COL6A3, COL6A6*) at multiple sites (Bushby et al. 2014; Hunter et al. 2015), and on the other hand to the inborn safety and ethical issues related to such approaches, ranging from the occurrence of off-targets, tumorigenesis, immunogenic responses potentially induced by the selected strategy, to the feasibility of modifying the gene pool in children or even embryos by *in utero* modification (Cox et al. 2015; DeWeerdt 2018). Alternative strategies focus on RNA, with the rationale of modifying the expression of a mutated allele by regulating mRNA splicing, stability, or translation (Fig. 6.4). In other terms, by introducing synthetic cues such as oligonucleotides, double-stranded RNAs, or specific drugs within cells, it is possible to force the inclusion/exclusion of mutated exons in the mature transcript to induce or abolish the decay of specific mRNA sequences, or to push the translational readthrough of premature stop codons (Le Roy et al. 2009).

On the basis of encouraging results obtained in clinical trials for Duchenne muscular dystrophy, in which antisense oligoribonucleotides (AON) designed to skip *DMD* exon 51 allowed the generation of an in-frame transcript and a partial restoration of dystrophin synthesis (Cirak et al. 2011), a similar strategy was designed for correcting specific forms of dominant-negative mutations in *COL6* genes. Considering that haploinsufficiency for one of the ColVI chains, despite causing a decrease in protein deposition, does not lead to any clinically relevant pathological phenotype (Camacho Vanegas et al. 2001; Foley et al. 2011), a study focused on targeting the mRNA produced by an allele presenting a dominant-negative heterozygous *COL6A2* deletion, known to induce a non-functional triple-helical association (Gualandi et al. 2012). Patient fibroblasts were transfected with a 2'-O-methyl phosphorothioate (2'OMePS) AON recognizing a single nucleotide polymorphism in cis with the mutation and designed to induce the skipping of exon 3 in the mutated *COL6A2* transcript. This causes a frameshift in the mRNA and ultimately leads to nonsense-mediated decay of the mutated transcript, but preserves the transcript encoded by the normal *COL6A2* allele. As predicted, this approach resulted in an enhanced secretion and ECM deposition of ColVI, with an increased microfibrillar network (Gualandi et al. 2012). Another study exploited a siRNA-based approach, targeting the most frequent mutation in Japanese UCMD patients, namely a dominant point mutation in the *COL6A1* gene that entails the complete absence of ColVI in the ECM. When patient fibroblasts were transfected with the designed siRNA, the mutant transcript appeared significantly downregulated when compared to the normal one, and secreted ColVI was detectable (Noguchi et al. 2014). Similarly, siRNA-mediated silencing of a *COL6A3* mutation causing the skipping of exon 16, one of the most common mutations of dominant UCMD patients, was found to elicit increased secretion and ECM deposition of ColVI in patient fibroblasts (Bolduc et al. 2014). Another UCMD-related heterozygous *COL6A3* deletion, in exon 15, was targeted by employing gapmer AONs (chimeric oligonucleotides containing a central block of deoxyribonucleotides) able to hybridize with the target mutant mRNA, thus leading to RNase H-mediated degradation of the RNA-DNA heteroduplex. This approach was able to selectively suppress the expression of the mutant *COL6A3* transcript, with increased ColVI deposition in the ECM (Marrosu et al. 2017).

The most recent example of such strategy came from the identification in a group of 35 UCMD patients of a deep intronic mutation in the *COL6A1* gene, generating a novel splice acceptor site and causing an in-frame insertion in mature mRNA (Cummings et al. 2017). Such insertion introduces an additional sequence in the triple-helical domain of α1(VI) chain, resulting in a dominant-negative effect on ColVI deposition and assembly in the ECM (Bolduc et al. 2019). The use of different combinations of AONs targeting either the splice junctions or pseudo-exon internal sites was shown to suppress pseudo-exon inclusion in a dose-dependent manner in patient-derived fibroblasts, leading to improved ColVI deposition in the ECM, with thicker and longer microfibrils. Interestingly, in this work, the authors also provided a proof of the applicability of the CRISPR/Cas9 technology to intronic mutations of *COL6* genes, by selecting specific gRNAs able to guide the excision of the pseudo-exon with an efficiency close to 80% and restoring an almost normal deposition of ColVI in patients' cells (Bolduc et al. 2019). Although gene editing is much attractive as a definitive and lasting cure, AON applications for ColVI-related myopathies may provide a more rapid route for translation in the clinical setting, given their favorable safety profile. Nonetheless, several open questions remain to be addressed. Thus far, the RNA-targeting strategies exploited for ColVI-related myopathies made use of patients' cells, but their efficacy for skeletal muscle remains to be proven. Moreover, it is still unknown whether the previously deposited ECM and/or the abnormally organized mutant ColVI may represent an obstacle for the proper deposition and assembly of the novel corrected ColVI in the muscle ECM.

6.5.2 Approaches Targeting Altered Molecular Pathways

Other strategies already entered a clinical setting, showing promising results in counteracting some major pathophysiological hallmarks of ColVI-related myopathies. These strategies belong to the second type of approach, aimed at targeting the cellular events triggered by ColVI deficiency (Fig. 6.4). In this context, the *Col6a1*$^{-/-}$ mouse model (Bonaldo et al. 1998) remains a valuable tool, which allowed over the years to unveil and dissect a number of pathophysiological in vivo defects, as well as to get insight into the molecular mechanisms underlying the muscle pathology caused ColVI deficiency. Studies carried out in the *Col6a1*$^{-/-}$ mice provided a wealth of information, which opened the field for further studies in patients' muscle biopsies and primary cultures.

6.5.2.1 Mitochondria

A first path taken to develop a therapy was provided by the unexpected discovery of a PTP-dependent mitochondrial dysfunction caused by ColVI deficiency. As described in detail in the previous section, the finding of remarkable ultrastructural alterations affecting mitochondria and SR in *Col6a1*$^{-/-}$ muscles and paralleled by an increased spontaneous apoptosis, triggered the interest in investigating the presence of a mitochondrial dysfunction linked to altered calcium load (Irwin et al. 2003). Indeed, while in basal conditions the resting $\Delta\psi_m$ of isolated *Col6a1*$^{-/-}$ myofibers did not show any difference when compared to that of wild-type myofibers, treatment with oligomycin, an inhibitor of the F_1F_O ATP synthase frequently used to study mitochondrial function in vitro (Antoniel et al. 2014; Li et al. 2014; Fisher-Wellman et al. 2018), caused a rapid mitochondrial depolarization in *Col6a1*$^{-/-}$ myofibers. This revealed that in the absence of ColVI, impaired mitochondrial respiration was inducing the consumption of ATP by F_1F_O ATP synthase working in a reverse manner as a hydrolase, in order to overcome the increased permeability and maintain a proper proton gradient (Bernardi et al. 2001; Irwin et al. 2003). The permeability transition causing mitochondrial depolarization is mediated by the increased opening of the PTP, which is known to be induced by mitochondrial matrix Ca^{2+} overload and potentiated by ROS (Bernardi et al. 2001; Šileikytė and Forte 2019). The oligomycin-induced mitochondrial depolarization displayed by isolated *Col6a1*$^{-/-}$ myofibers could be rescued by: (1) plating them on native ColVI, confirming the direct relationship between lack of extracellular ColVI and the mitochondrial defect; (2) treatment with Ca^{2+} chelators, such as the membrane-permeant BAPTA/AM, highlighting the involvement of Ca^{2+} dysregulation; (3) treatment with CsA, a well-known inhibitor of the PTP, but not with CsH, an analog unable to inhibit PTP, thus pointing at PTP opening as the first druggable target in ColVI pathobiology (Irwin et al. 2003). These ex vivo results triggered the interest in testing CsA efficacy in vivo. Indeed, CsA was highly effective in counteracting the myopathic defects of *Col6a1*$^{-/-}$ mice, with a remarkable amelioration of muscle structure and function (Irwin et al. 2003; Grumati et al. 2010).

The translation of mouse findings to patients required *in primis* the search for similar pathophysiological defects in biopsies and cell cultures from patients affected by ColVI-related myopathies. Towards this aim, a group of patients representative of the spectrum of UCMD severity was studied, revealing the increased occurrence of apoptosis in muscle biopsies and in primary muscle cultures, when compared to unaffected controls (Angelin et al. 2007). More importantly, the abnormal response to oligomycin observed in *Col6a1*$^{-/-}$ mouse myofibers was also displayed by patients' cells, showing a rapid mitochondrial depolarization that was rescued by plating cells on ColVI, as well as by treating cells with CsA, thus confirming a role for PTP involvement in ColVI deficiency (Angelin et al. 2007). The same study demonstrated that a non-immunosuppressive CsA analog, known as Debio 025 or alisporivir (Hansson et al. 2004), is able to counteract the mitochondrial dysfunction and to reduce the incidence of apoptosis in patients' cells (Angelin et al. 2007). This paved the way to the concept that the major drawback of targeting PTP in patients affected by ColVI-related myopathies, namely the unwanted CsA-induced immunosuppression, can be bypassed by desensitizing PTP opening with a more specific Cyp-D inhibitor, modified to prevent its interaction with calcineurin, and therefore with no effect on immune system surveillance.

These findings laid the foundation for the design of a first pilot clinical trial with CsA in ColVI-related myopathies. Taking into account that the number of BM/UCMD patients with known ColVI mutations is relatively low, together with a great variability in genotype-phenotype presentation and the potential risk of side effects induced by the use of an immunosuppressive drug, a short-term open trial with CsA was carried in one BM and four UCMD patients, three of pediatric age and the other two adults (Merlini et al. 2008a). Patients were administered 5 mg/kg CsA per day for one month—the selected dose being in the lower range of doses used for immunosuppressive treatment—with the aim of evaluating, as a major biological endpoint, mitochondrial function and apoptosis in biopsies taken before and after CsA administration. This required a careful optimization of muscle cell isolation from patients' biopsies and culture conditions, as well as of protocols for measuring the effect of CsA on $\Delta\psi_m$ ex vivo. This pilot clinical trial demonstrated the capability of CsA in rescuing the oligomycin-dependent mitochondrial depolarization and in lowering the incidence of apoptosis in patients' muscle biopsy (Merlini et al. 2008a). Additionally, significant signs of myofiber regeneration were detectable, especially in younger patients, suggesting a beneficial effect of CsA in promoting muscle regeneration (Merlini et al. 2008a), as indeed confirmed in subsequent studies in *Col6a1*$^{-/-}$ mice (Gattazzo et al. 2014a). This pilot trial represented a landmark achievement for ColVI research and for patients affected by ColVI-related myopathies, who saw for the first time the prospect of a treatment for their disease. Indeed, the parents of the three enrolled children asked to continue CsA administration after the conclusion of the one-month trial. Three other children were then additionally recruited in a second prospective long-term, open and non-comparative pilot clinical trial, where UCMD patients with an age ranging between 5 and 9 years swallowed 3–5mg/kg CsA daily for 1 to 3.2 years (Merlini et al. 2011). In this study, the primary endpoint was clinical, defined as an increase in muscle strength, and the secondary endpoint was the effect of CsA on the change in muscle mass, motor function, and respiratory function. Results were encouraging, since that five out of the six young UCMD patients enrolled in the study displayed a statistically significant increase in muscle strength. This was unfortunately not matched by a change in motor function, and despite the positive trend on limb muscle strength, CsA was not able to counteract the progressive decline in respiratory function (Merlini et al. 2011). The treatment was well tolerated by patients, with side effects related to renal dysfunction, hypertension, headache, gastrointestinal disturbances, and hirsutism/hypertrichosis. The results triggered the search of non-immunosuppressive alternatives to CsA, to be used as early as possible in UCMD patients when the diaphragm is less compromised and its regenerative potential is still high. In pursuing this objective, Debio 025 (alisporivir) and another non-immunosuppressive molecule derived from CsA modification, the NIM811 (Waldmeier et al. 2002), were tested in preclinical animal models. Debio 025 was able to recover the oligomycin-induced mitochondrial depolarization, as well as the spontaneous apoptosis and the ultrastructural defects of mitochondria and SR in *Col6a1*$^{-/-}$ mice (Tiepolo et al. 2009). Similarly, NIM811, was successfully tested in zebrafish *col6a1* exon 9 morphant embryos and in *Col6a1*$^{-/-}$ mice, showing its efficacy not only in recovering ultrastructural abnormalities, latent mitochondrial dysfunction, and muscle apoptosis, but also in significantly rescuing motor function in terms of coiling events in zebrafish morphants and of tetanic force in ColVI null mice (Zulian et al. 2014). The translation from bench to bedside of these therapeutic approaches for ColVI patients is relatively feasible, as the pharmacokinetics and safety of Debio 025 were already tested in a phase II clinical study in patients with chronic hepatitis C (Flisiak et al. 2009), and this drug has been more recently caught back in the spotlight for ColVI-related myopathies.

6.5.2.2 Oxidative Stress and Apoptosis

Although major efforts in the translation of therapeutic approaches to clinical use were focused on the desensitization of the PTP, other cellular targets were also considered. A large literature sustains that oxidative stress and ROS generation could contribute to the pathophysiology of muscular dystrophies (Terrill et al. 2013; Canton et al. 2014; Serra et al. 2018). As discussed in the previous section, muscles of *Col6a1*$^{-/-}$ mice display higher ROS levels, accompanied by increased expression and activity of MAOs. In vivo treatment with the MAO inhibitor pargyline was found to reduce ROS production and rescue myofiber apoptosis in *Col6a1*$^{-/-}$ mice, together with an amelioration of muscle strength and improved locomotor performance (Menazza et al. 2010). These data highlighted a significant role of ROS in the oxidation of myofibrillar proteins and in contributing to cell death in ColVI-deficient muscles, pointing at MAOs as candidate novel targets in counteracting disease progression in ColVI-related myopathies. Such concept was tested in myoblasts from BM and UCMD patients, showing that in vitro treatment with pargyline reduces cell susceptibility to oxidative damage, counteracts mitochondrial depolarization, and decrease the occurrence of apoptosis (Sorato et al. 2014). These results provided a rationale for targeting MAO activity as a prospective therapeutic strategy to counteract ColVI-related myopathies.

Another treatment, targeting the ColVI-apoptosis axis and suggested to have the opportunity of entering a clinical trial, features omigapil, a molecule produced by Santhera Pharmaceuticals. This molecule, displaying strong anti-apoptotic effects and previously found to be ineffective against Parkinson's disease and amyotrophic lateral sclerosis (Olanow et al. 2006; Miller et al. 2007), was then tested for drug repurposing in congenital muscular dystrophies. Indeed, omigapil was applied in animal models for laminin 2-deficient congenital muscular dystrophy MDC1A (Erb et al. 2009; Yu et al. 2013), and was found to be able to reduce apoptosis and ultrastructural alteration in myofibers of *Col6a1*$^{-/-}$ mice (unpublished data). Previous studies showed that this molecule interferes with the pro-apoptotic signaling pathway exerted by S-nitrosylation of GAPDH: S-nitrosylated GAPDH binds to the E3 ubiquitin ligase Siah1, which triggers its translocation into the nucleus where, through the engagement of CBP/p300 acetyltransferase, GAPDH activates the transcription of pro-apoptotic genes, including p53, PUMA, and p21 (Chuang et al. 2005; Sen et al. 2008). In 2013, a clinical study, named "CALLISTO," enrolled patients of ages ranging from 5 to 16 years and affected by ColVI-related myopathies or LAMA2-related MDC1A, with the objective of assessing the safety and tolerability of different doses of omigapil and its pharmacokinetics in pediatric patients. This study was completed in 2017 (ClinicalTrials.gov NCT01805024), but the omigapil results are not published yet. In spite of this, the study provided a number of clinical parameters in the cohort of patients observed for two to four years, which could be of relevance for estimating the efficacy of any other further treatment in pediatric ColVI patients (Meilleur et al. 2015; Bendixen et al. 2017; Nichols et al. 2018; Jain et al. 2019).

6.5.2.3 Autophagy

A different line of intervention, complementing the one directly targeting mitochondrial alterations, is focused on the autophagic pathway (Fig. 6.4). As described in the previous section, when investigating the mechanisms responsible for the persistence of altered mitochondria within ColVI-deficient myofibers, autophagy emerged as a major downregulated pathway. The defective autophagic flux of *Col6a1*$^{-/-}$ mice in basal conditions and upon 24-h starvation, restrained by an excessive phosphorylation of Akt and mediated by the defective levels of Beclin 1 and Bnip3, is paralleled in patients' biopsies by downregulated levels of these two autophagy effector proteins (Grumati et al. 2010). Besides this, a relevance for setting autophagy in the spotlight for therapeutic strategies counteracting BM and UCMD came from the evidence that this pathway can be modulated in the *Col6a1*$^{-/-}$ mouse model till the point of rescuing the myopathic phenotype (Grumati et al. 2011a). Indeed, in vivo treatment with the well-known pro-autophagic drug rapamycin was able to rescue autophagy induction, resulting in the amelioration of myofiber defects. Moreover, a prolonged 30-h starvation, as well as an LPD for four weeks, successfully restored all the myofiber features, including autophagic flux, organelle ultrastructure, mitochondrial dysfunction, and apoptosis, together with a marked increase of muscle strength (Grumati et al. 2010). Among the range of approaches exploited in *Col6a1*$^{-/-}$ mice and able to reactivate autophagy in skeletal muscle, nutritional-based ones result the most appealing for human therapies, as they should avoid the adverse immunosuppressive effects elicited by long-term treatments with rapamycin or CsA. These findings were translated into a pilot clinical trial aimed at testing the efficacy of a 1-year normocaloric LPD in activating autophagy in muscles of BM and UCMD patients' (ClinicalTrials.gov NCT01438788). The primary endpoint of the study was the increase in Beclin1 protein levels in muscle biopsies, compared to baseline, but the relevance of this study was much wider. The trial enrolled seven adult patients affected by BM or UCMD and resulted in safe and effective reactivation of autophagy (Castagnaro et al. 2016). This was paralleled by a decreased incidence of myofiber apoptosis, indicating benefits in muscle homeostasis, and by metabolic changes pointing at improved mitochondrial function. Of note, 1-year LPD displayed beneficial effects in counteracting the decline of functional parameters, with no significant worsening of any muscle strength parameter and improvement of motor and respiratory functions (Castagnaro et al. 2016). The setting up of this LPD-based clinical trial required as a prerequisite the detailed study of the nutritional and metabolic status of enrolled patients, which in turn enabled to deepen the understanding of the impact of energy metabolism on body composition and muscle strength in ColVI patients (Toni et al. 2014). Remarkably, the study also allowed the establishment of blood leukocytes as a non-invasive biomarker able to mirror and thus to reveal the autophagic status of patients' muscles, a valuable tool for monitoring the efficacy of future treatments targeting autophagy reactivation (Castagnaro et al. 2016). Incidentally, as underlined by the authors themselves, the significance of this pilot clinical trial was not to promote LPD as an ideal long-term treatment for ColVI pathologies—also considering the limitations due to the lack of a control group receiving a standard diet—but to demonstrate that modulating muscle autophagy through nutritional intervention is achievable and safe, not involving any specific side effect due to the administration of a synthetic drug (Castagnaro et al. 2016).

One of the major advantages of targeting autophagy relies on the fact that it is easily tunable by dietary means or by different nutraceuticals (Madeo et al. 2019). Indeed, the concept of treating human diseases by mean of nutraceutical food additives is particularly attractive in the light of different examples that already showed beneficial effects in the preclinical and clinical settings (Maiuri and Kroemer 2019). Among the most effective autophagy inducers found in natural foods and already applied to counteract human diseases are:

1. Stilbenoids, e.g., resveratrol: found in grapes, peanuts, berries, other plants and their derivatives including red wine and juices; its properties were particularly

highlighted in cardioprotection and neurodegenerative disorders (Sun et al. 2010; Kakoti et al. 2015).

2. Polyamines, e.g., spermidine: found in beans and other legumes, soy and derivatives, mushrooms, aged cheese; it was shown to enhance lifespan in different organisms and having protective properties in several tissues, including heart, brain, and muscle (Eisenberg et al. 2009, 2016; Kiechl et al. 2018).
3. The natural disaccharide trehalose: found in yeasts, sea algae, sunflower seeds, and mushrooms; it is considered a pro-autophagic inducer, acting on lysosomal biogenesis, with promising neuroprotective effects (Rusmini et al. 2019; Khalifeh et al. 2019).

Interestingly, as discussed in the previous section, spermidine administered to *Col6a1*$^{-/-}$ mice both *per os* and by intraperitoneal injection was successful in reactivating muscle autophagic flux, ameliorating organelle ultrastructure and myofiber defects, as well as in reducing the incidence of spontaneous apoptosis (Chrisam et al. 2015). Further studies in mice revealed strong beneficial effects of other autophagy-inducing nutraceuticals, such as stilbenoids, in ameliorating muscle structure and function and counteracting the myopathic pathology (Metti et al. 2020). These findings underline the remarkable opportunities offered by nutraceutical approaches in the treatment COL6 pathologies and support spermidine as a candidate for future trials in this context. Moreover, considering that spermidine and resveratrol do activate autophagy by modulating both the nuclear and the cytosolic acetylproteome through different mechanisms and in a converging manner (Morselli et al. 2011; Madeo et al. 2018), synergistic effects can be expected by combined nutraceutical treatments.

A huge effort has been made so far toward the identification of a safe cure for ColVI myopathies. The advancements achieved in the last decade, concerning the precision and efficacy of gene editing and molecular tools, together with the detailed knowledge about the biological pathways involved in the pathogenesis of ColVI-related myopathies, provide tremendous opportunities in the quest for a cure for patients. In the near future they may allow setting a brake to disease progression when combined with early diagnosis in infants.

6.6 Other Collagens Involved in Myopathies and Inherited Muscle Disorders

6.6.1 Collagen IV

Collagen IV (ColIV) is a major component of muscle basal lamina and it is present with different isoforms in the synaptic and extrasynaptic regions. Six genes (*COL4A1*-*COL4A6*) code for ColIV subunits in humans. The most abundant ColIV isoform in muscle basal lamina is [$\alpha1(IV)_2\alpha2(IV)$], whereas [$\alpha3(IV)\alpha4(IV)\alpha5(IV)$] and [$\alpha(IV)_2\alpha6(IV)$] are restricted to the NMJ (Singhal and Martin

2011; Gatseva et al. 2019). Mutations in *COL4A1* and *COL4A2* genes, coding for ColIV α1 and α2 chains, can cause a multisystemic disease characterized by the presence of cerebrovascular disorder with variable ocular, muscular, and renal involvement (see Chap. 5). Muscle pathology has been described in several individuals with *COL4A1* and *COL4A2* pathogenic variants (Jeanne and Gould 2017) and in animal models (i.e., mice and *Drosophila*) with mutations in *COL4A1* and *COL4A2* orthologs (Labelle-Dumais et al. 2011; Kelemen-Valkony et al. 2012; Kiss et al. 2019). In particular, mice carrying a heterozygous *Col4a1* splice site mutation that leads to the skipping of exon 41 ($Col4a1^{+/\Delta ex41}$) display a genetically complex skeletal myopathy with reduced grip strength and increased numbers of centrally nucleated cells (Labelle-Dumais et al. 2011) (for more details see also Chap. 3). Furthermore, $Col4a1^{+/\Delta ex41}$ mice also have ocular dysgenesis and neuronal localization defects, which, together with myopathy, are characteristic of the muscle-eye-brain disease and Walker-Warburg syndrome (Labelle-Dumais et al. 2011). Thus, this animal model is very useful for elucidating the pathophysiological mechanisms underlying *COL4A1*-related myopathy in humans. Recent work in the *Col4a1* mutant mice demonstrated that tissue-specific allelic and mechanistic heterogeneity contribute to the onset of the neuromuscular pathology caused by α1(IV*) deficiency* (Labelle-Dumais et al. 2019). These findings reveal the importance of the $[\alpha1(IV)]_2\alpha2(IV)$ network as an ECM platform that can evoke distinct tissue-specific responses and suggest that a potential therapeutic approach aimed at promoting $[\alpha1(IV)]_2\alpha2(IV)$ secretion may impinge on the myopathic pathology in a mutation-dependent way (Labelle-Dumais et al. 2019).

6.6.2 Collagen IX

A mild myopathy is also observed in some patients with multiple epiphyseal dysplasia (MED), a degenerative cartilage disorder caused by mutations in genes coding for collagen IX (ColIX), cartilage oligomeric matrix protein, and matrilin-3. In particular, mutations of *COL9A2* and *COL9A3* genes, coding for α2(IX) and α3 (IX) chains, are found in some individuals affected by a MED-related myopathy, with proximal muscle weakness and histological myopathic features, such as variation in myofiber size (Bönnemann et al. 2000; Jackson et al. 2009). Of note, ColIX is mainly expressed in cartilage ECM (see Chap. 4), and a role for this ECM protein in skeletal muscle was previously unreported. Thus, the findings in patients affected by MED with associated myopathy suggest that ColIX also plays a role in the musculoskeletal system (for a detailed overview see also Chap. 8).

6.6.3 Collagen XII

Human syndromes involving muscle and connective tissue are also associated with pathogenic variants in the *COL12A1* gene, coding for the α1 chain of collagen XII (ColXII (for more details see also Chap. 4). Indeed, heterozygous or homozygous mutations in *COL12A1* have been reported in patients with a mixed myopathic and Ehlers-Danlos syndrome (EDS) phenotype (Hicks et al. 2013; Zou et al. 2014; Punetha et al. 2017; Witting et al. 2018). The affected individuals display distal joint hypermobility and proximal joint contractures together with muscle hypotonia and weakness and delayed motor development. Furthermore, they are affected by scoliosis or kyphosis and abnormal scarring. Based on this characteristic phenotype, the condition was later classified as myopathic EDS (Malfait et al. 2017). Interestingly, *Col12a1* knockout mice display muscle weakness with decreased grip strength and connective tissue alterations, such as bone fragility, kyphoscoliosis, and short stature (Zou et al. 2014). Thanks to this mouse model, it was possible to highlight the function of ColXII in skeletal muscle, showing that its deficiency elicits structural changes in the muscle ECM and affects the transduction of force in the muscle-tendon-bone unit (Zou et al. 2014). A recent study by Malfait and coll. in 78 patients satisfying the clinical criteria for myopathic EDS identified among this cohort four novel heterozygous pathogenic *COL12A1* variants, one, *COL12A1* variant of unclear significance, and biallelic pathogenic variants in *COL6A1* (Delbaere et al. 2020). Altogether, these findings pointed out a significant clinical overlap between myopathic EDS and COL6-related myopathies, indicating that in the diagnosis of this spectrum of diseases the involvement of both ColVI and ColXII should be taken into account. These findings are also in agreement with independent studies indicating that ColVI and ColXII form functional complexes in the ECM (Izu et al. 2016).

6.6.4 Collagen V

ColV is a fibrillar collagen coded by three genes (*COL5A1-COL5A3*) and occurring in various isoforms in the interstitial ECM of different tissues, including tendons, ligaments, and muscles. It plays a key role in the modulation of collagen I (ColI) fibrillogenesis and in the regulation of fibril diameter (Birk et al. 1990). *Col5a1* null mice die early during embryogenesis due to impaired formation of collagen fibrils, whereas heterozygous *Col5a1*$^{+/-}$ mice display a reduction and altered size of ColI fibrils (Wenstrup et al. 2004). In humans, mutations of genes coding for COLV chains are causative for the classic forms of EDS, and a large number of patients with EDS carry heterozygous dominant mutations in *COL5A1* and *COL5A2* (Malfait et al. 2010; see also Chap. 3). Even if no major myopathic features were described so far in patients with ColV-related EDS, studies in mice showed that ColV is a critical component of the quiescent muscle SCs niche, as its depletion leads to anomalous cell cycle entry and gradual decrease of the stem cell pool, suggesting a potentially still neglected role of ColV also in the muscle (Baghdadi et al. 2018). Interestingly, a *COL5A1* mutation abolishing the N-propeptide cleavage and affecting collagen fibril organization was reported in a patient exhibiting muscular contractures and weakness, together with skin and nerve defects (Badara, 2016). Interestingly, a recent study also revealed ColV depletion in the ECM of patients with *COL6A1* defects (Delbaere et al. 2020), thus confirming the relevance of the interaction between ColVI and ColV (Symoens et al. 2011).

6.6.5 Collagen XV

A further type of collagen playing a role in maintaining muscle homeostasis is collagen XV (ColXV). Animal models generated to study the biological role of ColXV allowed to highlight that depletion of this ECM protein leads to muscle degeneration. Analysis of muscles from ColXV *knockout* (*Col15a1*$^{-/-}$) *mice* revealed a myopathic phenotype characterized by the presence of degenerating myofibers, centrally nucleated myofibers, and variations in myofiber size. These alterations were first detected at 3 months of age and became more evident in older mice. Furthermore, *Col15a1*$^{-/-}$ mice display increased vulnerability to exercise-induced muscle damage (Eklund et al. 2001). Zebrafish morphant embryos in which the expression of ColXV transcripts is downregulated show a severe impairment of muscle development. ColXV knockdown results in defective notochord differentiation and altered formation of slow and fast-twitch muscle in zebrafish, with the increased number of medial fast-twitch muscle fibers, thus suggesting that notochord-derived Hedgehog signals are enhanced by ColXV deficiency (Pagnon-Minot et al. 2008). Further work in zebrafish revealed that ColXV is involved in the slow muscle genetic program as a direct target of the Hedgehog/Gli signaling. The altered ColXV function causes pathfinding defects in motoneuron axons, resulting in muscle atrophy and compromised motility (Guillon et al. 2016).

6.6.6 Collagen XXII

A remarkable muscle phenotype was also detected in zebrafish models for collagen XXII (ColXXII). ColXXII is a distinctive ECM component of the MTJ, the highly specialized structure involved in the muscle force transmission to tendons (Koch et al. 2004). Knockdown of ColXXII in zebrafish *col22a1* morphant embryos causes a muscular dystrophy-like phenotype characterized by a strong reduction of MTJ folds and alterations in myoseptal structure, which ultimately cause a reduced contractile force and an increased susceptibility to contraction-induced myofiber detachment (Charvet et al. 2013).

6.6.7 Collagen XIII and Collagen Q

Two other components of collagen superfamily, namely collagen XIII (ColXIII) and collagen Q (ColQ), are instead linked to congenital myasthenic syndromes (CMS), a group of inherited diseases with compromised neuromuscular transmission, due to abnormal signal transmission at the motor endplate, and characterized by a muscle weakness that worsens upon exertion and recurrent respiratory illnesses. CMS represent a broad group of diverse disorders, and the age of onset and spectrum of symptoms (from variable fatigable weakness to disability) vary a lot among affected individuals (Rodríguez Cruz et al. 2018). ColXIII is a membrane-spanning and ECM shed protein with a critical role in the formation and function of NMJs (Heikkinen et al. 2019). In humans, homozygous recessive mutations of the *COL13A1* gene cause synaptic basal lamina-associated CMS type 19 (OMIM #616720) (Engel et al. 2016). The symptoms include hypotonia, muscle weakness, breathing and feeding difficulties, and respiratory problems soon after birth, but the severity of muscle weakness and of the disease progression is variable (Rodríguez Cruz et al. 2019). A number of studies were carried out by Pihlajaniemi and coll. in various engineered mouse models clearly underlined the critical role of ColXIII for the proper functionality of the musculoskeletal system. A transgenic mouse ($Col13a1^{N/N}$), producing an aberrant form of ColXIII lacking the short cytosolic and transmembrane domains, confirmed the crucial role of ColXIII for muscle integrity. In these mice, the mutant ColXIII protein is not correctly localized at the plasma membrane, but deposited in the adjacent ECM. $Col13a1^{N/N}$ mice display myopathic features, and their skeletal muscles display abnormal myofibers with a wavy interphase between sarcolemma and basement membrane, together with disorganization of myofilaments and Z-disks and vacuolization. These defects are progressive and exercise induces increased myofiber degeneration (Kvist et al. 2001). Additional work in ColXIII null ($Col13a1^{-/-}$) mice and in mutant mice expressing only the transmembrane form of ColXIII ($Col13^{tm/tm}$) demonstrated that ColXIII regulates the maturation of NMJs and it is required for NMJ regeneration and functional recovery after injury (Latvanlehto et al. 2010; Zainul et al. 2018). Further studies in $Col13a1^{-/-}$ mice and in transgenic mice overexpressing ColXIII confirmed the key role of ColXIII in NMJ and highlighted the importance of its correct expression and localization for the AChR clustering and for the proper formation and function of NMJs (Härönen et al. 2017; Härönen et al. 2019).

ColQ, corresponding to the collagen-like tail that anchors asymmetric acetylcholinesterase (AChE) to the basal lamina, is expressed predominantly in fast muscles at the level of NMJ. ColQ anchors AChE in the synaptic cleft and binds perlecan and muscle-specific kinase (MuSK), through which it exerts intracellular signaling and controls AChR clustering (Ohno et al. 1998). These properties make ColQ an important player in NMJ formation and stability and in muscle integrity. Mutations in the human *COLQ* gene are responsible for a CMS associated with AChE deficiency, classed as CMS type 5 (OMIM #603034) (Engel et al. 2016). The main symptoms include global and motor developmental delay, muscle hypotonia, fatigue, and respiratory failure (Mihaylova et al. 2008). Interestingly, mutations in the C-terminal domain of ColQ were found to compromise its interactions with MuSK and basement membrane components at the NMJ (Nakata et al. 2013; Arredondo et al. 2013). ColQ-deficient (*Colq*$^{-/-}$) mice are a model for this disease and display an upregulation of both embryonic γ-AChR and adult ε-AChR subunits, leading to mixed mature and immature AChRs at the NMJ of adult *Colq*$^{-/-}$ mice. In addition to myasthenia, *Colq*$^{-/-}$ animals display muscle atrophy, primarily affecting fast muscles, and defects in their development of fast muscle fibers (Sigoillot et al. 2016). Additional studies in ColQ-deficient focused on exploiting the efficacy of potential therapies, such as treatment with β2 adrenergic receptor agonists or AAV-mediated expression of ColQ (McMacken et al. 2019; Ito and Ohno 2018).

6.7 Conclusions

Altogether, the large amount of work carried out in the last decades highlighted the critical roles of different collagen types in skeletal muscle homeostasis and in the proper maintenance of muscle structure and function, as demonstrated by the causative contribution of mutations of several collagen genes in the pathogenesis of congenital muscular dystrophies and other muscle disorders. The increasing knowledge on the underlying cellular and molecular mechanisms and the identification of effective targets and approaches for therapy are of invaluable benefits not only for the accurate clinical diagnosis and the optimal management of patients, but also for developing effective therapeutic approaches able to counteract the onset and progression of clinical symptoms and disabilities. It can be foreseen that in the near future such increasing knowledge will be translated into further clinical studies and trials, with the aim of providing a safe and effective cure for such highly disabling and often life-threatening pathologies affecting muscle.

References

Aigner T, Hambach L, Söder S et al (2002) The C5 domain of Col6a3 is cleaved off from the Col6 fibrils immediately after secretion. Biochem Biophys Res Commun 290:743–748

Angelin A, Tiepolo T, Sabatelli P et al (2007) Mitochondrial dysfunction in the pathogenesis of Ullrich congenital muscular dystrophy and prospective therapy with cyclosporins. Proc Natl Acad Sci U S A 104:991–996

Antoniel M, Giorgio V, Fogolari F et al (2014) The oligomycin-sensitivity conferring protein of mitochondrial ATP synthase: emerging new roles in mitochondrial pathophysiology. Int J Mol Sci 15:7513–7536

Arredondo J, Lara M, Ng F et al (2013) COOH-terminal collagen Q (COLQ) mutants causing human deficiency of endplate acetylcholinesterase impair the interaction of ColQ with proteins of the basal lamina. Hum Genet 133:599–616

Aumailley M, Mann K, von der Mark H et al (1989) Cell attachment properties of collagen type VI and Arg-Gly-Asp dependent binding to its alpha 2(VI) and alpha 3(VI) chains. Exp Cell Res 181:463–474

Azzolin L, Basso E, Argenton F, Bernardi P (2010) Mitochondrial Ca^{2+} transport and permeability transition in zebrafish (*Danio rerio*). Biochim Biophys Acta Bioenerg 1797:1775–1779

Badara SN (2016) Une nouvelle mutation du collagène V conduisant, chez le patient, à des atteintes cutanées et musculaires. Biologie cellulaire. Université de Lyon, Français. ⟨NNT: 2016LYSEN064⟩

Baghdadi MB, Castel D, Machado L et al (2018) Reciprocal signalling by Notch-Collagen V-CALCR retains muscle stem cells in their niche. Nature 557:714–718

Bailey AJ, Restall DJ, Sims TJ et al (1979) Meat tenderness - immunofluorescent localization of the isomorphic forms of collagen in bovine muscles of varying texture. J Sci Food Agric 30:203–210

Baker NL, Mörgelin M, Peat R et al (2005) Dominant collagen VI mutations are a common cause of Ullrich congenital muscular dystrophy. Hum Mol Genet 14:279–293

Baker NL, Mörgelin M, Pace RA et al (2007) Molecular consequences of dominant Bethlem myopathy collagen VI mutations. Ann Neurol 62:390–405

Baldock C, Sherratt MJ, Shuttleworth CA, Kielty CM (2003) The supramolecular organization of collagen VI microfibrils. J Mol Biol 330:297–307

Ball SG, Baldock C, Kielty CM, Shuttleworth CA (2001) The role of the C1 and C2 A-domains in type VI collagen assembly. J Biol Chem 276:7422–7430

Beecher N, Roseman AM, Jowitt TA et al (2011) Collagen VI, conformation of A-domain arrays and microfibril architecture. J Biol Chem 286:40266–40275

Bendixen RM, Butrum J, Jain MS et al (2017) Upper extremity outcome measures for collagen VI-related myopathy and LAMA2-related muscular dystrophy. Neuromuscul Disord 27:278–285

Bernardi P, Bonaldo P (2013) Mitochondrial dysfunction and defective autophagy in the pathogenesis of collagen VI muscular dystrophies. Cold Spring Harb Perspect Biol 5:a011387

Bernardi P, Petronilli V, Di Lisa F, Forte M (2001) A mitochondrial perspective on cell death. Trends Biochem Sci 26:112–117

Bernardi P, Krauskopf A, Basso E et al (2006) The mitochondrial permeability transition from in vitro artifact to disease target. FEBS J 273:2077–2099

Bernardi P, Di Lisa F, Fogolari F et al (2015) From ATP to PTP and back: a dual function for the mitochondrial ATP synthase. Circ Res 116:1850–1862

Bethlem J, Wijngaarden GK (1976) Benign myopathy, with autosomal dominant inheritance. A report on three pedigrees. Brain 99:91–100

Bidanset DJ, Guidry C, Rosenberg LC et al (1992) Binding of the proteoglycan decorin to collagen type VI. J Biol Chem 267:5250–5256

Birk DE, Fitch JM, Babiarz JP et al (1990) Collagen fibrillogenesis in vitro: interaction of types I and V collagen regulates fibril diameter. J Cell Sci 95:649–657

Bolduc V, Zou Y, Ko D, Bönnemann CG (2014) SiRNA-mediated allele-specific silencing of a COL6A3 mutation in a cellular model of dominant Ullrich muscular dystrophy. Mol Ther Nucleic Acids 3:e147

Bolduc V, Reghan Foley A, Solomon-Degefa H et al (2019) A recurrent COL6A1 pseudoexon insertion causes muscular dystrophy and is effectively targeted by splice-correction therapies. JCI Insight 21:4(6). pii: 124403

Bonaldo P, Colombatti A (1989) The carboxyl terminus of the chicken α3 chain of collagen VI is a unique mosaic structure with glycoprotein Ib-like, fibronectin type III, and Kunitz modules. J Biol Chem 264:20235–20239

Bonaldo P, Russo V, Bucciotti F et al (1989) Alpha1 chain of chick type VI collagen. The complete cDNA sequence reveals a hybrid molecule made of one short collagen and three von Willebrand factor type A-like domains. J Biol Chem 264:5575–5580

Bonaldo P, Russo V, Bucciotti F et al (1990) Structural and functional features of the alpha 3 chain indicate a bridging role for chicken collagen VI in connective tissues. Biochemistry 29:1245–1254

Bonaldo P, Braghetta P, Zanetti M et al (1998) Collagen VI deficiency induces early onset myopathy in the mouse: an animal model for Bethlem myopathy. Hum Mol Genet 7:2135–2140

Bönnemann CG (2011a) The collagen VI-related myopathies. Ullrich congenital muscular dystrophy and Bethlem myopathy. Handb Clin Neurol 101:81–96

Bönnemann CG (2011b) The collagen VI-related myopathies: muscle meets its matrix. Nat Rev Neurol 7:379–390

Bönnemann CG, Cox GF, Shapiro F et al (2000) A mutation in the alpha 3 chain of type IX collagen causes autosomal dominant multiple epiphyseal dysplasia with mild myopathy. Proc Natl Acad Sci U S A 97:1212–1217

Bönnemann CG, Brockmann K, Hanefeld F (2003) Muscle ultrasound in Bethlem myopathy. Neuropediatrics 34:335–336

Bowe MA, Mendis DB, Fallon JR (2000) The small leucine-rich repeat proteoglycan biglycan binds to alpha-dystroglycan and is upregulated in dystrophic muscle. J Cell Biol 148:801–810

Braghetta P, Fabbro C, Piccolo S et al (1996) Distinct regions control transcriptional activation of the alpha1(VI) collagen promoter in different tissues of transgenic mice. J Cell Biol 135:1163–1177

Braghetta P, Vitale P, Piccolo S et al (1997) Tissue-specific expression of promoter regions of the alpha1(VI) collagen gene in cell cultures and transgenic mice. Eur J Biochem 247:200–208

Braghetta P, Ferrari A, Fabbro C et al (2008) An enhancer required for transcription of the Col6a1 gene in muscle connective tissue is induced by signals released from muscle cells. Exp Cell Res 314:3508–3518

Bretaud S, Nauroy P, Malbouyres M, Ruggiero F (2019) Fishing for collagen function: about development, regeneration and disease. Semin Cell Dev Biol 89:100–108

Brown JC, Golbik R, Mann K, Timpl R (1994) Structure and stability of the triple-helical domains of human collagen XIV. Matrix Biol 14:287–295

Bruns RR, Press W, Engvall E et al (1986) Type VI collagen in extracellular, 100-nm periodic filaments and fibrils: identification by immunoelectron microscopy. J Cell Biol 103:393–404

Burg MA, Tillet E, Timpl R et al (1996) Binding of the NG2 proteoglycan to type VI collagen and other extracellular matrix molecules. J Biol Chem 271:26110–26116

Bürgi J, Kunz B, Abrami L et al (2017) CMG2/ANTXR2 regulates extracellular collagen VI which accumulates in hyaline fibromatosis syndrome. Nat Commun 12:15861

Bushby KMD, Collins J, Hicks D (2014) Collagen type VI myopathies. Adv Exp Med Biol 802:185–199

Calve S, Odelberg SJ, Simon HG (2010) A transitional extracellular matrix instructs cell behavior during muscle regeneration. Dev Biol 344:259–271

Camacho Vanegas O, Bertini E, Zhang RZ et al (2001) Ullrich scleroatonic muscular dystrophy is caused by recessive mutations in collagen type VI. Proc Natl Acad Sci U S A 98:7516–7521. h

Canato M, Dal Maschio M, Sbrana F, et al (2010) Mechanical and electrophysiological properties of the sarcolemma of muscle fibers in two murine models of muscle dystrophy: Col6a1 −/− and mdx. J Biomed Biotechnol 2010:981945.

Canton M, Menazza S, Di Lisa F (2014) Oxidative stress in muscular dystrophy: from generic evidence to specific sources and targets. J Muscle Res Cell Motil 35:23–36

Capitanio D, Moriggi M, De Palma S et al (2017) Collagen VI null mice as a model for early onset muscle decline in aging. Front Mol Neurosci 10:337

Castagnaro S, Pellegrini C, Pellegrini M et al (2016) Autophagy activation in COL6 myopathic patients by a low-protein-diet pilot trial. Autophagy 12:2484–2495

Castagnaro S, Chrisam M, Cescon M et al (2018) Extracellular collagen VI has prosurvival and autophagy instructive properties in mouse fibroblasts. Front Physiol 9:1129

Cescon M, Gattazzo F, Chen P, Bonaldo P (2015) Collagen VI at a glance. J Cell Sci 128:3525–3531

Cescon M, Gregorio I, Eiber N et al (2018) Collagen VI is required for the structural and functional integrity of the neuromuscular junction. Acta Neuropathol 136:483–499

Charvet B, Ruggiero F, Le Guellec D (2012) The development of the myotendinous junction. A review. Muscles Ligaments Tendons J 2:53–63

Charvet B, Guiraud A, Malbouyres M et al (2013) Knockdown of col22a1 gene in zebrafish induces a muscular dystrophy by disruption of the myotendinous junction. Development 140:4602–4613

Chen P, Cescon M, Megighian A, Bonaldo P (2014) Collagen VI regulates peripheral nerve myelination and function. FASEB J 28:1145–1156

Chrisam M, Pirozzi M, Castagnaro S et al (2015) Reactivation of autophagy by spermidine ameliorates the myopathic defects of collagen VI-null mice. Autophagy 11:2142–2152

Chu ML, Pan TC, Conway D et al (1989) Sequence analysis of alpha 1(VI) and alpha 2(VI) chains of human type VI collagen reveals internal triplication of globular domains similar to the A domains of von Willebrand factor and two alpha 2(VI) chain variants that differ in the carboxy terminus. EMBO J 8:1939–1946

Chu ML, Zhang RZ, Pan TC et al (1990) Mosaic structure of globular domains in the human type VI collagen alpha 3 chain: similarity to von Willebrand factor, fibronectin, actin, salivary proteins and aprotinin type protease inhibitors. EMBO J 9:385–393

Chuang D-M, Hough C, Senatorov VV (2005) Glyceraldehyde-3-phosphate dehydrogenase, apoptosis, and neurodegenerative diseases. Annu Rev Pharmacol Toxicol 45:269–290

Cirak S, Arechavala-Gomeza V, Guglieri M et al (2011) Exon skipping and dystrophin restoration in patients with Duchenne muscular dystrophy after systemic phosphorodiamidate morpholino oligomer treatment: an open-label, phase 2, dose-escalation study. Lancet 378:595–605

Colombatti A, Bonaldo P (1987) Biosynthesis of chick type VI collagen. II. Processing and secretion in fibroblasts and smooth muscle cells. J Biol Chem 262:14461–14466

Colombatti A, Bonaldo P (1991) The superfamily of proteins with von Willebrand factor type A-like domains: one theme common to components of extracellular matrix, hemostasis, cellular adhesion, and defense mechanisms. Blood 77:2305–2315

Colombatti A, Bonaldo P, Ainger K et al (1987) Biosynthesis of chick type VI collagen. I. Intracellular assembly and molecular structure. J Biol Chem 262:14454–14460

Colombatti A, Bonaldo P, Doliana R (1993) Type A modules: interacting domains found in several non-fibrillar collagens and in other extracellular matrix proteins. Matrix 13:297–306

Colombatti A, Mucignat MT, Bonaldo P (1995) Secretion and matrix assembly of recombinant type VI collagen. J Biol Chem 270:13105–13111

Cox DBT, Platt RJ, Zhang F (2015) Therapeutic genome editing: prospects and challenges. Nat Med 21:121–131

Csapo R, Gumpenberger M, Wessner B (2020) Skeletal muscle extracellular matrix – what do we know about its composition, regulation, and physiological roles? A narrative review. Front Physiol 11:253

Cummings BB, Marshall JL, Tukiainen T et al (2017) Improving genetic diagnosis in Mendelian disease with transcriptome sequencing. Sci Transl Med 9:eaal5209

Dani C, Doglio A, Amri EZ et al (1989) Cloning and regulation of a mRNA specifically expressed in the preadipose state. J Biol Chem 264:10119–10125

De Palma S, Leone R, Grumati P et al (2013) Changes in muscle cell metabolism and mechanotransduction are associated with myopathic phenotype in a mouse model of collagen VI deficiency. PLoS One 8:e56716

Deconinck N, Dion E, Ben YR et al (2010) Differentiating Emery-Dreifuss muscular dystrophy and collagen VI-related myopathies using a specific CT scanner pattern. Neuromuscul Disord 20:517–523

Delbaere S, Dhooge T, Syx D et al (2020) Novel defects in collagen XII and VI expand the mixed myopathy/Ehlers-Danlos syndrome spectrum and lead to variant-specific alterations in the extracellular matrix. Genet Med 22:112–123

Demir E, Sabatelli P, Allamand V et al (2002) Mutations in COL6A3 cause severe and mild phenotypes of Ullrich congenital muscular dystrophy. Am J Hum Genet 70:1446–1458
DeWeerdt S (2018) Prenatal gene therapy offers the earliest possible cure. Nature 564:S6–S8
Doliana R, Bonaldo P, Colombatti A (1990) Multiple forms of chicken α3(VI) collagen chain generated by alternative splicing in type A repeated domains. J Cell Biol 111:2197–2205
Dziadek M, Kazenwadel JS, Hendrey A et al (2002) Alternative splicing of transcripts for the alpha3 chain of mouse collagen VI: identification of an abundant isoform lacking domains N7-N10 in mouse and human. Matrix Biol 21:227–241
Eisenberg T, Knauer H, Schauer A et al (2009) Induction of autophagy by spermidine promotes longevity. Nat Cell Biol 11:1305–1314
Eisenberg T, Abdellatif M, Schroeder S et al (2016) Cardioprotection and lifespan extension by the natural polyamine spermidine. Nat Med 22:1428–1438
Eklund L, Piuhola J, Komulainen J et al (2001) Lack of type XV collagen causes a skeletal myopathy and cardiovascular defects in mice. Proc Natl Acad Sci U S A 98:1194–11999
Engel J, Furthmayr H (1987) Electron microscopy and other physical methods for the characterization of extracellular matrix components: laminin, fibronectin, collagen IV, collagen VI, and proteoglycans. Methods Enzymol 145:3–78
Engel J, Furthmayr H, Odermatt E et al (1985) Structure and macromolecular organization of type VI collagen. Ann N Y Acad Sci 460:25–37
Engel AG, Shen XM, Selcen D et al (2016) Congenital myasthenic syndromes: pathogenesis, diagnosis, and treatment. Lancet Neurol 14:420–434
Engvall E, Hessle H, Klier G (1986) Molecular assembly, secretion, and matrix deposition of type VI collagen. J Cell Biol 102:703–710
Erb M, Meinen S, Barzaghi P et al (2009) Omigapil ameliorates the pathology of muscle dystrophy caused by laminin-α2 deficiency. J Pharmacol Exp Ther 331:787–795
Fabbro C, Braghetta P, Girotto D et al (1999) Cell type-specific transcription of the α1(VI) collagen gene. J Biol Chem 274:1759–1766
Fiaschi T, Cirelli D, Comito G et al (2009) Globular adiponectin induces differentiation and fusion of skeletal muscle cells. Cell Res 19:584–597
Finnis ML, Gibson MA (1997) Microfibril-associated glycoprotein-1 (MAGP-1) binds to the pepsin- resistant domain of the α3(VI) chain of type VI collagen. J Biol Chem 272:22817–22823
Fisher-Wellman KH, Davidson MT, Narowski TM et al (2018) Mitochondrial diagnostics: a multiplexed assay platform for comprehensive assessment of mitochondrial energy fluxes. Cell Rep 24:3593–3606.e10
Fitzgerald J, Rich C, Zhou FH, Hansen U (2008) Three novel collagen VI chains, α4(VI), α5(VI), and α6(VI). J Biol Chem 283:20170–20180
Flisiak R, Feinman SV, Jablkowski M et al (2009) The cyclophilin inhibitor Debio 025 combined with PEG IFNα2a significantly reduces viral load in treatment-naïve hepatitis C patients. Hepatology 49:1460–1468
Foley AR, Hu Y, Zou Y et al (2011) Large genomic deletions: a novel cause of Ullrich congenital muscular dystrophy. Ann Neurol 69:206–211
Furukawa T, Toyokura Y (1977) Congenital, hypotonic-sclerotic muscular dystrophy. J Med Genet 14:426–429
Gamberi T, Magherini F, Mannelli M et al (2019) Role of adiponectin in the metabolism of skeletal muscles in collagen VI-related myopathies. J Mol Med 97:793–801
Gara SK, Grumati P, Urciuolo A et al (2008) Three novel collagen VI chains with high homology to the α3 chain. J Biol Chem 283:10658–10670
Gara SK, Grumati P, Squarzoni S et al (2011) Differential and restricted expression of novel collagen VI chains in mouse. Matrix Biol 30:248–257
Gatseva A, Sin YY, Brezzo G, Van Agtmael T (2019) Basement membrane collagens and disease mechanisms. Essays Biochem 63:297–312

Gattazzo F, Molon S, Morbidoni V et al (2014a) Cyclosporin a promotes in vivo myogenic response in collagen VI-deficient myopathic mice. Front Aging Neurosci 6:244

Gattazzo F, Urciuolo A, Bonaldo P (2014b) Extracellular matrix: a dynamic microenvironment for stem cell niche. Biochim Biophys Acta 1840:2506–2519

Gillies AR, Lieber RL (2011) Structure and function of the skeletal muscle extracellular matrix. Muscle Nerve 44:318–331

Girotto D, Fabbro C, Braghetta P et al (2000) Analysis of transcription of the Col6a1 gene in a specific set of tissues suggests a new variant of enhancer region. J Biol Chem 275:17381–17390

Giusti B, Lucarini L, Pietroni V et al (2005) Dominant and recessive COL6A1 mutations in Ullrich scleroatonic muscular dystrophy. Ann Neurol 58:400–410

Grosu I, Truong D, Teodorescu S et al (2012) Anesthetic management of a child with Ullrich myopathy. J Anesth 26:636–637

Grumati P, Coletto L, Sabatelli P et al (2010) Autophagy is defective in collagen VI muscular dystrophies, and its reactivation rescues myofiber degeneration. Nat Med 16:1313–1320

Grumati P, Coletto L, Sandri M, Bonaldo P (2011a) Autophagy induction rescues muscular dystrophy. Autophagy 7:426–428

Grumati P, Coletto L, Schiavinato A et al (2011b) Physical exercise stimulates autophagy in normal skeletal muscles but is detrimental for collagen VI-deficient muscles. Autophagy 7:1415–1423

Gualandi F, Urciuolo A, Martoni E et al (2009) Autosomal recessive Bethlem myopathy. Neurology 73:1883–1891

Gualandi F, Manzati E, Sabatelli P et al (2012) Antisense-induced messenger depletion corrects a COL6A2 dominant mutation in Ullrich myopathy. Hum Gene Ther 23:1313–1318

Guillon E, Bretaud S, Ruggiero F et al (2016) Slow muscle precursors lay down a collagen XV matrix fingerprint to guide motor axon navigation. J Neurosci 36:2663–2676

Gupta VK, Scheunemann L, Eisenberg T et al (2013) Restoring polyamines protects from age-induced memory impairment in an autophagy-dependent manner. Nat Neurosci 16:1453–1460

Halfter W, Dong S, Schurer B et al (1998) Collagen XVIII is a basement membrane heparan sulfate proteoglycan. J Biol Chem 273:25404–25412

Hansen U, Allen JM, White R et al (2012) WARP interacts with collagen VI-containing microfibrils in the pericellular matrix of human chondrocytes. PLoS One 7:e52793

Hansson MJ, Mattiasson G, Månsson R et al (2004) The nonimmunosuppressive cyclosporin analogs NIM811 and UNIL025 display nanomolar potencies on permeability transition in brain-derived mitochondria. J Bioenerg Biomembr 36:407–413

Haq RU, Speer MC, Chu M-L, Tandan R (1999) Respiratory muscle involvement in Bethlem myopathy. Neurology 52:174–174

Härönen H, Zainul Z, Tu H et al (2017) Collagen XIII secures pre- and postsynaptic integrity of the neuromuscular synapse. Hum Mol Genet 26:2076–2090

Härönen H, Zainul Z, Naumenko N et al (2019) Correct expression and localization of collagen XIII are crucial for the normal formation and function of the neuromuscular system. Eur J Neurosci 49:1491–1511

Hatamochi A, Aumailley M, Mauch C et al (1989) Regulation of collagen VI expression in fibroblasts. Effects of cell density, cell-matrix interactions, and chemical transformation. J Biol Chem 264:3494–3499

Heikkinen A, Härönen H, Norman O et al (2019) Collagen XIII andother ECM components in the assembly and disease of the neuromuscular junction. Anat Rec (Hoboken). https://doi.org/10.1002/ar.24092

Heumüller SE, Talantikite M, Napoli M et al (2019) C-terminal proteolysis of the collagen VI α3 chain by BMP-1 and proprotein convertase(s) releases endotrophin in fragments of different sizes. J Biol Chem 294:13769–13780

Hicks D, Lampe AK, Barresi R et al (2008) A refined diagnostic algorithm for Bethlem myopathy. Neurology 70:1192–1199

Hicks D, Farsani GT, Laval S et al (2013) Mutations in the collagen XII gene define a new form of extracellular matrix-related myopathy. Hum Mol Genet 23:2353–2363
Higuchi I, Horikiri T, Niiyama T et al (2003) Pathological characteristics of skeletal muscle in Ullrich's disease with collagen VI deficiency. Neuromuscul Disord 13:310–316
Hunter JM, Ellen Ahearn M, Balak CD et al (2015) Novel pathogenic variants and genes for myopathies identified by whole exome sequencing. Mol Genet Genomic Med 3:283–301
Irwin WA, Bergamin N, Sabatelli P et al (2003) Mitochondrial dysfunction and apoptosis in myopathic mice with collagen VI deficiency. Nat Genet 35:367–371
Ishikawa H, Sugie K, Murayama K et al (2002) Ullrich disease: collagen VI deficiency: EM suggests a new basis for muscular weakness. Neurology 59:920–923
Ishikawa H, Sugie K, Murayama K et al (2004) Ullrich disease due to deficiency of collagen VI in the sarcolemma. Neurology 62:620–623
Ito M, Ohno K (2018) Protein-anchoring therapy to target extracellular matrix proteins to their physiological destinations. Matrix Biol 68-69:628–636
Iyengar P, Espina V, Williams TW et al (2005) Adipocyte-derived collagen VI affects early mammary tumor progression in vivo, demonstrating a critical interaction in the tumor/stroma microenvironment. J Clin Invest 115:1163–1176
Izu Y, Ezura Y, Koch M et al (2016) Collagens VI and XII form complexes mediating osteoblast interactions during osteogenesis. Cell Tissue Res 364:623–635
Jackson MT, Smith MM, Smith SM et al (2009) Activation of cartilage matrix metalloproteinases by activated protein C. Arthritis Rheum 60:780–791
Jain MS, Meilleur K, Kim E et al (2019) Longitudinal changes in clinical outcome measures in COL6-related dystrophies and LAMA2-related dystrophies. Neurology 93:E1932–E1943
Jeanne M, Gould DB (2017) Genotype-phenotype correlations in pathology caused by collagen type IV alpha 1 and 2 mutations. Matrix Biol 57-58:29–44
Jimenez-Mallebrera C, Maioli MA, Kim J et al (2006) A comparative analysis of collagen VI production in muscle, skin and fibroblasts from 14 Ullrich congenital muscular dystrophy patients with dominant and recessive COL6A mutations. Neuromuscul Disord 16:571–582
Jöbsis GJ, Keizers H, Vreijling JP et al (1996) Type VI collagen mutations in Bethlem myopathy, an autosomal dominant myopathy with contractures. Nat Genet 14:113–115
Jöbsis GJ, Boers JM, Barth PG, de Visser M (1999) Bethlem myopathy: a slowly progressive congenital muscular dystrophy with contractures. Brain 122:649–655
Jokela M, Lehtinen S, Palmio J et al (2019) A novel COL6A2 mutation causing late-onset limb-girdle muscular dystrophy. J Neurol 266:1649–1654
Kakoti BB, Hernandez-Ontiveros DG, Kataki MS et al (2015) Resveratrol and omega-3 fatty acid: its implications in cardiovascular diseases. Front Cardiovasc Med 2:38
Keene DR, Engvall E, Glanville RW (1988) Ultrastructure of type VI collagen in human skin and cartilage suggests an anchoring function for this filamentous network. J Cell Biol 107:1995–2006
Kelemen-Valkony I, Kiss M, Csiha J et al (2012) Drosophila basement membrane collagen col4a1 mutations cause severe myopathy. Matrix Biol 31:29–37
Khalifeh M, Barreto GE, Sahebkar A (2019) Trehalose as a promising therapeutic candidate for the treatment of Parkinson's disease. Br J Pharmacol 176:1173–1189
Kiechl S, Pechlaner R, Willeit P et al (2018) Higher spermidine intake is linked to lower mortality: a prospective population-based study. Am J Clin Nutr 108:371–380
Kirschner J, Hausser I, Zou Y et al (2005) Ullrich congenital muscular dystrophy: connective tissue abnormalities in the skin support overlap with Ehlers-Danlos syndromes. Am J Med Genet A 132A:296–301
Kiss AA, Somlyai-Popovics N, Kiss M et al (2019) Type IV collagen is essential for proper function of integrin-mediated adhesion in Drosophila muscle fibers. Int J Mol Sci 20. pii: E5124. doi: https://doi.org/10.3390/ijms20205124
Klionsky DJ, Abdelmohsen K, Abe A et al (2016) Guidelines for the use and interpretation of assays for monitoring autophagy (3rd edition). Autophagy 12:1–222

Koch M, Schulze J, Hansen U et al (2004) A novel marker of tissue junctions, collagen XXII. J Biol Chem 279(21):22514
Kok FO, Shin M, Ni CW et al (2015) Reverse genetic screening reveals poor correlation between morpholino-induced and mutant phenotypes in zebrafish. Dev Cell 32:97–108
Kuo H-J, Maslen CL, Keene DR, Glanville RW (1997) Type VI collagen anchors endothelial basement membranes by interacting with type IV collagen. J Biol Chem 272:26522–26529
Kvist AP, Latvanlehto A, Sund M et al (2001) Lack of cytosolic and transmembrane domains of type XIII collagen results in progressive myopathy. Am J Pathol 159:1581–1592
Labelle-Dumais C, Dilworth DJ, Harrington EP et al (2011) COL4A1 mutations cause ocular dysgenesis, neuronal localization defects, and myopathy in mice and Walker-Warburg syndrome in humans. PLoS Genet 7:e1002062
Labelle-Dumais C, Schuitema V, Hayashi G et al (2019) COL4A1 mutations cause neuromuscular disease with tissue-specific mechanistic heterogeneity. Am J Hum Genet 104:847–860
Lamandé SR, Bateman JF (2018) Collagen VI disorders: insights on form and function in the extracellular matrix and beyond. Matrix Biol 71–72:348–367
Lamandé SR, Sigalas E, Pan T et al (1998) The role of the α3(VI) chain in collagen VI assembly. J Biol Chem 273:7423–7430
Lamandé SR, Shields KA, Kornberg AJ et al (1999) Bethlem myopathy and engineered collagen VI triple helical deletions prevent intracellular multimer assembly and protein secretion. J Biol Chem 274:21817–21822
Lamandé SR, Mörgelin M, Selan C et al (2002) Kinked collagen VI tetramers and reduced microfibril formation as a result of Bethlem myopathy and introduced triple helical glycine mutations. J Biol Chem 277:1949–1956
Lampe AK, Bushby KMD (2005) Collagen VI related muscle disorders. J Med Genet 42:673–685
Lampe AK, Dunn DM, von Niederhausern AC et al (2005) Automated genomic sequence analysis of the three collagen VI genes: applications to Ullrich congenital muscular dystrophy and Bethlem myopathy. J Med Genet 42:108–120
Lampe AK, Zou Y, Sudano D et al (2008) Exon skipping mutations in collagen VI are common and are predictive for severity and inheritance. Hum Mutat 29:809–822
Latvanlehto A, Fox MA, Sormunen R et al (2010) Muscle-derived collagen XIII regulates maturation of the skeletal neuromuscular junction. J Neurosci 30:12230–12241
Le Roy F, Charton K, Lorson CL, Richard I (2009) RNA-targeting approaches for neuromuscular diseases. Trends Mol Med 15:580–591
Lee C, Kim M, Lee JH et al (2019) COL6A3-derived endotrophin links reciprocal interactions among hepatic cells in the pathology of chronic liver disease. J Pathol 247:99–109
Levine B, Kroemer G (2019) Biological functions of autophagy genes: a disease perspective. Cell 176:11–42
Li N, Oquendo E, Capaldi RA et al (2014) A systematic assessment of mitochondrial function identified novel signatures for drug-induced mitochondrial disruption in cells. Toxicol Sci 142:261–273
Li EW, McKee-Muir OC, Gilbert PM (2018) Cellular biomechanics in skeletal muscle regeneration. Curr Top Dev Biol 126:125–176
Lieschke GJ, Currie PD (2007) Animal models of human disease: zebrafish swim into view. Nat Rev Genet 8:353–367
Light N, Champion AE (1984) Characterization of muscle epimysium, perimysium and endomysium collagens. Biochem J 219:1017–1026
Listrat A, Lethias C, Hocquette JF et al (2000) Age-related changes and location of types I, III, XII and XIV collagen during development of skeletal muscles from genetically different animals. Histochem J 32:349–356
Liu J, Farmer JD, Lane WS et al (1991) Calcineurin is a common target of cyclophilin-cyclosporin A and FKBP-FK506 complexes. Cell 66:807–815

Lucarini L, Giusti B, Zhang RZ et al (2005) A homozygous COL6A2 intron mutation causes in-frame triple-helical deletion and nonsense-mediated mRNA decay in a patient with Ullrich congenital muscular dystrophy. Hum Genet 117:460–466

Lucioli S, Giusti B, Mercuri E et al (2005) Detection of common and private mutations in the COL6A1 gene of patients with Bethlem myopathy. Neurology 64:1931–1937

Madeo F, Eisenberg T, Pietrocola F, Kroemer G (2018) Spermidine in health and disease. Science 359:eaan2788

Madeo F, Carmona-Gutierrez D, Hofer SJ, Kroemer G (2019) Caloric restriction mimetics against age-associated disease: targets, mechanisms and therapeutic potential. Cell Metab 29:592–610

Maiuri MC, Kroemer G (2019) Therapeutic modulation of autophagy: which disease comes first? Cell Death Differ 26:680–689

Malfait F, Wenstrup RJ, De Paepe A (2010) Clinical and genetic aspects of Ehlers Danlos syndrome, classic type. Genet Med 12:597–605

Malfait F, Francomano C, Byers P et al (2017) The 2017 international classification of the Ehlers-Danlos syndromes. Am J Med Genet C Semin Med Genet 175:8–26

von der Mark H, Aumailley M, Wick G, et al (1984) Immunochemistry, genuine size and tissue localization of collagen VI. Eur J Biochem 142:493–502.

Marrosu E, Ala P, Muntoni F, Zhou H (2017) Gapmer antisense oligonucleotides suppress the mutant allele of COL6A3 and restore functional protein in Ullrich muscular dystrophy. Mol Ther Nucleic Acids 8:416–427

Martin DP, Tobias JD, Warhadpande S et al (2013) Perioperative care of a child with Ullrich congenital muscular dystrophy during posterior spinal fusion. S Afr J Anaesth Analg 19:73–76

Marvulli D, Volpin D, Bressan GM (1996) Spatial and temporal changes of type VI collagen expression during mouse development. Dev Dyn 206:447–454

Mashinchian O, Pisconti A, Le Moal E, Bentzinger CF (2018) The muscle stem cell niche in health and disease. Curr Top Dev Biol 126:23–65

Mazzucato M, Spessotto P, Masotti A et al (1999) Identification of domains responsible for von Willebrand factor type VI collagen interaction mediating platelet adhesion under high flow. J Biol Chem 274:3033–3041

McDevitt CA, Marcelino J, Tucker L (1991) Interaction of intact type VI collagen with hyaluronan. FEBS Lett 294:167–170

McMacken GM, Spendiff S, Whittaker RG et al (2019) Salbutamol modifies the neuromuscular junction in a mouse model of ColQ myasthenic syndrome. Hum Mol Genet 28:2339–2351

Meilleur KG, Jain MS, Hynan LS et al (2015) Results of a two-year pilot study of clinical outcome measures in collagen VI- and laminin alpha2-related congenital muscular dystrophies. Neuromuscul Disord 25:43–54

Menazza S, Blaauw B, Tiepolo T et al (2010) Oxidative stress by monoamine oxidases is causally involved in myofiber damage in muscular dystrophy. Hum Mol Genet 19:4207–4215

Mercuri E, Lampe A, Allsop J et al (2005) Muscle MRI in Ullrich congenital muscular dystrophy and Bethlem myopathy. Neuromuscul Disord 15:303–310

Mercuri E, Clements E, Offiah A et al (2010) Muscle magnetic resonance imaging involvement in muscular dystrophies with rigidity of the spine. Ann Neurol 67:201–208

Merlini L, Bernardi P (2008) Therapy of collagen VI-related myopathies (Bethlem and Ullrich). Neurotherapeutics 5:613–618

Merlini L, Morandi L, Granata C, Ballestrazzi A (1994) Bethlem myopathy: early-onset benign autosomal dominant myopathy with contractures. Description of two new families. Neuromuscul Disord 4:503–511

Merlini L, Angelin A, Tiepolo T et al (2008a) Cyclosporin A corrects mitochondrial dysfunction and muscle apoptosis in patients with collagen VI myopathies. Proc Natl Acad Sci U S A 105:5225–5229

Merlini L, Martoni E, Grumati P et al (2008b) Autosomal recessive myosclerosis myopathy is a collagen VI disorder. Neurology 71:1245–1253

Merlini L, Sabatelli P, Armaroli A et al (2011) Cyclosporine A in Ullrich congenital muscular dystrophy: long-term results. Oxid Med Cell Longev 2011:139194

Metti S, Gambarotto L, Chrisam M et al (2020) The polyphenol pterostilbene ameliorates the myopathic phenotype of Collagen VI deficient mice via autophagy induction. Front Cell Dev Biol 8:580933

Mihaylova V, Müller JS, Vilchez JJ et al (2008) Clinical and molecular genetic findings in COLQ-mutant congenital myasthenic syndromes. Brain 131:747–759

Miller R, Bradley W, Cudkowicz M et al (2007) Phase II/III randomized trial of TCH346 in patients with ALS. Neurology 69:776–784

Minois N, Carmona-Gutierrez D, Bauer MA et al (2012) Spermidine promotes stress resistance in *Drosophila melanogaster* through autophagy-dependent and -independent pathways. Cell Death Dis 3:e401–e401

Mizushima N, Levine B, Cuervo AM, Klionsky DJ (2008) Autophagy fights disease through cellular self-digestion. Nature 451:1069–1075

Mohire MD, Tandan R, Fries TJ et al (1988) Early-onset benign autosomal dominant limb-girdle myopathy with contractures (Bethlem myopathy). Neurology 38:573–580

Morselli E, Mariño G, Bennetzen MV et al (2011) Spermidine and resveratrol induce autophagy by distinct pathways converging on the acetylproteome. J Cell Biol 192:615–629

Myers JC, Dion AS, Abraham V et al (1996) Type XV collagen exhibits a widespread distribution in human tissues but a distinct localization in basement membrane zones. Cell Tissue Res 286:493–505

Nadeau A, Muntoni F (2008) Skin changes in Ullrich congenital muscular dystrophy. Neuromuscul Disord 18:982

Nadeau A, Kinali M, Main M et al (2009) Natural history of Ullrich congenital muscular dystrophy. Neurology 73:25–31

Nakata T, Ito M, Azuma Y et al (2013) Mutations in the C-terminal domain of ColQ in endplate acetylcholinesterase deficiency compromise ColQ-MuSK interaction. Hum Mutat 34:997–1004

Nanda A, Carson-Walter EB, Seaman S et al (2004) TEM8 interacts with the cleaved C5 domain of collagen alpha 3(VI). Cancer Res 64:817–820

Nichols C, Jain MS, Meilleur KG et al (2018) Electrical impedance myography in individuals with collagen 6 and laminin α-2 congenital muscular dystrophy: a cross-sectional and 2-year analysis. Muscle Nerve 57:54–60

Noguchi S, Ogawa M, Kawahara G et al (2014) Allele-specific gene silencing of mutant mRNA restores cellular function in Ullrich congenital muscular dystrophy fibroblasts. Mol Ther Nucleic Acids 3:e171

Nonaka I, Une Y, Ishihara T et al (1981) A clinical and histological study of Ullrich's disease (congenital atonic-sclerotic muscular dystrophy). Neuropediatrics 12:197–208

Ohno K, Brengman J, Tsujino A et al (1998) Human endplate acetylcholinesterase deficiency caused by mutations in the collagen-like tail subunit (ColQ) of the asymmetric enzyme. Proc Natl Acad Sci U S A 95:9654–9659

Okada M, Kawahara G, Noguchi S et al (2007) Primary collagen VI deficiency is the second most common congenital muscular dystrophy in Japan. Neurology 69:1035–1042

Olanow CW, Schapira AH, LeWitt PA et al (2006) TCH346 as a neuroprotective drug in Parkinson's disease: a double-blind, randomised, controlled trial. Lancet Neurol 5:1013–1020

Olsen DR, Peltonen J, Jaakkola S et al (1989) Collagen gene expression by cultured human skin fibroblasts. Abundant steady-state levels of type VI procollagen messenger RNA. J Clin Invest 83:791–795

Pace RA, Peat RA, Baker NL et al (2008) Collagen VI glycine mutations: perturbed assembly and a spectrum of clinical severity. Ann Neurol 64:294–303

Paco S, Kalko SG, Jou C et al (2013) Gene expression profiling identifies molecular pathways associated with collagen VI deficiency and provides novel therapeutic targets. PLoS One 8: e77430

Pagnon-Minot A, Malbouyres M, Haftek-Terreau Z et al (2008) Collagen XV, a novel factor in zebrafish notochord differentiation and muscle development. Dev Biol 316:21–35

Pallet N, Bouvier N, Legendre C et al (2008) Autophagy protects renal tubular cells against cyclosporine toxicity. Autophagy 4:783–791

Palma E, Tiepolo T, Angelin A et al (2009) Genetic ablation of cyclophilin D rescues mitochondrial defects and prevents muscle apoptosis in collagen VI myopathic mice. Hum Mol Genet 18:2024–2031

Pan TC, Zhang RZ, Sudano DG et al (2003) New molecular mechanism for Ullrich congenital muscular dystrophy: a heterozygous in-frame deletion in the COL6A1 gene causes a severe phenotype. Am J Hum Genet 73:355–369

Pan TC, Zhang R-Z, Markova D et al (2013) COL6A3 protein deficiency in mice leads to muscle and tendon defects similar to human collagen VI congenital muscular dystrophy. J Biol Chem 288:14320–14331

Pan TC, Zhang RZ, Arita M et al (2014) A mouse model for dominant collagen VI disorders heterozygous deletion of Col6a3 exon 16. J Biol Chem 289:10293–10307

Park J, Scherer PE (2012) Adipocyte-derived endotrophin promotes malignant tumor progression. J Clin Invest 122:4243–4256

Parola M, Marra F (2011) Adipokines and redox signaling: impact on fatty liver disease. Antioxid Redox Signal 15:461–483

Peat RA, Baker NL, Jones KJ et al (2007) Variable penetrance of COL6A1 null mutations: implications for prenatal diagnosis and genetic counselling in Ullrich congenital muscular dystrophy families. Neuromuscul Disord 17:547–557

Pepe G, Bertini E, Giusti B et al (1999a) A novel de novo mutation in the triple helix of the COL6A3 gene in a two-generation Italian family affected by Bethlem myopathy. A diagnostic approach in the mutations' screening of type VI collagen. Neuromuscul Disord 9:264–271

Pepe G, Giusti B, Bertini E et al (1999b) A heterozygous splice site mutation in COL6A1 leading to an in-frame deletion of the α1(VI) collagen chain in an Italian family affected by Bethlem myopathy. Biochem Biophys Res Commun 258:802–807

Pepe G, Lucarini L, Zhang R-Z et al (2006) COL6A1 genomic deletions in Bethlem myopathy and Ullrich muscular dystrophy. Ann Neurol 59:190–195

Petrini S, Tessa A, Stallcup WB et al (2005) Altered expression of the MCSP/NG2 chondroitin sulfate proteoglycan in collagen VI deficiency. Mol Cell Neurosci 30:408–417

Pfaff M, Aumailley M, Specks U et al (1993) Integrin and Arg-Gly-Asp dependence of cell adhesion to the native and unfolded triple helix of collagen type VI. Exp Cell Res 206:167–176

Piccolo S, Bonaldo P, Vitale P et al (1995) Transcriptional activation of the α1(VI) collagen gene during myoblast differentiation is mediated by multiple GA boxes. J Biol Chem 270:19583–19590

Pohl C, Dikic I (2019) Cellular quality control by the ubiquitin-proteasome system and autophagy. Science 366:818–822

Punetha J, Kesari A, Hoffman EP et al (2017) Novel Col12A1 variant expands the clinical picture of congenital myopathies with extracellular matrix defects. Muscle Nerve 55:277–281

Quarto R, Dozin B, Bonaldo P et al (1993) Type VI collagen expression is upregulated in the early events of chondrocyte differentiation. Development 117:245–251

Radev Z, Hermel JM, Elipot Y et al (2015) A TALEN-exon skipping design for a Bethlem myopathy model in zebrafish. PLoS One 10:e0133986

Rafii MS, Hagiwara H, Mercado ML et al (2006) Biglycan binds to alpha- and gamma-sarcoglycan and regulates their expression during development. J Cell Physiol 209:439–447

Ramanoudjame L, Rocancourt C, Lainé J et al (2015) Two novel COLVI long chains in zebrafish that are essential for muscle development. Hum Mol Genet 24:6624–6639

Robert V, Massimino ML, Tosello V et al (2001) Alteration in calcium handling at the subcellular level in mdx myotubes. J Biol Chem 276:4647–4651

Rodríguez Cruz PM, Palace J, Beeson D (2018) The neuromuscular junction and wide heterogeneity of congenital myasthenic syndromes. Int J Mol Sci 19. pii: E1677

Rodríguez Cruz PM, Cossins J, Estephan EP et al (2019) The clinical spectrum of the congenital myasthenic syndrome resulting from COL13A1 mutations. Brain 142:1547–1560
Rubinsztein DC, Mariño G, Kroemer G (2011) Autophagy and aging. Cell 146:682–695
Rusmini P, Cortese K, Crippa V et al (2019) Trehalose induces autophagy via lysosomal-mediated TFEB activation in models of motoneuron degeneration. Autophagy 15:631–651
Sabatelli P, Bonaldo P, Lattanzi G et al (2001) Collagen VI deficiency affects the organization of fibronectin in the extracellular matrix of cultured fibroblasts. Matrix Biol 20:475–486
Sabatelli P, Gara SK, Grumati P et al (2011) Expression of the collagen VI α5 and α6 chains in normal human skin and in skin of patients with collagen VI-related myopathies. J Invest Dermatol 131:99–107
Sabatelli P, Gualandi F, Gara SK et al (2012) Expression of collagen VI α5 and α6 chains in human muscle and in Duchenne muscular dystrophy-related muscle fibrosis. Matrix Biol 31:187–196
Sabatelli P, Sardone F, Traina F et al (2016) TGF-β1 differentially modulates the collagen VI α5 and α6 chains in human tendon cultures. J Biol Regul Homeost Agents 30:107–113
Saitta B, Stokes DG, Vissing H et al (1990) Alternative splicing of the human alpha 2(VI) collagen gene generates multiple mRNA transcripts which predict three protein variants with distinct carboxyl termini. J Biol Chem 265:6473–6480
Sanes JR (1982) Laminin, fibronectin, and collagen in synaptic and extrasynaptic portions of muscle fiber basement membrane. J Cell Biol 93:442–451
Sardone F, Traina F, Tagliavini F et al (2014) Effect of mechanical strain on the collagen VI pericellular matrix in anterior cruciate ligament fibroblasts. J Cell Physiol 229:878–886
Sardone F, Santi S, Tagliavini F et al (2016) Collagen VI–NG2 axis in human tendon fibroblasts under conditions mimicking injury response. Matrix Biol 55:90–105
Sasaki T, Göhring W, Pan TC et al (1995) Binding of mouse and human fibulin-2 to extracellular matrix ligands. J Mol Biol 254:892–899
Scacheri PC, Gillanders EM, Subramony SH et al (2002) Novel mutations in collagen VI genes: expansion of the Bethlem myopathy phenotype. Neurology 58:593–602
Schessl J, Goemans NM, Magold AI et al (2008) Predominant fiber atrophy and fiber type disproportion in early Ullrich disease. Muscle Nerve 38:1184–1191
Scotton C, Bovolenta M, Schwartz E et al (2016) Deep RNA profiling identified CLOCK and molecular clock genes as pathophysiological signatures in collagen VI myopathy. J Cell Sci 129:1671–1684
Sen N, Hara MR, Kornberg MD et al (2008) Nitric oxide-induced nuclear GAPDH activates p300/CBP and mediates apoptosis. Nat Cell Biol 10:866–873
Serra AJ, Prokić MD, Vasconsuelo A, Pinto JR (2018) Oxidative stress in muscle diseases: current and future therapy. Oxid Med Cell Longev 2018:6439138
Sigoillot SM, Bourgeois F, Karmouch J et al (2016) Neuromuscular junction immaturity and muscle atrophy are hallmarks of the ColQ-deficient mouse, a model of congenital myasthenic syndrome with acetylcholinesterase deficiency. FASEB J 30:2382–2399
Šileikytė J, Forte M (2019) The mitochondrial permeability transition in mitochondrial disorders. Oxid Med Cell Longev 2019:3403075
Singhal N, Martin PT (2011) Role of extracellular matrix proteins and their receptors in the development of the vertebrate neuromuscular junction. Dev Neurobiol 71:982–1005
Sipilä L, Ruotsalainen H, Sormunen R et al (2007) Secretion and assembly of type IV and VI collagens depend on glycosylation of hydroxylysines. J Biol Chem 282:33381–33388
Söderhäll C, Marenholz I, Kerscher T et al (2007) Variants in a novel epidermal collagen gene (COL29A1) are associated with atopic dermatitis. PLoS Biol 5:1952–1961
Sorato E, Menazza S, Zulian A et al (2014) Monoamine oxidase inhibition prevents mitochondrial dysfunction and apoptosis in myoblasts from patients with collagen VI myopathies. Free Radic Biol Med 75:40–47
Specks U, Mayer U, Nischt R et al (1992) Structure of recombinant N-terminal globule of type VI collagen alpha 3 chain and its binding to heparin and hyaluronan. EMBO J 11:4281–4290
Stokes DG, Saitta B, Timpl R, Chu ML (1991) Human α3(VI) collagen gene: Characterization of exons coding for the amino-terminal globular domain and alternative splicing in normal and tumor cells. J Biol Chem 266:8626–8633

Subramanian A, Schilling TF (2015) Tendon development and musculoskeletal assembly: emerging roles for the extracellular matrix. Development 142:4191–4204
Sun AY, Wang Q, Simonyi A, Sun GY (2010) Resveratrol as a therapeutic agent for neurodegenerative diseases. Mol Neurobiol 41:375–383
Sun K, Park J, Gupta OT et al (2014) Endotrophin triggers adipose tissue fibrosis and metabolic dysfunction. Nat Commun 5:3485
Symoens S, Renard M, Bonod-Bidaud C et al (2011) Identification of binding partners interacting with the α1-N-propeptide of type V collagen. Biochem J 433:371–381
Tagliavini F, Pellegrini C, Sardone F et al (2014) Defective collagen VI α6 chain expression in the skeletal muscle of patients with collagen VI-related myopathies. Biochim Biophys Acta 1842:1604–1612
Takahashi T, Cho HI, Kublin CL, Cintron C (1993) Keratan sulfate and dermatan sulfate proteoglycans associate with type VI collagen in fetal rabbit cornea. J Histochem Cytochem 41:1447–1457
Telfer WR, Busta AS, Bonnemann CG et al (2010) Zebrafish models of collagen VI-related myopathies. Hum Mol Genet 19:2433–2444
Terrill JR, Radley-Crabb HG, Iwasaki T et al (2013) Oxidative stress and pathology in muscular dystrophies: focus on protein thiol oxidation and dysferlinopathies. FEBS J 280:4149–4164
Thomas K, Engler AJ, Meyer GA (2015) Extracellular matrix regulation in the muscle satellite cell niche. Connect Tissue Res 56:1–8
Tiepolo T, Angelin A, Palma E et al (2009) The cyclophilin inhibitor Debio 025 normalizes mitochondrial function, muscle apoptosis and ultrastructural defects in Col6a1$^{-/-}$myopathic mice. Br J Pharmacol 157:1045–1052
Tillet E, Wiedemann H, Golbik R et al (1994) Recombinant expression and structural and binding properties of alpha 1(VI) and alpha 2(VI) chains of human collagen type VI. Eur J Biochem 221:177–185
Tillet E, Ruggiero F, Nishiyama A, Stallcup WB (1997) The membrane-spanning proteoglycan NG2 binds to collagens V and VI through the central nonglobular domain of its core protein. J Biol Chem 272:10769–10776
Tillet E, Gential B, Garrone R et al (2002) NG2 proteoglycan mediates beta1 integrin-independent cell adhesion and spreading on collagen VI. J Cell Biochem 86:726–736
Tonelotto V, Trapani V, Bretaud S et al (2019) Spatio-temporal expression and distribution of collagen VI during zebrafish development. Sci Rep 9:19851
Toni S, Morandi R, Busacchi M et al (2014) Nutritional status evaluation in patients affected by bethlem myopathy and Ullrich congenital muscular dystrophy. Front Aging Neurosci 6:315
Tooley LD, Zamurs LK, Beecher N et al (2010) Collagen VI microfibril formation is abolished by an a2(VI) von Willebrand factor type A domain mutation in a patient with Ullrich congenital muscular dystrophy. J Biol Chem 285:33567–33576
Tulla M, Pentikäinen OT, Viitasalo T et al (2001) Selective binding of collagen subtypes by integrin α1I, α2I, and α10I domains. J Biol Chem 276:48206–48212
Ullrich O (1930) Congenital, atonic–sclerotic muscular dystrophy, an additional type of heredodegenerative illness of the neuromuscular system [German]. Z Ges Neurol Psychiat 126:171–201
Urciuolo A, Quarta M, Morbidoni V et al (2013) Collagen VI regulates satellite cell self-renewal and muscle regeneration. Nat Commun 4:1964
Vitale P, Braghetta P, Volpin D et al (2001) Mechanisms of transcriptional activation of the col6a1 gene during Schwann cell differentiation. Mech Dev 102:145–156
Voit T (1998) Congenital muscular dystrophies: 1997 update. Brain Dev 20:65–74
Waldmeier PC, Feldtrauer J-J, Qian T, Lemasters JJ (2002) Inhibition of the mitochondrial permeability transition by the nonimmunosuppressive cyclosporin derivative NIM811. Mol Pharmacol 62:22–29
Wang YX, Dumont NA, Rudnicki MA (2014) Muscle stem cells at a glance. J Cell Sci 127:4543–4548
Wenstrup RJ, Florer JB, Brunskill EW et al (2004) Type V collagen controls the initiation of collagen fibril assembly. J Biol Chem 279:53331–53337

Werner E, Werb Z (2002) Integrins engage mitochondrial function for signal transduction by a mechanism dependent on Rho GTPases. J Cell Biol 158:357–368
Wiberg C, Hedbom E, Khairullina A et al (2001) Biglycan and decorin bind close to the N-terminal region of the collagen VI triple helix. J Biol Chem 276:18947–18952
Wiberg C, Heinegård D, Wenglén C et al (2002) Biglycan organizes collagen VI into hexagonal-like networks resembling tissue structures. J Biol Chem 277:49120–49126
Wiberg C, Klatt AR, Wagener R et al (2003) Complexes of matrilin-1 and biglycan or decorin connect collagen VI microfibrils to both collagen II and aggrecan. J Biol Chem 278:37698–37704
Witting N, Krag T, Werlauff U et al (2018) Collagen XII myopathy with rectus femoris atrophy and collagen XII retention in fibroblasts. Muscle Nerve 57:1026–1030
Yamauchi T, Kamon J, Minokoshi Y et al (2002) Adiponectin stimulates glucose utilization and fatty-acid oxidation by activating AMP-activated protein kinase. Nat Med 8:1288–1295
Yu Q, Sali A, van der Meulen J et al (2013) Omigapil treatment decreases fibrosis and improves respiratory rate in dy2J mouse model of congenital muscular dystrophy. PLoS One 8:e65468
Zainul Z, Heikkinen A, Koivisto H et al (2018) Collagen XIII is required for neuromuscular synapse regeneration and functional recovery after peripheral nerve injury. J Neurosci 38:4243–4258
Zanotti S, Negri T, Cappelletti C et al (2005) Decorin and biglycan expression is differentially altered in several muscular dystrophies. Brain 128:2546–2555
Zanussi S, Doliana R, Segat D, Bonaldo P (1992) The human type VI collagen gene. mRNA and protein variants of the alpha 3 chain generated by alternative splicing of an additional 5-end exon. J Biol Chem 267:24082–24089
Zech M, Lam DD, Francescatto L et al (2015) Recessive mutations in the α3 (VI) collagen gene COL6A3 cause early-onset isolated dystonia. Am J Hum Genet 96:883–893
Zelenski NA, Leddy HA, Sanchez-Adams J et al (2015) Type VI collagen regulates pericellular matrix properties, chondrocyte swelling, and mechanotransduction in mouse articular cartilage. Arthritis Rheumatol 67:1286–1294
Zorov DB, Filburn CR, Klotz LO et al (2000) Reactive oxygen species (ROS)-induced ROS release: A new phenomenon accompanying induction of the mitochondrial permeability transition in cardiac myocytes. J Exp Med 192:1001–1014
Zou Y, Zhang RZ, Sabatelli P et al (2008) Muscle interstitial fibroblasts are the main source of collagen VI synthesis in skeletal muscle: implications for congenital muscular dystrophy types Ullrich and Bethlem. J Neuropathol Exp Neurol 67:144–154
Zou Y, Zwolanek D, Izu Y et al (2014) Recessive and dominant mutations in COL12A1 cause a novel EDS/myopathy overlap syndrome in humans and mice. Hum Mol Genet 23:2339–2352
Zulian A, Rizzo E, Schiavone M et al (2014) NIM811, a cyclophilin inhibitor without immunosuppressive activity, is beneficial in collagen VI congenital muscular dystrophy models. Hum Mol Genet 23:5353–5363

Chapter 7
Skin Blistering and Collagens: From Bench to Therapies

Alexander Nyström, Dimitra Kiritsi, and Leena Bruckner-Tuderman

Abstract The skin is a multifunctional organ. Of its functions, the perhaps most evident one is that of being a flexible, mechanically robust barrier. This is achieved through a two-compartment arrangement with the skin's outermost compartment, the epidermis being the principal barrier and the inner dermis providing flexibility and tensile strength. Tight, durable joining of the two compartments is crucial and is achieved through the dermal–epidermal junction zone. Two specialized collagens, the large collagen VII and the transmembrane collagen XVII are essential for the mechano-resilience of this zone. The importance of these two collagens is evident from genetic and autoimmune skin blistering diseases occurring as consequence of their genetic deficiency or antibody targeting. Here, we will provide a detailed overview of the biology of collagen VII and XVII—from biosynthesis to their roles in human physiology and pathophysiology, which span beyond the skin.

Abbreviations

ASO	Antisense oligonucleotide
COPII	Conventional coat protein complex II
EB	Epidermolysis bullosa
ECM	Extracellular matrix
ER	Endoplasmic reticulum
DEB	Dystrophic EB
HMGB1	High mobility group protein B1
HSP47	Heath shock protein 47
IFM	Immunofluorescence mapping
JEB	Junctional EB
LH3	Lysyl hydroxylase 3
MSC	Mesenchymal stromal cell

A. Nyström · D. Kiritsi · L. Bruckner-Tuderman (✉)
Department of Dermatology, Medical Center, University of Freiburg, Freiburg, Germany
e-mail: bruckner-tuderman@uniklinik-freiburg.de

F. Ruggiero (ed.), *The Collagen Superfamily and Collagenopathies*, Biology of Extracellular Matrix 8, https://doi.org/10.1007/978-3-030-67592-9_7

NC	Non-collagenous
PTC	Premature termination codon
SCC	Squamous cell carcinoma
TANGO1	Transport and Golgi organization protein 1
TGM2	Transglutaminase 2
VWFA	Von Willebrand factor A

7.1 Introduction

The skin serves as the first barrier protecting the organism against external physical, chemical, biological, and frictional challenges and at the same time limiting the loss of fluids, salts, and proteins. The outermost cellular layer of the skin, the epidermis, is mainly responsible for the formation of a tight, chemically and biologically low-permeable barrier. Mechanical tautness of the skin is provided by the extracellular matrix (ECM) in its second layer, the dermis. To firmly attach the epidermal barrier to the mechano-resilient dermis the skin has developed an extended basement membrane zone, with numerous anchoring structures. This zone, collectively called the dermal–epidermal junction zone (DEJZ), encompasses the intracellular keratin intermediate filaments of basal keratinocytes, linked cell surface receptors that condense into hemidesmosomes or focal adhesions, the epidermal basement membrane and various anchoring structures protruding into the uppermost layers of the dermal ECM (Turcan and Jonkman 2015) (Fig. 7.1c). For the functionality of this zone allowing firm but yet flexible cohesion of the epidermis to the dermis, two collagens are essential: collagen VII that assembles into anchoring fibrils and collagen XVII that participates in the formation of hemidesmosomes and anchoring filaments (Burgeson 1993; Nishie 2020) (Fig. 7.1).

7.2 The Players

7.2.1 Collagen VII

Collagen VII in skin Collagen VII is a large homotypic collagen essential for mechano-robust attachment of stratified squamous epithelia to their underlying mesenchyme. In skin and other stratified squamous epithelia, collagen VII is distinctly deposited at the DEJZ, and arranged into large anchoring fibrils spanning from the level of the lamina densa over to the underlying pericellular ECM. Despite its tremendous importance for epithelium–mesenchyme cohesion, it is in these tissues a quantitatively minor collagen—of the total abundance of all collagens in the skin it makes up less than 0.001%. In addition to supporting skin integrity, collagen VII may also have an instructive role, e.g., promotion of wound healing

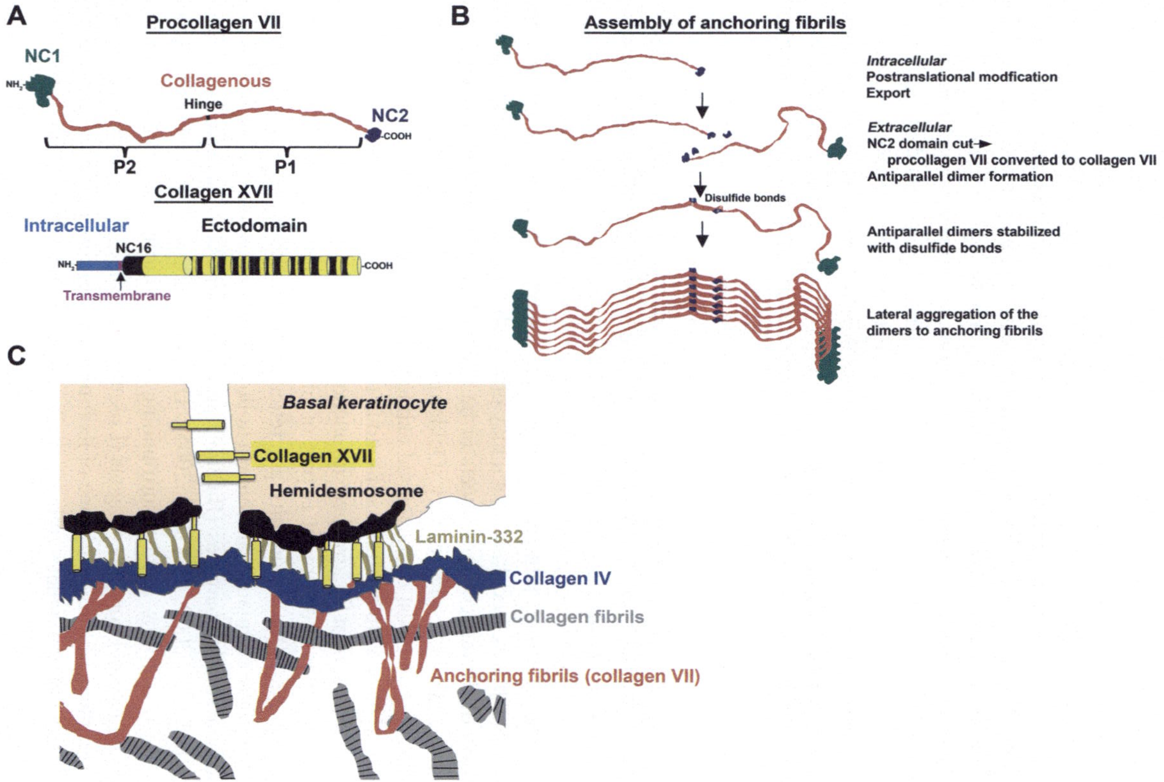

Fig. 7.1 Structure of collagen VII and collagen XVII and their interactions in the DEJZ. (**a**) Domain structure of collagen VII and collagen XVII. P1 and P2 indicate the pepsin-resistant fragments of the collagenous domain of collagen VII. (**b**) Assembly of anchoring fibrils. (**c**) Schematic representation of the interactions of collagen VII and collagen XVII in the DEJZ

directly or indirectly through pericellular ECM organization (Nyström et al. 2013a; Wang et al. 2013).

Extracutaneous expression of collagen VII Because genetic collagen VII deficiency in humans causes dystrophic epidermolysis bullosa (EB) (see below), with conspicuous skin blistering, most attention has been given on collagen VII in the skin. However, selected extracutaneous organs contain niches with notable collagen VII abundance. Its function in these organs remains in part elusive, but purposeful functions are likely to exist. Keeping or acquiring the specialized and highly energy consuming machinery for collagen VII biosynthesis as a mere bystander phenomenon seems not to make sense in an evolutionary and biological context.

Collagen VII has been shown to be essential for proper enamel formation of developing mammalian teeth (Umemoto et al. 2012). It supports differentiation of epithelial cells into enamelin- and amelogenin-secreting ameloblasts. Loss of collagen VII results in enamel malformation of teeth and a high prevalence of dental caries (Wright 2010). In addition to an instructing role, collagen VII seems also to be present in the enamel organic matrix and supports the connection of enamel and dentin (McGuire et al. 2014).

In the eye, collagen VII is not only essential for corneal attachment (Chen et al. 2018; Rashad et al. 2019), but is also present in the lens capsule, ciliary body, zonules, and even various retinal structures including the vitreoretinal interface (Wullink et al. 2015, 2018). In addition to retina, collagen VII is also found in other parts of the human nervous system, such as around pituitary and pineal gland cell nests, and under choroid plexus epithelial cells (Paulus et al. 1995).

Under the amniotic epithelium collagen VII assembles into enormous structures termed rivets. It was speculated that it may play such opposite roles as sites for nodes for growth of fibrous collagen sheets or be a site of enzymatic degradation during amnion weakening before delivery (Ockleford et al. 2013).

Intriguingly, the best characterized extracutaneous functions of collagen VII are in lymphoid organs. In thymus it is found in the subepithelial basement membrane of the capsule, where it is arranged in anchoring fibrils (Virtanen et al. 1996). In secondary lymphoid organs, it is part of immune conduits of the B cell follicles (Nyström et al. 2018). Conduits are reticular fibers of larger diameter that are connected to the circulation and have a hollow or perforated core allowing transport of fluid and smaller particles (Bajénoff and Germain 2009; Lokmic et al. 2008; Roozendaal et al. 2009; Sixt et al. 2005). They are composed of an outer basement membrane that forms a tube surrounding a microfibrillar layer and frequently contains an internal core of collagens (Lokmic et al. 2008). Our studies of collagen VII in the spleen disclosed that one of its functions is to support innate immunity by interacting with the innate immune cell activating protein cochlin (see below) (Nyström et al. 2018).

Biosynthesis and biochemical characteristics Collagen VII is synthesized in the skin by epidermal keratinocytes and dermal fibroblasts (Ryynänen et al. 1992) and in secondary lymphoid organs by lymphoid stromal cells (Nyström et al. 2018). The *COL7A1* gene encodes the 2944 amino acid large collagen VII proα1-chains

(Christiano et al. 1994) that while being translated translocate to the lumen of the endoplasmic reticulum (ER). Here, similar to other collagens, selected proline and lysine residues undergo enzyme-mediated hydroxylation and three polypeptide chains fold in an N-to-C terminal direction to a procollagen VII molecule (Canty and Kadler 2005). Structurally, the collagen VII proα1-chain is composed of three major domains (Fig. 7.1a). The approximately 1200 amino acid large NC1 domain contains one Von Willebrand Factor A (VWFA) domain at each end surrounding nine fibronectin type III repeats (FNIII). An approximately 1500 amino acid long collagenous domain with several imperfections follows the NC1 domain (Christiano et al. 1994). This represents the largest collagenous domain present in mammals (Nyström 2016). About midways is the longest stretch of interruption of the collagenous repeat (39 amino acids). This part is referred to as the hinge region and is suggested to provide some flexibility to the collagenous domain. The collagenous domain is flanked C-terminally by a short NC2 domain (160 amino acids) that contains a Kunitz/Bovine pancreatic trypsin inhibitor domain (Christiano et al. 1994), however, the collagen VII Kunitz domain does not possess protease inhibitor activity (Chen et al. 2001). In the biochemical literature, two fragments of collagen VII, the P1 and P2 fragments, are often mentioned. These pepsin digestion-derived fragments represent the major collagenous stretches on both sides of the hinge region; the P1 fragment is C-terminal and the P2 fragment N-terminal (Bentz et al. 1983) (Fig. 7.1a).

Intracellular transport and secretion Fully folded collagen VII is an extremely large molecule (>900 kDa and 450 nm), too large to fit into conventional coat protein complex II (COPII) complex-coated ER export vesicles (Jensen and Schekman 2011; Morris et al. 1986). Therefore, like other large collagens and ECM proteins, it requires specialized packing and transport machinery (for more details on collagen export we refer the reader to Chap. 5). Transport and Golgi organization protein 1 (TANGO1)—a large integral membrane protein—was first described as a facilitator of collagen VII packing at ER exit sites (Saito et al. 2009). Here, its luminal SH3 domain was described to directly interact with collagen VII. TANGO1 would thus act as a cargo receptor (Saito et al. 2009). However, in contrast to other cargo receptors, TANGO1 does not follow the cargo into export vesicles or structures, and remains at the ER exit site. While binding collagen VII at its luminal side the cytoplasmic part of TANGO1 promotes assembly of a large COPII structure through interaction with members of the COPII machinery SEC23/24 (Saito et al. 2009).

The proposed models and mechanisms of how TANGO1 would promote collagen VII secretion have subsequently become much more elaborate. The TANGO1 homologue cTAGE5, lacking the luminal SH3 domain, was shown to laterally interact with TANGO1 and also to be essential for collagen VII export (Saito et al. 2011). cTAGE5 would be involved in building the COPII complex through binding SEC12, which in turn recruits the small GTPase SAR1 that initiates vesicle budding (Jensen and Schekman 2011; Tanabe et al. 2016). TANGO1 recruits and aids fusion of ER–Golgi intermediate compartment (ERGIC) membranes to the ER to promote building of a mega carrier, or similar structure. For this direct or indirect interactions and activities of SLY-1 (a protein required for membrane fusion), various SNARE proteins and the NBAS, RINT1, ZW10 protein tether are needed (Nogueira et al. 2014; Raote et al. 2018; Santos et al. 2015). The fusion process depends on various SNARE proteins (Nogueira et al. 2014; Santos et al. 2015). Collectively through these studies multiple proteins more or less essential for collagen VII export have been identified. These include TANGO1, cTAGE5, SEC12, SAR1, SLY1, syntaxin 5, syntaxin 18, YKT6, BNIP1, USE1, NBAS, RINT1, and ZW10 (Saito et al. 2009, 2011; Nogueira et al. 2014; Raote et al. 2018; Santos et al. 2015).

Recent data suggest that multiple TANGO1 molecules are organized as a ring around ER exit sites. The creation of such structure is aided through lateral interactions with cTAGE5 and a TANGO1 in an alternative spliced form lacking the luminal domain (TANGO1-short) (Raote et al. 2018, 2017). The discovery of these structures has led to challenging of the dogma of ER export occurring via vesicles and that it may alternatively for large cargo occur through tunnels or similar extended structures between ER and Golgi (Raote and Malhotra 2019). However, the findings of TANGO1 being stationary at ER exit sites have been contested (Yuan et al. 2018).

Genetic studies have revealed TANGO1 to be an integral component for the secretion of many collagens (Lekszas et al. 2020; Wilson et al. 2011). Current data point to that the TANGO1 SH3 domain does not directly interact with collagens but this interaction is mediated via the collagen chaperone heat shock protein 47 (HSP47) with which the TANGO1 SH3 domain interacts with sub μM affinity (Ishikawa et al. 2016). If this is also the case for collagen VII has not been directly assessed, although collagen VII contains one potential high affinity and several middle-affinity HSP47 binding sequences, and HSP47 co-localize in the ER (Nogueira et al. 2014; Santos et al. 2015; Ishikawa et al. 2016; Widmer et al. 2012).

Wealth of in vitro data clearly implicates TANGO1 as a facilitator of collagen VII secretion, however, a heavy dependence on this mechanism for collagen VII deposition in vivo could be questioned based on phenotypes occurring in TANGO1 knockout mice and more recently reported for patients with TANGO1 deficiency (Lekszas et al. 2020; Wilson et al. 2011). TANGO1 deficiency does not mirror hallmarks of collagen VII deficiency, such as skin blistering. It rather leads to a generalized ECM disease due to impaired deposition of many collagens (Lekszas et al. 2020; Wilson et al. 2011). It is possible that a collagen VII deficiency-related phenotype is hidden among the other more severe clinical presentations of TANGO1 loss. In addition, as the TANGO1 knockout mice displayed neonatal lethality, spontaneous mechanically induced skin blistering cannot be investigated in this model (Wilson et al. 2011). However, more careful immunohistological and biochemical assessments of collagen VII secretion and deposition in TANGO1-deficient skin should bring more clarity to the in vivo dependence of TANGO1 for collagen VII secretion. It should be noted that human HSP47 deficiency also does not generally present overt skin blistering; localized self-resolving skin blistering was only observed in one newborn patient (Christiansen et al. 2010).

Both keratinocytes and fibroblasts may be able to deposit physiologically relevant levels of collagen VII in the skin. This is illustrated by improved skin stability in patients treated with gene-corrected epidermal skin grafts or conversely improved skin stability after spontaneous genetic correction of dermal fibroblasts—so-called revertant mosaicism (Siprashvili et al. 2016; Twaroski et al. 2019). Although some results have been confirmed in transformed keratinocytes, most work deciphering the role of TANGO1 in collagen VII secretion has been performed using fibroblasts or other cells with mesenchymal properties engineered to express supraphysiological collagen VII levels (Saito et al. 2009; Chen et al. 2002a). It is feasible that normo- vs. supraphysiological collagen VII synthesis may influence cellular secretion strategies. Furthermore, it cannot be excluded that keratinocytes and fibroblasts show differential dependence on various export mechanisms.

In the Golgi apparatus, after ER to Golgi transport, collagen VII may undergo modifications with N-linked oligosaccharides (Canty and Kadler 2005; Chen et al. 2000). The NC1 domain contains three potential sites for N-linked glycosylation (Christiano et al. 1994). Intriguingly, N-linked glycosylation of the NC1 domain has been suggested to be important for its secretion (Chen et al. 1997). In line with this, early studies have also disclosed that impaired glycosylation retards procollagen secretion (Housley et al. 1980). Glycosylation of hydroxylysines by lysyl hydroxylase 3 (LH3) appears to occur post-Golgi (Banushi et al. 2016). This glycosylation is important for secretion and supramolecular organization of tissue interface-stabilizing collagens including collagen IV, VI, and VII (Sipilä et al. 2007) (for more information on collagens IV and VI, please see Chaps. 3 and 7). Collagen VII is known to interact with LH3 (Küttner et al. 2014; Watt et al. 2015) and, importantly, LH3 deficiency in humans reduces collagen VII abundance in skin (Vahidnezhad et al. 2019).

Extracellular assembly and interactome Once exported, procollagen VII molecules align in an antiparallel manner with their C-terminal ends with an overlap of around 60 nm (Morris et al. 1986). This antiparallel dimer formation is mediated by the NC2 domain and the most proximal part the collagenous domain (Chen et al. 2001). Cysteine residues within the NC2 domain appear essential for the formation of an antiparallel dimer (Chen et al. 2001). Procollagen VII is converted to mature collagen VII by proteolytic release of about three-fourth of the NC2 domain by proteinases of the bone morphogenetic protein-1 (BMP-1)/Tolloid-like proteinase family that cut the NC2 domain between alanine 2821 and asparagine 2822 (Rattenholl et al. 2002) (Fig. 7.1b). However, other proteinases also participate in procollagen VII maturation (Rattenholl et al. 2002; Muir et al. 2016). Based on the cleavage site, two other members of the astacin family of metalloproteinases—meprin α and β—have been suggested to contribute to procollagen VII maturation (Muir et al. 2016). The procollagen-to-collagen conversion is essential for the functionality of collagen VII as reflected by skin blistering in patients with mutations of the cleavage site and retention of the NC2 domain (Rattenholl et al. 2002; Bruckner-Tuderman et al. 1995).

Antiparallel collagen VII dimers are around 800 nm in length and are stabilized by intermolecular disulfide bonds (Morris et al. 1986). In skin and other stratified squamous epithelia, multiple dimers condense laterally to form electron-dense striated structures, distinctly visible in transmission electron microscopy as anchoring fibrils (Burgeson 1993), which stretch from the level of the lamina densa in the basement membrane to around 300–400 nm into the papillary dermis (Sakai et al. 1986) (Fig. 7.1c).

The NC1 domains of collagen VII are predominantly located in and around the lamina densa of the basement membrane (Keene et al. 1987) meaning that the collagenous domains of the dimer loops down in the dermis. This organization is logical given that the NC1 domain binds laminin-332 and collagen IV with high affinity in the low nM range (Rousselle et al. 1997; Brittingham et al. 2006). The laminin β3 chain is considered to be the major collagen VII binding partner in the laminin-332 molecule (Chen et al. 1999), but the γ2 chain may also serve as an interaction partner of collagen VII (Rousselle et al. 1997; Chen et al. 1999).

Because the same region in the NC1 domain is used for binding of collagen IV and laminin-332, one NC1 domain cannot simultaneously interact with both (Brittingham et al. 2006). However, the trimeric arrangement of collagen VII, its folding and the size of the NC1 domains would allow one collagen VII molecule to interact with multiple laminin-332 and/or collagen IV molecules. Thus, collagen VII may also link the laminin and collagen IV networks in the epidermal basement membrane (Has and Nyström 2015). The affinity of the collagen VII NC1 domain for molecular collagen I is considerably lower than for laminin-332 and collagen IV (Brittingham et al. 2006), but collagen VII binds very tightly to dermal collagen I containing fibrils (Brittingham et al. 2006; Wegener et al. 2013). Furthermore, the dermal loops of the anchoring fibrils physically entrap collagen fibrils (Brittingham et al. 2006; Wegener et al. 2013). The end result of these interactions is an epidermal basement membrane that is firmly secured to the papillary dermal ECM (Fig. 7.1c).

Apart from collagen IV and laminin-332, additional intracellular and extracellular collagen-binding partners have been identified. Intracellular interactions have been reported to occur with TANGO1, thrombospondin-1, LH3, and transglutaminase 2 (TGM2) (Saito et al. 2009; Küttner et al. 2014; Watt et al. 2015; Atanasova et al. 2019), these interactions also occur extracellularly. Additional extracellular binding partners include fibronectin, BMP-1, thrombospondin-1, TGM2, and LH3 (Chen et al. 1997; Küttner et al. 2014; Watt et al. 2015; Rattenholl et al. 2002; Atanasova et al. 2019; Aho and Uitto 1998; Raghunath et al. 1996). Intriguingly, collagen VII deficiency correlates with reduced abundance of LH3 and TGM2, potentially widening the effect on the ECM evoked by collagen VII deficiency by reducing ECM-assembly-promoting glycosylation and ECM-stabilizing crosslinking, respectively (Sipilä et al. 2007; Bianchi et al. 2018; Ruotsalainen et al. 2006). In the dermis collagen VII shows close co-distribution with collagen XVI, but a direct interaction between the molecules has not been assessed (Grässel et al. 1999). We recently described the ECM protein and innate-immune activator cochlin as an extracutaneous binding partner of collagen VII in secondary lymphoid organs. Collagen VII interacted with the two cochlin VWFA domains, with a preference for the VWFA1 domain (Nyström et al. 2018).

7.2.2 Collagen XVII

Collagen XVII is an integral transmembrane protein in type II orientation. The intracellular N-terminal domain is relatively small, and the extracellular C-terminal collagenous domain extends from the plasma membrane like a flexible rod. Collagen XVII is synthesized by epithelial cells sitting on a basement membrane, most abundantly by basal keratinocytes of the epidermis and hair follicles, but also, e.g., by corneal epithelial cells. Its cellular localization is polarized, collagen XVII is found in the hemidesmosomes on the basal keratinocyte surface facing the basement membrane, but also as a non-hemidesmosomal protein on the basolateral cell surface facing neighboring keratinocytes (Natsuga et al. 2019).

Molecular structure and biosynthesis The collagen XVII molecule is a triple-helical trimer of three polypeptide chains. The *COL17A1* gene encodes an α1 (XVII) chain of 1497 amino acids and a molecular weight of 180 kDa. The intracellular domain contains 466 amino acids, the transmembrane domain 23 amino acid residues, and the collagenous extracellular domain 1008 amino acids. The latter consists of 15 collagenous subdomains interspersed with 16 non-collagenous (NC) subdomains. Like all collagens, collagen XVII undergoes a number of posttranslational modifications during its biosynthesis, including hydroxylation of prolyl and lysyl residues, glycosylation of hydroxylysine residues, and N-glycosylation of serine residues, disulfide bonding, and triple-helix folding (Kivirikko 1993; Franzke et al. 2003). These steps yield a triple-helical homotrimer of three α-chains. The export of the newly synthesized collagen XVII molecules onto the cell surface and their positioning in a type II transmembrane protein orientation still remain elusive (Malhotra and Erlmann 2015) (Fig. 7.1a).

Ectodomain shedding The ectodomain of collagen XVII is constitutively shed from the cell surface by ADAMs family of proteinases, mainly ADAMs-9 and -10 (Franzke et al. 2009). The cleavage product is a 120-kDa soluble triple-helical collagen. Depending on the biological context, other proteinases can also release the ectodomain from the cell surface and process it further to a 97-kDa fragment (Hofmann et al. 2009). The shedding is regulated by extracellular phosphorylation of collagen XVII (Zimina et al. 2007) and by its plasma membrane environment (Zimina et al. 2005), since collagen XVII resides in lipid rafts and the sheddases in non-raft membrane domains. The shedding is believed to regulate keratinocyte migration and differentiation under skin homeostasis and repair, such as wound healing, but also under pathological conditions like cancer progression (Liu et al. 2018; Galiger et al. 2018; Jacków et al. 2016).

The interactome The interactome of collagen XVII is not fully known, but its major intracellular binding partners include plectin, integrin β4, BP230, keratin 18, stratifin/14-3-3a, and adherence junction proteins like actinin-1 and -4, and p120 catenin (Natsuga et al. 2017, 2019; Koster et al. 2003; Gonzalez et al. 2001). The main extracellular ligands are laminin-332 and collagen IV at the basement membrane zone. In vitro, the presence of the latter ligands in the extracellular matrix enhances incorporation of the shed ectodomain into the matrix (Nishie et al. 2011).

Functions The major function of hemidesmosomal collagen XVII in skin homeostasis is the attachment of the epidermis to the dermis. Together with laminin 332, its ectodomain is a component of the anchoring filaments that protrude from the hemidesmosomes and traverse the basement membrane. The cohesive function is indirectly evidenced by collagen XVII mutations, which render it non-functional by abrogating its abundance or by changing its structure; such mutations lead to separation of the skin layers in response to mechanical forces and skin blistering in a genetic disorder called junctional epidermolysis bullosa (JEB). In a different skin blistering scenario, autoantibodies to collagen XVII ectodomain induce the autoimmune blistering disorder bullous pemphigoid (Nishie 2020).

An additional, intriguing role of collagen XVII is in the maintenance of hair follicle stem cells and melanocytic stem cells (Matsumura et al. 2016; Tanimura et al. 2011). First indications for these functions came from clinical observations of collagen XVII-deficient patients and mice, who also exhibit hair loss (Fig. 7.3) and hair greying at an early age (Tanimura et al. 2011; Has et al. 2020). It is now known that hair follicle stem cells express collagen XVII and that this collagen is an important constituent of the stem cell niche (Matsumura et al. 2016). The hair loss is explained by follicular stem cell exhaustion after loss of collagen XVII from the niche, e.g., via pathogenic *COL17A1* gene variants in junctional EB or via proteolytic degradation of collagen XVII by neutrophil elastase and other skin proteinases during aging. Hair greying is a consequence of the fact that follicular stem cell exhaustion leads to aberrant melanocyte stem cell maintenance (Natsuga et al. 2019).

Recent investigations have suggested that collagen XVII also plays a role in the homeostasis of the interfollicular epidermis by supporting the quiescence of epidermal stem cells there (Watanabe et al. 2017). Loss of the collagen and reduced epidermal–dermal cohesion destabilizes the maintenance of quiescent interfollicular epidermal stem cells. As a consequence the epidermis hyperproliferates, as seen in newborn collagen XVII-deficient mice (Tanimura et al. 2011) or, also, in UV-exposed aged human skin. These cell instructive and regulatory functions of collagen XVII are associated with different signaling pathways, e.g., TGF-β and Wnt signaling, but the mechanisms are likely to be context-dependent and complex and will need further elucidation (Watanabe et al. 2017).

7.3 Skin Fragility

The term skin fragility describes blister formation and easy breakability of the skin upon exposure to shearing forces or mechanical load. Skin fragility is typically seen in genetic and autoimmune skin blistering disorders and related conditions, for information on other collagens linked to skin fragility please see Chap. 4. Of all collagens, only collagen VII and collagen XVII are associated with skin blistering.

In both cases, pathogenic variants in the *COL7A1* and the *COL17A1* genes, but also autoantibodies to these collagens can render them non-functional and diminish cohesion of the skin layers, which manifests as skin blistering. The genetic blistering disorders, collectively called epidermolysis bullosa (EB), serve as a prototype for understanding skin fragility.

7.4 Dystrophic Epidermolysis Bullosa

Cause and genotype–phenotype correlations Dystrophic EB (DEB) is a genetic skin blistering disorder caused by functional deficiency of collagen VII as a consequence of pathogenic variants in the *COL7A1* gene. It can be inherited in a dominant or recessive manner. More than 1000 distinct mutations can be found in the databases (http://www.col7a1-database.info; http://www.hgmd.cf.ac.uk). However, the molecular disease mechanisms still need further elucidation. Collagen VII can be reduced or completely absent from the skin, and the anchoring fibrils structurally and functionally altered or totally missing, depending on the mutations. The human phenotypes range from occasional trauma-induced skin blisters and/or dystrophy of nails to severe generalized skin fragility (Fig. 7.2a), depending on the abundance of collagen VII in the skin, the rule of thumb being "the less collagen the more severe the phenotype."

Animal models Different animal models for collagen VII deficiency and DEB have been helpful in dissecting genetic and molecular disease mechanisms. A spontaneous *COL7A1* mutation in an inbred flock of sheep with severe DEB and complete lack of collagen VII helped establish a recessive inheritance pattern in an ECM disease (Bruckner-Tuderman et al. 1991), a novelty at a time when the dogma was that enzyme deficiencies are inherited in a recessive and disorders of structural proteins in a dominant manner. Both collagen VII knockout mice and different hypomorphic mouse, rat, and dog models for DEB (Webber et al. 2017; Heinonen et al. 1999; Fritsch et al. 2008; Gache et al. 2011; Palazzi et al. 2000; Nyström et al. 2013b) have facilitated dissection of molecular and cellular phenotypes and disease progression (Nyström et al. 2013a, 2015, 2018; Watt et al. 2015; Alexeev et al. 2016; Tamai et al. 2011; Vanden Oever et al. 2016) and testing of novel therapies, as delineated below.

Scarring and fibrosis As the major constituent of anchoring fibrils collagen VII is located below the basement membrane in the uppermost dermis and, therefore, in DEB tissue separation occurs at this level and healing takes place with scarring. In severe subtypes, the phenotype is progressive and remarkably complex. Injury-induced blistering and inflammation trigger scarring that, with advancing course of the disease, develops into wide-spread soft tissue fibrosis. As our understanding of the disease mechanisms increases, it becomes evident that inflammation and fibrosis are not limited to the skin, but all organ systems can be affected. In fact, severe DEB can be regarded as a systemic disease that exhibits deformities of the extremities,

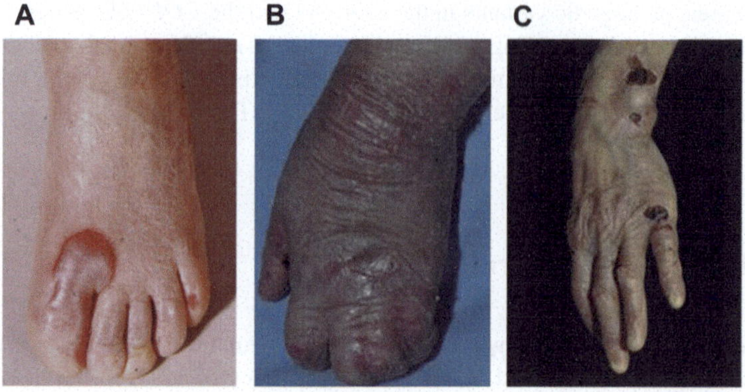

Fig. 7.2 Collagen VII-associated diseases. (**a**) The left foot of a 12-year patient with moderate DEB. Note the fresh blister on the big toe and the loss of nails as a consequence of scarring. (**b**) Pseudosyndactyly (mitten deformity) of the left hand of a 6-year-old patient with severe recessive DEB. The formation of pseudosyndactyly is driven by unremitting cycles of injury, inflammation, and healing with excessive scarring. (**c**) The left hand of a 69-year-old patient with EB acquisita. Note the blisters, scabs, and scars

joint contractures, limited mouth opening and mobility of the tongue, esophageal strictures, and consequently difficulties in swallowing, nutritional deficits, growth-retardation, and anemia (Reimer et al. 2020; Fine and Mellerio 2009a, b).

Mechanisms of fibrosis and skin cancer in DEB The mechanisms of fibrosis have been intensively investigated (Atanasova et al. 2019; Nyström et al. 2015; Odorisio et al. 2014; Mittapalli et al. 2016; Chacón-Solano et al. 2019). These studies, combined with unbiased global proteomics and biochemical and biological validation in human tissues and mouse models generate a picture of TGFβ-mediated derailment of tissue homeostasis. Unremitting cycles of mechanical injury, inflammation, and perturbed wound healing lead to altered tissue architecture in the dermis. This, in turn, causes excessive release and activity of TGF-β and development of a pro-fibrotic tissue microenvironment and increasing tissue stiffness.

Also, epidermal cells sense tissue changes in the dermis. Abnormal architecture and increased stiffness of the dermis and the DEJZ facilitate epithelial–mesenchymal transition and cancer initiation and progression (Butcher et al. 2009; Pickup et al. 2014). Indeed, aggressive squamous cell carcinoma (SCC) is the most feared complication of DEB and the leading cause of death in young adults with the disease (Has et al. 2020). DEB-associated SCCs (DEB-SCC) differ from common UV-light induced skin cancers in terms of mutation profiles. DEB-SCC exhibit endogenous mutation processes associated with APOBEC (apolipoprotein B mRNA-editing enzyme catalytic polypeptide-like), similar to mutation signatures in head and neck cancers (Cho et al. 2018). However, the APOBEC signature alone does not explain the aggressive behavior of the tumors. Rather, the most striking driver of DEB-SCC is the dermal microenvironment that supports cancer progression in multiple ways (Mittapalli et al. 2016). Apart from tissue stiffness and inflammatory processes, the environmental factors include the elevated bacterial colonization which, together with chronic inflammation and wounding have been shown to cooperate to trigger skin cancer in experimental animal models (Hoste et al. 2015).

Role in antibacterial immunity Drastically elevated bacterial colonization in DEB skin led to discovery of quite unexpected functions of collagen VII in antibacterial immunity (Nyström et al. 2018). As discussed above, collagen VII is also a structural component of the lymphoid ECM in the spleen, specifically of the conduits. In the lumen of the conduits, a multi-protein complex that supports innate immunity harbors collagen VII that binds and sequesters the innate immune activator cochlin. In response to bacterial infection, the LCCL domain of cochlin is released from the complex and transported into the circulation. Systemic increase of LCCL levels activates macrophages and neutrophils in the periphery to combat bacterial challenges at infection sites. In DEB, genetic loss of collagen VII evokes concomitant loss of cochlin from lymphoid conduits. This results in an inability to activate innate immune cells in the skin and in a subsequent increase of the bacterial load, as seen in both recessive DEB patients and mouse models (Nyström et al. 2018).

7.5 Junctional Epidermolysis Bullosa

Junctional EB (JEB) is a clinically and genetically heterogeneous subtype of EB characterized by trauma-induced tissue separation at the level of the lamina lucida of the DEJZ. It can be caused by pathogenic variants in multiple genes encoding proteins of the hemidesmosome-anchoring filament complex at the DEJZ (Has et al. 2020; Pfendner and Lucky 1993), most commonly collagen XVII and laminin-332. It is inherited in an autosomal recessive manner (Pfendner and Lucky 1993). The immunofluorescence mapping (IFM) of the skin, which is used to diagnose the disease, shows collagen XVII at the blister roof in the patient's skin, whereas laminin 332 is found on the blister floor. Depending on the mutation, these proteins may be reduced or be completely absent, corresponding to diminished or missing hemidesmosomes in transmission electron microscopy (Shinkuma et al. 2011; Kiritsi et al. 2013).

Genotype–phenotype correlations The JEB disease severity ranges from localized to severe forms. The most severe and early lethal subtype, severe generalized JEB, is caused by null mutations in the laminin 332 genes *LAMA3*, *LAMB3*, or *LAMC2*. Other clinical subtypes, coined intermediate or localized forms, can be caused by mutations both in the laminin 332 genes or in the collagen XVII gene, *COL17A1* (Has et al. 2020; Fine et al. 2008). Although patients with mutations in these genes have similarities in their clinical presentation, there are also differences that hint to the molecular and genetic background. Specifically, the presence of exuberant granulation tissue appears to point to deficiency of laminin 332 rather than of collagen XVII (Kiritsi et al. 2013; McLean et al. 2003).

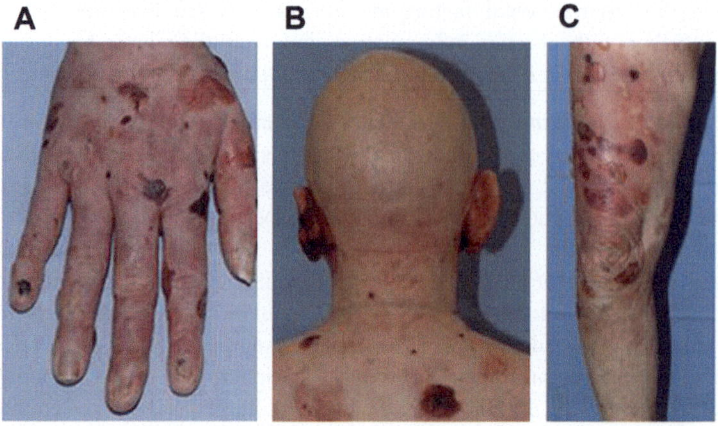

Fig. 7.3 Collagen XVII-associated diseases. (**a**) The right hand of a 26-year-old patient with JEB due to collagen XVII deficiency. Widespread blistering, scabs, skin atrophy, hypopigmentation, and nail dystrophy are seen. (**b**) The same patient as in A exhibits complete loss of hair and disseminated blisters. (**c**) The right leg of a 75-year-old patient with bullous pemphigoid due to autoantibodies against collagen XVII. Typical signs are disseminated blisters, erosions, and itchy inflammatory plaques

The phenotype of patients with *COL17A1* mutations is characterized by variable degrees of skin blistering, dystrophy or loss of nails, mucosal involvement, enamel defects, and hair involvement (Fig. 7.3). The severity of the clinical manifestations correlates inversely with the abundance of collagen XVII (Pasmooij et al. 2007; Varki et al. 2006). Patients with complete lack of collagen XVII in the skin exhibit congenital, generalized blistering with severe mucosal involvement, nail dystrophy and, later, universal loss of hair starting in the teens and amelogenesis imperfecta. As shown by a comprehensive analysis of a cohort of 43 patients with *COL17A1* mutations, about 12–14% of the physiological collagen XVII levels are sufficient to render the phenotype rather mild and the life span long (Kiritsi et al. 2011). In patients with splice-site mutations allowing low expression of collagen XVII mild disease manifestations were observed, with some cases being diagnosed only at advanced age, when cofounding disorders, e.g., diabetes or immobility aggravated the skin fragility (Kiritsi et al. 2011). If part of the collagen XVII molecule is spliced out, both the percentage of transcripts not causing premature termination of protein translation and the exact location of the mutation play roles in the outcome. Mutations within the collagenous domains will perturb the G-X-Y repeat motif and therefore affect the folding and stability of the triple-helical ectodomain, and in some cases lead to intracellular retention, aberrant post-translational modifications, and perturbed export to the cell surface, in which case extracellular collagen XVII abundance is lower. However, these mutations are associated with a milder phenotype than null mutations (Huilaja et al. 2009).

The JEB subtype, JEB late onset, is caused by the specific *COL17A1* mutation p. Arg1303Gln (Has et al. 2020; Kiritsi et al. 2011; Yuen et al. 2011; Schumann et al.

1997). Blistering appears not to be congenital, but starts in childhood and results in scarring and pronounced skin atrophy. Nails are either lost or dystrophic. Loss of dermatoglyphs, sclerosis, contractures of fingers, palmoplantar keratoderma, microstomia, esophageal stenosis, corneal erosions, ectropion, and lacrimal duct obstruction are further features of the phenotype (Kiritsi et al. 2011; Has et al. 2014). IFM shows a disorganized and duplicated basement membrane, whereas in transmission electron microscopy reduplication of the lamina densa, as well as deposition of amorphous material in the superficial dermis is found. Thus, this specific missense mutation, which is located in the NC-4 domain of collagen XVII has been suggested to cause structural changes in the molecule. It was further speculated that the structural changes result in altered interactions with laminin-332 by exposing a cryptic/hidden? glutamine residue and de novo crosslinking between collagen XVII and laminin-332 (Has et al. 2014).

Revertant mosaicism A therapeutically interesting feature of JEB due to collagen XVII deficiency is the frequent presence of spontaneous correction of the deficiency in skin patches, also known as revertant mosaicism or "natural gene therapy" (Pasmooij et al. 2005, 2012). This correction occurs through a second—somatic—mutation that restores protein expression. Thus, a mosaic of clinically healthy (revertant) patches surrounded by fragile (mutant) skin exists. The clinical observation that the mutant skin was hypopigmented, as compared to the revertant patches, led to the identification of the importance of collagen XVII for melanocyte supply to the epidermis (Gostynski et al. 2014).

7.6 Autoimmune Blistering Disorders Associated with Autoantibodies Against Collagen XVII or Collagen VII

Skin fragility occurs not only through a missing skin protein due to genetic variants, but also through production of autoantibodies against the respective proteins in autoimmune blistering disorders. These diseases are mediated mainly by IgG autoantibodies that are deposited along the DEJZ and that target collagen VII in EB acquisita (EBA) or collagen XVII in bullous pemphigoid (BP) and mucous membrane pemphigoid (MMP) (Ujiie et al. 2019).

Epidermolysis bullosa acquisita EBA was initially described in 1985 as an acquired blistering disease resembling genetic DEB (Elliott 1895) (Fig. 7.2). It is caused by autoantibodies against collagen VII (Woodley et al. 1984). Based on the pathogenetic mechanism and the clinical picture, two main clinical EBA manifestations have been described: (a) the mechanobullous type, where blisters are induced by direct antibody impact and (b) the non-mechanobullous/inflammatory type, where blistering occurs through antibody-induced local inflammation (Koga et al. 2018). A subtype of the latter includes MMP-like inflammatory EBA that is clinically defined by predominant involvement of the mucosa. Interestingly, the clinical presentation may change in the same patient over time, thus patients may switch from a BP-like form to a mechanobullous form or mucosal lesions may appear later on (Koga et al. 2018). In the mechanobullous EBA variant, skin fragility with blisters, scarring, and milia primarily at trauma-prone sites can be accompanied by nail involvement (Iwata et al. 2018). In the non-mechanobullous type, widespread blistering is typically accompanied by pruritus (Iwata et al. 2018).

Many autoimmune disorders have been reported to be associated with EBA, but the strongest association is with chronic inflammatory bowel diseases, since around 25% of the EBA patients were found to suffer from Crohn's disease, too (Chen et al. 2002b).

The immunodominant part of collagen VII has been described to be the NC1 domain (Lapiere et al. 1993; Vorobyev et al. 2017), but antibodies against the NC2 and the collagenous domains have also been identified in several cases (Ishii et al. 2009; Saleh et al. 2011). No clear correlation between the target regions of the autoantibodies and the clinical picture or prognosis exists yet (Vorobyev et al. 2015). It has been postulated that the disruption of the collagen VII interaction with its binding partners explains the variation of the disease manifestations.

Bullous pemphigoid group Collagen XVII was initially identified as a structural component of the hemidesmosomes using IgG antibodies from BP sera (Giudice et al. 1992). Typically, BP affects patients in their late 70s or older, and the clinical presentation includes urticarial, erythematous plaques, and tense blisters, accompanied by intense pruritus (Schmidt and Zillikens 2013), although atypical presentations also occur (Cozzani et al. 2015) (Fig. 7.3c). The mucous membranes might show erosions, especially the oral mucosa. In cases of sole mucosal involvement, the oral mucosa is most commonly affected, but ocular and nasal epithelia, as well as the larynx and the genital area can also be impaired (La Placa et al. 2019). Patients suffering from MMP are also frequently of higher age, the disease appears to be more treatment refractory than classical BP and more often paraneoplastic (La Placa et al. 2019). If untreated, MMP may result in functional impairment of the visual acuity, the ability to speak and breath, or urinary and sexual dysfunction through the development of mucosal synechiae (Kiritsi and Schauer 2019).

Several authors have reported a higher rate of neurological diseases (e.g., cerebral infarction, dementia, Parkinson's disease, epilepsy) in patients with BP than in the general population (Ujiie et al. 2019; Papakonstantinou et al. 2019; Messingham et al. 2019a), however, a causal relationship has not been found yet.

The immunodominant region of collagen XVII is the NC16A domain, with more than 80% of the patients having antibodies against it during the active phase (Ujiie et al. 2019; Has and Kern 2010). In pemphigoid gestationis, an autoimmune blistering disorder associated with pregnancy (Huilaja et al. 2014), more than 90% of the women have antibodies against epitopes of the NC16a domain (Di Zenzo et al. 2007). It has been postulated that after the autoantibodies bind to the target antigen, inflammatory cell infiltration is locally induced through complement activation and blister formation through proteolytic enzymes, e.g., plasmin, neutrophil elastase, and

MMP-9 (Nishie 2020; Nishie 2014). But other pathogenetic mechanisms have also been discussed, e.g., that IgG binding to collagen XVII results in reduced hemidesmosomal collagen XVII content and thus weakened DEJZ stability (Iwata et al. 2009). Hurskainen et al. have sought to provide an answer to why the NC16A domain is specifically immunodominant and produced a genetically manipulated mouse lacking it. Since this is the region where shedding of the molecule takes place, this mouse model is a shedding-deficient one. Interestingly, the mice are prone to suffer from itch and spontaneously develop autoantibodies against collagen XVII, even though no blistering is observed (Hurskainen et al. 2015). The molecular and cellular mechanisms of this interesting observation need further elucidation.

In patients with MMP, in addition to autoantibodies to the NC16a domain, autoantibodies to the C-terminal domain of collagen XVII (Balding et al. 1996; Schmidt et al. 2001) and frequently also to BP230, laminin 332, collagen VII or integrin α6β4 are found (Kamaguchi and Iwata 2019).

Linear IgA disease This autoimmune blistering disorder is caused by IgA class autoantibodies targeting hemidesmosomal proteins (Lamberts et al. 2019), as visualized by the deposition of IgA along the DEJZ. It is the most common autoimmune blistering disorder in childhood (Kiritsi and Schauer 2019). The antigens in most cases are the 120-kDa ectodomain of collagen XVII or LAD-1 antigen, or its cleavage product, the 97-kDa LABD97 antigen, which is generated from the ectodomain by plasmin or other proteinases (Hofmann et al. 2009). The clinical hallmarks include blisters arranged in a circular manner (as crowns of jewels) and urticarial plaques; the mucosa might also be affected (Juratli and Sárdy 2019; Genovese et al. 2019). Most patients suffer from pruritus. In rare cases collagen VII is the autoantigen, a subtype termed IgA-EBA. The disease appears to be self-limiting in childhood within a few years, whereas in adults it tends to be chronic and treatment refractory (Lamberts et al. 2019).

Diagnostics The autoimmune blistering disorders are diagnosed by demonstration of the presence of tissue bound or circulating autoantibodies to collagen VII or XVII. Around 80–90% of BP patients have autoantibodies that react with the extracellular non-collagenous 16A domain (NC16A), with commercial ELISA systems recognizing autoantibodies against this domain (MBL, Nagoya, Japan and Euroimmun, Lübeck, Germany; both have a sensitivity ranging from 80% to 95% (Di Zenzo et al. 2008)). Recently, a noninflammatory phenotype of BP was associated with antibodies against the midportion of collagen XVII, whereas these patients had less erythema and less eosinophils in skin lesions. Surprisingly, the majority of them had received dipeptidyl peptidase-IV inhibitors for the treatment of diabetes (Izumi et al. 2016; Varpuluoma et al. 2018). For EBA, a specific pattern in direct immunofluorescence, the so-called serration pattern of linear immunoglobulin deposits at the DEJZ is almost pathognomonic (Koga et al. 2018). In contrast, indirect immunofluorescence staining with the patient's serum is not very specific (with 74.7%), thus ELISA or immunoblotting should be used for detection of antibodies (Schmidt et al. 2017). There are two commercial ELISA systems for detection of autoantibodies to collagen VII: one recognizing both the NC1 and the NC2 domains (MBL, Nagoya,

Japan) that has the highest sensitivity (97.9%), and another recognizing only the NC1 domain (Euroimmun, Lübeck, Germany with a sensitivity of 89.5%). Immunoblotting with antibodies against the NC1 domain has a sensitivity of 85.3% (Schmidt et al. 2017).

7.7 Evidence-Based and Regenerative Therapies

Despite EB-related collagenopathies being rare diseases, there have been rather extraordinary efforts toward developing better treatment options for the disease, (for therapy approaches on other collagenopathies we refer the reader to Chaps. 3, 5 and 7). As of February 2020, there were more than 20 recruiting clinical trials listed on ClinicalTrials.gov (Tables 7.1 and 7.2). A few more trials are ongoing but no longer recruiting. The aims of the trials span from cure by means of collagen VII or XVII replacement, to providing symptom relief by reducing, e.g., itch or pain, improving wound healing, reducing fibrosis, or targeting cancer. Some trials, like certain cell therapies, fall in both categories by using medical products that have the theoretical capability to relieve symptoms, e.g., by suppressing inflammation, while at the same time providing neo-synthesis of the protein at fault. Another distinction is the area or body site of treatment. Since EB is a systemic disease, systemic treatments should be strived for. However, systemic delivery is more challenging than topical delivery and, therefore, many trials first focus on achieving efficient topical treatment of particularly problematic skin areas. This is true both for curative and for symptom-relief approaches and dependent on drug modalities.

Disease progression as a therapeutic challenge It is important to consider that both DEB and collagen XVII-deficient JEB are progressive diseases. In DEB, disease progression is a chain of events that increasingly worsens the phenotype and adds complications: skin fragility leads to chronic damage and chronic inflammation, which promote fibrosis and the establishment of a tumor promoting and subsequently tumor-accommodating microenvironment. The progressive nature of the disease also means that it is challenging to reverse changes and alleviate them once progressed; one obvious example is dermal fibrosis. This means that affected individuals should be treated as young as possible; breaking the chain early could effectively reduce fibrosis and cancer burden in DEB. However, depending on the therapeutic modality, also older patients can benefit from certain symptom-relief therapies (Table 7.2). One consequence of the progressive changes in the skin microenvironment is that these can hamper the response to curative therapies by providing a poor ground, e.g., gene-corrected keratinocyte grafts. Consequently, the most effective treatment would likely be achieved through combinatory cycles of symptom-relief and curative therapies, with the treatment regimens titrated for the patient's disease severity and progression.

Gene therapies with intent to cure Curative therapies that are in clinical trials encompass localized injection of gene-corrected dermal fibroblasts, grafting of

Table 7.1 Actively recruiting clinical trials on EB for therapies with the intent to cure as of February 2020

Therapies with intent- to-cure			
Drug type	Investigational drug	EB type	Trial Identification Nr.
Gene therapy	Transplantation surgery of genetically corrected cultured epidermal autograft	JEB *COL17A1* mutations	ClinicalTrials.gov NCT03490331 Austria, Salzburg
Gene therapy	Genetically corrected cultured epidermal autograft	Recessive DEB	ClinicalTrials.gov NCT02984085 Austria, Salzburg
Gene therapy	EB-101 autologous, gene-corrected keratinocyte sheets	Recessive DEB	ClinicalTrials.gov NCT04227106 USA, Stanford
Gene therapy	KB103, topically applied non-integrating, replication-incompetent herpes simplex virus vector expressing human collagen VII protein	DEB	ClinicalTrials.gov NCT03536143 USA, Stanford
Gene therapy	Genetically corrected skin equivalents	Recessive DEB	EudraCT Number 2016-002790-35 France, Paris
Antisense oligos	QR-313, topically applied antisense oligonucleotide	DEB with *COL7A1* exon 73 mutations	ClinicalTrials.gov NCT03605069 USA, multicenter
PTC readthrough	Gentamicin, intravenous	Recessive DEB	ClinicalTrials.gov NCT03392909 USA, Los Angeles
PTC readthrough	Gentamicin, topical	JEB	ClinicalTrials.gov NCT03526159 USA, Los Angeles
PTC readthrough	Optimizing i.v. gentamicin in JEB	JEB	ClinicalTrials.gov NCT04140786 USA, Los Angeles
Protein therapy	PTR-01, recombinant human collagen VII	Recessive DEB	ClinicalTrials.gov NCT03752905 USA, Stanford

gene-corrected epidermal sheets, grafting of gene-corrected full-thickness skin grafts, and in vivo gene therapy (Siprashvili et al. 2016; Lwin et al. 2019; Gaucher et al. 2020; Krystal Biotech Announces Positive Results from Phase 2 Clinical Trial 2020) (Table 7.1). Of the published trials on curative therapies for DEB so far, grafting of autologous retroviral-mediated gene-corrected epidermal sheets have produced most robust effects on improved skin durability (Siprashvili et al. 2016; Eichstadt et al. 2019). However, sustained transgene expression appears to decline with time (Siprashvili et al. 2016; Eichstadt et al. 2019). This is remarkable, in particular when compared to epidermal grafts for correction of laminin-332-deficient JEB that seem to stay stable for at least 2 years or so (Eichstadt et al. 2019; De Rosa et al. 2014; Marinkovich and Tang 2019). There are many potential reasons for this

Table 7.2 Actively recruiting clinical trials on EB for regenerative and symptom-relief therapies as of February 2020

Regenerative cell-based therapies			
Investigational drug		EB type	Trial identification Nr
Serial mesenchymal stem cell (MSC) infusions from a related donor		All EB types	ClinicalTrials.gov NCT02582775 USA, Minnesota
Allogeneic stem cell transplantation and "off-the-shelf" mesenchymal stem cells		All EB types	ClinicalTrials.gov NCT01033552 USA, Minnesota
Allogeneic ABCB5-positive stem cells		Recessive DEB	ClinicalTrials.gov NCT03529877 Germany, international
Epidermal grafts generated using the cellutome system		EB after hematopoietic cell transplant	ClinicalTrials.gov NCT02670837 USA, Minnesota
Bone marrow-derived MSC infusions (MisenSistem-EB)		Recessive DEB	EudraCT Number 2017–000606-37 Spain
Symptom-relief therapies			
Target aim	Investigational drug	EB type	Trial Identification Nr.
Anti-inflammatory	Pharmacokinetics, safety of diacerein after maximum use	EBS	ClinicalTrials.gov NCT03472287 USA, international
Anti-inflammatory	Oleogel-S-10, topical	All EB types	ClinicalTrials.gov NCT03068780 International
Accelerator of wound healing	RGN-137, a thymosin beta-4 gel, topical	JEB, DEB	ClinicalTrials.gov NCT03578029 USA
Analgesic and anti-pruritic	Pregabalin for itch and pain	Recessive DEB	ClinicalTrials.gov NCT03928093 Canada, Toronto
Anticancer	Rigosertib for advanced EB cancer	Recessive DEB	ClinicalTrials.gov NCT03786237 EudraCT number: 2016-0036832-19 Salzburg

difference including the preparation and targeting of epidermal stem cells (Marinkovich and Tang 2019), but also the fact that laminin-332 promotes epidermal stem cell maintenance (De Rosa et al. 2019). It would be interesting to assess, if the changes in the dermal microenvironment also contribute. The use of gene-corrected whole skin grafts is currently being investigated (Gaucher et al. 2020) and could bring some insights toward this.

An additional factor that should be considered with regard to the declining collagen VII abundance is the strength of the transgene expression in gene-corrected cells. Collagen VII is physiologically a low-abundant protein, and to support skin stability low levels of collagen VII are needed. In other words, abundance is not a good measure to predict therapeutic efficacy—"less could be more." Supraphysiological expression of collagen VII that may occur after expressing *COL7A1* under the control of the strong promoter may cause cellular stress due to limited capacity of the intricate machinery for time- and energy-consuming collagen VII biosynthesis and export. This cellular stress may result in subclinical inflammation reducing therapeutic durability. The secreted collagen VII may also be of lower functionality or, alternatively, supraphysiological levels of collagen VII could affect microenvironmental homeostasis. Therefore, it is conceivable that gene-corrected keratinocytes expressing physiological levels of collagen VII, e.g., through genome editing, could yield longer-lasting benefit.

Nonpermanent restoration of collagen Different strategies aim to provide temporal symptom relief by nonpermanent restoration of collagen presence at the DEJZ. One approach is through the direct injection of recombinant procollagen VII, which has shown some evidence of not only working as a localized therapy but also being systemically effective after infusion in small animal models (Woodley et al. 2004; Hou et al. 2015).

Premature termination codons (PTC) may be amendable by drugs evoking PTC readthrough. This would lead to collagen VII or collagen XVII being translated from patient-endogenous mRNA transcript but with, depending on the mutation, alternative amino acids inserted at the PTC codon. Topical delivery or intradermal injections of high concentrations of the aminoglycoside gentamicin has in one clinical study shown promise to increase collagen VII abundance, promote wound healing, and reduced skin blistering in DEB (Woodley et al. 2017). Both collagen VII and collagen XVII are composed of large repeats of exons with an in-frame arrangement, allowing these to be skipped without disturbing the reading frame (Kowalewski et al. 2016; Bornert et al. 2016). Several clinical trials are currently testing this approach for JEB and DEB (Table 7.1).

Exclusion of mutation-carrying exons at the pre-mRNA level could lead to the translation of an internally shortened protein but with sufficiently retained functionality to provide tissue stabilization. Observations from natural skipping of mutation-carrying exons occurring in DEB provide support that this can, depending on the exon, indeed result in phenotypic amelioration (Bremer et al. 2019). Pharmacologically, exon skipping can be achieved through antisense oligonucleotides (ASOs) binding to the exon targeted for exclusion. ASO-mediated skipping has been shown for multiple exons in *COL7A1,* spanning exons encoding segments of the NC1 domain to the C-terminal part of the collagenous domain (Bornert et al. 2016; Goto et al. 2006; Turczynski et al. 2016). A major challenge with this approach is the delivery to achieve sufficient intracellular ASO uptake yielding therapeutically relevant levels of exon skipping. This favors topical delivery. ASOs would not readily pass through the epidermal barrier—application on wounds would circumvent this challenge. At the moment topical delivery on wounds of an ASO targeting *COL7A1* exon 73 is being investigated in a clinical trial (NCT03605069) as a wound healing therapy for exon 73 mutated DEB patients (Haisma et al. 2018) (Table 7.1).

Antifibrotic therapies We recently provided preclinical evidence that the angiotensin II type 1 receptor inhibitor losartan effectively delayed fibrosis and alleviated inflammation in chronically injured skin in DEB model mice (Nyström et al. 2015). Losartan was used because of its suitable safety profile, its capability of systemic delivery, and its suggested efficacy in another genetically predisposed fibroproliferative disease—Marfan Syndrome (Habashi et al. 2006; Brooke et al.

2008). A point of intersection between the two diseases was altered TGFβ bioavailability and losartan had been shown to indirectly lower TGFβ activity (Nyström et al. 2015; Habashi et al. 2006; Neptune et al. 2003). Based on the effect on reducing inflammation and delaying dermal fibrosis in our DEB models, we initiated an investigator-initiated trial to assess the safety and tolerability of losartan in children (EudraNr. 2015-003670-32).

Another systemic symptom-relief therapy that is currently being evaluated for DEB is injection with a modified high mobility group protein B1 (HGMB1) peptide. The study is based on the finding that injured epithelia release HGMB1, which mobilizes tissue-regenerating cells, among them epithelial progenitors, from the bone marrow (Tamai et al. 2011, 2017). The rationale is that injections of the peptide would increase in the skin the mobilization of such tissue-regenerating progenitor cells that promote healing. Another benefit could be that mobilization of mesenchymal stromal cells (MSCs), which suppress inflammation, may also occur.

Regenerative cell therapies After initial enthusiasm about bone marrow transplantation as a curative therapy for DEB, long-term monitoring revealed only limited clinical benefits and a high ratio of adverse effects (Wagner et al. 2010). Thereafter, most cell therapy studies for EB deal with MSCs, which have dual potential. They have anti-inflammatory properties, i.e., they can provide symptom relief, but they can also synthesize functional collagen VII (Kühl et al. 2015). They have and are being evaluated as therapies for DEB; both as sole MSC infusions but also jointly with allogeneic bone marrow transplantation (Rashidghamat et al. 2019; Ebens et al. 2019) (Table 7.2). Suppression of inflammation can be achieved at a lower number of cells than what would be needed for effective restoration of collagen VII abundance in tissue (Kühl et al. 2015; Rashidghamat et al. 2019). In addition, the effect on inflammation can be mediated on both systemic and tissue levels. Therefore, the clinical benefit of MSC infusions is generally due to reduction of inflammation and damage response, rather than due to collagen VII replacement.

Therapies for autoimmune blistering skin diseases For autoimmune disorders, no approved pathogenesis-specific therapies exist yet. It is important to diagnose the disease and start treatment with immunosuppressive drugs as early as possible. Based on the skin and mucosal involvement, potent topical steroids will be initially used in most cases, followed by systemic corticosteroids in dosages adapted to the patients' weight. If insufficient to control disease, other immunosuppressants (e.g., azathioprine, mycophenolate mofetil, and others) or immunomodulatory agents, e.g., dapsone will be required. In severe or treatment-resistant cases, plasma exchange, intravenous immunoglobulin therapy, or use of anti-CD20 antibodies, e.g., rituximab is employed (Ujiie et al. 2019; Eming et al. 2015).

Lately, several different pathogenesis-derived approaches have emerged. For example, the pathogenetic role of IgE, eosinophils, and a Th2-oriented immune response in autoimmune blistering disorders, especially in BP, has been investigated and their relevance for therapy strategies evaluated (Kamata et al. 2019; Cozzani et al. 2018; Saniklidou et al. 2018). Elevated serum IgE has been observed in around 80% of BP patients, while the incidence of BP180-specific IgE antibodies varies between 22 and 100% (Messingham et al. 2019b). The numbers of both circulating and lesional eosinophils closely correlate with the BP severity (Kamata et al. 2019). The therapeutic effects of IgE antibody inhibition were proven by the use of omalizumab, a humanized monoclonal antibody that binds the Fc portion of IgE (Yu et al. 2014). Interestingly, IgE autoantibodies have also been disclosed in MMP and EBA (Natsuga et al. 2010; Koga et al. 2019; Ludwig 2019), but their pathogenetic relevance needs still to be evaluated. Another approach that is currently being sought is to block eotaxin and its receptor CCR3. Both were identified to be highly abundant in BP lesions (Frezzolini et al. 2002), being associated with activation of the Th2 immune system and correlating with the number of tissue eosinophils (Griffith et al. 2014; Wakugawa et al. 2000). Finally, blocking interleukins 4 and 13, which are crucial cytokines in Th2 response, might also show therapeutic potential in these disorders. To that end, dupilumab, the recombinant IgG4 monoclonal antibody which binds to the interleukin-4 receptor IL-4Rα, has recently been employed in treatment-refractory BP and showed satisfactory results (Abdat et al. 2020). These treatments focus on specific signaling pathways in autoimmune blistering disorders and are thus associated with fewer side effects than the classical immunosuppressants.

7.8 Outlook

Collagen VII and collagen XVII are disparate in terms of their structure, biochemical properties, interactome, and location. Still, both collagens are essential for supporting skin integrity. This is evidenced by skin blistering diseases arising from genetic deficiency of the two proteins. Importantly, what we learned from studying these two genetic collagenopathies has also been applied to more common, acquired autoimmune blistering diseases with which pathomechanisms are shared. Research on all these has aided our understanding of the physiological functions of collagens VII and XVII that take on in health, regeneration, and disease. Consequently, the new knowledge has allowed for continuously improved management of the diseases—both symptomatic and curative. It should, however, be underscored that there are still many aspects of the two collagens that are insufficiently understood, spanning from their biosynthesis to extracutaneous functions. A better understanding of these will facilitate the translation of the new knowledge into evidence-based, biologically valid therapies. On a higher level, studies of two specialized collagens have allowed for and continue to promote an improved understanding of the ECM in health and disease.

Acknowledgments Work in the authors' laboratory and EB Center is supported by grants from the German Research Foundation (DFG) NY90/2-1 (AN), NY90/3-2 (AN), NY90/5-1 (AN), SFB850 project B11 (AN), SFB1160 project B03 (AN, DK), KI1795/1-1 (DK), KI 1795/2-1 (DK), the Fritz-Thyssen Foundation (DK, AN), the Berta-Ottenstein Advanced Clinician Scientist Program of the Medical Faculty, University of Freiburg (DK) and research grants from the patient advocacy organization "DEBRA International" Nyström-Bruckner-Tuderman 1 (LBT, AN) and Bruckner-Tuderman 5 (LBT, DK).

References

Abdat R, Waldman RA, de Bedout V et al (2020) Dupilumab as a novel therapy for bullous pemphigoid: a multicenter case series. J Am Acad Dermatol 83:46

Aho S, Uitto J (1998) Two-hybrid analysis reveals multiple direct interactions for thrombospondin 1. Matrix Biol J Int Soc Matrix Biol 17:401–412

Alexeev V, Donahue A, Uitto J, Igoucheva O (2016) Chemotaxis-driven disease-site targeting of therapeutic adult stem cells in dystrophic epidermolysis bullosa. Stem Cell Res Ther 7:124

Atanasova VS, Russell RJ, Webster TG et al (2019) Thrombospondin-1 is a major activator of TGF-β signaling in recessive dystrophic epidermolysis bullosa fibroblasts. J Invest Dermatol 139:1497–1505.e5

Bajénoff M, Germain RN (2009) B-cell follicle development remodels the conduit system and allows soluble antigen delivery to follicular dendritic cells. Blood 114:4989–4997

Balding SD, Prost C, Diaz LA et al (1996) Cicatricial pemphigoid autoantibodies react with multiple sites on the BP180 extracellular domain. J Invest Dermatol 106:141–146

Banushi B, Forneris F, Straatman-Iwanowska A et al (2016) Regulation of post-Golgi LH3 trafficking is essential for collagen homeostasis. Nat Commun 7:12111

Bentz H, Morris NP, Murray LW et al (1983) Isolation and partial characterization of a new human collagen with an extended triple-helical structural domain. Proc Natl Acad Sci U S A 80:3168–3172

Bianchi N, Beninati S, Bergamini CM (2018) Spotlight on the transglutaminase 2 gene: a focus on genomic and transcriptional aspects. Biochem J 475:1643–1667

Bornert O, Kühl T, Bremer J et al (2016) Analysis of the functional consequences of targeted exon deletion in COL7A1 reveals prospects for dystrophic epidermolysis bullosa therapy. Mol Ther J Am Soc Gene Ther 24:1302–1311

Bremer J, der Heijden EHV, Eichhorn DS et al (2019) Natural exon skipping sets the stage for antisense oligonucleotide-mediated exon skipping as therapy for dystrophic epidermolysis bullosa. Mol Ther – Nucleic Acids 18:465

Brittingham R, Uitto J, Fertala A (2006) High-affinity binding of the NC1 domain of collagen VII to laminin 5 and collagen IV. Biochem Biophys Res Commun 343:692–699

Brooke BS, Habashi JP, Judge DP et al (2008) Angiotensin II blockade and aortic-root dilation in Marfan's syndrome. N Engl J Med 358:2787–2795

Bruckner-Tuderman L, Guscetti F, Ehrensperger F (1991) Animal model for dermolytic mechanobullous disease: sheep with recessive dystrophic epidermolysis bullosa lack collagen VII. J Invest Dermatol 96:452–458

Bruckner-Tuderman L, Nilssen O, Zimmermann DR et al (1995) Immunohistochemical and mutation analyses demonstrate that procollagen VII is processed to collagen VII through removal of the NC-2 domain. J Cell Biol 131:551–559

Burgeson RE (1993) Type VII collagen, anchoring fibrils, and epidermolysis bullosa. J Invest Dermatol 101:252–255

Butcher DT, Alliston T, Weaver VM (2009) A tense situation: forcing tumour progression. Nat Rev Cancer 9:108–122

Canty EG, Kadler KE (2005) Procollagen trafficking, processing and fibrillogenesis. J Cell Sci 118:1341–1353

Chacón-Solano E, León C, Díaz F et al (2019) Fibroblast activation and abnormal extracellular matrix remodelling as common hallmarks in three cancer-prone genodermatoses. Br J Dermatol 181:512–522

Chen M, Marinkovich MP, Veis A et al (1997) Interactions of the amino-terminal noncollagenous (NC1) domain of type VII collagen with extracellular matrix components. A potential role in epidermal-dermal adherence in human skin. J Biol Chem 272:14516–14522
Chen M, Marinkovich MP, Jones JC et al (1999) NC1 domain of type VII collagen binds to the beta3 chain of laminin 5 via a unique subdomain within the fibronectin-like repeats. J Invest Dermatol 112:177–183
Chen M, O'Toole EA, Muellenhoff M et al (2000) Development and characterization of a recombinant truncated type VII collagen "minigene". Implication for gene therapy of dystrophic epidermolysis bullosa. J Biol Chem 275:24429–24435
Chen M, Keene DR, Costa FK et al (2001) The carboxyl terminus of type VII collagen mediates antiparallel dimer formation and constitutes a new antigenic epitope for epidermolysis bullosa acquisita autoantibodies. J Biol Chem 276:21649–21655
Chen M, Kasahara N, Keene DR et al (2002a) Restoration of type VII collagen expression and function in dystrophic epidermolysis bullosa. Nat Genet 32:670–675
Chen M, O'Toole EA, Sanghavi J et al (2002b) The epidermolysis bullosa acquisita antigen (type VII collagen) is present in human colon and patients with crohn's disease have autoantibodies to type VII collagen. J Invest Dermatol 118:1059–1064
Chen VM, Shelke R, Nyström A et al (2018) Collagen VII deficient mice show morphologic and histologic corneal changes that phenotypically mimic human dystrophic epidermolysis bullosa of the eye. Exp Eye Res 175:133–141
Cho RJ, Alexandrov LB, den Breems NY et al (2018) APOBEC mutation drives early-onset squamous cell carcinomas in recessive dystrophic epidermolysis bullosa. Sci Transl Med 10: eaas9668
Christiano AM, Greenspan DS, Lee S, Uitto J (1994) Cloning of human type VII collagen. Complete primary sequence of the alpha 1(VII) chain and identification of intragenic polymorphisms. J Biol Chem 269:20256–20262
Christiansen HE, Schwarze U, Pyott SM et al (2010) Homozygosity for a missense mutation in SERPINH1, which encodes the collagen chaperone protein HSP47, results in severe recessive osteogenesis imperfecta. Am J Hum Genet 86:389–398
Cozzani E, Gasparini G, Burlando M et al (2015) Atypical presentations of bullous pemphigoid: clinical and immunopathological aspects. Autoimmun Rev 14:438–445
Cozzani E, Gasparini G, Di Zenzo G, Parodi A (2018) Immunoglobulin E and bullous pemphigoid. Eur J Dermatol 28:440–448
De Rosa L, Carulli S, Cocchiarella F et al (2014) Long-term stability and safety of transgenic cultured epidermal stem cells in gene therapy of junctional epidermolysis bullosa. Stem Cell Rep 2:1–8
De Rosa L, Secone Seconetti A, De Santis G et al (2019) Laminin 332-dependent YAP dysregulation depletes epidermal stem cells in junctional epidermolysis bullosa. Cell Rep 27:2036–2049.e6
Di Zenzo G, Calabresi V, Grosso F et al (2007) The intracellular and extracellular domains of BP180 antigen comprise novel epitopes targeted by pemphigoid gestationis autoantibodies. J Invest Dermatol 127:864–873
Di Zenzo G, Thoma-Uszynski S, Fontao L et al (2008) Multicenter prospective study of the humoral autoimmune response in bullous pemphigoid. Clin Immunol Orlando Fla 128:415–426
Ebens CL, McGrath JA, Tamai K et al (2019) Bone marrow transplant with post-transplant cyclophosphamide for recessive dystrophic epidermolysis bullosa expands the related donor pool and permits tolerance of nonhaematopoietic cellular grafts. Br J Dermatol 181:1238–1246
Eichstadt S, Barriga M, Ponakala A et al (2019) Phase 1/2a clinical trial of gene-corrected autologous cell therapy for recessive dystrophic epidermolysis bullosa. JCI Insight 4
Elliott GT (1895) Two cases of epidermolysis bullosa 13:10
Eming R, Sticherling M, Hofmann SC et al (2015) S2k guidelines for the treatment of pemphigus vulgaris/foliaceus and bullous pemphigoid. J Dtsch Dermatol Ges J Ger Soc Dermatol 13:833–844

Fine J-D, Mellerio JE (2009a) Extracutaneous manifestations and complications of inherited epidermolysis bullosa: part I. Epithelial associated tissues. J Am Acad Dermatol 61:367–384; quiz 385–386
Fine J-D, Mellerio JE (2009b) Extracutaneous manifestations and complications of inherited epidermolysis bullosa: part II. Other organs. J Am Acad Dermatol 61:387–402; quiz 403–404
Fine JD, Eady RA, Bauer EA et al (2008) The classification of inherited epidermolysis bullosa (EB): report of the third international consensus meeting on diagnosis and classification of EB. J Am Acad Dermatol 58:931–950
Franzke C-W, Tasanen K, Schumann H, Bruckner-Tuderman L (2003) Collagenous transmembrane proteins: collagen XVII as a prototype. Matrix Biol J Int Soc Matrix Biol 22:299–309
Franzke C-W, Bruckner-Tuderman L, Blobel CP (2009) Shedding of collagen XVII/BP180 in skin depends on both ADAM10 and ADAM9. J Biol Chem 284:23386–23396
Frezzolini A, Teofoli P, Cianchini G et al (2002) Increased expression of eotaxin and its specific receptor CCR3 in bullous pemphigoid. Eur J Dermatol 12:27–31
Fritsch A, Loeckermann S, Kern JS et al (2008) A hypomorphic mouse model of dystrophic epidermolysis bullosa reveals mechanisms of disease and response to fibroblast therapy. J Clin Invest 118:1669–1679
Gache Y, Pin D, Gagnoux-Palacios L et al (2011) Correction of dog dystrophic epidermolysis bullosa by transplantation of genetically modified epidermal autografts. J Invest Dermatol 131:2069–2078
Galiger C, Löffek S, Stemmler MP et al (2018) Targeting of cell surface proteolysis of collagen XVII impedes squamous cell carcinoma progression. Mol Ther J Am Soc Gene Ther 26:17–30
Gaucher S, Lwin SM, Titeux M et al (2020) EBGene trial: patient preselection outcomes for the European GENEGRAFT ex vivo phase I/II gene therapy trial for recessive dystrophic epidermolysis bullosa. Br J Dermatol 182:794–797
Genovese G, Venegoni L, Fanoni D et al (2019) Linear IgA bullous dermatosis in adults and children: a clinical and immunopathological study of 38 patients. Orphanet J Rare Dis 14:115
Giudice GJ, Emery DJ, Diaz LA (1992) Cloning and primary structural analysis of the bullous pemphigoid autoantigen BP180. J Invest Dermatol 99:243–250
Gonzalez AM, Otey C, Edlund M, Jones JC (2001) Interactions of a hemidesmosome component and actinin family members. J Cell Sci 114:4197–4206
Gostynski A, Pasmooij AM, Del Rio M et al (2014) Pigmentation and melanocyte supply to the epidermis depend on type XVII collagen. Exp Dermatol 23:130–132
Goto M, Sawamura D, Nishie W et al (2006) Targeted skipping of a single exon harboring a premature termination codon mutation: implications and potential for gene correction therapy for selective dystrophic epidermolysis bullosa patients. J Invest Dermatol 126:2614–2620
Grässel S, Unsöld C, Schäcke H et al (1999) Collagen XVI is expressed by human dermal fibroblasts and keratinocytes and is associated with the microfibrillar apparatus in the upper papillary dermis. Matrix Biol J Int Soc Matrix Biol 18:309–317
Griffith JW, Sokol CL, Luster AD (2014) Chemokines and chemokine receptors: positioning cells for host defense and immunity. Annu Rev Immunol 32:659–702
Habashi JP, Judge DP, Holm TM et al (2006) Losartan, an AT1 antagonist, prevents aortic aneurysm in a mouse model of Marfan syndrome. Science 312:117–121
Haisma I, Bornert O, Schuijt M et al (2018) 1086 QR-313, an antisense oligonucleotide, restores expression of functional type VII collagen in DEB patient cells. J Invest Dermatol 138:S184
Has C, Kern JS (2010) Collagen XVII. Dermatol Clin 28:61–66
Has C, Nyström A (2015) Epidermal basement membrane in health and disease. Curr Top Membr 76:117–170
Has C, Kiritsi D, Mellerio JE et al (2014) The missense mutation p.R1303Q in type XVII collagen underlies junctional epidermolysis bullosa resembling Kindler syndrome. J Invest Dermatol 134:845–849
Has C, Bauer JW, Bodemer C et al (2020) Consensus re-classification of inherited epidermolysis bullosa and other disorders with skin fragility. Br J Dermatol 183:614

Heinonen S, Männikkö M, Klement JF et al (1999) Targeted inactivation of the type VII collagen gene (Col7a1) in mice results in severe blistering phenotype: a model for recessive dystrophic epidermolysis bullosa. J Cell Sci 112(Pt 21):3641–3648
Hofmann SC, Voith U, Schönau V et al (2009) Plasmin plays a role in the in vitro generation of the linear IgA dermatosis antigen LADB97. J Invest Dermatol 129:1730–1739
Hoste E, Arwert EN, Lal R et al (2015) Innate sensing of microbial products promotes wound-induced skin cancer. Nat Commun 6:5932
Hou Y, Guey LT, Wu T et al (2015) Intravenously administered recombinant human type VII collagen derived from Chinese hamster ovary cells reverses the disease phenotype in recessive dystrophic epidermolysis bullosa mice. J Invest Dermatol 135:3060–3067.
Housley TJ, Rowland FN, Ledger PW et al (1980) Effects of tunicamycin on the biosynthesis of procollagen by human fibroblasts. J Biol Chem 255:121–128
Huilaja L, Hurskainen T, Autio-Harmainen H et al (2009) Glycine substitution mutations cause intracellular accumulation of collagen XVII and affect its post-translational modifications. J Invest Dermatol 129:2302–2306
Huilaja L, Mäkikallio K, Tasanen K (2014) Gestational pemphigoid. Orphanet J Rare Dis 9:136
Hurskainen T, Kokkonen N, Sormunen R et al (2015) Deletion of the major bullous pemphigoid epitope region of collagen XVII induces blistering, autoimmunization, and itching in mice. J Invest Dermatol 135:1303–1310
Ishii N, Yoshida M, Ishida-Yamamoto A et al (2009) Some epidermolysis bullosa acquisita sera react with epitopes within the triple-helical collagenous domain as indicated by immunoelectron microscopy. Br J Dermatol 160:1090–1093
Ishikawa Y, Ito S, Nagata K et al (2016) Intracellular mechanisms of molecular recognition and sorting for transport of large extracellular matrix molecules. Proc Natl Acad Sci USA 113: E6036–E6044
Iwata H, Kamio N, Aoyama Y et al (2009) IgG from patients with bullous pemphigoid depletes cultured keratinocytes of the 180-kDa bullous pemphigoid antigen (type XVII collagen) and weakens cell attachment. J Invest Dermatol 129:919–926
Iwata H, Vorobyev A, Koga H et al (2018) Meta-analysis of the clinical and immunopathological characteristics and treatment outcomes in epidermolysis bullosa acquisita patients. Orphanet J Rare Dis 13:153
Izumi K, Nishie W, Mai Y et al (2016) Autoantibody profile differentiates between inflammatory and noninflammatory bullous pemphigoid. J Invest Dermatol 136:2201–2210
Jacków J, Löffek S, Nyström A et al (2016) Collagen XVII shedding suppresses re-epithelialization by directing keratinocyte migration and dampening mTOR signaling. J Invest Dermatol 136:1031–1041
Jensen D, Schekman R (2011) COPII-mediated vesicle formation at a glance. J Cell Sci 124:1–4
Juratli HA, Sárdy M (2019) Linear IgA bullous dermatosis. Hautarzt Z Dermatol Venerol Verwandte Geb 70:254–259
Kamaguchi M, Iwata H (2019) The diagnosis and blistering mechanisms of mucous membrane pemphigoid. Front Immunol 10:34
Kamata A, Kurihara Y, Funakoshi T et al (2019) Basement membrane zone IgE deposition is associated with bullous pemphigoid disease severity and treatment results. Br J Dermatol 182:1221
Keene DR, Sakai LY, Lunstrum GP et al (1987) Type VII collagen forms an extended network of anchoring fibrils. J Cell Biol 104:611–621
Kiritsi D, Schauer F (2019) Autoimmune blistering dermatoses in children. Hautarzt Z Dermatol Venerol Verwandte Geb 70:277–282
Kiritsi D, Kern JS, Schumann H et al (2011) Molecular mechanisms of phenotypic variability in junctional epidermolysis bullosa. J Med Genet 48:450–457
Kiritsi D, Has C, Bruckner-Tuderman L (2013) Laminin 332 in junctional epidermolysis bullosa. Cell Adhes Migr 7:135–141

Kivirikko KI (1993) Collagens and their abnormalities in a wide spectrum of diseases. Ann Med 25:113–126
Koga H, Prost-Squarcioni C, Iwata H et al (2018) Epidermolysis bullosa acquisita: the 2019 update. Front Med 5:362
Koga H, Teye K, Yamashita K et al (2019) Detection of anti-type VII collagen IgE antibodies in epidermolysis bullosa acquisita. Br J Dermatol 180:1107–1113
Koster J, Geerts D, Favre B et al (2003) Analysis of the interactions between BP180, BP230, plectin and the integrin alpha6beta4 important for hemidesmosome assembly. J Cell Sci 116:387–399
Kowalewski C, Bremer J, Gostynski A et al (2016) Amelioration of junctional epidermolysis bullosa due to exon skipping. Br J Dermatol 174:1375–1379
Krystal Biotech Announces Positive Results from Phase 2 Clinical Trial ("GEM-2 study") of KB103 and Receives Regenerative Medicine Advanced Therapy ("RMAT") Designation from FDA for KB103. Kryst Biotech (2020). Accessed 5 Mar 2020
Kühl T, Mezger M, Hausser I et al (2015) High local concentrations of intradermal MSCs restore skin integrity and facilitate wound healing in dystrophic epidermolysis bullosa. Mol Ther J Am Soc Gene Ther 23:1368–1379
Küttner V, Mack C, Gretzmeier C et al (2014) Loss of collagen VII is associated with reduced transglutaminase 2 abundance and activity. J Invest Dermatol 134:2381–2389
La Placa M, Balestri R, Tartari F et al (2019) Mucous membrane pemphigoid-associated malignancies: case series and a brief overview of the literature. Dermatol Pract Concept 9:119–125
Lamberts A, Rashid H, Pas HH et al (2019) Pemphigoid variants affecting the skin. Clin Exp Dermatol 44:721–727
Lapiere JC, Woodley DT, Parente MG et al (1993) Epitope mapping of type VII collagen. Identification of discrete peptide sequences recognized by sera from patients with acquired epidermolysis bullosa. J Clin Invest 92:1831–1839
Lekszas C, Foresti O, Raote I et al (2020) Biallelic TANGO1 mutations cause a novel syndromal disease due to hampered cellular collagen secretion. eLife 9
Liu C-C, Lin J-H, Hsu T-W et al (2018) Collagen XVII/laminin-5 activates epithelial-to-mesenchymal transition and is associated with poor prognosis in lung cancer. Oncotarget 9:1656–1672
Lokmic Z, Lämmermann T, Sixt M et al (2008) The extracellular matrix of the spleen as a potential organizer of immune cell compartments. Semin Immunol 20:4–13
Ludwig RJ (2019) Type VII collagen IgE autoantibodies in epidermolysis bullosa acquisita: more common than suspected. Br J Dermatol 180:981–983
Lwin SM, Syed F, Di W-L et al (2019) Safety and early efficacy outcomes for lentiviral fibroblast gene therapy in recessive dystrophic epidermolysis bullosa. JCI Insight 4
Malhotra V, Erlmann P (2015) The pathway of collagen secretion. Annu Rev Cell Dev Biol 31:109–124
Marinkovich MP, Tang JY (2019) Gene therapy for epidermolysis bullosa. J Invest Dermatol 139:1221–1226
Matsumura H, Mohri Y, Binh NT et al (2016) Hair follicle aging is driven by transepidermal elimination of stem cells via COL17A1 proteolysis. Science 351:aad4395
McGuire JD, Walker MP, Mousa A et al (2014) Type VII collagen is enriched in the enamel organic matrix associated with the dentin-enamel junction of mature human teeth. Bone 63:29–35
McLean WH, Irvine AD, Hamill KJ et al (2003) An unusual N-terminal deletion of the laminin alpha3a isoform leads to the chronic granulation tissue disorder laryngo-onycho-cutaneous syndrome. Hum Mol Genet 12:2395–2409
Messingham KN, Miller AD, Narayanan NS et al (2019a) Demographics and autoantibody profiles of pemphigoid patients with underlying neurologic diseases. J Invest Dermatol 139:1860–1866.e1
Messingham KN, Crowe TP, Fairley JA (2019b) The intersection of IgE autoantibodies and eosinophilia in the pathogenesis of bullous pemphigoid. Front Immunol 10:2331
Mittapalli VR, Madl J, Löffek S et al (2016) Injury-driven stiffening of the dermis expedites skin carcinoma progression. Cancer Res 76:940–951

Morris NP, Keene DR, Glanville RW et al (1986) The tissue form of type VII collagen is an antiparallel dimer. J Biol Chem 261:5638–5644

Muir AM, Massoudi D, Nguyen N et al (2016) BMP1-like proteinases are essential to the structure and wound healing of skin. Matrix Biol J Int Soc Matrix Biol 56:114–131

Natsuga K, Nishie W, Shinkuma S et al (2010) Circulating IgA and IgE autoantibodies in antilaminin-332 mucous membrane pemphigoid. Br J Dermatol 162:513–517

Natsuga K, Nishie W, Nishimura M et al (2017) Loss of interaction between plectin and type XVII collagen results in epidermolysis bullosa simplex. Hum Mutat 38:1666–1670

Natsuga K, Watanabe M, Nishie W, Shimizu H (2019) Life before and beyond blistering: the role of collagen XVII in epidermal physiology. Exp Dermatol 28:1135–1141

Neptune ER, Frischmeyer PA, Arking DE et al (2003) Dysregulation of TGF-beta activation contributes to pathogenesis in Marfan syndrome. Nat Genet 33:407–411

Nishie W (2014) Update on the pathogenesis of bullous pemphigoid: an autoantibody-mediated blistering disease targeting collagen XVII. J Dermatol Sci 73:179–186

Nishie W (2020) Collagen XVII processing and blistering skin diseases. Acta Derm Venereol 100: adv00054

Nishie W, Kiritsi D, Nyström A et al (2011) Dynamic interactions of epidermal collagen XVII with the extracellular matrix: laminin 332 as a major binding partner. Am J Pathol 179:829–837

Nogueira C, Erlmann P, Villeneuve J et al (2014) SLY1 and Syntaxin 18 specify a distinct pathway for procollagen VII export from the endoplasmic reticulum. eLife 3:e02784

Nyström A (2016) Collagens in wound healing 9. Wound Heal Biomater-Vol 2 Funct Biomater 171

Nyström A, Velati D, Mittapalli VR et al (2013a) Collagen VII plays a dual role in wound healing. J Clin Invest 123:3498–3509

Nyström A, Buttgereit J, Bader M et al (2013b) Rat model for dominant dystrophic epidermolysis bullosa: glycine substitution reduces collagen VII stability and shows gene-dosage effect. PLoS One 8:e64243

Nyström A, Thriene K, Mittapalli V et al (2015) Losartan ameliorates dystrophic epidermolysis bullosa and uncovers new disease mechanisms. EMBO Mol Med 7:1211–1228

Nyström A, Bornert O, Kühl T et al (2018) Impaired lymphoid extracellular matrix impedes antibacterial immunity in epidermolysis bullosa. Proc Natl Acad Sci USA 115:E705–E714

Ockleford CD, McCracken SA, Rimmington LA et al (2013) Type VII collagen associated with the basement membrane of amniotic epithelium forms giant anchoring rivets which penetrate a massive lamina reticularis. Placenta 34:727–737

Odorisio T, Di Salvio M, Orecchia A et al (2014) Monozygotic twins discordant for recessive dystrophic epidermolysis bullosa phenotype highlight the role of TGF-β signalling in modifying disease severity. Hum Mol Genet 23:3907–3922

Palazzi X, Marchal T, Chabanne L et al (2000) Inherited dystrophic epidermolysis bullosa in inbred dogs: a spontaneous animal model for somatic gene therapy. J Invest Dermatol 115:135–137

Papakonstantinou E, Limberg MM, Gehring M et al (2019) Neurological disorders are associated with bullous pemphigoid. J Eur Acad Dermatol Venereol 33:925–929

Pasmooij AM, Pas HH, Deviaene FC et al (2005) Multiple correcting COL17A1 mutations in patients with revertant mosaicism of epidermolysis bullosa. Am J Hum Genet 77:727–740

Pasmooij AM, Pas HH, Jansen GH et al (2007) Localized and generalized forms of blistering in junctional epidermolysis bullosa due to COL17A1 mutations in the Netherlands. Br J Dermatol 156:861–870

Pasmooij AM, Nijenhuis M, Brander R, Jonkman MF (2012) Natural gene therapy may occur in all patients with generalized non-Herlitz junctional epidermolysis bullosa with COL17A1 mutations. J Invest Dermatol 132:1374–1383

Paulus W, Baur I, Liszka U et al (1995) Expression of type VII collagen, the major anchoring fibril component, in normal and neoplastic human nervous system. Virchows Arch Int J Pathol 426:199–202

Pfendner EG, Lucky AW (1993) Junctional epidermolysis bullosa. In: Adam MP, Ardinger HH, Pagon RA et al (eds) GeneReviews®. University of Washington, Seattle

Pickup MW, Mouw JK, Weaver VM (2014) The extracellular matrix modulates the hallmarks of cancer. EMBO Rep 15:1243–1253

Raghunath M, Höpfner B, Aeschlimann D et al (1996) Cross-linking of the dermo-epidermal junction of skin regenerating from keratinocyte autografts. Anchoring fibrils are a target for tissue transglutaminase. J Clin Invest 98:1174–1184

Raote I, Malhotra V (2019) Protein transport by vesicles and tunnels. J Cell Biol 218:737–739

Raote I, Ortega Bellido M, Pirozzi M et al (2017) TANGO1 assembles into rings around COPII coats at ER exit sites. J Cell Biol 216:901–909

Raote I, Ortega-Bellido M, Santos AJ et al (2018) TANGO1 builds a machine for collagen export by recruiting and spatially organizing COPII, tethers and membranes. eLife 7

Rashad R, Weed MC, Quinn N, Chen VM (2019) Extended wear bandage contact lenses decrease pain and preserve vision in patients with epidermolysis bullosa: case series and review of literature. Ocul Immunol Inflamm 28:1–5

Rashidghamat E, Kadiyirire T, Ayis S et al (2019) Phase I/II open-label trial of intravenous allogeneic mesenchymal stromal cell therapy in adults with recessive dystrophic epidermolysis bullosa. J Am Acad Dermatol 83:447

Rattenholl A, Pappano WN, Koch M et al (2002) Proteinases of the bone morphogenetic protein-1 family convert procollagen VII to mature anchoring fibril collagen. J Biol Chem 277:26372–26378

Reimer A, Hess M, Schwieger-Briel A et al (2020) Natural history of growth and anaemia in children with epidermolysis bullosa: a retrospective cohort study. Br J Dermatol 182:1437–1448

Roozendaal R, Mempel TR, Pitcher LA et al (2009) Conduits mediate transport of low-molecular-weight antigen to lymph node follicles. Immunity 30:264–276

Rousselle P, Keene DR, Ruggiero F et al (1997) Laminin 5 binds the NC-1 domain of type VII collagen. J Cell Biol 138:719–728

Ruotsalainen H, Sipilä L, Vapola M et al (2006) Glycosylation catalyzed by lysyl hydroxylase 3 is essential for basement membranes. J Cell Sci 119:625–635

Ryynänen J, Sollberg S, Parente MG et al (1992) Type VII collagen gene expression by cultured human cells and in fetal skin. Abundant mRNA and protein levels in epidermal keratinocytes. J Clin Invest 89:163–168

Saito K, Chen M, Bard F et al (2009) TANGO1 facilitates cargo loading at endoplasmic reticulum exit sites. Cell 136:891–902

Saito K, Yamashiro K, Ichikawa Y et al (2011) cTAGE5 mediates collagen secretion through interaction with TANGO1 at endoplasmic reticulum exit sites. Mol Biol Cell 22:2301–2308

Sakai LY, Keene DR, Morris NP, Burgeson RE (1986) Type VII collagen is a major structural component of anchoring fibrils. J Cell Biol 103:1577–1586

Saleh MA, Ishii K, Kim Y-J et al (2011) Development of NC1 and NC2 domains of type VII collagen ELISA for the diagnosis and analysis of the time course of epidermolysis bullosa acquisita patients. J Dermatol Sci 62:169–175

Saniklidou AH, Tighe PJ, Fairclough LC, Todd I (2018) IgE autoantibodies and their association with the disease activity and phenotype in bullous pemphigoid: a systematic review. Arch Dermatol Res 310:11–28

Santos AJM, Raote I, Scarpa M et al (2015) TANGO1 recruits ERGIC membranes to the endoplasmic reticulum for procollagen export. eLife 4

Schmidt E, Zillikens D (2013) Pemphigoid diseases. Lancet Lond Engl 381:320–332

Schmidt E, Skrobek C, Kromminga A et al (2001) Cicatricial pemphigoid: IgA and IgG autoantibodies target epitopes on both intra- and extracellular domains of bullous pemphigoid antigen 180. Br J Dermatol 145:778–783

Schmidt T, Hoch M, Lotfi Jad SS et al (2017) Serological diagnostics in the detection of IgG autoantibodies against human collagen VII in epidermolysis bullosa acquisita: a multicentre analysis. Br J Dermatol 177:1683–1692

Schumann H, Hammami-Hauasli N, Pulkkinen L et al (1997) Three novel homozygous point mutations and a new polymorphism in the COL17A1 gene: relation to biological and clinical phenotypes of junctional epidermolysis bullosa. Am J Hum Genet 60:1344–1353

Shinkuma S, McMillan JR, Shimizu H (2011) Ultrastructure and molecular pathogenesis of epidermolysis bullosa. Clin Dermatol 29:412–419

Sipilä L, Ruotsalainen H, Sormunen R et al (2007) Secretion and assembly of type IV and VI collagens depend on glycosylation of hydroxylysines. J Biol Chem 282:33381–33388

Siprashvili Z, Nguyen NT, Gorell ES et al (2016) Safety and wound outcomes following genetically corrected autologous epidermal grafts in patients with recessive dystrophic epidermolysis bullosa. JAMA 316:1808–1817

Sixt M, Kanazawa N, Selg M et al (2005) The conduit system transports soluble antigens from the afferent lymph to resident dendritic cells in the T cell area of the lymph node. Immunity 22:19–29

Tamai K, Yamazaki T, Chino T et al (2011) PDGFRalpha-positive cells in bone marrow are mobilized by high mobility group box 1 (HMGB1) to regenerate injured epithelia. Proc Natl Acad Sci USA 108:6609–6614

Tamai K, Yamazaki S, Wang X et al (2017) 179 systemic administration of HMGB1 peptide drastically improves survival of the RDEB model mice by mobilizing multipotent stem/progenitor cells from bone marrow. J Invest Dermatol 137:S223

Tanabe T, Maeda M, Saito K, Katada T (2016) Dual function of cTAGE5 in collagen export from the endoplasmic reticulum. Mol Biol Cell 27:2008–2013

Tanimura S, Tadokoro Y, Inomata K et al (2011) Hair follicle stem cells provide a functional niche for melanocyte stem cells. Cell Stem Cell 8:177–187

Turcan I, Jonkman MF (2015) Blistering disease: insight from the hemidesmosome and other components of the dermal-epidermal junction. Cell Tissue Res 360:545–569

Turczynski S, Titeux M, Tonasso L et al (2016) Targeted exon skipping restores type VII collagen expression and anchoring fibril formation in an in vivo RDEB model. J Invest Dermatol 136:2387–2395

Twaroski K, Eide C, Riddle MJ et al (2019) Revertant mosaic fibroblasts in recessive dystrophic epidermolysis bullosa. Br J Dermatol 181:1247–1253

Ujiie H, Iwata H, Yamagami J et al (2019) Japanese guidelines for the management of pemphigoid (including epidermolysis bullosa acquisita). J Dermatol 46:1102–1135

Umemoto H, Akiyama M, Domon T et al (2012) Type VII collagen deficiency causes defective tooth enamel formation due to poor differentiation of ameloblasts. Am J Pathol 181:1659–1671

Vahidnezhad H, Youssefian L, Saeidian AH et al (2019) Mutations in PLOD3, encoding lysyl hydroxylase 3, cause a complex connective tissue disorder including recessive dystrophic epidermolysis bullosa-like blistering phenotype with abnormal anchoring fibrils and type VII collagen deficiency. Matrix Biol J Int Soc Matrix Biol 81:91–106

Vanden Oever M, Muldoon D, Mathews W et al (2016) miR-29 regulates type VII collagen in recessive dystrophic epidermolysis bullosa. J Invest Dermatol 136:2013–2021

Varki R, Sadowski S, Pfendner E, Uitto J (2006) Epidermolysis bullosa. I. Molecular genetics of the junctional and hemidesmosomal variants. J Med Genet 43:641–652

Varpuluoma O, Försti A-K, Jokelainen J et al (2018) Vildagliptin significantly increases the risk of bullous pemphigoid: a Finnish Nationwide registry study. J Invest Dermatol 138:1659–1661

Virtanen I, Lohi J, Tani T et al (1996) Laminin chains in the basement membranes of human thymus. Histochem J 28:643–650

Vorobyev A, Ujiie H, Recke A et al (2015) Autoantibodies to multiple epitopes on the non-collagenous-1 domain of type VII collagen induce blisters. J Invest Dermatol 135:1565–1573

Vorobyev A, Ludwig RJ, Schmidt E (2017) Clinical features and diagnosis of epidermolysis bullosa acquisita. Expert Rev Clin Immunol 13:157–169

Wagner JE, Ishida-Yamamoto A, McGrath JA et al (2010) Bone marrow transplantation for recessive dystrophic epidermolysis bullosa. N Engl J Med 363:629–639

Wakugawa M, Nakamura K, Hino H et al (2000) Elevated levels of eotaxin and interleukin-5 in blister fluid of bullous pemphigoid: correlation with tissue eosinophilia. Br J Dermatol 143:112–116

Wang X, Ghasri P, Amir M et al (2013) Topical application of recombinant type VII collagen incorporates into the dermal-epidermal junction and promotes wound closure. Mol Ther J Am Soc Gene Ther 21:1335–1344

Watanabe M, Natsuga K, Nishie W et al (2017) Type XVII collagen coordinates proliferation in the interfollicular epidermis. eLife 6:e26635

Watt SA, Dayal JHS, Wright S et al (2015) Lysyl hydroxylase 3 localizes to epidermal basement membrane and is reduced in patients with recessive dystrophic epidermolysis bullosa. PLoS One 10:e0137639

Webber BR, O'Connor KT, McElmurry RT et al (2017) Rapid generation of Col7a1(−/−) mouse model of recessive dystrophic epidermolysis bullosa and partial rescue via immunosuppressive dermal mesenchymal stem cells. Lab Investig J Tech Methods Pathol 97:1218–1224

Wegener H, Leineweber S, Seeger K (2013) The vWFA2 domain of type VII collagen is responsible for collagen binding. Biochem Biophys Res Commun 430:449–453

Widmer C, Gebauer JM, Brunstein E et al (2012) Molecular basis for the action of the collagen-specific chaperone Hsp47/SERPINH1 and its structure-specific client recognition. Proc Natl Acad Sci 109:13243–13247

Wilson DG, Phamluong K, Li L et al (2011) Global defects in collagen secretion in a Mia3/TANGO1 knockout mouse. J Cell Biol 193:935–951

Woodley DT, Briggaman RA, O'Keefe EJ et al (1984) Identification of the skin basement-membrane autoantigen in epidermolysis bullosa acquisita. N Engl J Med 310:1007–1013

Woodley DT, Keene DR, Atha T et al (2004) Injection of recombinant human type VII collagen restores collagen function in dystrophic epidermolysis bullosa. Nat Med 10:693–695

Woodley DT, Cogan J, Hou Y et al (2017) Gentamicin induces functional type VII collagen in recessive dystrophic epidermolysis bullosa patients. J Clin Invest 127:3028–3038

Wright JT (2010) Oral manifestations in the epidermolysis bullosa spectrum. Dermatol Clin 28:159–164

Wullink B, Pas HH, Van der Worp RJ et al (2015) Type VII collagen expression in the human vitreoretinal Interface, corpora Amylacea and inner retinal layers. PLoS One 10:e0145502

Wullink B, Pas HH, Van der Worp RJ et al (2018) Type VII collagen in the human accommodation system: expression in ciliary body, zonules, and lens capsule. Invest Ophthalmol Vis Sci 59:1075–1083

Yu KK, Crew AB, Messingham KAN et al (2014) Omalizumab therapy for bullous pemphigoid. J Am Acad Dermatol 71:468–474

Yuan L, Kenny SJ, Hemmati J et al (2018) TANGO1 and SEC12 are copackaged with procollagen I to facilitate the generation of large COPII carriers. Proc Natl Acad Sci USA 115:E12255–E12264

Yuen WY, Pas HH, Sinke RJ, Jonkman MF (2011) Junctional epidermolysis bullosa of late onset explained by mutations in COL17A1. Br J Dermatol 164:1280–1284

Zimina EP, Bruckner-Tuderman L, Franzke C-W (2005) Shedding of collagen XVII ectodomain depends on plasma membrane microenvironment. J Biol Chem 280:34019–34024

Zimina EP, Fritsch A, Schermer B et al (2007) Extracellular phosphorylation of collagen XVII by ecto-casein kinase 2 inhibits ectodomain shedding. J Biol Chem 282:22737–22746

Chapter 8
Collagens as New Players in Nervous System Diseases

Anne Heikkinen, Michael A. Fox, and Taina Pihlajaniemi

Abstract The binomial nervous system involves the central nervous system (CNS), comprising the brain and the spinal cord, and the peripheral nervous system (PNS). Both divisions of the nervous system contain electrically excitable neurons as well as a number of supporting neuroglial cells, which include oligodendrocytes, astrocytes, microglia, choroid plexus ependymal cells in the CNS, and satellite and Schwann cells in the PNS. Connective tissues rich in fibrillar collagens form the outermost cover for the nervous system proper. Moreover, there are rich basement membranes (BM) surrounding all nervous system tissues and vessels within these structures. BMs compartmentalize nervous tissues and contribute to selective barrier and filtration functions essential for brain homeostasis. While BMs are absent from the brain parenchyma, there are extracellular matrices (ECM) in these regions that remain under-explored. The composition and types of matrices differ substantially in different parts of the nervous system. ECMs are significantly abundant during development, guiding cellular migration and differentiation as well as axon navigation and synaptogenesis. Additionally, ECMs promote neuronal health, contribute to synaptic homeostasis and plasticity, and are upregulated in response to disease and

A. Heikkinen
Faculty of Biochemistry and Molecular Medicine, Center for Cell-Matrix Research, University of Oulu, Oulu, Finland

Biocenter Oulu, University of Oulu, Oulu, Finland
e-mail: anne.heikkinen@oulu.fi

M. A. Fox
School of Neuroscience, Virginia Tech, Blacksburg, VA, USA

Fralin Biomedical Research Institute, Virginia Tech Carilion, Roanoke, VA, USA

Department of Biological Sciences, Virginia Tech, Blacksburg, VA, USA

Department of Pediatrics, Virginia Tech Carilion School of Medicine, Roanoke, VA, USA
e-mail: mafox1@vt.edu

T. Pihlajaniemi (✉)
Faculty of Biochemistry and Molecular Medicine, Center for Cell-Matrix Research, University of Oulu, Oulu, Finland
e-mail: taina.pihlajaniemi@oulu.fi

F. Ruggiero (ed.), *The Collagen Superfamily and Collagenopathies*, Biology of Extracellular Matrix 8, https://doi.org/10.1007/978-3-030-67592-9_8

trauma. It is for these reasons that the mutation and malfunction of collagens have been linked to neurodevelopmental, degenerative, and psychiatric disorders as well as motor and sensory dysfunction. In recent years, it has become clear that collagens, constituting a major family of ECM proteins, and other extracellular components are generated not only by glial cells but also by neurons. The functions and expression patterns of the collagen superfamily members in the nervous system are summarized in this chapter, where we focus on their roles in vitro and in vivo in a number of animal models, and in human diseases of the nervous system.

Abbreviations

Aβ	Amyloid beta
AChR	Acetylcholine receptor
AD	Alzheimer's disease
ALS	Amyotrophic lateral sclerosis
APP	Amyloid precursor protein
BBB	Blood–brain barrier
BM	Basement membrane
BP	Bullous pemphigoid
COL	Collagenous domain
CLAC	Collagen-like Alzheimer amyloid plaque component
CLAC-P	CLAC precursor
CNS	Central nervous system
CMS	Congenital myasthenic syndrome
DDR	Discoidin domain receptor
DRG	Dorsal root ganglion
ECM	Extracellular matrix
FACIT	Fibril-associated collagens with interrupted triple helices
MACIT	Membrane-associated collagens with interrupted triple helices
Multiplexin	Multiple triple-helix domains with interruptions
NC	Non-collagenous domain
NMJ	Neuromuscular junction
PNN	Perineuronal net
PNS	Peripheral nervous system
PTP	Protein tyrosine phosphatase
ROS	Reactive oxygen species
SVCT2	Sodium-dependent vitamin C transporter 2

8.1 Extracellular Matrix of the Nervous System

Animals sense changes in their environment through a variety of sensory organs or receptor cells, which then send signals into the brain through afferent sensory neurons of either the PNS or CNS. For example, in the case of the somatosensory and pain systems, the somas of these first-order neurons lie outside the CNS in the dorsal root ganglia (DRG), where they are embedded and shielded by peripheral glial satellite cells and extracellular matrices (ECM). In the case of the visual system, however, the detection of light photons occurs in the retina, a portion of the CNS that resides in the back of the eye. Signals from the retina are sent to the brain along the optic nerve, which is also a component of the CNS. Regardless of whether sensory signals propagate along components of the PNS or CNS, once they reach the brain they are processed by a complex network of central excitatory and inhibitory interneurons. The neurons of the brain are supported by astrocytes and ensheathed by oligodendrocytes. The astrocytes further function in limiting other structures such as the blood–brain barrier (BBB) and meninges, and they also guide neuronal migration. The astrocytes also contribute to neurotransmission by maintaining the homeostasis of the brain's extracellular compounds. In response to internal and external stimuli, central neurons conduct decisions to command efferent neurons of the PNS. Efferent axons of motor nerves run from the spinal cord nuclei, where their somas reside, to innervate their peripheral targets, muscles and glands. In the peripheral nerve, Schwann cells generate support and myelin around axons to promote neural actions. In this chapter, we go through various ECMs of the PNS and CNS with an established link with collagens. Despite collagens being relatively broadly expressed in the retina, retinas are not discussed here as we chose to mainly focus on the brain.

8.1.1 ECMs of the Central Nervous System

In adults, the CNS is encapsulated by the BM, a specialized ECM chiefly built of laminin and collagen IV networks interconnected by nidogens and perlecan, providing comprehensive boundaries to tissue structures. The BM is further covered by ECM that is especially rich in fibrillar collagens, and the two are interconnected by a complexity of molecules (Hubert et al. 2009; Gordon and Hahn 2010; Ricard-Blum 2011). Together with cellular components, these ECMs form the outermost, protecting layer of the CNS, the meninges (Fig. 8.1a). All three layers of the meninges, the pia mater, arachnoid mater, and dura mater, derive from the meningeal mesenchyme and are of neural crest origin. Meningeal cells produce many ECM constituents, including BM components and interstitial fibrillar collagens, and together these extracellular cues promote the development of radial glial cells, a key component of the developing brain that plays critical and necessary roles in directing neuronal migration (Sievers et al. 1994; Hartmann et al. 1998). At the edge

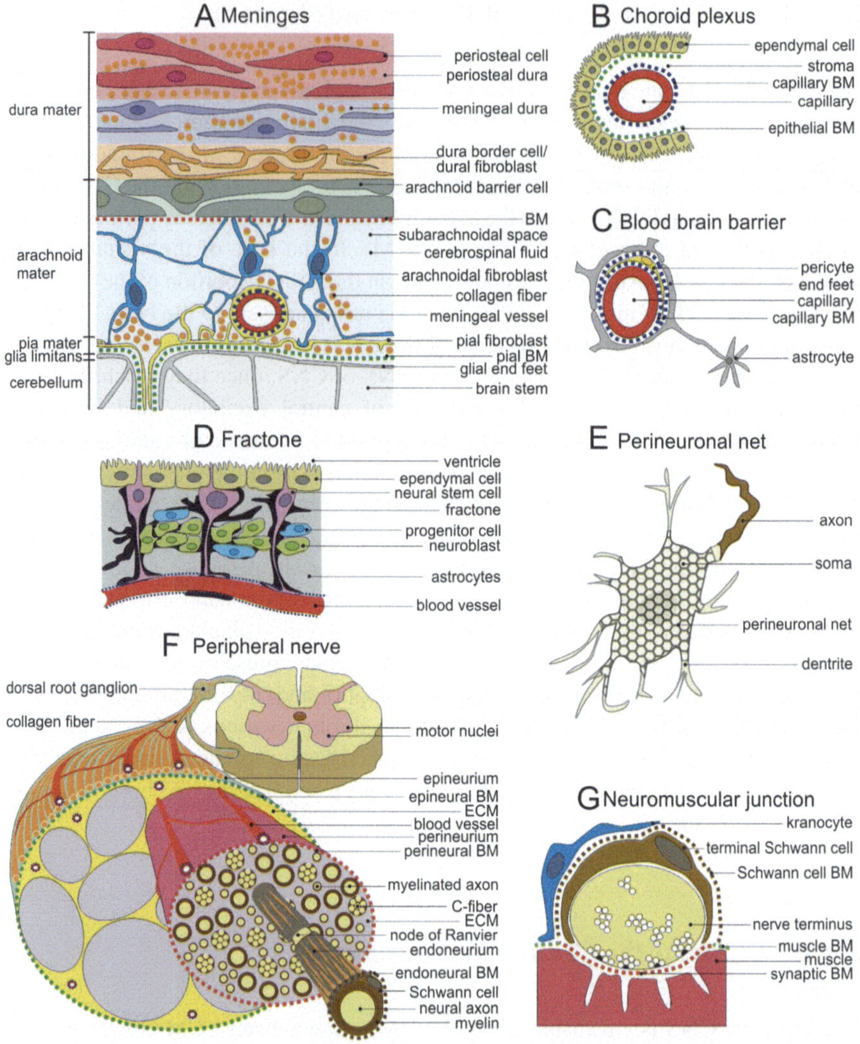

Fig. 8.1 Extracellular matrices of the nervous system containing collagens. (**a–e**) Structures in the CNS; meninges (**a**), choroid plexus (**b**), blood brain barrier (**c**), fractone (**d**) and perineuronal net (**e**). (**f–g**) Multicellular assemblies in the PNS; peripheral nerve (**f**), and neuromuscular junction (**g**)

of brain tissues, astrocyte end-feet produce components that contribute to the pial BM and establish the cortical glial limitans (Sievers et al. 1994; Heck et al. 2003). The pia mater separates the brain from the cerebrospinal fluid circulating within the subarachnoid space. Choroid plexuses filter and secrete the cerebrospinal fluid that both protects and nourishes the brain and the spinal cord. The BMs of both ependymal epithelia and capillaries participate in this filtration process from blood to ventricles (Fig. 8.1b) and compromise in this function may result in hydrocephalus (Utriainen et al. 2004). In addition, the BM plays a relevant role in the CNS blood vessels where it contributes to the BBB, a selective semipermeable segregator between blood and brain. Endothelial cells, pericytes, and astrocyte foot projections, together with the ECM produced by these cell types, all constitute the four layers of the BBB (Fig. 8.1c), which is critical for brain homeostasis. Different from peripheral pericytes, which originate from the mesoderm, BBB pericytes are of neural crest origin. Compromises in the barrier function may result in hemorrhage as well as a wide range of other neurological disorders (Sharif et al. 2018; Xu et al. 2018).

The CNS parenchyma has low amounts of BMs and fibrillar collagens and therefore it relies on the bony skull and vertebra for structural support. Long ago, electron microscopy on brain tissue, which requires highly fixed and dehydrated samples before embedding and sectioning, gave the impression that there is very little extracellular space in the brain. However, this may be an artifact of these imaging techniques (Korogod et al. 2015), and increasing evidence indicates that the brain is in fact richer in ECMs than previously thought. Specialized brain ECMs exist in several distinct forms and several precise locations. Unlike outside of the nervous system, where BMs are largely composed of collagens, laminins, nidogens, and proteoglycans, the primary ECM constituents in the brain include proteoglycans and glycosaminoglycans (Novak and Kaye 2000; Zimmermann and Dours-Zimmermann 2008; Krishnaswamy et al. 2019). One specialized brain ECM rich in these components is the fractone (Fig. 8.1d), which is associated with neural stem cell niches adjacent to the walls of the lateral ventricles (Kerever et al. 2007). A second, specialized brain ECM is the perineuronal net (PNN), a robust structure that ensheaths the soma and proximal processes of selective interneurons dendrites and the initial segment of axons (Fig. 8.1e). Not only do PNNs provide support and promote neuronal health, they contribute to synaptic plasticity as they emerge at the end of the critical period of brain development. In addition to these well-defined ECMs, it is becoming abundantly clear that astroglial processes that contact pre- and postsynaptic elements (and therefore contribute to the tripartite synapse) generate and deposit ECM into and around the synaptic cleft, which is critical for synapse formation and function (Rohrbough et al. 2007; Ferrer-Ferrer and Dityatev 2018; Song and Dityatev 2018). In fact, the role of the ECM is so important at brain synapses (like those of the neuromuscular junction in the PNS) that it may be better to re-term the central synapse as a tetrapartite synapse. In the developing brain, the extracellular space is even more substantial and during this period the brain ECM plays important roles in axon guidance, synaptogenesis, neurotransmission, and plasticity. Reduced brain ECM is connected with complex neurological diseases, but on the other hand, accumulation of ECM aggregates also promotes cellular dysfunction and neurodegeneration (Frischknecht and Gundelfinger 2012; Mouw et al. 2014). As we learn more about the roles of ECM in brain development, function, and disease, there is growing evidence that non-fibrillar collagens, such as collagens VI, XVII, XVIII, XIX, and XXV (Hashimoto et al. 2002; Seppänen et al. 2006; Su et al. 2010; Cescon et al. 2016) are not only generated by neural cells but also that brain development and function requires them and/or their proteolytically released, functionally active fragments, collectively termed as matricryptins (Fox 2008; Su et al. 2010, 2012, 2016, 2017; Cescon et al. 2016).

8.1.2 ECMs of the Peripheral Nervous System

While the ECM is relatively sparse in the CNS proper, the PNS is rich in ECMs. In a peripheral nerve, all motor axons, as well as a portion of sensory and autonomic axons, are individually encapsulated by myelinating Schwann cells. Small caliber, unmyelinated sensory axons, called C-fibers, are enwrapped by non-myelinating Schwann cells. The BM surrounding axons, and its adjacent ECM containing thin collagen fibrils (Osawa and Ide 1986), are both produced by the Schwann cells and constitute together the endoneurium which provides insulation. Axons, vessels, supporting cells, and the interstitial ECM together form a nerve fascicle that is covered by a tight perineurium, which is produced by perineural fibroblasts and serves as a diffusion barrier. Several fascicles constitute the peripheral nerve, covered by epineurial BM which is further ensheathed by ECM that is rich in fibrillar collagens, like the CNS (Fig. 8.1f). The epineurium constructs the nerve trunk and provides mechanical strength. The epineurial collagen fibrils are as thick as those in the dermis of the skin, while the endoneurial collagen fibrils generated by the Schwann cells are thin (Osawa and Ide 1986). Although collagen IV is a typical and essential component of endo-, peri-, and epineural BMs, several other, nontypical collagens associate with the peripheral nerve BMs, and these are discussed here. Schwann cells, for their part, express critical receptors on their cell membrane, through which collagen-rich/dependent ECMs transmit signals to affect cell behavior such as adhesion and migration (Chernousov et al. 2008). As described above for the brain, during PNS development the ECM plays essential roles in neural migration, axon outgrowth, and guidance, and in the formation of peripheral synapses, such as the neuromuscular junction (NMJ) (Fox et al. 2007). In the PNS, the ECM surrounding nerve fibers is so robust that it is preserved following nerve injury (such as nerve crushing or cutting), and in these cases, the preserved endoneurium and perineurium serve as a hollow tube, studded with extrinsic guidance factors, that not only provide a path for regenerating axons to regrow, but also facilitate regrowth (Sanes and Lichtman 1999; Nguyen et al. 2002).

At innervation, the neural axon forms synapses with the muscle to form the NMJ (Fig. 8.1g). Non-myelinating terminal Schwann cells enshield the axon terminus and are themselves encapsulated by BM. The postnatal motor synapse is further covered by mesenchymal cells, kranocytes, which are suggested to contribute to synapse regeneration (Court et al. 2008). The BM that passes through the synapse, between the nerve and muscle, is unique in its ECM composition and contains specific synaptic isoforms different from the muscle BM counterparts (Fig. 8.1g). The synaptic BM is primarily and prior to innervation secreted by the developing postsynaptic muscle, but later also other cellular parties of the NMJ contribute to the synaptic BM. These molecules maintain the homeostasis and enable plasticity by both retro- and anterograde signaling (Sanes and Lichtman 1999, 2001; Kummer et al. 2006; Heikkinen et al. 2020). Moreover, the ECMs embed "hidden" potential in regulating cell behavior; for example, many ECM components are capable of releasing matricryptins which, once liberated from the ECM, can contribute to distinct bio-activities such as acting as guidance or synaptogenic cues (Ackley et al. 2001; Fox et al. 2007; Meyer and Moussian 2009; Ricard-Blum and Ballut 2011).

8.2 Distribution of Collagen Subtypes in the Nervous System

The collagen superfamily comprises 28 distinct collagen types that are divided into subfamilies according to the three-dimensional assemblies they form. Due to their trimeric characteristics, each distinct collagen type can be encoded by a single gene or by multiple genes. The distribution of collagens in the CNS and PNS is summarized in Tables 8.1 and 8.2, respectively.

8.2.1 Fibrillar Collagens and Other Fibril-Forming Collagens

Fibrillar collagens, comprised of collagen types I, II, III, V, XI, XXIV, and XXVII, form the best-known collagen subfamily as they are assembled into structural fibers that densely populate connective tissues throughout the body. Fibrillar collagens form heterotypic fibers where collagens III/V and XI preferentially incorporate collagen I and II fibers, respectively. The former collagen types nucleate the fibers, but they also limit the fiber diameter by occasional N-peptides retained. Through the latter property, fibers achieve unique features that are important for tissue-specific functions (Myllyharju and Kivirikko 2004; Gordon and Hahn 2010; Ricard-Blum 2011). As mentioned, fibrillar collagens form the outermost layer of both the CNS and PNS. Pia Mater is rich in collagens I and III while collagen V colocalizes with collagen IV at the pial BM (Sievers et al. 1994), and is expressed already in the neuroepithelium (Roulet et al. 2007). Collagen V is present in the ECM of small CNS capillaries, but collagen I/III fibrils are absent (Maxwell et al. 1984; Munji et al. 2019). Collagen II, typically found in cartilaginous tissues, is transiently expressed in the developing choroid plexus and meninges, temporally coinciding with active tissue remodeling (Cheah et al. 1991; Andrikopoulos et al. 1992; Sandberg et al. 1993; Lui et al. 1995a; Yoshioka et al. 1995). Collagen XI coexists with collagen II and is similarly transiently present in developing meninges and the brain cortex (Nah et al. 1992; Sandberg et al. 1993; Lui et al. 1995b; Yoshioka et al. 1995). Collagens XXIV and XXVII are either absent or present at very low levels in the brain

Table 8.1 Expression of collagens in the CNS

Subfamily	Type	Expression	References
Fibrillar	I/III	Meninges	Sievers et al. (1994)
	II/XI	Developing meninges	Cheah et al. (1991), Andrikopoulos et al. (1992), Nah et al. (1992), Sandberg et al. (1993), Lui et al. (1995a, b), Yoshioka et al. (1995)
	V	Neuroepithelium, pial BM, small capillaries	Maxwell et al. (1984), Sievers et al. (1994), Roulet et al. (2007), Munji et al. (2019)
FACIT	IX	With collagen II in developing meninges	Ring et al. (1995)
	XII	With collagen I in meninges	Oh et al. (1993)
	XVI	Hippocampal neurons	Hubert et al. (2009)
	XIX	PNN-rich telencephalic interneurons	Sumiyoshi et al. (2001), Su et al. (2016)
Network-forming	IV $[\alpha 1]_2\alpha 2$	BMs of pia, choroid plexus, BBB, and other vessels	Shellswell et al. (1979), Roggendorf et al. (1988), Kleppel et al. (1989), Sievers et al. (1994), Halfter et al. (1998)
	IV $\alpha 3\alpha 4\alpha 5$	BMs of choroid plexus	Urabe et al. (2002)
	IV $[\alpha 5]_2\alpha 6$	Pial BM	Urabe et al. (2002)
	VI $\alpha 1\alpha 2\alpha 3$	Pial BM, vessel adventitia, choroid plexus, hippocampal neurons, and corpus callosum	Roggendorf et al. (1988), Kamei et al. (1992), Dziadek et al. (1996), Cescon et al. (2016)
	VIII	Meninges, spinal cord white matter	Kapoor et al. (1988), Muragaki et al. (1992)
Multiplexin	XV	Developing CNS capillaries, vessels of meninges, and choroid plexus	Muona et al. (2002)
	XVIII	BMs of pia, choroid plexus and most brain vessels, Purkinje neurons	Halfter et al. (1998), Miosge et al. (2003), Utriainen et al. (2004), Su et al. (2012), Caglayan et al. (2014)
Transmembrane	XIII	Developing CNS neurons, meninges, brain vessels	Sandberg-Lall et al. (2000), Sund et al. (2001a)
	XXIII	Developing dura mater, neurons of olfactory bulb	Koch et al. (2006), Monavarfeshani et al. (2017)
	XXV (CLAC-P)	Developing spinal cord, CNS neurons, retina	Hashimoto et al. (2002), Kay et al. (2011), Tanaka et al. (2014), Monavarfeshani et al. (2017)
	XVII	Cortical, hippocampal, and amygdaloid neurons	Seppänen et al. (2006, 2007, 2009)
Others	VII	Choroid plexus, pineal and pituitary glands	Paulus et al. (1995)

Table 8.2 Expression of collagens in the peripheral nervous system

Subfamily	Type	Expression	References
Fibrillar	I	Endo-, peri-, and epineurium	Shellswell et al. (1979), Osawa and Ide (1986), Wälchli et al. (1994)
	II	Schwann cells, nodes of Ranvier	D'Antonio et al. (2006)
	III	Endo- and perineurium	Shellswell et al. (1979), Osawa and Ide (1986), Wälchli et al. (1994)
	V	Endoneurial Schwann cell BM	Shellswell et al. (1979), Chernousov et al. (2006)
FACIT	XII	Developing endo- and perineurium	Oh et al. (1993)
	XIV	With collagen I in peri- and epineurium	Wälchli et al. (1994)
	XVI	ECM and neurons of DRG	Lai and Chu (1996), Hubert et al. (2007)
	XIX	Maturing muscle and gastro-esophageal junction smooth muscle	Myers et al. (1997), Sumiyoshi et al. (2001, 2004)
Network-forming	IV $[\alpha1]_2\alpha2$	Endo- and perineurium, DRG capsule BM, all muscle BMs	Shellswell et al. (1979), Halfter et al. (1998), Miner and Sanes (1994), Sund et al. (2001a)
	IV $\alpha3\alpha4\alpha5$	Postnatal synaptic BM at the NMJ	Miner and Sanes (1994), Fox et al. (2007)
	IV $[\alpha5]_2\alpha6$	Postnatal synaptic BM at the NMJ	Miner and Sanes (1994), Fox et al. (2007)
	VI $\alpha1\alpha2\alpha3$	Endo-, peri-, and epineural BM, myelinating Schwann cells, neural resident macrophages, synaptic BM at the NMJ	Keene et al. (1988), Peltonen et al. (1990), Allen et al. (2009), Chen et al. (2014), Cescon et al. (2018)
Multiplexin	XV	Developing endo- and perineural BM zone, Schwann cell BM at the NMJ, ECM of developing DRG	Muona et al. (2002)
	XVIII	BMs of DRG, endoneurium and muscle, Schwann cell BM at the NMJ	Saarela et al. (1998), Miosge et al. (2003), Su et al. (2012)
Transmembrane	XIII	Developing neurons, neural vessels, postsynaptic muscle, and synaptic BM at the NMJ	Hägg et al. (2001), Latvanlehto et al. (2010), Logan et al. (2015), Härönen et al. (2017), Zainul et al. (2018), Härönen et al. (2019)
	XXV (CLAC-P)	Developing muscle, spinal motor, and DRG neurons	Tanaka et al. (2014), Goncalves et al. (2019), Munezane et al. (2019)
Others	XXVIII	Developing non-myelinating Schwann cells in DRG, nerve and NMJ, nodes of Ranvier	Veit et al. (2006), Grimal et al. (2010)

(Boot-Handford et al. 2003; Matsuo et al. 2008). Although collagen VII is definitively not a fibrillar collagen, it forms anchoring fibers that are especially critical in anchoring of the epidermis to the dermis in the skin (see Chap. 7 for further details). In the brain, collagen VII underlies choroid plexus epithelia and surrounds pineal gland and pituitary gland cell nests (Paulus et al. 1995).

In the PNS, collagen I is generally abundant in endo-, peri-, and epineurium, while collagens III and V are rather enriched in endo- and perineurium (Shellswell et al. 1979; Osawa and Ide 1986; Wälchli et al. 1994). Different from collagen III, in that it exclusively encompasses delicate endoneurial collagen fibrils, collagen V is additionally present at the Schwann cell BM (Shellswell et al. 1979; Chernousov et al. 2006). The incorporation of collagen V limits the diameter of collagen fibers (Birk et al. 1990) and speculatively may therefore motivate the presence of small-caliber fibers in the endoneurium (Osawa and Ide 1986). Collagen II is expressed during development by Schwann cell precursors, later by both non-myelinating and myelinating Schwann cells, and in adulthood, it accumulates at the nodes of Ranvier (D'Antonio et al. 2006). The expression of collagen II is regulated by nerve-derived signals (D'Antonio et al. 2006). Since deficiency of major fibril-forming collagens results in devastating connective tissue diseases, such as osteogenesis imperfecta and Ehlers–Danlos syndromes (see Chaps. 2 and 3), assessing their roles specifically in the nervous system is challenging.

8.2.2 FACIT Collagens

As the name implies, FACIT collagens (fibril-associated collagens with interrupted triple helices) associate with fibrils with a preference for a major collagen type, I or II. FACITs contribute to fibrillogenesis by limiting the fibril diameter and participate in forming supramolecular aggregates within the ECM. Collagen types IX, XII, XIV, XVI, XIX, XX, XXI, and XXII constitute the FACIT subfamily (Wälchli et al. 1994; Gordon and Hahn 2010). The abundance of some of these FACITs varies within the nervous tissue, with some being present broadly in the nervous system and some being restricted to very localized brain regions.

Collagen IX is bound to collagen II fibrils and in developing chicken it is found to be present in the meninges (Ring et al. 1995). Collagen XII stabilizes collagen I fibers and by antibody staining on mouse tissues, it is defined in the dense connective tissue of meninges, but also in the endoneurium and perineurium, especially during embryonic development (Oh et al. 1993). Collagen XIV limits lateral fibril fusion, and in situ hybridizations and immunofluorescence staining of chicken embryos show that collagen XIV colocalizes with collagen I in the epineurium and perineurium (Wälchli et al. 1994). Collagen XVI modulates ECM aggregates and is present both in the CNS and PNS. In situ hybridizations reveal adult mouse hippocampus positive for *Col16a1* signal and in cultured hippocampal neurons, collagen XVI locates in the soma, along neurites and at the axonal growth cone (Hubert et al. 2009), resembling findings on DRG neurons in the PNS (Hubert et al. 2007). Studies in mice indicate that collagen XVI is an ECM component of the DRG where it surrounds neuronal cell bodies (Lai and Chu 1996). Its expression peaks at around birth and reverts again if exposed to nerve injury (Lai and Chu 1996; Hubert et al. 2007). Increasing *Col16a1* gene expression in the DRG in the latter half of gestation suggests that it may participate in the final steps of DRG maturation. In culture, besides the soma, axotomized DRG neurons, preconditioned for regenerative growth, exhibit collagen XVI along neurites but also at their tips, representing the growth cone. Besides neurons, collagen XVI is highly induced in Schwann/fibroblastic cells growing out from axotomized DRG explants (Hubert et al. 2007). A combination of neuronal and glial cell deposition of collagen XVI in vivo is thus highly probable.

Based on its primary structure, collagen XIX resembles FACIT collagens (Khaleduzzaman et al. 1997) but immunohistochemistry suggests it associates in PNS with BMs (Myers et al. 1997; Sumiyoshi et al. 2001, 2004). Collagen XIX expression peaks transiently during embryonic muscle development, following myogenic regulatory factor Myf-5 expression and declines as myogenin emerges in differentiating skeletal muscle progenitors (Sumiyoshi et al. 2001). Studies in zebrafish have revealed its expression at intermediate targets of navigating motor axons (Beattie et al. 2000; Hilario et al. 2010). Moreover, collagen XIX is expressed by developing and postnatal smooth muscle cells of the gastroesophageal junction prior to their transdifferentiation into skeletal muscle cells (Sumiyoshi et al. 2001). In the brain, collagen XIX expression increases postnatally, peaks during periods of synaptogenesis, and in the adult brain it is largely confined to telencephalic interneurons (Sumiyoshi et al. 2001; Su et al. 2010, 2016).

8.2.3 Network-Forming Collagens

Collagens IV, VI, VIII, and X form networks and sheath-like structures. Collagen X acts primarily outside the nervous system. With the aid of amino-terminal 7S domain and carboxyterminal NC1 (non-collagenous) domain, collagen IV self-assembles into a network to constitute BMs together with laminins, nidogens, and perlecan (Xu et al. 2018; Gatseva et al. 2019). Collagen IV presents in three heterotrimeric isoforms; $[\alpha 1(IV)]_2\alpha 2(IV)$, $\alpha 3(IV)\alpha 4(IV)\alpha 5(IV)$, and $[\alpha 5(IV)]_2\alpha 6(IV)$, giving rise to chain-dependent matricryptins. The classical $[\alpha 1]_2\alpha 2(IV)$ isoform is broadly expressed and it is the main isoform in BMs of both CNS and PNS (Miner and Sanes 1994; Khoshnoodi et al. 2008). Due to its abundance, a broad clinical spectrum of human diseases arises from *COL4A1* and *COL4A2* mutations (see also Chap. 5). Various immunohistochemical and transcript-level studies on human and animal tissues have shown that $[\alpha 1(IV)]_2\alpha 2(IV)$ colonizes all CNS BMs. Collagen IV appears in the developing BMs very early on in embryonic development (Dziadek and Timpl 1985). At mid-gestation in mice, collagen IV locates beneath the neuroepithelium (Sund et al. 2001a). In the adult brain, it shields the glia limitans at pial BM, ependymal, and vascular BMs of the choroid plexus and

BMs of meningeal and intraparenchymal vessels, as well as gray and white matter BBB capillaries (Shellswell et al. 1979; Roggendorf et al. 1988; Halfter et al. 1998; Kleppel et al. 1989; Sievers et al. 1994). Together with some other BM proteins, $[\alpha 1(IV)]_2\alpha 2(IV)$ is a component of fractones (Kerever et al. 2007). Collagen IV is synthesized at the glia limitans by both meningeal cells and astrocytes (Urabe et al. 2002), and by both endothelial cells and pericytes at the BBB (Jeanne et al. 2015). Besides the major $[\alpha 1(IV)]_2\alpha 2(IV)$ isotype, α3(IV)α4(IV)α5(IV) is found at the choroid plexus and $[\alpha 5(IV)]_2\alpha 6(IV)$ at the glia limitans (Urabe et al. 2002), reflecting context-dependent functions of these collagens IV.

In the peripheral nerve, $[\alpha 1(IV)]_2\alpha 2(IV)$ heterotrimers inhabit the endoneurial and perineural BMs, and collagen IV also occurs in the capsule surrounding the DRG (Shellswell et al. 1979; Miner and Sanes 1994; Halfter et al. 1998; Sund et al. 2001a). Schwann cells are capable of synthesizing and secreting collagen IV (Carey et al. 1983). While $[\alpha 1(IV)]_2\alpha 2(IV)$ heterotrimers inhabit all muscle BMs, α3-α6 chains are present exclusively in the synaptic BM at the NMJ where they appear in the third postnatal week in mice (Miner and Sanes 1994; Fox et al. 2007).

Although six distinct collagen VI alpha chains exist in mammals, typically collagen VI molecules are composed of α1, α2, and α3 chains, with the secretion of collagen VI molecules being regulated by the latter (Lamande et al. 1998). Collagen VI molecules form antiparallel dimers and further tetramers, and in the ECM they are deposited as end-to-end beaded microfilaments (Gregorio et al. 2018). Collagen VI was originally reported to be present in the brain, both in the BMs and nervous tissue proper (Roggendorf et al. 1988). Collagen VI expression is defined within the superficial glia and the sheath of cranial nerves, the adventitia of meningeal and larger intraparenchymal vessels and choroid plexus, starting at the embryonic stage and persisting into adulthood (Roggendorf et al. 1988; Kamei et al. 1992; Dziadek et al. 1996; Cescon et al. 2016). Moreover, nerve cell-related transcripts for α1-α3 chains are identified in the brain (Karkheiran et al. 2013; Zech et al. 2015; Cescon et al. 2016) and collagen VI protein is detected in the hippocampus and corpus callosum (Cescon et al. 2016). Cultured cortical and hippocampal neurons produce collagen VI, which retains primarily intracellular (Cescon et al. 2016). Transcripts for α4-α6 chains are also present in both developing and adult mouse brain (Gara et al. 2008), but their distribution remains to be resolved.

In the adult peripheral nerve, collagen VI inhabits the BMs at the endo-, peri-, and epineurium (Keene et al. 1988; Peltonen et al. 1990; Allen et al. 2009; Chen et al. 2014) where it is deposited by Schwann cells, perineural cells, and fibroblasts (Peltonen et al. 1990; Muona et al. 1993; Chen et al. 2014). During development, collagen VI expression is induced in Schwann cells by nerve-derived neuregulin at the time of differentiation toward a myelinating phenotype and is downregulated in mature myelinating Schwann cells (Vitale et al. 2001). Furthermore, resident macrophages in the peripheral nerve are capable of producing collagen VI (Chen et al. 2014), similar to cultured human macrophages (Schnoor et al. 2008). Moreover, collagen VI inhabits the synaptic BM at the NMJ (Cescon et al. 2018). Primarily *COL6A1/COL6A2/COL6A3* mutations affect the muscle, causing severe Ullrich congenital muscular dystrophy and milder Bethlem myopathy. Mouse models have allowed to segregate disease phenotypes and helped in defining a wide range of collagen VI functions (see also Chap. 6). Collagen VI in the nervous system was thoroughly reviewed recently (Gregorio et al. 2018).

Structurally, collagen XXVIII, the latest member of the collagen superfamily, resembles collagen VI, although it does not form networks. Its expression is highly restricted to special structures in the PNS. Collagen XXVIII surrounds non-myelinating glial cells of the PNS; satellite cells of the DRG, non-myelinating Schwann cells of peripheral nerves, and terminal Schwann cells of sensory end organs and NMJs, but also nodes of Ranvier at myelinated axons (Veit et al. 2006; Grimal et al. 2010).

Yet another network-forming collagen, collagen VIII, is expressed by several ectoderm-derived ECMs, although its overall distribution is limited. In the CNS it appears in fibrillar arrays at the meninges surrounding the brain, spinal cord, and optic nerve, and in the white matter of the spinal cord (Kapoor et al. 1988; Muragaki et al. 1992). Proteolytic processing of collagen VIII gives rise to matricryptin vastatin (Ricard-Blum and Ballut 2011).

8.2.4 Multiplexin Collagens

Multiplexin (multiple triple-helix domains with interruptions) collagens XV and XVIII share structural similarities, owing to their common evolutional origin. Collagen XVIII presents in three amino terminally different isoforms; the short, medium, and long, which share common collagenous portions and a carboxyterminal endostatin domain. This proteolytically releasable matricryptin conveys distinct biological activities, and a corresponding domain also exists in collagen XV. Collagen XVIII is a heparan sulfate proteoglycan, while collagen XV bears both heparan and chondroitin sulfate side chains (Bretaud et al. 2020). Collagen XVIII forms are differentially deposited in various BMs where they orient in a polarized manner, while collagen XV bridges the rear face of BM with surroundings (Seppinen and Pihlajaniemi 2011; Heljasvaara et al. 2017; Gatseva et al. 2019). Collagen XV exists both in the CNS and PNS. It is transiently expressed in the developing CNS capillaries, while in adults collagen XV is downregulated in BBB vessels but confined primarily to meningeal and choroid plexus blood vessels (Muona et al. 2002). In the PNS in mice, collagen XV appears in the skeletal muscle upon primary myotube formation. It is deposited to the developing BM, delineating individual muscle fibers and persisting into adulthood. Moreover, collagen XV is enriched at the NMJ, but not in the synaptic BM but instead in the terminal Schwann cell-capping BM. Collagen XV is a component of the developing endoneurial and perineural BM zones and it is also transiently expressed in the ECM of developing DRG (Muona et al. 2002). In vivo mouse studies emphasize the significance of collagen XV in Schwann cell differentiation.

Early studies with chicken embryos show that collagen XVIII delineates the pial BM, DRG, and peripheral nerve, and it is also detected in post-hatch chicken peripheral nerves and muscle fibers (Halfter et al. 1998). Further immunolabelling studies on mouse and human tissues confined collagen XVIII at the developing and adult BMs of pia (Miosge et al. 2003; Utriainen et al. 2004; Caglayan et al. 2014). Besides pial BM, collagen XVIII is present at the BMs of most brain blood vessels and those associated with the choroid plexus (Miosge et al. 2003; Utriainen et al. 2004; Caglayan et al. 2014). In the brain parenchyma, all three *Col18a1* transcripts are detected exclusively in the cerebellum, expressed by Purkinje neurons and temporally coinciding with synaptogenesis. The presence of endostatin in extracted synaptosomes suggests its involvement in synaptogenesis (Su et al. 2012). The polarized orientation of collagen XVIII in BMs was originally identified in the Bruch's membrane beneath the retinal pigment epithelium (Marneros et al. 2004). In the developing mouse PNS, the ganglial BM and endoneurium stain positively for collagen XVIII (Miosge et al. 2003). Similar to collagen XV, collagen XVIII is a component of the terminal Schwann cell BM at the NMJ (Su et al. 2012) and it also inhabits extrasynaptic muscle BM (Saarela et al. 1998).

8.2.5 Transmembrane Collagens

Types XIII, XVII, XXIII, and XXV are type II transmembrane collagens, composed largely of collagenous sequences. All four membrane-spanning collagens are shed by proteases to release pericellular ectodomains (Franzke et al. 2005; Heikkinen et al. 2012). Collagen XVII is unique in primary structure while collagens XIII, XXIII, and XXV share structural similarities due to their evolution through gene duplication of a common ancestor and are collectively called membrane-associated collagens with interrupted triple helices (MACIT) (Tu et al. 2015).

An early in situ hybridization study did not detect *COL13A1* transcripts in the fetal human brain (Sandberg et al. 1989). Immunostaining on mouse embryos, however, suggests a transient expression of collagen XIII during mouse development in neurons of the CNS as well as in the PNS. In the late embryonic period meninges also become positive for collagen XIII (Sund et al. 2001a). Postnatally, collagen XIII expression in the brain becomes radically reduced and *Col13a1* transcripts are hardly detectable in distinct brain regions (Monavarfeshani et al. 2017). Instead, meninges as well as blood vessels in the brain (Fig. 8.2) and peripheral nerve exhibit collagen XIII protein in adult mice (Zainul et al. 2018). Furthermore, collagen XIII colonizes the NMJ (Hägg et al. 2001; Latvanlehto et al. 2010; Logan et al. 2015; Härönen et al. 2017, 2019; Zainul et al. 2018) where it is expressed postsynaptically, shed, and incorporated into the synaptic cleft BM (Latvanlehto et al. 2010). Collagen XIII deficiency causes congenital myasthenia, highlighting its importance at the NMJ.

In situ hybridization confines collagen XXIII expression in the developing dura mater of the brain and spinal cord in the mouse embryo (Koch et al. 2006). A detailed examination of different CNS regions reveals *Col23a1* transcripts selectively in the accessory olfactory bulb (Monavarfeshani et al. 2017). According to Western

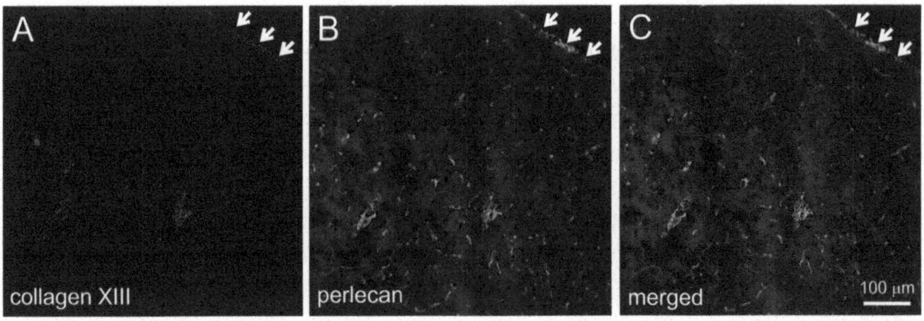

Fig. 8.2 Expression of collagen XIII in the cerebral cortex. Adult mouse brain stained (**a**) for collagen XIII with an anti-human collagen XIII/NC3 antibody (Hägg et al. 1998), (**b**) for perlecan, and (**c**) merged. Arrows point to pia mater

blotting, in the adult mouse brain collagen XXIII protein is primarily in proteolytically processed forms (Koch et al. 2006). To date, no human disease or study on mutant animals has been published on collagen XXIII.

The shed form of collagen XXV (originally named collagen-like Alzheimer amyloid plaque component, CLAC) was originally found to associate with senile plaques in the Alzheimer's disease (AD) brain and its precursor, collagen XXV (CLAC-P), expressed by CNS neurons (Hashimoto et al. 2002). In the mouse brain transcripts for *Col25a1* are present both during embryonic development and in adulthood (Tanaka et al. 2014; Monavarfeshani et al. 2017). The spinal cord demonstrates higher collagen XXV expression during development compared to the adult situation (Tanaka et al. 2014).

In the brain, collagen XXV becomes increasingly expressed postnatally, being far more prevalent than the other MACITs (Monavarfeshani et al. 2017). Collagen XXV is found broadly in distinct subsets of both inhibitory and excitatory CNS neurons, with the hippocampus exhibiting the highest expression (Monavarfeshani et al. 2017). Collagen XXV is enriched in regions of the brain innervated by axons of retinal ganglion cells, and it can also be found in the retina per se (Kay et al. 2011; Monavarfeshani et al. 2017).

Besides being present both in the developing and adult CNS, collagen XXV is transiently expressed early in the developing muscle, coinciding with myogenic differentiation and becoming depressed by nerve-induced excitation (Tanaka et al. 2014; Goncalves et al. 2019; Munezane et al. 2019). In addition, developing spinal motor neurons and DRG neurons are positive for *Col25a1* transcripts, as demonstrated at the time of myoblast fusion (Tanaka et al. 2014). Muscle-derived collagen XXV is important in the development of the neuromuscular system.

Collagen XVII is the fourth transmembrane collagen with a unique structure and is different from the MACIT collagens XIII/XXIII/XXV. Similar to the latter, its ectodomain can also become proteolytically shed (Franzke et al. 2005). Collagen XVII is a component of dermal hemidesmosomes and genetic mutations lead to junctional epidermolysis bullosa while autoantibodies targeting shed collagen XVII cause a blistering disease bullous pemphigoid (BP), whereby collagen XVII is also known as BP180 (see Chap. 7 for further details). Collagen XVII is also broadly although variably present among different neuroanatomical regions of the CNS which share a common developmental origin with the ectoderm (Seppänen et al. 2007). Therefore, it is not surprising that BP patients exhibit variable neurological symptoms. Collagen XVII is expressed in human cortical, hippocampal and amygdaloid neurons, most intensively in motor nuclei and Betz cells, and pyramidal neurons (Seppänen et al. 2006, 2007, 2009). Collagen XVII colonizes the soma and proximal axons and it accumulates intracellularly in neuronal lipofuscin granules (Seppänen et al. 2007, 2010) but does not cumulate over time (Seppänen et al. 2007).

8.3 Collagen Functions in the Nervous System

Collagens are critical constituents of the nervous system ECMs. Since many of the different collagen types often occupy ECMs in several tissues, collagen-related diseases typically involve several organ systems. Collagens and their proteolytic products are involved in multiple signaling pathways and provide extracellular cues that are important in guiding developmental processes. Such actions are reutilized in tissue repair in response to injury, and consequently, several collagens typically become upregulated. Further in adulthood, collagen assemblies provide support and flexibility in stabilizing cellular structures such as synapses. Aging impairs cellular metabolism, which also affects the extracellular environment and results in protein accumulation, neurotoxicity, and neurodegeneration. Mutations in genes coding for collagens result in a broad range of symptoms affecting nervous tissues and thereby have a major impact on patients' life. Neuropathologies entailing collagens are reviewed in the following paragraphs. In addition, the involvement of collagens in inherited CNS and PNS diseases and gene-specific animal models are summarized in Tables 8.3 and 8.4, respectively.

8.3.1 Neuronal Migration

Cell migration and differentiation lay the foundations for the development of specialized organs in multicellular organisms. ECM provides qualitative, temporal, and spatial information for migrating and differentiating cells during development for them to properly form a functional tissue. ECMs of the developing meninges promote radial glia organization and together these deposit cues and routes for migrating neural cells, respectively (Fig. 8.3a). Several members of the collagen superfamily, most importantly fibrillar and BM collagens, support morphogenesis during development, applying in the nervous system. Collagen I further promotes neural maturation. Collagen I actions are exemplified by a broad range of neuropathological findings in Osteogenesis imperfecta patients (see also Chap. 2). Those

Table 8.3 Collagens in inherited CNS diseases and disease models

Subfamily	Type	Gene affected	Phenotype	References
Fibrillar	I	*COL1A1/A2*	Developmental and various other brain defects in Osteogenesis imperfecta	Charnas and Marini (1995), Emery et al. (1999)
	II	*Col2a1*	Loss of head tissues and holoprosencephaly	Leung et al. (2010)
FACIT	XIX	*Col19a1*	Increased transcription of proteases involved in PNN homeostasis, reduced number of PNNs in the cerebral cortex and hippocampus, compromised establishment of perisynaptic synapses in a subset of hippocampal and cortical inhibitory interneurons, and expression of seizures and schizophrenia-like behaviors	Su et al. (2010, 2016, 2017)
Network-forming	IV $[\alpha1]_2\alpha2$	*COL4A1/A2*	Dominant mutations integrate with cerebrovascular disease and intracerebral hemorrhage that may result in porencephaly (OMIM #614519, #175780, #614483)	Gould et al. (2005), Breedveld et al. (2006), Favor et al. (2007), Volonghi et al. (2010), Jeanne et al. (2012)
		Col4a1	Focal disturbances in pial BM and concurrent misplacement of migrating neurons	Pöschl et al. (2004)
	VI $\alpha1\alpha2\alpha3$	*Col6a1*	Increased ROS and apoptosis in aged brain, impaired motor coordination, and spatial memory	Cescon et al. (2016)
Multiplexin	XV	*Col15a1*	Deficiency beneficial in experimental ischemic stroke	Dhungana et al. (2017)
	XVIII	*COL18A1*	Knobloch syndrome (OMIM #267750) associating with CNS malformations	Suzuki et al. (2002), Kliemann et al. (2003), Keren et al. (2007), Caglayan et al. (2014), Charsar and Goldberg (2017), Corbett et al. (2017), White et al. (2017)
		Col18a1	Structural alterations in the choroid plexus	Utriainen et al. (2004)

(continued)

Table 8.3 (continued)

Subfamily	Type	Gene affected	Phenotype	References
			resulting in hydrocephalus	
			Defected climbing fiber synaptogenesis through interaction of presynaptic α3β1 integrin and target-derived endostatin on Purkinje neurons	Su et al. (2012)
Transmembrane	XIII	*COL13A1*	Mild mental retardation among CNS19 patients	Rodriguez Cruz et al. (2019)

include brainstem compression but also macrocephaly and ventriculomegaly, impaired neuroblast migration, hippocampal malrotation, agyria, and abnormal neuronal lamination, among others (Charnas and Marini 1995; Emery et al. 1999). Mice deficient in collagen II display a partially penetrant phenotype showing loss of head tissue and holoprosencephaly (Leung et al. 2010). Furthermore, pial BM constituent collagen IV also contributes to brain development. Due to BBB breakage and resultant hemorrhage, collagen IV mutations (detailed later) promote the development of secondary porencephalic lesions. Alternatively, this may directly result from impaired neuronal ventricular-to-cortex migration that is dependent on radial glial support (Fig. 8.3a). Collagen IV mutations may compromise the interaction of glial end feet projections with the pial BM. In fact, *Col4a1/a2*-deficient embryos demonstrate focal disturbances in pial BM and concurrent misplacement of migrating neurons (Pöschl et al. 2004). Additionally, collagen XIII may affect neural migration since transgenic overexpression of truncated molecules delays brain development. Ectopic collagen XIII enhances axon outgrowth in rat hippocampal neuron cultures, suggesting a means by which collagen XIII may influence brain formation. These collagen XIII mutant mouse embryos die due to defects in angiogenesis, which provides an alternative explanation (Sund et al. 2001b).

Loss-of-function mutations in *COL18A1* resulting in deficiency of one or all collagen XVIII isoforms cause Knobloch syndrome (OMIM #267750), a growingly heterogeneous rare disorder defined by high myopia, vitreoretinal degeneration associating with retinal detachment and often associating with occipital scalp defect causing encephalocele (Sertie et al. 2000; Suzuki et al. 2002, 2009; Kliemann et al. 2003; Passos-Bueno et al. 2006; Keren et al. 2007; Mahajan et al. 2010; Caglayan et al. 2014; Haghighi et al. 2014; Zhang et al. 2018). Immunolabeling studies on mouse and human tissues indicate that collagen XVIII delineates the developing and adult BMs of pia (Utriainen et al. 2004; Caglayan et al. 2014). Deposition of collagen XVIII provokes novel interactions in the ECM which subsequently contribute to guiding the neuroectodermal morphogenesis and cell migration in the closure of the neural tube. Moreover, the existence of neuronal migratory and circuit

Table 8.4 Collagens in inherited PNS diseases and disease models

Subfamily	Type	Gene affected	Phenotype	References
FACIT	XIX	*Col19a1*	Prohibited sphincter muscle transdifferentiation, altered muscle ECM and disturbed relaxation of the sphincter muscle leading to malnutrition and lethality	Sumiyoshi et al. (2004)
		Zebrafish *Stumpy*	Impaired motor axon pathfinding and branching resulting in incomplete motor axon synaptogenesis	Beattie et al. (2000), Hilario et al. (2010)
Network-forming	IV $[\alpha1]_2\alpha2$	*COL4A1/A2*	Myopathy in a portion of multisystem disorder patients	Labelle-Dumais et al. (2011), Mao et al. (2015), Jeanne and Gould (2017)
		Col4a1/a2	Thickened myelin and impaired radial sorting, reduced nerve conduction velocities and peripheral neuropathy recapitulating general and progressive human myopathy	Kuo et al. (2014), Labelle-Dumais et al. (2011)
		Col4a1	Transient delay in neonatal presynaptic differentiation at the NMJ	Fox et al. (2007)
		C. elegans emb-9 [col (IV)α1]	Ectopic presynaptic boutons at the NMJ	Qin et al. (2014)
	IV α3α4α5, IV $[\alpha5]_2\alpha6$	*Col5a4*	Compromised synaptic integrity at the NMJ	Fox et al. (2007)
	VI α1α2α3	*Col6a1, COL6A1/A2/A3*	AChR cluster fragmentation and subunit switch, altered expression of several synaptic and postsynaptic proteins, compromised neurotransmission, and corresponding alterations in Ullrich congenital muscular dystrophy patients	Cescon et al. (2018)
		Col6a1	Hypermyelination, impaired nerve conduction properties and motor coordination,	Chen et al. (2014)

(continued)

Table 8.4 (continued)

Subfamily	Type	Gene affected	Phenotype	References
			thickened C-fibers and slowdown in nociception	
			Compromised post injury macrophage polarization and peripheral nerve regeneration	Chen et al. (2014)
Multiplexin	XV	*Col15a1*	Loosely packed axons of C-fibers and compromised radial sorting resulting in polyaxonal myelination	Rasi et al. (2010)
		Zebrafish *col15a1b*	Defected axon path-finding and branching, and formation of the neuromuscular system	Guillon et al. (2016)
	XV/XVIII multiplexin orthologue	*C. elegans cle-1*, *Drosophila dmp*	Compromised axon growth cone navigation, reduced number of synapses, and defected postsynaptic and presynaptic structures	Ackley et al. (2001, 2003), Meyer and Moussian (2009), Qin et al. (2014)
	XVIII	Zebrafish *col18a1*	Defected neural crest cell migration through somites	Banerjee et al. (2013)
Transmembrane	XIII	*COL13A1*	Congenital myasthenic syndrome type 19 (OMIM #616720)	Logan et al. (2015), Dusl et al. (2019), Marquardt and Li (2019), Rodriguez Cruz et al. (2019)
		Col13a1	Terminal sprouting, axon detachment, retracting and invaginating terminal Schwann cells, and erroneous spreading of synaptic vesicles in the preterminal axon at the NMJ	Latvanlehto et al. (2010), Härönen et al. (2017)
			Delayed and incomplete regeneration of the NMJ post nerve injury	Zainul et al. (2018)

(continued)

Table 8.4 (continued)

Subfamily	Type	Gene affected	Phenotype	References
	XXV	*COL25A1*	Congenital cranial dysinnervation disorder (OMIM #616219) due to compromised interaction between muscular collagen XXV and axonal PTP σ/δ	Shinwari et al. (2015), Munezane et al. (2019)
		Col25a1	Defected muscle development, total block of motor axon invasion into the muscle, innervation and NMJ formation leading to motor nerve degeneration and neonatal lethality due to a respiratory failure	Tanaka et al. (2014), Goncalves et al. (2019), Munezane et al. (2019)
	XIII/XXIII/ XXV orthologue	*C. elegans col-99*	Impaired peripheral axon guidance	Taylor et al. (2018)

formation defects in Knobloch syndrome are evident and exemplified by various CNS malformations, such as pachygyria and polymicrogyria associated with epilepsy (Suzuki et al. 2002; Kliemann et al. 2003; Keren et al. 2007; Caglayan et al. 2014; Charsar and Goldberg 2017; Corbett et al. 2017; White et al. 2017). In collagen XVIII-deficient mice, the pial BM appears outwardly normal, and obvious CNS malformations are not observed (Utriainen et al. 2004), evidencing species specificity in this respect. Studies in zebrafish have shown that collagen XVIII is involved in neural migration in the PNS. In this model, collagen XVIII transcripts are transiently expressed in a subset of muscle cell precursors (adaxial cells) in close contact with the path taken by motor axon and neural crest cells around the onset of motor axon outgrowth and neural crest cell migration (Schneider and Granato 2006). Morpholino knockdown of *col18a1* in zebrafish results in both motor axon extension and neural crest cell migration defects which both normally follow the same path through the muscle territory in the somites (Schneider and Granato 2006; Banerjee et al. 2013). Interestingly, this literature hypothesizes that collagen XVIII affects motor axon extension and neural crest cell migration via its glycosylated side chains, added by the glycosyltransferase Lhr3, as the *col18a1* morphants and the *lhr3* mutant *diwanka* display similar motor axon and neural crest cell phenotypes. However, this still remains an untested hypothesis which will be among the next upcoming technical challenges in the field.

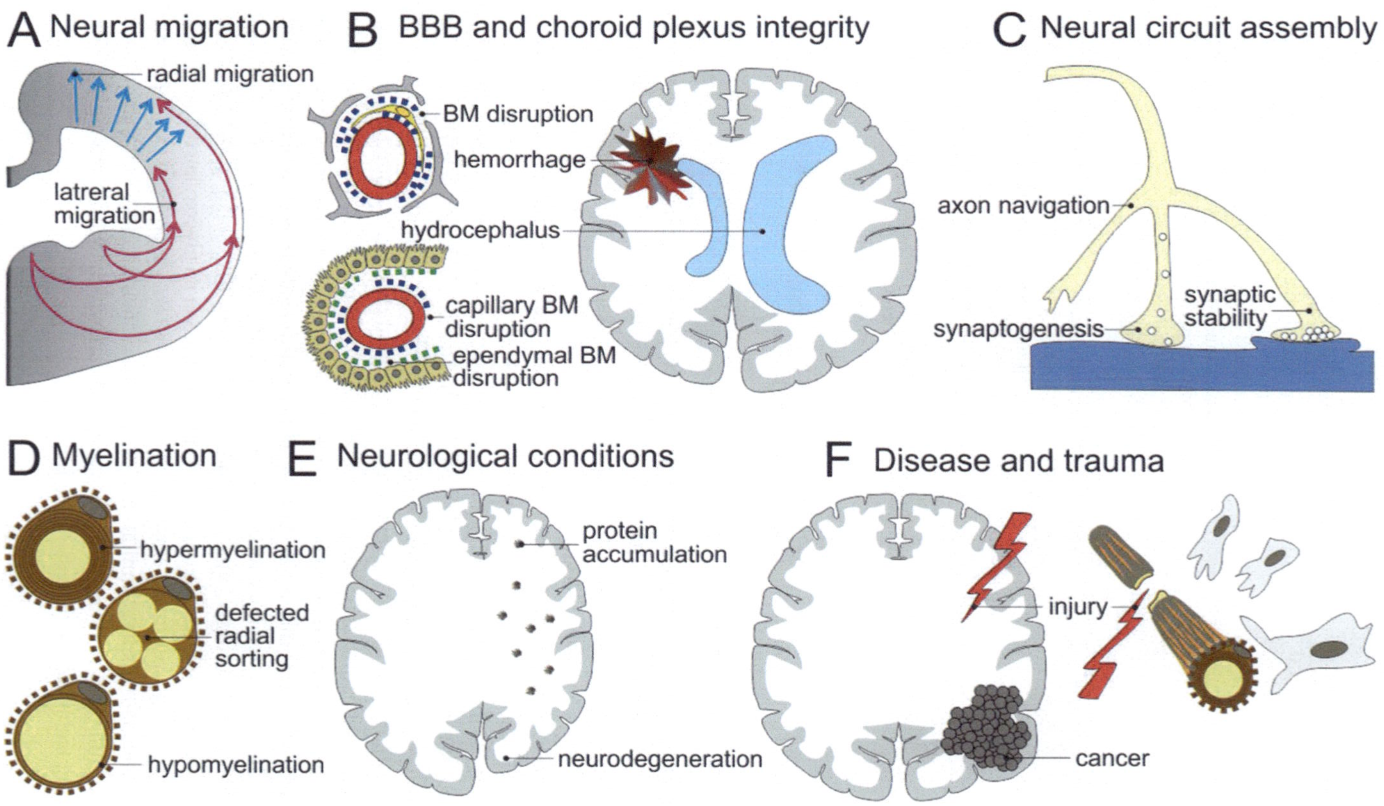

Fig. 8.3 Pathophysiological conditions of the nervous system involving collagens. Defects (**a**) in neural migration may result in malformations, (**b**) in integrity of barrier and filtration functions result in hemorrhage and hydrocephalus, respectively, (**c**) in neural circuit development, synaptogenesis and synaptic

homeostasis lead to malfunctioning, **(d)** in myelination compromises peripheral nerve structure and function, **(e)** in the nervous system ECM integrity may result in neurodevelopmental/psychiatric/degenerative conditions, and other processes involving collagens include **(f)** regeneration from disease or traumatic injury

8.3.2 Barrier and Filtration Functions

The formation and maintenance of a fully functional BBB requires seamless cooperation of all four constituent layers; astrocytes, pericytes, endothelial cells, and the extracellular BM. Collagen IV is a critical component of BMs and in mice *Col4a1/a2* deficiency results in embryonic disruption of BM structures, causing lethality at embryonic day (E) 10.5–11.5 (Pöschl et al. 2004). Owing to dominant mutations in *COL4A1/A2* genes, collagen IV integrates with cerebrovascular disease and intracerebral hemorrhage (Fig. 8.3b) that may result in porencephaly (OMIM #614519, #175780, #614483). Mutations often represent missense mutations of glycines and cause impaired secretion of collagen IV, proposedly due to a dominant-negative effect (Gould et al. 2005; Breedveld et al. 2006; Volonghi et al. 2010; Jeanne et al. 2012). Similarly, in mutant mice, heterozygous *Col4a1/a2* mutations compromise vascular BM development, cause local disruption of vascular BM, and result in cerebral hemorrhage at birth, which in turn leads to the development of porencephaly in survivors (Gould et al. 2005; Favor et al. 2007). In preclinical studies, postnatal treatment with chemical chaperones diminished intracellular accumulation of mutant collagen IV in endothelial cells and pericytes, and resulted in reduced disease severity (Jeanne et al. 2015; Hayashi et al. 2018), suggesting a potential frame to develop a therapy to treat human patients (Detailed in Chap. 5).

Interestingly, Dhungana and colleagues showed that collagen XV deficiency is beneficial in ischemic stroke generated in the middle cerebral artery in knock-out mice. Collagen XV is dispensable for cerebrovascular development and integrity in physiology, but changes in the molecular composition of the vascular BM prior to stoke proposedly contribute to beneficial events, such as an increase in VEGF-A. The effect is similar to recombinant tissue plasminogen activator treatment, indeed increasing circulating collagen XV levels in wild-type mice and therefore suggestively acting at least partly through a proteolysis of collagen XV-dependent matrix (Dhungana et al. 2017).

Additionally, some other collagens have been genetically linked to the integrity of BBB. In line with the expression of collagen XIII in intracerebral vessels (Fig. 8.2) and collagen VI at meningeal vessels, genetic polymorphisms of *COL13A1* and *COL6A3* genes in the Japanese population were found to associate with intracerebral and subarachnoid hemorrhage, respectively (Yoshida et al. 2010). Post-irreversible ischemic damage, collagen XVII immunoreactivity depletes which indicates a dynamic injury response (Seppänen et al. 2007). More recently, genetic polymorphisms in the *COL17A1* gene have also been linked to subarachnoid hemorrhage (Yamada et al. 2017).

As collagen XVIII normally occupies both the ependymal and capillary BMs, phenotypical findings in the choroid plexus in collagen XVIII deficiency is not a surprise. Collagen XVIII knock-out mice exhibit broadened BMs, the thickness associating with deteriorated epithelial cells and tight junctions in choroid plexuses. This results in hydrocephalus in collagen XVIII deficient mice, but only within a certain mixed genetic C57 background, indicating that the composition of other factors is critical for the choroid plexus function and cerebrospinal fluid homeostasis within the collagen XVIII context (Utriainen et al. 2004). In support of this, the dilation of ventricles is also reported in Knobloch syndrome patients (Kliemann et al. 2003). BMs are critical actors of biological barriers and filters. This is also evidenced by the involvement of BM and BM-associated collagens, specifically, in these functions.

8.3.3 Neural Circuit Assembly, Maturation, and Maintenance

Studies on animal models have primarily revealed the importance of several collagens in neural circuit formation, synaptogenesis, and synaptic homeostasis both in the CNS and PNS (Fig. 8.3c). For example, CNS malformations in Knobloch syndrome predominate in the forebrain, but collagen XVIII is also expressed by cerebellar Purkinje neurons concomitantly with synaptogenesis, and endostatin is found in synaptic-rich extracts. Although the cerebellum of collagen XVIII-deficient mice is morphologically normal, the assembly of climbing fiber synapses (which are the synapses formed between presynaptic axons from the inferior olivary nucleus and postsynaptic dendrites of Purkinje cells) was impaired in the absence of collagen XVIII. In mice, lack of target-derived endostatin reduces synaptic contacts and compromises presynaptic differentiation. This concomitantly influences the motor performance in a rotarod test, which was confirmed to derive from cerebral defects, as NMJs appear morphologically normal in these mice (Su et al. 2012). In vitro, endostatin induces accumulation of synaptic vesicle proteins, indicative of presynaptic differentiation, and such an effect is reversed by the administration of anti-$\alpha3\beta1$ integrin function-blocking antibodies (Su et al. 2012). $\alpha3\beta1$ integrin has been identified at presynaptic active zones (Carlson et al. 2010) and further binding experiments confirm this integrin as a presynaptic ligand for the target-derived endostatin on Purkinje cells (Su et al. 2012). Collagen XVIII also acts in the PNS circuit assembly, a function that is also attributed to the closely related collagen XV. Mutants of the single orthologue in worms and flies, as well as for individual multiplexin collagens in fish, all exhibit peripheral axon guidance defects, although such associations are sparse in higher animals (reviewed in Bretaud et al. 2020). Mutants of the collagen XV/XVIII orthologs, *cle-1* in *C. elegans* (Ackley et al. 2001) and *dmp (multiplexin)* in *Drosophila* (Meyer and Moussian 2009), fail to establish a proper innervation. CLE-1 accumulates at nerve cords and often at the anterior–posterior axon tracks in a punctate manner in regions rich in synapses. The deletion of the restin/endostatin-embedding NC1 domain results in low penetrance migration and axon guidance defects that are rescued by trimeric NC1 (Ackley et al. 2001), shown to induce cell motility (Kuo et al. 2001). On the contrary, the monomeric restin/endostatin domain is incapable of doing the same, but instead, it exaggerates the navigation defect, which suggests a role as a feedback inhibitor (Ackley et al. 2001). The vertebrate zebrafish genome harbors distinct genes for collagens XV and XVIII and evidence suggests that both proteins (isoforms Col15a1b and Col18a1a) contribute to neural circuit formation. In zebrafish *diwanka* mutants the myotomal glycosyltransferase activity was found to be essential for axon growth cone migration and motor innervation (Schneider and Granato 2006). Presumably its modifications on collagen XVIII-carbohydrates at muscle precursor ECM may be a prerequisite for interaction with neuronal receptor protein tyrosine phosphatases (PTP) and/or kinases, known to contribute to axon pathfinding and also to bind collagens/heparan sulfate proteoglycans (Vogel et al. 1997; Aricescu et al. 2002; Unsoeld et al. 2013; Munezane et al. 2019).

In a study with mutant zebrafish models, the deposition of collagen XV-B, one of the two collagen XV forms present in zebrafish, in the slow muscle ECM constitutes the path and branching milestones for growing axons in order for them to establish an intact neuromuscular system (Guillon et al. 2016). In the developing slow muscle, adaxial cells respond to Hedgehog/Gli and MuSK signaling by expressing *col15a1b* and depositing specifically the collagen XV-B protein in the motor path (Bretaud et al. 2011; Guillon et al. 2016). Reminiscent of what was suggested for collagen XVIII, collagen XV-B might also interact with discoidin domain receptors (DDR) to guide motor axon navigation as such a direct interaction between collagen XV and DDR1 is identified on tumor cells (Clementz et al. 2013). Both vertebrate multiplexin homologs, collagens XV and XVIII, are also enriched at the NMJ in the BM capping terminal Schwann cells (Muona et al. 2002; Su et al. 2012). In worms, CLE-1 concentrates at the NMJ near synaptic zones and serves as a synaptic organizer (Ackley et al. 2003). Mutant worms exhibit a high penetrance phenotype of reduced number of synapses and compromised integrity of both postsynaptic and presynaptic structures, such as sprouting (Ackley et al. 2003; Qin et al. 2014). Functional analyses indicate that synaptic transmission in cholinergic synapses is impaired (Ackley et al. 2003). In mice, neonatal examination shows that the motor nerve innervation is correctly patterned in the absence of collagen XVIII. Developmental establishment and postnatal maturation of the NMJ are also normal (Su et al. 2012). Assembly and maintenance of the NMJ are thus far not explored in collagen XV knock-out mice. This would, however, be of interest since collagen XVIII appears not to convey such functions in mammals. Still, clear defects, especially in axon navigation but also in synaptogenesis, exist in mutants of other species. Moreover, studies on collagen XV/XVIII double mutant mice confirm a lack of major compensation between the two multiplexins (Ylikärppä et al. 2003). Yet, collagen XV-deficient mice develop myopathy, and potential impairment in circuit formation could in part contribute to the evolvement of such phenotypes (Eklund et al. 2001).

Collagen IV and its NC1 domain extracted from the electric organ of *Torpedo californica* was initially identified to promote synaptogenesis on cultured motor neurons, and the recombinant NC1 domains of human α2 (canstatin), α3 (tumstatin), and α6 (hexastatin) chains were found to share this property (Fox et al. 2007). *C. elegans* genome contains only two collagen IV genes, *emb-9* and *let-2*, encoding respectively the α1(IV) and α2(IV) chains. Mutations in *emb-9* result in intracellular persistence of synaptic collagen IV, which promotes ectopic presynaptic boutons at the NMJ (Qin et al. 2014). *Col4a1* mutant mice lacking $[\alpha1(IV)]_2\alpha2$

(IV) heterotrimers exhibit transient delay in presynaptic differentiation neonatally. The α3(IV)α4(IV)α5(IV) and $[\alpha5(IV)]_2\alpha6(IV)$ heterotrimers synergistically maintain synaptic integrity when it is compromised in *Col5a4* mutants, while loss of either of the two heterotrimers alone has no detectable effect on the NMJ architecture (Fox et al. 2007).

At the moment, at least 30 distinct genes have been identified to cause congenital myasthenic syndrome (CMS). One of these, CMS19 (OMIM #616720), is caused by loss-of-function mutations in *COL13A1,* and 16 different mutations have been reported since the first discoveries in 2015 (Logan et al. 2015; Dusl et al. 2019; Marquardt and Li 2019; Rodriguez Cruz et al. 2019). Intriguingly, anti-collagen XIII autoimmune antibodies were recently identified in sera of myasthenia gravis patients, although their disease-causing potential remains to be proven (Tu et al. 2018). Typically, muscle weakness in CMS19 predominates in facial, bulbar, respiratory, and axial muscles and consequently patients suffer from neonatal respiratory distress and severe dysphagia. The critical symptoms often resolve early in infancy, and the disease continues to gradually improve, suggesting a relevant role for collagen XIII specifically in early phases of NMJ maturation postnatally. Along this line, defects in acetylcholine receptor (AChR) clustering and maturation thereafter are obvious in collagen XIII knock-out mice at 2 weeks of age and onward (Latvanlehto et al. 2010). Defects at the NMJ include sprouting axons, retracting terminal Schwann cells, and erroneous spreading of synaptic vesicles in the preterminal axon (Latvanlehto et al. 2010; Härönen et al. 2017). Occasionally, the nerve terminus is detached, allowing Schwann cell invagination (Latvanlehto et al. 2010). On the other hand, knock-in mice incapable of collagen XIII shedding exhibit enhanced transsynaptic adhesion and overwhelming neurotransmission capacity, suggestive of collagen XIII motivating transsynaptic adhesion and/or acting as a receptor for nerve-derived ligand(s). However, NMJs of the knock-in mice are not completely normal (Härönen et al. 2017), implying a relevant role also for the shed form, which was indeed shown to enhance AChR cluster maturation in cultured myotubes (Latvanlehto et al. 2010). Moreover, collagen XIII was found to bind ColQ, the collagenous tail of acetylcholinesterase, and contribute to the fine-localization of them both at the NMJ (Härönen et al. 2017). Interestingly, CMS19 patients do not respond well to anti-cholinesterase treatment in general but do respond to 3,4-Diaminopyridine, a blocker of presynaptic potassium channels and salbutamol, β2-adrenergic receptor agonist (Logan et al. 2015; Dusl et al. 2019; Rodriguez Cruz et al. 2019). Likewise, 3,4-Diaminopyridine improved muscle performance of knock-out mice in the short term (Härönen et al. 2017). Mouse models have demonstrated their value in revealing details of the disease that would not have been achievable by studying human patients solely. For example, muscle histology is relatively normal both in patients and knock-out mice, but mouse studies reveal fiber-type change at the expense of slow muscle fibers (Härönen et al. 2017), especially those dependent on neurotrophic support. Regeneration after injury is delayed in knock-out mice, further pointing to its functions in NMJ maturation. Proposedly, due to preexisting defects at the NMJ, structural and functional recovery of the NMJ in knock-out mice never reaches the pre-injury status (Zainul et al. 2018).

In mice, collagen XXV deficiency interferes with the development of the neuromuscular system far earlier than collagen XIII deficiency. Based on in vitro studies, shed collagen XXV is capable of inducing myoblast fusion into myotubes during primary myogenesis. This process does not involve transcriptional regulation, but collagen XXV rather aids in myoblast adhesion and/or recognition (Goncalves et al. 2019). In line with such a suggested role, muscle development is compromised in collagen XXV knock-out mice (Tanaka et al. 2014; Goncalves et al. 2019). This leads to a total block of motor axon invasion into the muscle, innervation and NMJ formation, and subsequent motor nerve degeneration. Loss of muscle innervation causes neonatal lethality due to respiratory failure (Tanaka et al. 2014). Tissue-specific depletion confirms that muscle-derived collagen XXV is essential for innervation, while nerve-derived collagen XXV appears dispensable (Munezane et al. 2019). Similarly, *COL25A1* mutations also affect peripheral innervation in humans, but in a milder way. One form of congenital cranial dysinnervation disorder, congenital fibrosis of extra-ocular muscles-5 (CFEOM5; OMIM #616219), is caused by recessive mutations in *COL25A1* influencing the structure of the collagen XXV protein (Shinwari et al. 2015; Munezane et al. 2019). Shinwari and colleagues demonstrated collagen XXV expression in human extra-ocular muscles and decreased levels of axon guidance molecules in an affected patient, suggestive of the disease evolvement from defected innervation of extra-ocular muscles during development (Shinwari et al. 2015). Collagen XXV directly binds with PTP σ and δ on invading axons, thereby inducing cell–cell interaction and axon elongation whereas mutant collagen XXV molecules are incapable of doing the same (Munezane et al. 2019). Interestingly, mutations in the *C. elegans* MACIT orthologue, *col-99* (Tu et al. 2015), compromise peripheral axon guidance suggesting a role for COL-99 as a target-derived cue for axon outgrowth in worms (Taylor et al. 2018). Mouse models in which either collagen XIII and XVII shedding is prevented show that the increased ratio of the transmembrane form over the shed one improves selected physiological events (Jacków et al. 2016a, b; Härönen et al. 2017). Therefore, it would be of interest to evaluate to what extent preventing shedding of collagen XXV would rescue mice from dysinnervation.

Moreover, collagen VI is found to be important for motor synapse integrity and function (see also Chap. 6). It regulates various critical synaptic components through ETS and/or bHLH transcription factor family-dependent pathways and thereby affects AChR cluster homeostasis. In line, collagen VI deficiency in mice results in compromised neurotransmission, due to AChR cluster fragmentation and subunit switch, and altered expression of several synaptic and postsynaptic proteins. Similar alterations were observed in Ullrich congenital muscular dystrophy patients (Cescon et al. 2018). Furthermore, zebrafish morphants for *col6a2* and *col6a4–6* mammalian orthologue present altered axonal growth and a muscular disease (Ramanoudjame et al. 2015).

Animal studies and in vitro experiments both reveal that collagen XIX self-aggregates into oligomers (Myers et al. 2003). It acts as a pericellular matrix/BM zone organizer by affecting the composition of signaling cues that contribute to intercellular communication. Collagen XIX knock-out mice die by 3 weeks of age from malnourishment due to compromised swallowing and failing relaxation of the sphincter muscle. Lack of collagen XIX prevents the smooth-to-skeletal muscle transdifferentiation in the abdominal segment of the esophagus because of failed myogenin transcription induction and expression in the region that normally expresses collagen XIX. The sphincter muscle innervation is normal, but electron microscopy indicates irregular muscle ECM and thick BM, which compromises NO-dependent relaxation of the sphincter muscle. Related to its association with BM, collagen XIX may act as a BM organizer at the motor synapse as well and contribute to the function of the neuromuscular junction in sphincter muscle (Sumiyoshi et al. 2004). In zebrafish, Stumpy, an orthologue for mammalian collagen XIX, guides motor axon pathfinding and branching exemplified by a halt in trunk motor axons at their intermediate targets in *stumpy* mutants (Beattie et al. 2000; Hilario et al. 2010). Co-immunoprecipitation and the analysis of the mutant fish that phenocopy incomplete motor axon synaptogenesis observed in *stumpy* mutants identified the axonal chondrolectin as an interaction partner for muscle-derived collagen XIX through their C-type lectin domain and collagenous repeats, respectively (Oprisoreanu et al. 2019). Collectively, several non-fibrillar collagens act as target-derived organizers in neural circuit development. This is pronounced in the PNS, implying the importance of ECM cues for axon guidance and synaptogenesis.

8.3.4 Myelination

The formation of endoneurium starts at E15 in mice when BM and thin collagen fibrils appear around Schwann cells (Osawa and Ide 1986). Developing nerves at this stage are devoid of fibroblasts (Osawa and Ide 1986) and it has been shown that Schwann cells themselves are capable of producing BM and delicate collagen-fibril matrix (Bunge et al. 1980; Eldridge et al. 1987, 1989; Chernousov et al. 1998, 2006). Specifically, the production of the BM is critical for complete lay-down of the endoneurial ECM and subsequent steps in Schwann cell differentiation and myelination (Eldridge et al. 1989; Chernousov et al. 1998) (Fig. 8.3d). Oligodendrocyte myelination does not require collagen-dependent ECM and therefore collagens are dispensable for the CNS myelination (Eldridge et al. 1989). Altered Schwann cell function and myelination in the PNS is associated with a variety of neuropathies affecting both motor and sensory functions.

In in vitro cultures, the presence of neurons significantly induces Schwann cells to deposit BM (Bunge et al. 1980; Chernousov et al. 1998), indicating that interactions with both axons and ECM are critical for myelination. Early on, the essence of collagen IV in myelination is demonstrated in vitro (Eldridge et al. 1989). The multisystem disorder caused by *COL4A1/A2* mutations involves myopathy in a proportion of patients (Labelle-Dumais et al. 2011; Mao et al. 2015; Jeanne and Gould 2017). Symptoms are partially explained by vascular defects but since collagen IV is such an abundant BM component, myopathy obviously evolves from various locations and levels of the BM. Collagen IV mutant mice recapitulate general and progressive human myopathy (Labelle-Dumais et al. 2011; Kuo et al. 2014). During postnatal maturation, sciatic nerves exhibit thickened myelin and impaired radial sorting. This results in reduced nerve conduction velocities and peripheral neuropathy in collagen IV mutant mice. Resembling favorable effects on vasculature, administration of the chemical chaperone 4PBA increases collagen IV secretion and consequently prevents peripheral nerve hypomyelination in mutant mice (Labelle-Dumais et al. 2019).

The significance of collagens in the peripheral nerve myelination is further highlighted by the study of sodium-dependent vitamin C transporter 2 (SVCT2)-deprived mice (Gess et al. 2011). As vitamin C is crucial for collagen synthesis (Booth and Uitto 1981; Murad et al. 1981), the deposition of collagen types IV, V, and XXVIII were found to be noticeably reduced in the peripheral nerve ECM and the peripheral nerve showed decrease myelin thickness in *SVCT2*$^{+/-}$ mice. Surprisingly, the authors showed that collagen synthesis is regulated at the transcriptional level and not at the posttranslational level revealing a specific mechanism of action of Vitamin C in the PNS. Interestingly, *SVCT2* knockdown in Schwann cell cultures resulted in a significant reduction in collagen V synthesis (Gess et al. 2011). In vitro, collagen V promotes axon fasciculation and affiliation of axons in Schwann cells (Chernousov et al. 2001). Atypical for mature fibrillar collagen trimers, collagen V retains N-terminal non-collagenous domains (Chernousov et al. 2000), mediating an autocrine interaction with Schwann cell heparan sulfate receptor glypican-1 and not with collagen-binding integrins $\alpha1\beta1$ or $\alpha2\beta1$, to regulate myelination (Chernousov et al. 2001, 2006).

Collagen VI also promotes Schwann cell differentiation and regulates myelination and studies on animal models confirm its involvement in peripheral nerve integrity and function. Chen and colleagues showed that the reduction in collagen VI in *Col6a1*$^{-/-}$ mice does not alter the number of axons or axon fiber type distribution, but it results in hypermyelination of all fiber types later in life. Lack of collagen VI in the ECM deteriorates the regulation of signaling pathways controlling myelination in Schwann cells. This compromises nerve conduction properties and motor coordination of mice. Moreover, C-fibers are thickened and concomitantly the withdrawal response of mutant mice to both thermal and mechanical stimuli is compromised, indicating alteration of nociception (Chen et al. 2014). Collagen VI is thus proven important in Schwann cell homeostasis and function.

Axon segregation and myelination is not affected by the lack of collagen XV in mice, although the endoneurial BM occasionally exhibits mild disorganization. However, this phenomenon is associated with loosely packed axons of C-fibers and compromised radial sorting, resulting in polyaxonal myelination (Rasi et al. 2010). Double deficiency for collagen XV and laminin $\alpha4$ generates additional phenotypes, further witnessing the importance of Schwann cell BM for proper peripheral nerve maturation (Rasi et al. 2010).

8.3.5 Neurodevelopmental/Psychiatric/Degenerative Diseases

An increasing body of evidence supports the involvement of ECM molecules in the development of neural systems and neurodevelopmental diseases (Jovanov Milosevic et al. 2014) (Fig. 8.3e). *COL13A1* mutations cause congenital myasthenia, but in line with the collagen XIII expression pattern, patients also suffer from skeletal defects and occasionally from mild mental retardation (Rodriguez Cruz et al. 2019). Collagen XIX is defined to regions rich in PNNs, and in line, in knock-out mice the number of PNNs and levels of their constituents in distinct locations of the cerebral cortex and hippocampus normally positive for collagen XIX are reduced. This is associated with an increase in the transcripts from several protease-encoding genes involved in PNN homeostasis, exemplified by an ectopic neuronal ADAMTS4 expression in the collagen XIX-deficient cortex (Su et al. 2017). Studies of knock-out mice further show that collagen XIX is required for the establishment of axosomatic synapses in a subset of hippocampal and cortical inhibitory interneurons (Su et al. 2010, 2016). PNNs and inhibitory interneurons are involved in the emergence of schizophrenia and interestingly, collagen XIX-deficient mice express seizures and schizophrenia-like behaviors (Su et al. 2016). In humans, the deletion of a chromosomal region containing *COL19A1*, but also several other genes, has been associated with schizophrenia (Liao et al. 2012). Intracellularly collagen XIX-dependent cues affect transcriptional regulation and consecutive protein levels (Sumiyoshi et al. 2004; Su et al. 2017). Collagen XIX contributes to the formation of inhibitory nerve terminals in a paracrine fashion through $\alpha_5\beta_1$ integrin signaling (Su et al. 2016). Correct collagen XIX expression appears to be critical for proper neuronal pericellular matrix homeostasis, since its increase associates with faster progression and higher mortality among patients suffering from Amyotrophic lateral sclerosis (ALS) (Calvo et al. 2019).

Later in life, ECM molecules provide survival signals and support cellular health. Studies of mouse models and human patients both suggest a neuroprotective role for collagen VI. In cultured primary neurons *Col6a1* deficiency induces spontaneous apoptosis through defective autophagy and higher susceptibility to oxidative stress. Collagen VI expression is induced in the brain of aged mice and concordantly reactive oxygen species (ROS) and apoptosis was found to increase in aged *Col6a1*$^{-/-}$ brains, suggesting that collagen VI protects against age-induced oxidative damage. The performance of knock-out mice in motor coordination and spatial memory tests was compromised (Cescon et al. 2016) although no such cognitive impairment has thus far been reported in patients with collagen VI-related skeletal muscle diseases. Interestingly, compound heterozygous *COL6A3* mutation carriers however demonstrate early-onset segmental isolated dystonia without muscular involvement (Zech et al. 2015) and *COL6A2* mutations expose to progressive myoclonus epilepsy syndrome (Karkheiran et al. 2013). An additional line of evidence comes from AD mouse models and human patients. Transcripts coding for collagen α1-α3(VI) chains are particularly elevated in the brain of mice and patients with AD (Cheng et al. 2009). Neurodegenerative diseases typically involve neurotoxic intracellular accumulation and extracellular protein aggregation with aging. Senile plaques in AD are formed by extracellular accumulation of amyloid-beta (Aβ) peptides (Aβ40 and Aβ42) that are γ-secretase-cleaved proteolytic products of amyloid precursor protein (APP) (O'Brien and Wong 2011). Incubation of primary neuron cultures with Aβ42 dose-dependently promotes collagen VI protein expression through TβRII and Smad3 signaling. Wild-type neurons beat *Col6a1*$^{-/-}$ counterparts in resistance to Aβ42-mediated neurotoxicity and ectopic collagen VI augments protection against Aβ42-induced cell death. Experiments with primary neuronal cultures suggest that neuroprotection is mediated, at least partly, by preventing Aβ42 oligomer interaction with neurons (Cheng et al. 2009). Combined, these results make collagen VI a considerable instrument when designing novel therapies to treat neurodegenerative diseases.

Collagen XVIII accumulates in amyloid-positive vasculature and senile plaques in AD brains (van Horssen et al. 2002). Collagen XVIII is a collagen/proteoglycan hybrid and it may promote and stabilize senile plagues through its heparan sulfate proteoglycans. Moreover, the collagen XVIII matricryptin endostatin is increased in the cerebrospinal fluid of AD patients, providing a potential diagnostic tool in AD (Salza et al. 2015). Additionally, collagen XXV is associated with senile plaques in AD disease (Hashimoto et al. 2002; Kowa et al. 2004) and its involvement in the dynamics of β-amyloidogenesis has been demonstrated. In vitro studies are, however, even conflicting to some extent, most likely due to the utilization of variable Aβ peptides, Aβ40 or Aβ42, and different methods. Hashimoto and colleagues showed that both transmembrane and shed furin-mediated proteolytic product CLAC are capable of binding fibrillized Aβ (Hashimoto et al. 2002) whereas Söderberg et al. further showed that shed collagen XXV also interacts with the non-Aβ components of the AD amyloid plaques (Söderberg et al. 2005). Two different studies identified distinct positively charged sequences, locating either at the COL1 (collagenous) or NC2 domains of collagen XXV, important for Aβ42 and Aβ40 binding, respectively, through possible ionic interaction with the negatively charged amino acids in Aβ peptides (Kakuyama et al. 2005; Osada et al. 2005; Söderberg et al. 2005). In immunohistochemical staining of AD brains, however, Aβ40-positive plaques remain negative for collagen XXV and its accumulation to loose, rather than typical, Aβ42-positive fibrils implies early involvement in plaque formation and rather suggests an inhibitory role for collagen XXV in senile plaque maturation (Kowa et al. 2004). Indeed, shed collagen XXV inhibits Aβ fibril elongation in vitro (Kakuyama et al. 2005; Osada et al. 2005), while on the other hand, it enhances Aβ fibril aggregation and resistance against proteases (Söderberg et al. 2005). Collagen XXV may thus contribute to regulating fibril stabilization into aggregates and disease progression.

In line with in vitro results, overexpression of human collagen XXV in mouse CNS neurons exposes transgenic mice to increased Aβ accumulation, although not its fibrillization. With an unknown mechanism, excess collagen XXV appears to elevate p35/p25 expression levels, which in turn results in Cdk5 activation and subsequent increase of BACE1, a β-secretase involved in APP processing and AD pathology (Tong et al. 2010). In these transgenic mice, induced AD pathogenesis involves a decline in the level of the synaptic vesicle protein synaptophysin and an activation of astrocytes, associated with behavioral disabilities typical for AD (Tong et al. 2010). Besides a clear involvement of collagen XXV in AD pathogenesis in the transgenic mouse model and AD patients, only one study supports such a relationship at the genetic level, as Forsell and colleagues confirmed that certain *COL25A1* gene polymorphisms neighboring exon 9 expose patients to AD development (Forsell et al. 2010). Such SNPs could speculatively influence AD evolvement by in/out-splicing of the NC2 Aβ binding site or affect collagen XXV protein levels. In addition to AD, collagen XXV aggregates exist in Down syndrome (Kowa et al. 2004) but not in cognitive impairment at senescence (Gal et al. 2018). Increased levels of shed collagen XIII and XXIII ectodomains are detected in cancer patients' blood or urea due to increased expression and following proteolytic processing (Spivey et al. 2010; Miyake et al. 2017). It can be speculated whether circulating collagen XXV ectodomain/matricryptin is similarly increased in certain neurodegenerative diseases, and if so, could it be utilized as a diagnostic or prognostic marker?

Collagen XVII is one of the two major auto-antigens in BP, a subepidermal blistering skin disease (see Chap. 7 for further details). An exponentially growing body of evidence implies that BP, which typically raises in later life, predisposes to neurological comorbidities such as dementia, Parkinson's disease, cerebrovascular disorders, multiple sclerosis, and epilepsy, and to psychiatric diseases such as schizophrenia, uni- and bipolar disorder, schizotypal and delusional disorders, and personality disorders (Foureur et al. 2001; Parker et al. 2008; Marazza et al. 2009; Brick et al. 2014; Cai et al. 2014; Ren et al. 2017). On the other hand, neuropsychiatric disorders increase the risk of development of anti-collagen XVII autoantibodies and BP (Chosidow et al. 2000; Bastuji-Garin et al. 2011; Teixeira et al. 2014; Kibsgaard et al. 2017; Kokkonen et al. 2017; Langan et al. 2011; Yu Phuan et al. 2017; Katisko et al. 2018). Due to its existence in both skin and CNS, both of which are of ectodermal origin, collagen XVII may act as a shared autoantigen, but also neuroinflammation or neurodegeneration could trigger the autoimmunization (Amber et al. 2017). Since knock-out mice die soon after birth because of dermal and glomerular defects, it is not possible to utilize these mice to investigate the role of collagen XVII in neurological diseases (Hurskainen et al. 2012). Taken together, several non-fibrillar collagens, many of which are of neural origin, contribute to the homeostasis of CNS neurons, and thereby to brain physiology and function.

8.3.6 *Reactive Collagen Production/Distribution Following Disease or Trauma*

Tissue repair typically involves matrix remodeling to mediate signaling and support regrowth. Genes active during development are often reactivated during repair due to similarities shared by developmental and repair processes. Dysfunction of the BBB is linked to various neurological and neurodegenerative diseases and several collagens are induced in injured BBB vessels (Christov et al. 2008; Munji et al. 2019). For example, levels of collagens I and III are elevated at BBB in experimental mouse models for seizure, multiple sclerosis, stroke, and traumatic brain injury (Munji et al. 2019). It is suggested that post-hemorrhagic meningeal fibrosis is the cause for following hydrocephalus both in humans and in a rat model (Sajanti et al. 1999). In a cerebrovascular disease, stroke-like episodes induce accumulation of the fibrillar collagens I, III, and V in the wall of intracerebral arterioles (Zhang et al. 1994). The accumulation of fibrillar collagens and subsequent vascular fibrosis is a response to prolonged repair signaling. In hypertensive patients, vascular BMs generally become thickened but additionally, collagen VI is atypically deposited in the broadened vascular wall of larger arteries and cortical vessels, indicative of ECM remodeling in response to hypertension (Roggendorf et al. 1988). In the brain vessels of AD patients, the levels of collagens I and III increase while collagen IV level is reduced, suggestive of BM disruption and postinjury response (Christov et al. 2008). Expression of some additional collagens, such as type VI, VIII, and XVIII either in neurons or in reactive microglia/astrocytes is reported to be induced after brain injuries (Hirano et al. 2004; Zhang et al. 2007; Cheng et al. 2011).

Unlike in CNS, injured nerves in the PNS do regenerate (Perrin et al. 2005). Recruited macrophages, exhibiting M2 regenerative phenotype, engulf debris at the injured site and secrete signals responsible for the induction of axonal regrowth (Mokarram et al. 2012). Thereafter, new projections grow along the preexisting path to reestablish innervation and synapses (Fig. 8.3f). Laminin $\alpha2\beta1\gamma1$ is the most important BM player of nerve regeneration (Chen and Strickland 2003) but fibrillar collagens also support axon growth. Also, collagen VI appears important in several reactive processes in the nervous system. Its expression is induced in injured nerves and collagen VI is shown to promote monocyte adhesion (Schnoor et al. 2008) and macrophage migration (Chen et al. 2014). In line, in collagen VI-deficient injured nerves myelin clearance is delayed due to defective recruitment of macrophages (Chen et al. 2014). Regenerative M2 macrophages produce more collagen VI over proinflammatory M1 macrophages (Schnoor et al. 2008) and collagen VI was found to be important for macrophage M2 polarization. Collagen VI regulates macrophage migration and polarization through the regulation of AKT and PKA pathways. Concomitantly, collagen VI deficiency in mice compromises peripheral nerve regeneration (Chen et al. 2014). In diabetic neuropathy patients, collagen VI accumulates at the epineurium and perineurium, and in in vitro experiments its expression is induced by glucose (Muona et al. 1993; Hill 2009). As collagen VI expression normally ceases following the completion of myelination (Vitale et al. 2001), a surplus of collagen VI may thus become a pathogenic signal. Besides collagen VI, collagen V is also found at increased levels in the endoneurium of diabetic neuropathy but collagen IV remains unaltered (Hill 2009). Moreover, collagen types I, III, IV, V, and VI are upregulated, especially in Charcot–Marie–Tooth type 1 disease but also in some other inherited demyelinating neuropathies caused by mutations in genes expressed by myelinating Schwann cells (Palumbo et al. 2002). Collagen

XXVIII is also induced in the dysmyelinated nerves in a Charcot–Marie–Tooth disease mouse model (Grimal et al. 2010).

Tumors that develop in both the peripheral and central nervous systems contain distinct ECM from those described throughout this chapter. Collagens as part of the tumor stroma may support tumor growth and invasion or counteractingly limit tumor growth. Several collagens are differentially, typically increasingly, expressed between the normal nervous tissue cells and the tumoral counterparts. Abnormal levels of at least on collagens I, III, IV, VI, VII, VIII, XI, XVI, XVII, and XXVIII have been reported in brain tumors or cancer cells (Sage et al. 1984; Paulus et al. 1991; Fujita et al. 2008; Senner et al. 2008; An et al. 2009; Di Rosa et al. 2015; Ishihara et al. 2019; Yang et al. 2019). Abnormal expression of collagens can be utilized as prognostic markers or therapeutic targets in neuro-oncology. One of those collagens differentially expressed in brain cancer is collagen XVI since *COL16A1* expression is induced in glioblastoma (Senner et al. 2008). Suggestively, collagen XVI affects cell adhesion through its recognized interaction with $\alpha 1\beta 1$ integrin (Eble et al. 2006) and promotes tumor invasiveness (Bauer et al. 2011). In vitro *COL16A1* knockdown results in a reduction of focal adhesion number and $\beta 1$-integrin activation (Senner et al. 2008; Bauer et al. 2011). Collagen XVI promotes the invasion of oral cancer through the induction of *MMP9* expression (Bedal et al. 2014) and this may also apply to brain tumors, since MMP9 is a recognized organizer of the PNN (Ethell and Ethell 2007). Although collagen XVII is confined to neurons, transcriptome sequencing identifies a novel *PTEN-COL17A1* fusion gene transcribed in glioblastoma, that results in a significant increase in collagen XVII expression and association with recurrence (Yan et al. 2017). In glioma cell lines, collagen XVII induction promotes invasiveness while its knockdown concomitantly decreases MMP9 expression and suppresses invasiveness (Yan et al. 2017), suggesting a regulatory role for collagen XVII in glioblastoma malignance. Angiogenesis is a prerequisite for tumor growth. Collagen VII becomes neo-expressed in scattered abnormal vessels in astrocytic and ependymal tumors (Paulus et al. 1995). Similarly, collagen VIII is not expressed in the normal brain parenchyma or nonneoplastic cerebral disorders, but it is highly induced in blood vessels of distinct brain tumors (Paulus et al. 1991). Interestingly, ectopic expression of the collagen $\alpha 1$ (VIII) matricryptin vastatin results in reduced angiogenesis and glioblastoma inhibition in an orthotopic model in mice (Li et al. 2017), suggesting a role reminiscent of endostatin (O'Reilly et al. 1997). Injuries result in changes in cellular signaling and responding alteration in ECMs to support regeneration. Variable collagens are either neo- or re-deposited in injured or cancerous nervous tissue stroma. The accumulation of fibrillar collagens concomitant with the appearance of fibrotic tissues is indicative of adverse repair response, while the assembly of variable collagens in the tumorous stroma may promote or limit cancer growth on a case-by-case basis.

8.4 Conclusion

While the presence of fibrillar and BM collagens in the nervous tissue was established decades ago, exciting emerging data suggest the existence of previously unknown types of ECMs in the CNS. Novel roles for matrix molecules in the nervous tissues have been identified and, for example, it is now known that neuronal cells themselves express several types of collagens. An increasing number of distinct collagens are involved in the assembly of neural circuits, signifying the importance of the collagen-dependent ECMs, especially in founding platforms and in providing target-derived signals for navigating axons, thus enabling proper synaptogenesis. In recent years with the aid of next-generation sequencing, several novel collagens have been linked to human nervous system diseases and their number is expected to still increase. Moreover, animal models have provided valuable information on the functional mechanisms of collagens. The expansion of "omics" analyses will certainly lead to the identification of more alterations in collagens regarding their structure or expression levels in various animal models and cell types, and in patients. Confidently this chapter will not only serve as an introduction to the diverse roles of collagens in the nervous system but will also inspire researchers not to overlook collagens as mere artifacts in their datasets. The field of collagens in the brain is only in its infancy, and it will be interesting to see how the field evolves in the next decade or even sooner.

References

Ackley BD, Crew JR, Elamaa H, Pihlajaniemi T, Kuo CJ, Kramer JM (2001) The NC1/endostatin domain of *Caenorhabditis elegans* type XVIII collagen affects cell migration and axon guidance. J Cell Biol 152:1219–1232

Ackley BD, Kang SH, Crew JR, Suh C, Jin Y, Kramer JM (2003) The basement membrane components nidogen and type XVIII collagen regulate organization of neuromuscular junctions in *Caenorhabditis elegans*. J Neurosci 23:3577–3587

Allen JM, Zamurs L, Brachvogel B, Schlotzer-Schrehardt U, Hansen U, Lamande SR, Rowley L, Fitzgerald J, Bateman JF (2009) Mice lacking the extracellular matrix protein WARP develop normally but have compromised peripheral nerve structure and function. J Biol Chem 284:12020–12030

Amber KT, Zikry J, Hertl M (2017) A multi-hit hypothesis of bullous pemphigoid and associated neurological disease: is HLA-DQB1*03:01, a potential link between immune privileged antigen exposure and epitope spreading? HLA 89:127–134

An JH, Lee SY, Jeon JY, Cho KG, Kim SU, Lee MA (2009) Identification of gliotropic factors that induce human stem cell migration to malignant tumor. J Proteome Res 8:2873–2881

Andrikopoulos K, Suzuki HR, Solursh M, Ramirez F (1992) Localization of pro-alpha 2 (V) collagen transcripts in the tissues of the developing mouse embryo. Dev Dyn 195:113–120

Aricescu AR, McKinnell IW, Halfter W, Stoker AW (2002) Heparan sulfate proteoglycans are ligands for receptor protein tyrosine phosphatase sigma. Mol Cell Biol 22:1881–1892

Banerjee S, Isaacman-Beck J, Schneider VA, Granato M (2013) A novel role for Lh3 dependent ECM modifications during neural crest cell migration in zebrafish. PLoS One 8:e54609

Bastuji-Garin S, Joly P, Lemordant P, Sparsa A, Bedane C, Delaporte E, Roujeau JC, Bernard P, Guillaume JC, Ingen-Housz-Oro S, Maillard H, Pauwels C, Picard-Dahan C, Dutronc Y, Richard MA, French Study Group for Bullous Diseases (2011) Risk factors for bullous pemphigoid in the elderly: a prospective case-control study. J Invest Dermatol 131:637–643

Bauer R, Ratzinger S, Wales L, Bosserhoff A, Senner V, Grifka J, Grassel S (2011) Inhibition of collagen XVI expression reduces glioma cell invasiveness. Cell Physiol Biochem 27:217–226

Beattie CE, Melancon E, Eisen JS (2000) Mutations in the stumpy gene reveal intermediate targets for zebrafish motor axons. Development 127:2653–2662

Bedal KB, Grassel S, Oefner PJ, Reinders J, Reichert TE, Bauer R (2014) Collagen XVI induces expression of MMP9 via modulation of AP-1 transcription factors and facilitates invasion of oral squamous cell carcinoma. PLoS One 9:e86777

Birk DE, Fitch JM, Babiarz JP, Doane KJ, Linsenmayer TF (1990) Collagen fibrillogenesis in vitro: interaction of types I and V collagen regulates fibril diameter. J Cell Sci 95(Pt 4):649–657

Booth BA, Uitto J (1981) Collagen biosynthesis by human skin fibroblasts. III. The effects of ascorbic acid on procollagen production and prolyl hydroxylase activity. Biochim Biophys Acta 675:117–122. 0304-4165(81)90076-3 [pii]

Boot-Handford RP, Tuckwell DS, Plumb DA, Rock CF, Poulsom R (2003) A novel and highly conserved collagen (pro(alpha)1(XXVII)) with a unique expression pattern and unusual molecular characteristics establishes a new clade within the vertebrate fibrillar collagen family. J Biol Chem 278:31067–31077

Breedveld G, de Coo IF, Lequin MH, Arts WF, Heutink P, Gould DB, John SW, Oostra B, Mancini GM (2006) Novel mutations in three families confirm a major role of COL4A1 in hereditary porencephaly. J Med Genet 43:490–495. jmg.2005.035584 [pii]

Bretaud S, Pagnon-Minot A, Guillon E, Ruggiero F, Le Guellec D (2011) Characterization of spatial and temporal expression pattern of Col15a1b during zebrafish development. Gene Expr Patterns 11:129–134

Bretaud S, Guillon E, Karppinen S, Pihlajaniemi T, Ruggiero F (2020) Collagen XV, a multifaceted multiplexin present across tissues and species. Matrix Biol Plus 6–7

Brick KE, Weaver CH, Savica R, Lohse CM, Pittelkow MR, Boeve BF, Gibson LE, Camilleri MJ, Wieland CN (2014) A population-based study of the association between bullous pemphigoid and neurologic disorders. J Am Acad Dermatol 71:1191–1197

Bunge MB, Williams AK, Wood PM, Uitto J, Jeffrey JJ (1980) Comparison of nerve cell and nerve cell plus Schwann cell cultures, with particular emphasis on basal lamina and collagen formation. J Cell Biol 84:184–202

Caglayan AO, Baranoski JF, Aktar F, Han W, Tuysuz B, Guzel A, Guclu B, Kaymakcalan H, Aktekin B, Akgumus GT, Murray PB, Erson-Omay EZ, Caglar C, Bakircioglu M, Sakalar YB, Guzel E, Demir N, Tuncer O, Senturk S, Ekici B, Minja FJ, Sestan N, Yasuno K, Bilguvar K, Caksen H, Gunel M (2014) Brain malformations associated with Knobloch syndrome—review of literature, expanding clinical spectrum, and identification of novel mutations. Pediatr Neurol 51:806–813.e8

Cai SC, Allen JC, Lim YL, Chua SH, Tan SH, Tang MB (2014) Mortality of bullous pemphigoid in Singapore: risk factors and causes of death in 359 patients seen at the National Skin Centre. Br J Dermatol 170:1319–1326

Calvo AC, Cibreiro GA, Merino PT, Roy JF, Galiana A, Rufian AJ, Cano JM, Martin MA, Moreno L, Larrode P, Vazquez PC, Galan L, Mora J, Munoz-Blanco JL, Munoz MJ, Zaragoza P, Pegoraro E, Soraru G, Mora M, Lunetta C, Penco S, Tarlarini C, Esteban J, Osta R, Redondo AG (2019) Collagen XIX alpha 1 improves prognosis in amyotrophic lateral sclerosis. Aging Dis 10:278–292

Carey DJ, Eldridge CF, Cornbrooks CJ, Timpl R, Bunge RP (1983) Biosynthesis of type IV collagen by cultured rat Schwann cells. J Cell Biol 97:473–479

Carlson SS, Valdez G, Sanes JR (2010) Presynaptic calcium channels and alpha3-integrins are complexed with synaptic cleft laminins, cytoskeletal elements and active zone components. J Neurochem 115:654–666

Cescon M, Chen P, Castagnaro S, Gregorio I, Bonaldo P (2016) Lack of collagen VI promotes neurodegeneration by impairing autophagy and inducing apoptosis during aging. Aging (Albany NY) 8:1083–1101

Cescon M, Gregorio I, Eiber N, Borgia D, Fusto A, Sabatelli P, Scorzeto M, Megighian A, Pegoraro E, Hashemolhosseini S, Bonaldo P (2018) Collagen VI is required for the structural and functional integrity of the neuromuscular junction. Acta Neuropathol 136:483

Charnas LR, Marini JC (1995) Neurologic profile in osteogenesis imperfecta. Connect Tissue Res 31:23

Charsar BA, Goldberg EM (2017) Polymicrogyria and intractable epilepsy in siblings with Knobloch syndrome and homozygous mutation of COL18A1. Pediatr Neurol 76:91–92. S0887-8994(17)30804-4 [pii]

Cheah KS, Lau ET, Au PK, Tam PP (1991) Expression of the mouse alpha 1(II) collagen gene is not restricted to cartilage during development. Development 111:945–953

Chen ZL, Strickland S (2003) Laminin gamma1 is critical for Schwann cell differentiation, axon myelination, and regeneration in the peripheral nerve. J Cell Biol 163:889–899

Chen P, Cescon M, Megighian A, Bonaldo P (2014) Collagen VI regulates peripheral nerve myelination and function. FASEB J 28:1145–1156

Cheng JS, Dubal DB, Kim DH, Legleiter J, Cheng IH, Yu GQ, Tesseur I, Wyss-Coray T, Bonaldo P, Mucke L (2009) Collagen VI protects neurons against Abeta toxicity. Nat Neurosci 12:119–121

Cheng IH, Lin YC, Hwang E, Huang HT, Chang WH, Liu YL, Chao CY (2011) Collagen VI protects against neuronal apoptosis elicited by ultraviolet irradiation via an Akt/phosphatidylinositol 3-kinase signaling pathway. Neuroscience 183:178–188

Chernousov MA, Stahl RC, Carey DJ (1998) Schwann cells use a novel collagen-dependent mechanism for fibronectin fibril assembly. J Cell Sci 111(Pt 18):2763–2777

Chernousov MA, Rothblum K, Tyler WA, Stahl RC, Carey DJ (2000) Schwann cells synthesize type V collagen that contains a novel alpha 4 chain. Molecular cloning, biochemical characterization, and high affinity heparin binding of alpha 4(V) collagen. J Biol Chem 275:28208–28215

Chernousov MA, Stahl RC, Carey DJ (2001) Schwann cell type V collagen inhibits axonal outgrowth and promotes Schwann cell migration via distinct adhesive activities of the collagen and noncollagen domains. J Neurosci 21:6125–6135

Chernousov MA, Rothblum K, Stahl RC, Evans A, Prentiss L, Carey DJ (2006) Glypican-1 and alpha4(V) collagen are required for Schwann cell myelination. J Neurosci 26:508–517

Chernousov MA, Yu WM, Chen ZL, Carey DJ, Strickland S (2008) Regulation of Schwann cell function by the extracellular matrix. Glia 56:1498–1507

Chosidow O, Doppler V, Bensimon G, Joly P, Salachas F, Lacomblez L, Prost C, Camu W, Frances C, Herson S, Meininger V (2000) Bullous pemphigoid and amyotrophic lateral sclerosis: a new clue for understanding the bullous disease? Arch Dermatol 136:521–524

Christov A, Ottman J, Hamdheydari L, Grammas P (2008) Structural changes in Alzheimer's disease brain microvessels. Curr Alzheimer Res 5:392–395

Clementz AG, Mutolo MJ, Leir SH, Morris KJ, Kucybala K, Harris H, Harris A (2013) Collagen XV inhibits epithelial to mesenchymal transition in pancreatic adenocarcinoma cells. PLoS One 8:e72250

Corbett MA, Turner SJ, Gardner A, Silver J, Stankovich J, Leventer RJ, Derry CP, Carroll R, Ha T, Scheffer IE, Bahlo M, Jackson GD, Mackey DA, Berkovic SF, Gecz J (2017) Familial epilepsy with anterior polymicrogyria as a presentation of COL18A1 mutations. Eur J Med Genet 60:437–443

Court FA, Gillingwater TH, Melrose S, Sherman DL, Greenshields KN, Morton AJ, Harris JB, Willison HJ, Ribchester RR (2008) Identity, developmental restriction and reactivity of extralaminar cells capping mammalian neuromuscular junctions. J Cell Sci 121:3901–3911

D'Antonio M, Michalovich D, Paterson M, Droggiti A, Woodhoo A, Mirsky R, Jessen KR (2006) Gene profiling and bioinformatic analysis of Schwann cell embryonic development and myelination. Glia 53:501–515

Dhungana H, Huuskonen MT, Pihlajaniemi T, Heljasvaara R, Vivien D, Kanninen KM, Malm T, Koistinaho J, Lemarchant S (2017) Lack of collagen XV is protective after ischemic stroke in mice. Cell Death Dis 8:e2541

Di Rosa M, Sanfilippo C, Libra M, Musumeci G, Malaguarnera L (2015) Different pediatric brain tumors are associated with different gene expression profiling. Acta Histochem 117:477–485

Dusl M, Moreno T, Munell F, Macaya A, Gratacos M, Abicht A, Strom TM, Lochmuller H, Senderek J (2019) Congenital myasthenic syndrome caused by novel COL13A1 mutations. J Neurol 266:1107–1112

Dziadek M, Timpl R (1985) Expression of nidogen and laminin in basement membranes during mouse embryogenesis and in teratocarcinoma cells. Dev Biol 111:372–382

Dziadek M, Darling P, Bakker M, Overall M, Zhang RZ, Pan TC, Tillet E, Timpl R, Chu ML (1996) Deposition of collagen VI in the extracellular matrix during mouse embryogenesis correlates with expression of the alpha 3(VI) subunit gene. Exp Cell Res 226:302–315

Eble JA, Kassner A, Niland S, Morgelin M, Grifka J, Grassel S (2006) Collagen XVI harbors an integrin alpha1 beta1 recognition site in its C-terminal domains. J Biol Chem 281:25745–25756

Eklund L, Piuhola J, Komulainen J, Sormunen R, Ongvarrasopone C, Fässler R, Muona A, Ilves M, Ruskoaho H, Takala TE, Pihlajaniemi T (2001) Lack of type XV collagen causes a skeletal myopathy and cardiovascular defects in mice. Proc Natl Acad Sci USA 98:1194–1199

Eldridge CF, Bunge MB, Bunge RP, Wood PM (1987) Differentiation of axon-related Schwann cells in vitro. I. Ascorbic acid regulates basal lamina assembly and myelin formation. J Cell Biol 105:1023–1034

Eldridge CF, Bunge MB, Bunge RP (1989) Differentiation of axon-related Schwann cells in vitro: II. Control of myelin formation by basal lamina. J Neurosci 9:625–638

Emery SC, Karpinski NC, Hansen L, Masliah E (1999) Abnormalities in central nervous system development in osteogenesis imperfecta type II. Pediatr Dev Pathol 2:124–130

Ethell IM, Ethell DW (2007) Matrix metalloproteinases in brain development and remodeling: synaptic functions and targets. J Neurosci Res 85:2813–2823

Favor J, Gloeckner CJ, Janik D, Klempt M, Neuhauser-Klaus A, Pretsch W, Schmahl W, Quintanilla-Fend L (2007) Type IV procollagen missense mutations associated with defects of the eye, vascular stability, the brain, kidney function and embryonic or postnatal viability in the mouse, *Mus musculus*: an extension of the Col4a1 allelic series and the identification of the first two Col4a2 mutant alleles. Genetics 175:725–736

Ferrer-Ferrer M, Dityatev A (2018) Shaping synapses by the neural extracellular matrix. Front Neuroanat 12:40

Forsell C, Björk BF, Lilius L, Axelman K, Fabre SF, Fratiglioni L, Winblad B, Graff C (2010) Genetic association to the amyloid plaque associated protein gene COL25A1 in Alzheimer's disease. Neurobiol Aging 31:409–415

Foureur N, Descamps V, Lebrun-Vignes B, Picard-Dahan C, Grossin M, Belaich S, Crickx B (2001) Bullous pemphigoid in a leg affected with hemiparesia: a possible relation of neurological diseases with bullous pemphigoid? Eur J Dermatol 11:230–233

Fox MA (2008) Novel roles for collagens in wiring the vertebrate nervous system. Curr Opin Cell Biol 20:508–513

Fox MA, Sanes JR, Borza DB, Eswarakumar VP, Fässler R, Hudson BG, John SW, Ninomiya Y, Pedchenko V, Pfaff SL, Rheault MN, Sado Y, Segal Y, Werle MJ, Umemori H (2007) Distinct target-derived signals organize formation, maturation, and maintenance of motor nerve terminals. Cell 129:179–193

Franzke CW, Bruckner P, Bruckner-Tuderman L (2005) Collagenous transmembrane proteins: recent insights into biology and pathology. J Biol Chem 280:4005–4008

Frischknecht R, Gundelfinger ED (2012) The brain's extracellular matrix and its role in synaptic plasticity. Adv Exp Med Biol 970:153–171

Fujita A, Sato JR, Festa F, Gomes LR, Oba-Shinjo SM, Marie SK, Ferreira CE, Sogayar MC (2008) Identification of COL6A1 as a differentially expressed gene in human astrocytomas. Genet Mol Res 7:371–378

Gal J, Chen J, Katsumata Y, Fardo DW, Wang WX, Artiushin S, Price D, Anderson S, Patel E, Zhu H, Nelson PT (2018) Detergent insoluble proteins and inclusion body-like structures immunoreactive for PRKDC/DNA-PK/DNA-PKcs, FTL, NNT, and AIFM1 in the amygdala of cognitively impaired elderly persons. J Neuropathol Exp Neurol 77:21–39

Gara SK, Grumati P, Urciuolo A, Bonaldo P, Kobbe B, Koch M, Paulsson M, Wagener R (2008) Three novel collagen VI chains with high homology to the alpha3 chain. J Biol Chem 283:10658–10670

Gatseva A, Sin YY, Brezzo G, Van Agtmael T (2019) Basement membrane collagens and disease mechanisms. Essays Biochem 63:297–312

Gess B, Rohr D, Fledrich R, Sereda MW, Kleffner I, Humberg A, Nowitzki J, Strecker JK, Halfter H, Young P (2011) Sodium-dependent vitamin C transporter 2 deficiency causes hypomyelination and extracellular matrix defects in the peripheral nervous system. J Neurosci 31:17180–17192

Goncalves TJM, Boutillon F, Lefebvre S, Goffin V, Iwatsubo T, Wakabayashi T, Oury F, Armand AS (2019) Collagen XXV promotes myoblast fusion during myogenic differentiation and muscle formation. Sci Rep 9:5878–5876

Gordon MK, Hahn RA (2010) Collagens. Cell Tissue Res 339:247–257

Gould DB, Phalan FC, Breedveld GJ, van Mil SE, Smith RS, Schimenti JC, Aguglia U, van der Knaap, M S, Heutink P, John SW (2005) Mutations in Col4a1 cause perinatal cerebral hemorrhage and porencephaly. Science 308:1167–1171

Gregorio I, Braghetta P, Bonaldo P, Cescon M (2018) Collagen VI in healthy and diseased nervous system. Dis Model Mech 11

Grimal S, Puech S, Wagener R, Venteo S, Carroll P, Fichard-Carroll A (2010) Collagen XXVIII is a distinctive component of the peripheral nervous system nodes of ranvier and surrounds nonmyelinating glial cells. Glia 58:1977–1987

Guillon E, Bretaud S, Ruggiero F (2016) Slow muscle precursors lay down a collagen XV matrix fingerprint to guide motor axon navigation. J Neurosci 36:2663–2676

Hägg P, Rehn M, Huhtala P, Väisänen T, Tamminen M, Pihlajaniemi T (1998) Type XIII collagen is identified as a plasma membrane protein. J Biol Chem 273:15590–15597

Hägg P, Väisänen T, Tuomisto A, Rehn M, Tu H, Huhtala P, Eskelinen S, Pihlajaniemi T (2001) Type XIII collagen: a novel cell adhesion component present in a range of cell-matrix adhesions and in the intercalated discs between cardiac muscle cells. Matrix Biol 19:727–742

Haghighi A, Tiwari A, Piri N, Nurnberg G, Saleh-Gohari N, Haghighi A, Neidhardt J, Nurnberg P, Berger W (2014) Homozygosity mapping and whole exome sequencing reveal a novel homozygous COL18A1 mutation causing Knobloch syndrome. PLoS One 9:e112747

Halfter W, Dong S, Schurer B, Cole GJ (1998) Collagen XVIII is a basement membrane heparan sulfate proteoglycan. J Biol Chem 273:25404–25412

Härönen H, Zainul Z, Tu H, Naumenko N, Sormunen R, Miinalainen I, Shakirzyanova A, Oikarainen T, Abdullin A, Martin P, Santoleri S, Koistinaho J, Silman I, Giniatullin R, Fox MA, Heikkinen A, Pihlajaniemi T (2017) Collagen XIII secures pre- and postsynaptic integrity of the neuromuscular synapse. Hum Mol Genet 26:2076–2090

Härönen H, Zainul Z, Naumenko N, Sormunen R, Miinalainen I, Shakirzyanova A, Santoleri S, Kemppainen AV, Giniatullin R, Pihlajaniemi T, Heikkinen A (2019) Correct expression and localization of collagen XIII are crucial for the normal formation and function of the neuromuscular system. Eur J Neurosci 49:1491–1511

Hartmann D, Ziegenhagen MW, Sievers J (1998) Meningeal cells stimulate neuronal migration and the formation of radial glial fascicles from the cerebellar external granular layer. Neurosci Lett 244:129–132

Hashimoto T, Wakabayashi T, Watanabe A, Kowa H, Hosoda R, Nakamura A, Kanazawa I, Arai T, Takio K, Mann DM, Iwatsubo T (2002) CLAC: a novel Alzheimer amyloid plaque component derived from a transmembrane precursor, CLAC-P/collagen type XXV. EMBO J 21:1524–1534

Hayashi G, Labelle-Dumais C, Gould DB (2018) Use of sodium 4-phenylbutyrate to define therapeutic parameters for reducing intracerebral hemorrhage and myopathy in Col4a1 mutant mice. Dis Model Mech 11

Heck N, Garwood J, Schutte K, Fawcett J, Faissner A (2003) Astrocytes in culture express fibrillar collagen. Glia 41:382–392

Heikkinen A, Tu H, Pihlajaniemi T (2012) Collagen XIII: a type II transmembrane protein with relevance to musculoskeletal tissues, microvessels and inflammation. Int J Biochem Cell Biol 44:714–717

Heikkinen A, Härönen H, Norman O, Pihlajaniemi T (2020) Collagen XIII and other ECM components in the assembly and disease of the neuromuscular junction. Anat Rec (Hoboken)

Heljasvaara R, Aikio M, Ruotsalainen H, Pihlajaniemi T (2017) Collagen XVIII in tissue homeostasis and dysregulation – lessons learned from model organisms and human patients. Matrix Biol 57–58:55–75

Hilario JD, Wang C, Beattie CE (2010) Collagen XIXa1 is crucial for motor axon navigation at intermediate targets. Development 137:4261–4269

Hill R (2009) Extracellular matrix remodelling in human diabetic neuropathy. J Anat 214:219–225

Hirano S, Yonezawa T, Hasegawa H, Hattori S, Greenhill NS, Davis PF, Sage EH, Ninomiya Y (2004) Astrocytes express type VIII collagen during the repair process of brain cold injury. Biochem Biophys Res Commun 317:437–443

Hubert T, Grimal S, Ratzinger S, Mechaly I, Grassel S, Fichard-Carroll A (2007) Collagen XVI is a neural component of the developing and regenerating dorsal root ganglia extracellular matrix. Matrix Biol 26:206–210

Hubert T, Grimal S, Carroll P, Fichard-Carroll A (2009) Collagens in the developing and diseased nervous system. Cell Mol Life Sci 66:1223–1238

Hurskainen T, Moilanen J, Sormunen R, Franzke CW, Soininen R, Löffek S, Huilaja L, Nuutinen M, Bruckner-Tuderman L, Autio-Harmainen H, Tasanen K (2012) Transmembrane collagen XVII is a novel component of the glomerular filtration barrier. Cell Tissue Res 348:579–588

Ishihara E, Takahashi S, Fukaya R, Ohta S, Yoshida K, Toda M (2019) Identification of KLRC2 as a candidate marker for brain tumor-initiating cells. Neurol Res 41:1043–1049

Jacków J, Löffek S, Nyström A, Bruckner-Tuderman L, Franzke CW (2016a) Collagen XVII shedding suppresses re-epithelialization by directing keratinocyte migration and dampening mTOR signaling. J Invest Dermatol 136:1031–1041

Jacków J, Schlosser A, Sormunen R, Nyström A, Sitaru C, Tasanen K, Bruckner-Tuderman L, Franzke CW (2016b) Generation of a functional non-shedding collagen XVII mouse model: relevance of collagen XVII shedding in wound healing. J Invest Dermatol 136:516–525

Jeanne M, Gould DB (2017) Genotype-phenotype correlations in pathology caused by collagen type IV alpha 1 and 2 mutations. Matrix Biol 57-58:29–44

Jeanne M, Labelle-Dumais C, Jorgensen J, Kauffman WB, Mancini GM, Favor J, Valant V, Greenberg SM, Rosand J, Gould DB (2012) COL4A2 mutations impair COL4A1 and COL4A2 secretion and cause hemorrhagic stroke. Am J Hum Genet 90:91–101

Jeanne M, Jorgensen J, Gould DB (2015) Molecular and genetic analyses of collagen type IV mutant mouse models of spontaneous intracerebral hemorrhage identify mechanisms for stroke prevention. Circulation 131:1555–1565

Jovanov Milosevic N, Judas M, Aronica E, Kostovic I (2014) Neural ECM in laminar organization and connectivity development in healthy and diseased human brain. Prog Brain Res 214:159–178

Kakuyama H, Soderberg L, Horigome K, Winblad B, Dahlqvist C, Naslund J, Tjernberg LO (2005) CLAC binds to aggregated Abeta and Abeta fragments, and attenuates fibril elongation. Biochemistry 44:15602–15609

Kamei A, Houdou S, Mito T, Konomi H, Takashima S (1992) Developmental change in type VI collagen in human cerebral vessels. Pediatr Neurol 8:183–186
Kapoor R, Sakai LY, Funk S, Roux E, Bornstein P, Sage EH (1988) Type VIII collagen has a restricted distribution in specialized extracellular matrices. J Cell Biol 107:721–730
Karkheiran S, Krebs CE, Makarov V, Nilipour Y, Hubert B, Darvish H, Frucht S, Shahidi GA, Buxbaum JD, Paisan-Ruiz C (2013) Identification of COL6A2 mutations in progressive myoclonus epilepsy syndrome. Hum Genet 132:275–283
Katisko K, Kokkonen N, Kruger J, Hartikainen P, Koivisto AM, Helisalmi S, Korhonen VE, Kokki M, Tuusa J, Herukka SK, Solje E, Haapasalo A, Tasanen K, Remes AM (2018) The association between frontotemporal lobar degeneration and bullous pemphigoid. J Alzheimers Dis 66:743–750
Kay JN, De la Huerta I, Kim IJ, Zhang Y, Yamagata M, Chu MW, Meister M, Sanes JR (2011) Retinal ganglion cells with distinct directional preferences differ in molecular identity, structure, and central projections. J Neurosci 31:7753–7762
Keene DR, Engvall E, Glanville RW (1988) Ultrastructure of type VI collagen in human skin and cartilage suggests an anchoring function for this filamentous network. J Cell Biol 107:1995–2006
Keren B, Suzuki OT, Gerard-Blanluet M, Bremond-Gignac D, Elmaleh M, Titomanlio L, Delezoide AL, Passos-Bueno MR, Verloes A (2007) CNS malformations in Knobloch syndrome with splice mutation in COL18A1 gene. Am J Med Genet A 143A:1514–1518
Kerever A, Schnack J, Vellinga D, Ichikawa N, Moon C, Arikawa-Hirasawa E, Efird JT, Mercier F (2007) Novel extracellular matrix structures in the neural stem cell niche capture the neurogenic factor fibroblast growth factor 2 from the extracellular milieu. Stem Cells 25:2146–2157
Khaleduzzaman M, Sumiyoshi H, Ueki Y, Inoguchi K, Ninomiya Y, Yoshioka H (1997) Structure of the human type XIX collagen (COL19A1) gene, which suggests it has arisen from an ancestor gene of the FACIT family. Genomics 45:304–312
Khoshnoodi J, Pedchenko V, Hudson BG (2008) Mammalian collagen IV. Microsc Res Tech 71:357–370
Kibsgaard L, Rasmussen M, Lamberg A, Deleuran M, Olesen AB, Vestergaard C (2017) Increased frequency of multiple sclerosis among patients with bullous pemphigoid: a population-based cohort study on comorbidities anchored around the diagnosis of bullous pemphigoid. Br J Dermatol 176:1486–1491
Kleppel MM, Santi PA, Cameron JD, Wieslander J, Michael AF (1989) Human tissue distribution of novel basement membrane collagen. Am J Pathol 134:813–825
Kliemann SE, Waetge RT, Suzuki OT, Passos-Bueno MR, Rosemberg S (2003) Evidence of neuronal migration disorders in Knobloch syndrome: clinical and molecular analysis of two novel families. Am J Med Genet A 119A:15–19
Koch M, Veit G, Stricker S, Bhatt P, Kutsch S, Zhou P, Reinders E, Hahn RA, Song R, Burgeson RE, Gerecke DR, Mundlos S, Gordon MK (2006) Expression of type XXIII collagen mRNA and protein. J Biol Chem 281:21546–21557
Kokkonen N, Herukka SK, Huilaja L, Kokki M, Koivisto AM, Hartikainen P, Remes AM, Tasanen K (2017) Increased levels of the bullous pemphigoid BP180 autoantibody are associated with more severe dementia in Alzheimer's disease. J Invest Dermatol 137:71–76
Korogod N, Petersen CC, Knott GW (2015) Ultrastructural analysis of adult mouse neocortex comparing aldehyde perfusion with cryo fixation. Elife 4
Kowa H, Sakakura T, Matsuura Y, Wakabayashi T, Mann DM, Duff K, Tsuji S, Hashimoto T, Iwatsubo T (2004) Mostly separate distributions of CLAC- versus Abeta40- or thioflavin S-reactivities in senile plaques reveal two distinct subpopulations of beta-amyloid deposits. Am J Pathol 165:273–281
Krishnaswamy VR, Benbenishty A, Blinder P, Sagi I (2019) Demystifying the extracellular matrix and its proteolytic remodeling in the brain: structural and functional insights. Cell Mol Life Sci 76:3229–3248

Kummer TT, Misgeld T, Sanes JR (2006) Assembly of the postsynaptic membrane at the neuromuscular junction: paradigm lost. Curr Opin Neurobiol 16:74–82

Kuo CJ, LaMontagne KR, Garcia-Cardena G, Ackley BD, Kalman D, Park S, Christofferson R, Kamihara J, Ding YH, Lo KM, Gillies S, Folkman J, Mulligan RC, Javaherian K (2001) Oligomerization-dependent regulation of motility and morphogenesis by the collagen XVIII NC1/endostatin domain. J Cell Biol 152:1233–1246

Kuo DS, Labelle-Dumais C, Mao M, Jeanne M, Kauffman WB, Allen J, Favor J, Gould DB (2014) Allelic heterogeneity contributes to variability in ocular dysgenesis, myopathy and brain malformations caused by Col4a1 and Col4a2 mutations. Hum Mol Genet 23:1709–1722

Labelle-Dumais C, Dilworth DJ, Harrington EP, de Leau M, Lyons D, Kabaeva Z, Manzini MC, Dobyns WB, Walsh CA, Michele DE, Gould DB (2011) COL4A1 mutations cause ocular dysgenesis, neuronal localization defects, and myopathy in mice and Walker-Warburg syndrome in humans. PLoS Genet 7:e1002062

Labelle-Dumais C, Schuitema V, Hayashi G, Hoff K, Gong W, Dao DQ, Ullian EM, Oishi P, Margeta M, Gould DB (2019) COL4A1 mutations cause neuromuscular disease with tissue-specific mechanistic heterogeneity. Am J Hum Genet 104:847–860

Lai CH, Chu ML (1996) Tissue distribution and developmental expression of type XVI collagen in the mouse. Tissue Cell 28:155–164

Lamande SR, Sigalas E, Pan TC, Chu ML, Dziadek M, Timpl R, Bateman JF (1998) The role of the alpha3(VI) chain in collagen VI assembly. Expression of an alpha3(VI) chain lacking N-terminal modules N10-N7 restores collagen VI assembly, secretion, and matrix deposition in an alpha3(VI)-deficient cell line. J Biol Chem 273:7423–7430

Langan SM, Groves RW, West J (2011) The relationship between neurological disease and bullous pemphigoid: a population-based case-control study. J Invest Dermatol 131:631–636

Latvanlehto A, Fox MA, Sormunen R, Tu H, Oikarainen T, Koski A, Naumenko N, Shakirzyanova A, Kallio M, Ilves M, Giniatullin R, Sanes JR, Pihlajaniemi T (2010) Muscle-derived collagen XIII regulates maturation of the skeletal neuromuscular junction. J Neurosci 30:12230–12241

Leung AW, Wong SY, Chan D, Tam PP, Cheah KS (2010) Loss of procollagen IIA from the anterior mesendoderm disrupts the development of mouse embryonic forebrain. Dev Dyn 239:2319–2329

Li Y, Li J, Woo YM, Shen Z, Yao H, Cai Y, Lin MC, Poon WS (2017) Enhanced expression of Vastatin inhibits angiogenesis and prolongs survival in murine orthotopic glioblastoma model. BMC Cancer 17:126–128

Liao HM, Chao YL, Huang AL, Cheng MC, Chen YJ, Lee KF, Fang JS, Hsu CH, Chen CH (2012) Identification and characterization of three inherited genomic copy number variations associated with familial schizophrenia. Schizophr Res 139:229–236

Logan CV, Cossins J, Rodriguez Cruz PM, Parry DA, Maxwell S, Martinez-Martinez P, Riepsaame J, Abdelhamed ZA, Lake AV, Moran M, Robb S, Chow G, Sewry C, Hopkins PM, Sheridan E, Jayawant S, Palace J, Johnson CA, Beeson D (2015) Congenital myasthenic syndrome type 19 is caused by mutations in COL13A1, encoding the atypical non-fibrillar collagen type XIII alpha1 chain. Am J Hum Genet 97:878–885

Lui VC, Ng LJ, Nicholls J, Tam PP, Cheah KS (1995a) Tissue-specific and differential expression of alternatively spliced alpha 1(II) collagen mRNAs in early human embryos. Dev Dyn 203:198–211

Lui VC, Kong RY, Nicholls J, Cheung AN, Cheah KS (1995b) The mRNAs for the three chains of human collagen type XI are widely distributed but not necessarily co-expressed: implications for homotrimeric, heterotrimeric and heterotypic collagen molecules. Biochem J 311(Pt 2):511–516

Mahajan VB, Olney AH, Garrett P, Chary A, Dragan E, Lerner G, Murray J, Bassuk AG (2010) Collagen XVIII mutation in Knobloch syndrome with acute lymphoblastic leukemia. Am J Med Genet A 152A:2875–2879

Mao M, Alavi MV, Labelle-Dumais C, Gould DB (2015) Type IV collagens and basement membrane diseases: cell biology and pathogenic mechanisms. Curr Top Membr 76:61–116

Marazza G, Pham HC, Scharer L, Pedrazzetti PP, Hunziker T, Trueb RM, Hohl D, Itin P, Lautenschlager S, Naldi L, Borradori L, Autoimmune bullous disease Swiss study group (2009) Incidence of bullous pemphigoid and pemphigus in Switzerland: a 2-year prospective study. Br J Dermatol 161:861–868

Marneros AG, Keene DR, Hansen U, Fukai N, Moulton K, Goletz PL, Moiseyev G, Pawlyk BS, Halfter W, Dong S, Shibata M, Li T, Crouch RK, Bruckner P, Olsen BR (2004) Collagen XVIII/endostatin is essential for vision and retinal pigment epithelial function. EMBO J 23:89–99

Marquardt RJ, Li Y (2019) Congenital myasthenic syndrome type 19 due to a novel mutation in the COL13A1 GENE. Muscle Nerve 60:E3–E4

Matsuo N, Tanaka S, Yoshioka H, Koch M, Gordon MK, Ramirez F (2008) Collagen XXIV (Col24a1) gene expression is a specific marker of osteoblast differentiation and bone formation. Connect Tissue Res 49:68–75

Maxwell WL, Duance VC, Lehto M, Ashurst DE, Berry M (1984) The distribution of types I, III, IV and V collagens in penetrant lesions of the central nervous system of the rat. Histochem J 16:1215–1229

Meyer F, Moussian B (2009) Drosophila multiplexin (Dmp) modulates motor axon pathfinding accuracy. Develop Growth Differ 51:483–498

Miner JH, Sanes JR (1994) Collagen IV alpha 3, alpha 4, and alpha 5 chains in rodent basal laminae: sequence, distribution, association with laminins, and developmental switches. J Cell Biol 127:879–891

Miosge N, Simniok T, Sprysch P, Herken R (2003) The collagen type XVIII endostatin domain is co-localized with perlecan in basement membranes in vivo. J Histochem Cytochem 51:285–296

Miyake M, Hori S, Morizawa Y, Tatsumi Y, Toritsuka M, Ohnishi S, Shimada K, Furuya H, Khadka VS, Deng Y, Ohnishi K, Iida K, Gotoh D, Nakai Y, Inoue T, Anai S, Torimoto K, Aoki K, Tanaka N, Konishi N, Fujimoto K (2017) Collagen type IV alpha 1 (COL4A1) and collagen type XIII alpha 1 (COL13A1) produced in cancer cells promote tumor budding at the invasion front in human urothelial carcinoma of the bladder. Oncotarget 8:36099–36114

Mokarram N, Merchant A, Mukhatyar V, Patel G, Bellamkonda RV (2012) Effect of modulating macrophage phenotype on peripheral nerve repair. Biomaterials 33:8793–8801

Monavarfeshani A, Knill CN, Sabbagh U, Su J, Fox MA (2017) Region- and cell-specific expression of transmembrane collagens in mouse brain. Front Integr Neurosci 11:20

Mouw JK, Ou G, Weaver VM (2014) Extracellular matrix assembly: a multiscale deconstruction. Nat Rev Mol Cell Biol 15:771–785

Munezane H, Oizumi H, Wakabayashi T, Nishio S, Hirasawa T, Sato T, Harada A, Yoshida T, Eguchi T, Yamanashi Y, Hashimoto T, Iwatsubo T (2019) Roles of collagen XXV and its putative receptors PTPsigma/delta in intramuscular motor innervation and congenital cranial dysinnervation disorder. Cell Rep 29:4362–4376.e6

Munji RN, Soung AL, Weiner GA, Sohet F, Semple BD, Trivedi A, Gimlin K, Kotoda M, Korai M, Aydin S, Batugal A, Cabangcala AC, Schupp PG, Oldham MC, Hashimoto T, Noble-Haeusslein LJ, Daneman R (2019) Profiling the mouse brain endothelial transcriptome in health and disease models reveals a core blood-brain barrier dysfunction module. Nat Neurosci 22:1892–1902

Muona P, Jaakkola S, Zhang RZ, Pan TC, Pelliniemi L, Risteli L, Chu ML, Uitto J, Peltonen J (1993) Hyperglycemic glucose concentrations up-regulate the expression of type VI collagen in vitro. Relevance to alterations of peripheral nerves in diabetes mellitus. Am J Pathol 142:1586–1597

Muona A, Eklund L, Väisänen T, Pihlajaniemi T (2002) Developmentally regulated expression of type XV collagen correlates with abnormalities in Col15a1(−/−) mice. Matrix Biol 21:89–102

Murad S, Grove D, Lindberg KA, Reynolds G, Sivarajah A, Pinnell SR (1981) Regulation of collagen synthesis by ascorbic acid. Proc Natl Acad Sci USA 78:2879–2882

Muragaki Y, Shiota C, Inoue M, Ooshima A, Olsen BR, Ninomiya Y (1992) alpha 1(VIII)-collagen gene transcripts encode a short-chain collagen polypeptide and are expressed by various

epithelial, endothelial and mesenchymal cells in newborn mouse tissues. Eur J Biochem 207:895–902
Myers JC, Li D, Bageris A, Abraham V, Dion AS, Amenta PS (1997) Biochemical and immunohistochemical characterization of human type XIX defines a novel class of basement membrane zone collagens. Am J Pathol 151:1729–1740
Myers JC, Li D, Amenta PS, Clark CC, Nagaswami C, Weisel JW (2003) Type XIX collagen purified from human umbilical cord is characterized by multiple sharp kinks delineating collagenous subdomains and by intermolecular aggregates via globular, disulfide-linked, and heparin-binding amino termini. J Biol Chem 278:32047–32057
Myllyharju J, Kivirikko KI (2004) Collagens, modifying enzymes and their mutations in humans, flies and worms. Trends Genet 20:33–43
Nah HD, Barembaum M, Upholt WB (1992) The chicken alpha 1 (XI) collagen gene is widely expressed in embryonic tissues. J Biol Chem 267:22581–22586
Nguyen QT, Sanes JR, Lichtman JW (2002) Pre-existing pathways promote precise projection patterns. Nat Neurosci 5:861–867
Novak U, Kaye AH (2000) Extracellular matrix and the brain: components and function. J Clin Neurosci 7:280–290
O'Brien RJ, Wong PC (2011) Amyloid precursor protein processing and Alzheimer's disease. Annu Rev Neurosci 34:185–204
Oh SP, Griffith CM, Hay ED, Olsen BR (1993) Tissue-specific expression of type XII collagen during mouse embryonic development. Dev Dyn 196:37–46
Oprisoreanu AM, Smith HL, Arya S, Webster R, Zhong Z, Wehner D, Cardozo MJ, Becker T, Talbot K, Becker CG (2019) Interaction of axonal chondrolectin with collagen XIXa1 is necessary for precise neuromuscular junction formation. Cell Rep 29:1082–1098.e10
O'Reilly MS, Boehm T, Shing Y, Fukai N, Vasios G, Lane WS, Flynn E, Birkhead JR, Olsen BR, Folkman J (1997) Endostatin: an endogenous inhibitor of angiogenesis and tumor growth. Cell 88:277–285
Osada Y, Hashimoto T, Nishimura A, Matsuo Y, Wakabayashi T, Iwatsubo T (2005) CLAC binds to amyloid beta peptides through the positively charged amino acid cluster within the collagenous domain 1 and inhibits formation of amyloid fibrils. J Biol Chem 280:8596–8605
Osawa T, Ide C (1986) Changes in thickness of collagen fibrils in the endo- and epineurium of the mouse sciatic nerve during development. Acta Anat (Basel) 125:245–251
Palumbo C, Massa R, Panico MB, Di Muzio A, Sinibaldi P, Bernardi G, Modesti A (2002) Peripheral nerve extracellular matrix remodeling in Charcot-Marie-Tooth type I disease. Acta Neuropathol 104:287–296
Parker SR, Dyson S, Brisman S, Pennie M, Swerlick RA, Khan R, Manos S, Korman BD, Xia Z, Korman NJ (2008) Mortality of bullous pemphigoid: an evaluation of 223 patients and comparison with the mortality in the general population in the United States. J Am Acad Dermatol 59:582–588
Passos-Bueno MR, Suzuki OT, Armelin-Correa LM, Sertie AL, Errera FI, Bagatini K, Kok F, Leite KR (2006) Mutations in collagen 18A1 and their relevance to the human phenotype. An Acad Bras Cienc 78:123–131
Paulus W, Sage EH, Liszka U, Iruela-Arispe ML, Jellinger K (1991) Increased levels of type VIII collagen in human brain tumours compared to normal brain tissue and non-neoplastic cerebral disorders. Br J Cancer 63:367–371
Paulus W, Baur I, Liszka U, Drlicek M, Leigh I, Bruckner-Tuderman L (1995) Expression of type VII collagen, the major anchoring fibril component, in normal and neoplastic human nervous system. Virchows Arch 426:199–202
Peltonen J, Jaakkola S, Hsiao LL, Timpl R, Chu ML, Uitto J (1990) Type VI collagen. In situ hybridizations and immunohistochemistry reveal abundant mRNA and protein levels in human neurofibroma, schwannoma and normal peripheral nerve tissues. Lab Investig 62:487–492

Perrin FE, Lacroix S, Aviles-Trigueros M, David S (2005) Involvement of monocyte chemoattractant protein-1, macrophage inflammatory protein-1alpha and interleukin-1beta in Wallerian degeneration. Brain 128:854–866

Pöschl E, Schlotzer-Schrehardt U, Brachvogel B, Saito K, Ninomiya Y, Mayer U (2004) Collagen IV is essential for basement membrane stability but dispensable for initiation of its assembly during early development. Development 131:1619–1628

Qin J, Liang J, Ding M (2014) Perlecan antagonizes collagen IV and ADAMTS9/GON-1 in restricting the growth of presynaptic boutons. J Neurosci 34:10311–10324

Ramanoudjame L, Rocancourt C, Laine J, Klein A, Joassard L, Gartioux C, Fleury M, Lyphout L, Kabashi E, Ciura S, Cousin X, Allamand V (2015) Two novel COLVI long chains in zebrafish that are essential for muscle development. Hum Mol Genet 24:6624–6639

Rasi K, Hurskainen M, Kallio M, Staven S, Sormunen R, Heape AM, Avila RL, Kirschner D, Muona A, Tolonen U, Tanila H, Huhtala P, Soininen R, Pihlajaniemi T (2010) Lack of collagen XV impairs peripheral nerve maturation and, when combined with laminin-411 deficiency, leads to basement membrane abnormalities and sensorimotor dysfunction. J Neurosci 30:14490–14501

Ren Z, Hsu DY, Brieva J, Silverberg NB, Langan SM, Silverberg JI (2017) Hospitalization, inpatient burden and comorbidities associated with bullous pemphigoid in the U.S.A. Br J Dermatol 176:87–99

Ricard-Blum S (2011) The collagen family. Cold Spring Harb Perspect Biol 3:a004978

Ricard-Blum S, Ballut L (2011) Matricryptins derived from collagens and proteoglycans. Front Biosci (Landmark Ed) 16:674–697

Ring C, Lemmon V, Halfter W (1995) Two chondroitin sulfate proteoglycans differentially expressed in the developing chick visual system. Dev Biol 168:11–27

Rodriguez Cruz PM, Cossins J, Estephan EP, Munell F, Selby K, Hirano M, Maroofin R, Mehrjardi MYV, Chow G, Carr A, Manzur A, Robb S, Munot P, Wei Liu W, Banka S, Fraser H, De Goede C, Zanoteli E, Conti Reed U, Sage A, Gratacos M, Macaya A, Dusl M, Senderek J, Topf A, Hofer M, Knight R, Ramdas S, Jayawant S, Lochmuller H, Palace J, Beeson D (2019) The clinical spectrum of the congenital myasthenic syndrome resulting from COL13A1 mutations. Brain 142:1547–1560

Roggendorf W, Opitz H, Schuppan D (1988) Altered expression of collagen type VI in brain vessels of patients with chronic hypertension. A comparison with the distribution of collagen IV and procollagen III. Acta Neuropathol 77:55–60

Rohrbough J, Rushton E, Woodruff E, Fergestad T, Vigneswaran K, Broadie K (2007) Presynaptic establishment of the synaptic cleft extracellular matrix is required for post-synaptic differentiation. Genes Dev 21:2607–2628

Roulet M, Ruggiero F, Karsenty G, LeGuellec D (2007) A comprehensive study of the spatial and temporal expression of the col5a1 gene in mouse embryos: a clue for understanding collagen V function in developing connective tissues. Cell Tissue Res 327:323–332

Saarela J, Rehn M, Oikarinen A, Autio-Harmainen H, Pihlajaniemi T (1998) The short and long forms of type XVIII collagen show clear tissue specificities in their expression and location in basement membrane zones in humans. Am J Pathol 153:611–626

Sage H, Balian G, Vogel AM, Bornstein P (1984) Type VIII collagen. Synthesis by normal and malignant cells in culture. Lab Investig 50:219–231

Sajanti J, Björkstrand AS, Finnila S, Heikkinen E, Peltonen J, Majamaa K (1999) Increase of collagen synthesis and deposition in the arachnoid and the dura following subarachnoid hemorrhage in the rat. Biochim Biophys Acta 1454:209–216

Salza R, Oudart JB, Ramont L, Maquart FX, Bakchine S, Thoannes H, Ricard-Blum S (2015) Endostatin level in cerebrospinal fluid of patients with Alzheimer's disease. J Alzheimers Dis 44:1253–1261

Sandberg M, Tamminen M, Hirvonen H, Vuorio E, Pihlajaniemi T (1989) Expression of mRNAs coding for the alpha 1 chain of type XIII collagen in human fetal tissues: comparison with expression of mRNAs for collagen types I, II, and III. J Cell Biol 109:1371–1379

Sandberg MM, Hirvonen HE, Elima KJ, Vuorio EI (1993) Co-expression of collagens II and XI and alternative splicing of exon 2 of collagen II in several developing human tissues. Biochem J 294 (Pt 2):595–602

Sandberg-Lall M, Hägg PO, Wahlström I, Pihlajaniemi T (2000) Type XIII collagen is widely expressed in the adult and developing human eye and accentuated in the ciliary muscle, the optic nerve and the neural retina. Exp Eye Res 70:401–410

Sanes JR, Lichtman JW (1999) Development of the vertebrate neuromuscular junction. Annu Rev Neurosci 22:389–442

Sanes JR, Lichtman JW (2001) Induction, assembly, maturation and maintenance of a postsynaptic apparatus. Nat Rev Neurosci 2:791–805

Schneider VA, Granato M (2006) The myotomal diwanka (lh3) glycosyltransferase and type XVIII collagen are critical for motor growth cone migration. Neuron 50:683–695

Schnoor M, Cullen P, Lorkowski J, Stolle K, Robenek H, Troyer D, Rauterberg J, Lorkowski S (2008) Production of type VI collagen by human macrophages: a new dimension in macrophage functional heterogeneity. J Immunol 180:5707–5719

Senner V, Ratzinger S, Mertsch S, Grassel S, Paulus W (2008) Collagen XVI expression is upregulated in glioblastomas and promotes tumor cell adhesion. FEBS Lett 582:3293–3300

Seppänen A, Autio-Harmainen H, Alafuzoff I, Särkioja T, Veijola J, Hurskainen T, Bruckner-Tuderman L, Tasanen K, Majamaa K (2006) Collagen XVII is expressed in human CNS neurons. Matrix Biol 25:185–188

Seppänen A, Suuronen T, Hofmann SC, Majamaa K, Alafuzoff I (2007) Distribution of collagen XVII in the human brain. Brain Res 1158:50–56

Seppänen A, Pikkarainen M, Hartikainen P, Hofmann SC, Majamaa K, Alafuzoff I (2009) Expression of collagen XVII and ubiquitin-binding protein p62 in motor neuron disease. Brain Res 1247:171–177

Seppänen A, Miettinen R, Alafuzoff I (2010) Neuronal collagen XVII is localized to lipofuscin granules. Neuroreport 21:1090–1094

Seppinen L, Pihlajaniemi T (2011) The multiple functions of collagen XVIII in development and disease. Matrix Biol 30:83–92

Sertie AL, Sossi V, Camargo AA, Zatz M, Brahe C, Passos-Bueno MR (2000) Collagen XVIII, containing an endogenous inhibitor of angiogenesis and tumor growth, plays a critical role in the maintenance of retinal structure and in neural tube closure (Knobloch syndrome). Hum Mol Genet 9:2051–2058

Sharif Y, Jumah F, Coplan L, Krosser A, Sharif K, Tubbs RS (2018) Blood brain barrier: a review of its anatomy and physiology in health and disease. Clin Anat 31:812–823

Shellswell GB, Restall DJ, Duance VC, Bailey AJ (1979) Identification and differential distribution of collagen types in the central and peripheral nervous systems. FEBS Lett 106:305–308

Shinwari JM, Khan A, Awad S, Shinwari Z, Alaiya A, Alanazi M, Tahir A, Poizat C, Al Tassan N (2015) Recessive mutations in COL25A1 are a cause of congenital cranial dysinnervation disorder. Am J Hum Genet 96:147–152

Sievers J, Pehlemann FW, Gude S, Berry M (1994) Meningeal cells organize the superficial glia limitans of the cerebellum and produce components of both the interstitial matrix and the basement membrane. J Neurocytol 23:135–149

Söderberg L, Kakuyama H, Moller A, Ito A, Winblad B, Tjernberg LO, Naslund J (2005) Characterization of the Alzheimer's disease-associated CLAC protein and identification of an amyloid beta-peptide-binding site. J Biol Chem 280:1007–1015

Song I, Dityatev A (2018) Crosstalk between glia, extracellular matrix and neurons. Brain Res Bull 136:101–108

Spivey KA, Banyard J, Solis LM, Wistuba II, Barletta JA, Gandhi L, Feldman HA, Rodig SJ, Chirieac LR, Zetter BR (2010) Collagen XXIII: a potential biomarker for the detection of primary and recurrent non-small cell lung cancer. Cancer Epidemiol Biomarkers Prev 19:1362–1372

Su J, Gorse K, Ramirez F, Fox MA (2010) Collagen XIX is expressed by interneurons and contributes to the formation of hippocampal synapses. J Comp Neurol 518:229–253
Su J, Stenbjorn RS, Gorse K, Su K, Hauser KF, Ricard-Blum S, Pihlajaniemi T, Fox MA (2012) Target-derived matricryptins organize cerebellar synapse formation through alpha3beta1 integrins. Cell Rep 2:223–230
Su J, Chen J, Lippold K, Monavarfeshani A, Carrillo GL, Jenkins R, Fox MA (2016) Collagen-derived matricryptins promote inhibitory nerve terminal formation in the developing neocortex. J Cell Biol 212:721–736
Su J, Cole J, Fox MA (2017) Loss of interneuron-derived collagen XIX leads to a reduction in Perineuronal nets in the mammalian telencephalon. ASN Neuro 9:1759091416689020
Sumiyoshi H, Laub F, Yoshioka H, Ramirez F (2001) Embryonic expression of type XIX collagen is transient and confined to muscle cells. Dev Dyn 220:155–162
Sumiyoshi H, Mor N, Lee SY, Doty S, Henderson S, Tanaka S, Yoshioka H, Rattan S, Ramirez F (2004) Esophageal muscle physiology and morphogenesis require assembly of a collagen XIX-rich basement membrane zone. J Cell Biol 166:591–600
Sund M, Väisänen T, Kaukinen S, Ilves M, Tu H, Autio-Harmainen H, Rauvala H, Pihlajaniemi T (2001a) Distinct expression of type XIII collagen in neuronal structures and other tissues during mouse development. Matrix Biol 20:215–231
Sund M, Ylönen R, Tuomisto A, Sormunen R, Tahkola J, Kvist AP, Kontusaari S, Autio-Harmainen H, Pihlajaniemi T (2001b) Abnormal adherence junctions in the heart and reduced angiogenesis in transgenic mice overexpressing mutant type XIII collagen. EMBO J 20:5153–5164
Suzuki OT, Sertie AL, Der Kaloustian VM, Kok F, Carpenter M, Murray J, Czeizel AE, Kliemann SE, Rosemberg S, Monteiro M, Olsen BR, Passos-Bueno MR (2002) Molecular analysis of collagen XVIII reveals novel mutations, presence of a third isoform, and possible genetic heterogeneity in Knobloch syndrome. Am J Hum Genet 71:1320–1329
Suzuki O, Kague E, Bagatini K, Tu H, Heljasvaara R, Carvalhaes L, Gava E, de Oliveira G, Godoi P, Oliva G, Kitten G, Pihlajaniemi T, Passos-Bueno MR (2009) Novel pathogenic mutations and skin biopsy analysis in Knobloch syndrome. Mol Vis 15:801–809
Tanaka T, Wakabayashi T, Oizumi H, Nishio S, Sato T, Harada A, Fujii D, Matsuo Y, Hashimoto T, Iwatsubo T (2014) CLAC-P/collagen type XXV is required for the intramuscular innervation of motoneurons during neuromuscular development. J Neurosci 34:1370–1379
Taylor J, Unsoeld T, Hutter H (2018) The transmembrane collagen COL-99 guides longitudinally extending axons in *C. elegans*. Mol Cell Neurosci 89:9–19
Teixeira VB, Cabral R, Brites MM, Vieira R, Figueiredo A (2014) Bullous pemphigoid and comorbidities: a case-control study in Portuguese patients. An Bras Dermatol 89:274–278
Tong Y, Xu Y, Scearce-Levie K, Ptacek LJ, Fu YH (2010) COL25A1 triggers and promotes Alzheimer's disease-like pathology in vivo. Neurogenetics 11:41–52
Tu H, Huhtala P, Lee HM, Adams JC, Pihlajaniemi T (2015) Membrane-associated collagens with interrupted triple-helices (MACITs): evolution from a bilaterian common ancestor and functional conservation in *C. elegans*. BMC Evol Biol 15:281–283
Tu H, Pirskanen-Matell R, Heikkinen A, Oikarainen T, Risteli J, Pihlajaniemi T (2018) Autoimmune antibodies to collagen XIII in myasthenia gravis patients. Muscle Nerve 57:506–510
Unsoeld T, Park JO, Hutter H (2013) Discoidin domain receptors guide axons along longitudinal tracts in *C. elegans*. Dev Biol 374:142–152
Urabe N, Naito I, Saito K, Yonezawa T, Sado Y, Yoshioka H, Kusachi S, Tsuji T, Ohtsuka A, Taguchi T, Murakami T, Ninomiya Y (2002) Basement membrane type IV collagen molecules in the choroid plexus, pia mater and capillaries in the mouse brain. Arch Histol Cytol 65:133–143
Utriainen A, Sormunen R, Kettunen M, Carvalhaes LS, Sajanti E, Eklund L, Kauppinen R, Kitten GT, Pihlajaniemi T (2004) Structurally altered basement membranes and hydrocephalus in a type XVIII collagen deficient mouse line. Hum Mol Genet 13:2089–2099

van Horssen J, Wilhelmus MM, Heljasvaara R, Pihlajaniemi T, Wesseling P, de Waal RM, Verbeek MM (2002) Collagen XVIII: a novel heparan sulfate proteoglycan associated with vascular amyloid depositions and senile plaques in Alzheimer's disease brains. Brain Pathol 12:456–462
Veit G, Kobbe B, Keene DR, Paulsson M, Koch M, Wagener R (2006) Collagen XXVIII, a novel von Willebrand factor A domain-containing protein with many imperfections in the collagenous domain. J Biol Chem 281:3494–3504
Vitale P, Braghetta P, Volpin D, Bonaldo P, Bressan GM (2001) Mechanisms of transcriptional activation of the col6a1 gene during Schwann cell differentiation. Mech Dev 102:145–156
Vogel W, Gish GD, Alves F, Pawson T (1997) The discoidin domain receptor tyrosine kinases are activated by collagen. Mol Cell 1:13–23
Volonghi I, Pezzini A, Del Zotto E, Giossi A, Costa P, Ferrari D, Padovani A (2010) Role of COL4A1 in basement-membrane integrity and cerebral small-vessel disease. The COL4A1 stroke syndrome. Curr Med Chem 17:1317–1324
Wälchli C, Koch M, Chiquet M, Odermatt BF, Trueb B (1994) Tissue-specific expression of the fibril-associated collagens XII and XIV. J Cell Sci 107(Pt 2):669–681
White RJ, Wang Y, Tang P, Montezuma SR (2017) Knobloch syndrome associated with Polymicrogyria and early onset of retinal detachment: two case reports. BMC Ophthalmol 17:214-z
Xu L, Nirwane A, Yao Y (2018) Basement membrane and blood-brain barrier. Stroke Vasc Neurol 4:78–82
Yamada Y, Sakuma J, Takeuchi I, Yasukochi Y, Kato K, Oguri M, Fujimaki T, Horibe H, Muramatsu M, Sawabe M, Fujiwara Y, Taniguchi Y, Obuchi S, Kawai H, Shinkai S, Mori S, Arai T, Tanaka M (2017) Identification of six polymorphisms as novel susceptibility loci for ischemic or hemorrhagic stroke by exome-wide association studies. Int J Mol Med 39:1477–1491
Yan X, Zhang C, Liang T, Yang F, Wang H, Wu F, Wang W, Wang Z, Cheng W, Xu J, Jiang T, Chen J, Ding Y (2017) A PTEN-COL17A1 fusion gene and its novel regulatory role in collagen XVII expression and GBM malignance. Oncotarget 8:85794–85803
Yang H, Jin L, Sun X (2019) A thirteengene set efficiently predicts the prognosis of glioblastoma. Mol Med Rep 19:1613–1621
Ylikärppä R, Eklund L, Sormunen R, Muona A, Fukai N, Olsen BR, Pihlajaniemi T (2003) Double knockout mice reveal a lack of major functional compensation between collagens XV and XVIII. Matrix Biol 22:443–448
Yoshida T, Kato K, Yokoi K, Oguri M, Watanabe S, Metoki N, Yoshida H, Satoh K, Aoyagi Y, Nozawa Y, Yamada Y (2010) Association of genetic variants with hemorrhagic stroke in Japanese individuals. Int J Mol Med 25:649–656
Yoshioka H, Iyama K, Inoguchi K, Khaleduzzaman M, Ninomiya Y, Ramirez F (1995) Developmental pattern of expression of the mouse alpha 1 (XI) collagen gene (Col11a1). Dev Dyn 204:41–47
Yu Phuan CZ, Yew YW, Tey HL (2017) Bullous pemphigoid and antecedent neurological diseases: an association with dementia. Indian J Dermatol Venereol Leprol 83:457–461
Zainul Z, Heikkinen A, Koivisto H, Rautalahti I, Kallio M, Lin S, Härönen H, Norman O, Rüegg MA, Tanila H, Pihlajaniemi T (2018) Collagen XIII is required for neuromuscular synapse regeneration and functional recovery after peripheral nerve injury. J Neurosci 38:4243–4258
Zech M, Lam DD, Francescatto L, Schormair B, Salminen AV, Jochim A, Wieland T, Lichtner P, Peters A, Gieger C, Lochmuller H, Strom TM, Haslinger B, Katsanis N, Winkelmann J (2015) Recessive mutations in the alpha3 (VI) collagen gene COL6A3 cause early-onset isolated dystonia. Am J Hum Genet 96:883–893
Zhang WW, Ma KC, Andersen O, Sourander P, Tollesson PO, Olsson Y (1994) The microvascular changes in cases of hereditary multi-infarct disease of the brain. Acta Neuropathol 87:317–324
Zhang ZY, Zhang Z, Fauser U, Artelt M, Burnet M, Schluesener HJ (2007) Dexamethasone transiently attenuates up-regulation of endostatin/collagen XVIII following traumatic brain injury. Neuroscience 147:720–726

Zhang LS, Li HB, Zeng J, Yang Y, Ding C (2018) Knobloch syndrome caused by homozygous frameshift mutation of the COL18A1 gene in a Chinese pedigree. Int J Ophthalmol 11:918–922
Zimmermann DR, Dours-Zimmermann MT (2008) Extracellular matrix of the central nervous system: from neglect to challenge. Histochem Cell Biol 130:635–653

The manufacturer's authorised representative in the EU is Springer Nature Customer Service Centre GmbH, Europaplatz 3, 69115 Heidelberg, Germany. If you have any concerns regarding our products, please contact ProductSafety@springernature.com

Printed and bound by CPI Group (UK) Ltd, Croydon, CR0 4YY
30/09/2026
02225691-0007